# INTRODUCTION TO MODERN METHODS OF QUANTUM MANY-BODY THEORY AND THEIR APPLICATIONS

Series on Advances in Quantum Many-Body Theory – Vol. 7

# INTRODUCTION TO MODERN METHODS OF QUANTUM MANY-BODY THEORY AND THEIR APPLICATIONS


Editors

**Adelchi Fabrocini**
*University of Pisa, Italy*

**Stefano Fantoni**
*National Institute for Advanced Studies, Italy*

**Eckhard Krotscheck**
*Johannes-Kepler University, Austria*



World Scientific
*New Jersey • London • Singapore • Hong Kong*

*Published by*

World Scientific Publishing Co. Pte. Ltd.

P O Box 128, Farrer Road, Singapore 912805

*USA office:* Suite 1B, 1060 Main Street, River Edge, NJ 07661

*UK office:* 57 Shelton Street, Covent Garden, London WC2H 9HE

**British Library Cataloguing-in-Publication Data**
A catalogue record for this book is available from the British Library.

**INTRODUCTION TO MODERN METHODS OF QUANTUM MANY-BODY THEORY AND THEIR APPLICATIONS**



ISBN 981-238-069-8

Printed in Singapore by Mainland Press

## CONTENTS


PREFACE xi

**Chapter 1 DENSITY FUNCTIONAL THEORY** 1

1. Introduction 1
   1.1. Units and notation 3
   1.2. Hartree-Fock theory 3
   1.3. Homogeneous electron gas 5
      1.3.1. Free electrons 6
      1.3.2. Exchange energy 7
2. What is density functional theory? 8
   2.1. Hohenberg-Kohn theorem 8
   2.2. A simple example: the Thomas-Fermi theory 9
      2.2.1. Variational equation of Thomas-Fermi theory 10
      2.2.2. Thomas-Fermi atom 10
      2.2.3. An example 12
3. Kohn-Sham theory 13
   3.1. Local density approximation 13
   3.2. Spin and the local spin density approximation 14
   3.3. The generalized gradient approximation 15
4. Numerical methods for the Kohn-Sham equation 16
      4.0.1. Exact exchange 20
      4.0.2. $\mathcal{O}(N)$ methods 20
5. Some applications and limitations of DFT 21
   5.1. Two examples of condensed matter 21
   5.2. Vibrations 22
   5.3. NMR chemical shifts 23
6. Limitations of DFT 24
7. Time-dependent density functional theory: the equations 25
   7.1. Optical properties 27
      7.1.1. $f$-sum rule 28
   7.2. Methods to solve the TDDFT equations 29
      7.2.1. Linear response formula 31
   7.3. Dynamic polarizability 32
   7.4. Dielectric function 33

8. TDDFT: numerical aspects 34
8.1. Configuration matrix method 35
8.2. Linear response method 36
8.3. Sternheimer method 37
8.4. Real time method 37
9. Applications of TDDFT 39
9.1. Simple metal clusters 39
9.2. Carbon structures 41
9.3. Diamond 42
9.4. Other applications 44
9.5. Limitations 45
References 46

**Chapter 2 MICROSCOPIC DESCRIPTION OF QUANTUM LIQUIDS 49**

1. Introduction 49
1.1. General properties 51
2. Microscopic description 54
3. Hypernetted–chain equations 63
3.1. Results for liquid $^4$He 75
4. Minimization of the energy: Optimal two–body correlation functions 81
4.1. Asymptotic behavior 85
5. Low excited states 87
6. A $^3$He impurity in liquid $^4$He 93
6.1. The $^3$He impurity as a probe in liquid $^4$He 97
6.2. The excitation spectrum of the $^3$He impurity 98
6.3. Correlated perturbative approach 99
7. Variational description of Fermi systems 102
7.1. Diagrammatic rules and Fermi hypernetted chain equations 109
7.2. Results for $^3$He 112
7.3. Excited states and the dynamic structure function 114
8. Summary 116
9. Acknowledgments 117
References 118

**Chapter 3 THE COUPLED CLUSTER METHOD AND ITS APPLICATIONS 121**

1. Introduction 122
2. The Coupled Cluster formalism 123
2.1. The exponential form of the wave function 123
2.2. The Configuration Interaction Method (CIM) 126
2.3. The Coupled Cluster equations 129
2.4. The reference state 131

2.5. The Bra Ground state 133
3. Approximation schemes 134
3.1. The SUB(n) or CCn approximation 134
3.2. Applications to Coulomb interacting systems 138
3.3. The HCSUB(n) approximation 142
4. Applications to light nuclei 143
4.1. The TICC2 approximation in configuration representation 145
4.2. TICC2 in coordinate representation 150
4.3. The nuclei $^4$He and $^{16}$O in the TICI2 approximation 156
4.4. Beyond TICI2 161
5. Helium droplets 163
5.1. The J-TICI3 approximation 164
5.2. Ground state of $^4$He droplets 166
5.3. Collective states of $^4$He droplets 168
5.4. $^3$He droplets 173
References 176

**Chapter 4 EXPERIMENTS WITH A RUBIDIUM BOSE-EINSTEIN CONDENSATE 179**
1. Introduction 179
2. Micromotion 181
3. BEC in 1D optical lattice 182
4. Condensate photoionization 186
5. Conclusions and acknowledgments 188
References 188

**Chapter 5 THEORETICAL ASPECTS OF BOSE–EINSTEIN CONDENSATION 191**
1. Bosons and condensation 191
2. BEC in $^4$He 194
3. BEC in dilute systems 197
4. Conclusions 201
5. Acknowledgments 202
References 202

**Chapter 6 ELEMENTARY EXCITATIONS AND DYNAMIC STRUCTURE OF QUANTUM FLUIDS 205**
1. Introduction 205
2. Ground state of a quantum Bose fluid 208
2.1. Optimized ground state 208
2.2. Euler equation with the Jastrow wave function 210
3. Equation of motion method 214
3.1. Linear response 214

3.2. Time–dependent correlation functions 216
3.3. Action integral 217
3.4. Least action principle 218
3.5. Many–particle densities and currents 219
3.6. One– and two–particle continuity equations 220
4. Solving the continuity equations 221
4.1. Feynman approximation 222
5. CBF–approximation 224
5.1. Convolution approximation 225
5.2. Two–particle equation 226
5.3. One–particle equation 228
5.4. The self–energy and the linear response function 229
5.4.1. Numerical evaluation of the self–energy 230
5.5. Analytic structure of the self–energy 231
5.5.1. Anomalous dispersion in liquid $^4$He 233
5.5.2. Absolute minimum in the spectrum 234
5.6. Dynamic structure function in the CBF–approximation 236
5.7. Summary 239
6. The full solution 239
6.1. Continuity equations 239
6.2. Continuity equations in momentum space 242
6.3. Phonon–roton spectrum 244
6.4. Sum rules 246
6.5. Results on two–particle currents 250
6.6. Precursor of the liquid–solid transition 251
7. Dynamics of a single impurity 253
7.1. Continuity equations 253
7.2. Linear response and self–energy 257
7.3. Hydrodynamic effective mass 260
8. Summary 262
References 262

**Chapter 7 THEORY OF CORRELATED BASIS FUNCTIONS 265**
1. Introduction 265
2. Basics of CBF theory 268
2.1. Motivations, basic concepts, and definitions 268
2.2. Finite–order correlated basis functions theory 271
3. Techniques for matrix elements 272
3.1. Definitions and notations 272
3.2. Techniques for matrix elements. I. Diagonal quantities 275
3.3. Techniques for matrix elements. II. Off–diagonal quantities 278
4. Interpretation of effective interactions 283

4.1. Quasiparticle interaction 283
4.2. Variational BCS theory 289
4.3. CBF, the optimization problem, and momentum–dependent correlations 295
5. Infinite order CBF theory 298
5.1. Introduction 298
5.2. Correlated coupled cluster theory 299
5.3. CBF ring diagrams 308
6. Dynamics in correlated basis functions 315
6.1. Equations of motion for correlated states 317
6.2. Coherence and the Feynman theory of excitations 319
6.3. Diagrammatic reduction 320
6.4. Local approximations 322
6.5. Averaged CRPA equations 324
6.6. Response function and dynamic structure function 325
References 326

**Chapter 8 THE MAGNETIC SUSCEPTIBILITY OF LIQUID $^3$He 329**

1. Introduction 329
2. The susceptibility of bulk liquid $^3$He 330
3. Liquid $^3$He confined in aerogel 334
4. Two-dimensional liquid $^3$He 335
5. Conclusions 337
6. Acknowledgements 337
References 337

**Chapter 9 THE HYPERSPHERICAL HARMONIC METHOD: A REVIEW AND SOME RECENT DEVELOPMENTS 339**

1. Introduction 339
2. Microscopic systems 341
3. Jacobi coordinates 342
4. Hyperspherical coordinates 345
5. Hyperspherical functions 347
6. The coupled equations 351
7. The hyperspherical harmonic expansion in momentum space 352
8. Results for the $A = 3$, 4 nuclei with the hyperspherical harmonic expansion 354
9. Modified hyperspherical harmonic expansions 356
9.1. The potential basis 357
9.2. The correlated expansion 357
9.3. The adiabatic approximation 360

9.4. The extended hyperspherical harmonic expansion 361
9.5. Application of the EHH expansion to the helium atom 362
10. Variational calculations for three- and four-nucleon scattering processes 365
10.1. $N-d$ scattering 365
10.2. Results for the $N-d$ scattering 367
10.3. Results for the low energy $n-^{3}H$ and $p-^{3}He$ scattering 368
11. Electro-weak reaction on few-nucleon systems 369
11.1. The $p-d$ radiative capture 370
11.2. The *hep* reaction 371
12. Conclusions 374
References 376

**Chapter 10 THE NUCLEAR MANY–BODY PROBLEM 379**
1. Introduction 379
1.1. The nuclear interaction 381
1.2. Quantum simulations 383
1.3. Plan of the paper 385
2. The Hamiltonian 385
2.1. Two–body potential 385
2.2. Three–body interaction 387
3. The AFDMC method 388
3.1. The auxiliary field breakup for a $v_6$ two–body potential 389
3.2. Break–up for the three–body potential 389
3.3. The spin-orbit propagator 390
3.4. Trial wave function and path constraint 391
3.5. Tail corrections 393
3.6. The AFDMC algorithm 393
3.7. Finite size effects: The periodic box FHNC method 395
4. AFDMC applications to nucleon matter 396
4.1. Equation of state of neutron matter 396
4.2. Symmetry energy of nuclear matter 399
4.3. Spin susceptibility of neutron matter 400
5. Outlook and Conclusions 402
References 405

INDEX 407

# PREFACE

This volume of the series *Advances in Quantum–Many–Body Theories* contains the lecture notes of the second European summer school on *Microscopic Quantum Many–Body Theories and their Applications,* which was hosted during the time of Sept. 3– Sept. 14 by the Abdus Salam Center for Theoretical Physics in Miramare, Trieste (Italy). The aim of this school was to introduce a selected group of graduate students and young postdoctoral researchers to the dominant and most successful techniques of modern microscopic many–body theory. This summer school is the sequel of a preceding one that was held in September 1997 at the Universidad Internacional Menéndez Pelayo (UIMP) in Valencia. [a]

Modern quantum many-body theory (QMBT) had its birth some 50 years ago with the pioneering work of Brueckner, Gell–Mann, Feynman, Landau, Noziéres, Pines, to name only a few. It has since grown to become one of the most fundamental and exciting areas of modern theoretical physics. Its aims are to understand and predict those properties of macroscopic matter that have their origins in the underlying interactions between, and the quantum-mechanical nature of, the elementary constituents at the most microscopic level relevant to the energy range under consideration. The field is naturally multi–disciplinary within physics. Hence, QMBT has become an essential tool for researchers working in several, and apparently different, fields of physics, chemistry, and other disciplines. Among the specific areas of application we may count condensed matter; nuclear and high-energy physics; dense matter astrophysics; atoms and molecules; and elementary particles.

The variety of current approaches to the microscopic many-body problem includes density functional theory, the hypernetted chain/correlated basis functions formalism, the coupled cluster method, and numerical simulation methods. An important point that must be stressed is that the rapid evolution of the different formulations of QMBT over the last decade has provided valuable new insights regarding their intimate interrelationships. An appreciation of this underlying identity will surely provide students and researchers with a much deeper understanding of the physical content of QMBT itself and will offer a broader variety of practical theoretical tools. In this sense, familiarity with the basics of many-body theories should be part of the background knowledge of many researchers.

[a] *Microscopic Quantum Many–Body Theories and their Applications,* eds. J. Navarro and A. Polls, in "Lecture Notes in Physics" Vol. 510 (1998).

This book contains pedagogical introductions to the above–mentioned dominant techniques of modern many–body theory, leading up to today's front–line research. These techniques have their roots in the standard analytical methods of theoretical physics: perturbation theory, scattering theory, and stationarity principles. Moreover, the interplay between these methods and the computational ones has led to fruitful and novel insights into the physics of many–particle systems that deserve to be brought to the attention of the students.

A series of lectures on numerical simulation techniques has been delivered by Gaetano Senatore. They followed relatively closely the published notes by D. Ceperley[b] and it was felt that this adequate reference material on the subject did not justify an independent re–writing.

Two aspects can be broadly distinguished in QMBT: the *methods* or *techniques* used to study QMB systems, and the *specific fields of application* as well as experimental verifications. Mindful of these primary objectives, this volume contains also reviews on *modern developments* and the applications include hypersherical expansion methods, the theory of highly dynamic systems, as well as some key experiments that address questions containing direct challenges to many–body theorists. The contributions by E. Arimondo and H. Godfrin are included; two more seminars were given by P. Martin on metallic clusters and A. F. G. Wyatt on quantum evaporation.

We have selected authors for all of the above subjects with particular care both on the basis of outstanding reputations in the fields they represent and of their recognized research experience and knowledge of generic many–body theory. This volume, together with the lecture notes of the Valencia school, addresses the striking lack of a related pedagogical literature that would allow researchers to acquire the requisite physical insight and technical skills. While trying to avoid too much overlap with the Valencia lecture notes, we have nevertheless tried to make the articles contained in this volume self–contained.

The school was facilitated by a grant from the European Community (HPCF-CT-1999-00197) as well as support from the ICTP and SISSA. The organizers would like to thank Mrs. Doreen Sauleek for her efficient and courteous management.

A. Fabrocini
S. Fantoni
E. Krotscheck

[b]see, for example, `http://archive.ncsa.uiuc.edu/Apps/CMP/papers/cep96b.ps` and `http://www.mcc.uiuc.edu/SummerSchool/David%20Ceperley/ceperley.pdfit`

## CHAPTER 1

# DENSITY FUNCTIONAL THEORY

George F. Bertsch
*Institute for Nuclear Theory*
*and Department of Physics and Astronomy*
*University of Washington Seattle WA 98195 USA*
*E-mail: bertsch@u.washington.edu*

Kazuhiro Yabana
*Institute of Physics*
*University of Tsukuba*
*Tsukuba 305-8577 Japan*

Density functional theory is a remarkably successful theory of ordinary matter, despite its *ad hoc* origins. These lectures describe the theory and its applications starting from an elementary level. The practical theory uses the Kohn-Sham equations, well-chosen energy functionals, and efficient numerical methods for solving the Schroedinger equation. The time-dependent version of the theory is also useful for describing excitations. These notes are based on courses given by one of the authors (GFB) at the Graduiertenkolleg in Rostock, Germany in March 2001 and the Summer School on Microscopic Quantum Many-Body Theories in Trieste, Italy in September 2001.

## 1. Introduction

The density functional theory is now widely applied in all areas of physics and chemistry, wherever properties of systems of electrons need to be calculated. The theory is very successful in calculating certain properties–hence its popularity. This is reason enough for a student of theory to learn what it is all about. However, it is quite different in philosophy to other many-body approaches that you will hear about. The tried-and-true path in theoretical physics is to look for systematic expansions for calculating the properties of interest, finding controlled approximations that be refined to achieve greater accuracy. The density functional theory is not at all systematic, and in the end its justification is only the quality of its predictions. However, it is rightly described as an *ab initio* framework, giving theories whose parameters are determined *a priori* by general considerations. These lectures will present the theory and its applications at a pace that I hope is understandable with a minimum of prior formal training in advanced quantum mechanics. In the first lecture today, I will set the stage by deriving Hartree-Fock theory, presenting some results on the homogeneous electron gas, and finally presenting the Hohenberg-Kohn theorem, which has motivated the density functional approach.

I will begin the next lecture with a simple example of a density functional theory which can be worked out, ending up with the Thomas-Fermi theory of many-electron systems. Unfortunately, the Thomas-Fermi theory has very limited validity, and it has not been possible to make useful improvements despite many attempts. The DFT became useful only after Kohn and Sham introduced electron orbitals into the functional. In their theory the variables are the single-particle wave functions of electrons in occupied orbitals as well as the electron density. The theory then has a structure very close to mean-field theories such as the Hartree theory. The emphasis on using the density variable wherever possible leads directly to a version of the theory called the Local Density Approximation (LDA). The LDA is a significant improvement over Hartree-Fock (in ways we shall discuss), but at the same time one can see deficiencies inherent in that scheme. A more complicated implementation of the theory, called the Generalized Gradient Approximation, makes it surprising accurate for calculating structures and binding energies, and in this form the theory is widely applied.

The Kohn-Sham theory requires solving the 3-dimensional Schroedinger equation many times, and questions of algorithms and numerical methods are important in making applications of the theory. There are several well-developed methods to solve the equations, and each has its advocates. In my third lecture I will discuss some of these numerical aspects. I will also survey some of the applications, noting where the DFT is reliable and where its accuracy is problematic. I will also mention some directions that have been taken to make more accurate theories, going beyond the DFT.

All of this so far is a theory of matter in its ground state. We are of course also very interested in the excitations of many-body systems, and the DFT can also be applied to dynamics, where it is called time-dependent density functional theory (TDDFT). In my fourth lecture I will derive the equations to be solved and the algorithms used to solve the equations. The time-dependent theory is quite computationally intensive, and much progress can be made by finding more efficient numerical techniques. Finally, in the last lecture, I will show you some state-of-the-art applications of the TDDFT.

Although it is not really necessary for my lectures, I will use a second-quantized field operator notation because it is the most efficient way to write down expectation values in many-particle spaces. Let us start with the basic Hamiltonian, which can be taken as the sum of three terms,

$$H = \int d^3r \frac{\hbar^2}{2m} \nabla\psi^\dagger(\mathbf{r}) \cdot \nabla\psi(\mathbf{r}) + \frac{1}{2}\int d^3r \int d^3r' \frac{e^2}{|\mathbf{r}-\mathbf{r}'|}\psi^\dagger(\mathbf{r})\psi^\dagger(\mathbf{r}')\psi(\mathbf{r}')\psi(\mathbf{r}) + \int d^3r V_{ext}(\mathbf{r})\psi^\dagger(\mathbf{r})\psi(\mathbf{r}). \tag{1.1}$$

The terms represent the electron kinetic energy, the electron-electron interaction, and the interaction of the electrons with an external field, respectively. The $\psi^\dagger$ and $\psi$ are field operators with the Fermion anticommutation relations, $\{\psi^\dagger(\mathbf{r}), \psi(\mathbf{r}')\} = \delta(\mathbf{r} - \mathbf{r}')$. I will explain what one needs to know about these as we go along. As a warm-up to the theory, I will derive the Hartree-Fock theory. But before that, some issues of notation and units should be clarified.

### 1.1. *Units and notation*

In eq. 1.1 we used units in which $e^2$ has dimensions of energy-length. If you are used to the MKS system, you can convert formulas by the substitution $e^2 \rightarrow e^2_{MKS}/4\pi\epsilon_0$. One often sees formulas quoted in atomic units, with no explicit dimensional quantities. In atomic units, lengths are expressed in units of the Bohr, $a_0 = \hbar^2/me^2 = 0.529..$ Å and energies in units of the Hartree, $e^2/a_0 = 27.2.$ eV. Confusingly, one also sees energies quoted in Rydbergs, $e^2/2a_0 = 13.6..$ eV. Personally, I do not care for implicit atomic units because they hide the functional dependence on mass and charge. It is also common to express densities in terms of the parameter $r_s$, defined as the radius in atomic units of a sphere whose volume is the reciprocal density. Thus $r_s = (3n/4\pi)^{1/3}\hbar^2/me^2$, where $n$ is the density of electrons. In presenting numerical results, I will often use "practical atomic units", taking eV for energy and Å (0.1 nm) for length.

### 1.2. *Hartree-Fock theory*

Hartree-Fock theory is very simple to describe: it is the variational theory obtained by the expectation value of the Hamiltonian, allowing all wave functions that can be represented as Slater determinants. Let's see how this comes about. Using second-quantized notation, the Slater determinants constructed from a set of orthonormal single-particle wave functions $\{a\}$ are represented by a product of creation operators $c^\dagger$ acting on the vacuum. An $N$-particle state is thus

$$|N\rangle = \sum_a^N c_a^\dagger|\rangle.$$

The operators $c\dagger$ and $c$ satisfy the anticommutation relation $\{c_a^\dagger, c_{a'}\} = \delta_{a,a'}$. To get back the orbital wave function in position space, i.e. to reveal the spatial wave function $\phi_a(r)$, we apply the field operator $\psi(\mathbf{r})$ to the state $a$. The anticommutator gives the sought amplitude,

$$\{\psi(\mathbf{r}), c_a^\dagger\} = \phi_a(\mathbf{r}).$$

We now take the expectation value of $H$ in the state $|N\rangle$ and reduce the operator expectation values by moving annihilation operators to the right and creation operators to the left with the help of the above anticommutators. The result at the end is

$$\langle N|H|N\rangle = \sum_a^N \frac{\hbar^2}{2m}\int d^3r \nabla\phi_a^* \cdot \nabla\phi_a + \sum_{a<b}\int d^3r \int d^3r' \frac{e^2}{|\mathbf{r}-\mathbf{r}'|}|\phi_a|^2|\phi_b|^2$$
$$-\sum_{a<b}\int d^3r \int d^3r' \frac{e^2}{|\mathbf{r}-\mathbf{r}'|}\phi_a^*(\mathbf{r})\phi_a(\mathbf{r}')\phi_b^*(\mathbf{r}')\phi_b(\mathbf{r}) + \sum_a \int d^3r V_{ext}(\mathbf{r})|\phi_a|^2 . \quad (1.2)$$

The result looks very similar to eq. (1.1) with respect to the kinetic energy and the external potential energy terms. But the electron-electron interaction has given rise to two terms, the direct (or Hartree) energy, and the exchange (or Fock) energy. Notice also that the factor of 1/2 in eq. (1.1) has disappeared; instead one has a double sum over the $N(N-1)/2$ orbital pairs $(a,b)$. It is often convenient to rewrite eq. (1.2) rearranging the sums slightly. Let us add terms with $a=b$ to the direct and exchange sums. This won't affect the result, because the direct and exchange cancel if the two orbitals are the same. The direct term can then be written as an independent sum over the $a$ and $b$ orbitals. Defining the single particle density $n(\mathbf{r}) = \langle N|\psi^\dagger(\mathbf{r})\psi(\mathbf{r})|N\rangle = \sum_a^N |\phi_a(r)|^2$, the direct and external field terms are seen to depend directly on $n(r)$. The full expectation value becomes

$$\langle N|H|N\rangle = \sum_a^N \frac{\hbar^2}{2m}\int d^3r \nabla\phi_a^* \cdot \nabla\phi_a + \frac{1}{2}\int d^3r \int d^3r' \frac{e^2}{|\mathbf{r}-\mathbf{r}'|}n(\mathbf{r})n(\mathbf{r}')$$
$$-\sum_{a<b}\int d^3r \int d^3r' \frac{e^2}{|\mathbf{r}-\mathbf{r}'|}\phi_a^*(\mathbf{r})\phi_a(\mathbf{r}')\phi_b^*(\mathbf{r}')\phi_b(\mathbf{r})$$
$$+\int d^3r V_{ext}(\mathbf{r})n(\mathbf{r}) . \quad (1.3)$$

Now that we have the Hartree-Fock energy function, the next task is to find the minimum within the allowed variational space. First let us recall quickly how variational principles work. If we have a integral expression $\int F(\phi)dx$ that depends on a function $\phi(x)$, the condition that the value is stationary with respect to variations in $\phi$ is

$$\frac{dF}{d\phi} = 0. \quad (1.4)$$

This must be satisfied for all values of $x$. If there is a constraint that some other integral $\int G(\phi)dx$ has a fixed value, the stationary condition contains the constraint as a Lagrange multiplier,

$$\frac{dF}{d\phi} + \mu\frac{dG}{d\phi} = 0. \quad (1.5)$$

We now apply this to the Hartree-Fock energy, eq. (1.2), varying with respect to a wave function amplitude $\phi_a^*$. Remembering that the wave functions were assumed to be normalized, we impose the constraint $\int \phi_a^*\phi_a d^3r = 1$ with a Lagrange multiplier. The multiplier will be denoted $\epsilon_a$; it looks exactly like the energy in the Schrödinger equation. The wave functions also have to be orthogonal as well, but it turns out

Table 1. Atomization energies of selected molecules

| | $Li_2$ | $C_2H_2$ | 20 simple molecules (mean absolute error) |
|---|---|---|---|
| Experimental | 1.04 eV | 17.6 eV | - |
| Theoretical errors: | | | |
| Hartree-Fock | -0.94 | -4.9 | 3.1 |
| LDA | -0.05 | 2.4 | 1.4 |
| GGA | -0.2 | 0.4 | 0.35 |
| $\tau$ | -0.05 | -0.2 | 0.13 |

that it is not necessary to put in Lagrange multipliers to satisfy that condition. There is one more technical point in carrying out the variation. When the gradient of a function is varied, one first integrates by parts to move the gradient elsewhere in the expression. One must impose suitable boundary conditions on the function to carry out the integration by parts, and that must be remembered in solving the differential equations that result from the variation.

Without going through the steps I will just quote the result here. One obtains $N$ equations for the amplitudes $\phi_a$,

$$-\frac{\hbar^2}{2m}\nabla^2\phi_a(\mathbf{r}) + \int \frac{e^2}{|\mathbf{r}-\mathbf{r}'|}n(\mathbf{r}')d^3r'\phi_a(\mathbf{r}) - \sum_b \int d^3r' \frac{e^2}{|\mathbf{r}-\mathbf{r}'|}\phi_a(\mathbf{r}')\phi_b^*(\mathbf{r}')\phi_b(\mathbf{r}) + V_{ext}(\mathbf{r})\phi_a(\mathbf{r}) = \epsilon_a\phi_a(\mathbf{r})\,. \tag{1.6}$$

These are the Hartree-Fock equations. It is interesting to see how well they do in making a theory of matter. In Table I is shown some energies calculated with eq. (1.6), taken from Refs. 1, 2. The entries in the table are atomization energies, which is the energy require to pull the cluster or molecule apart into individual atoms. Results are given for a simple atomic cluster, a simple molecule, and a set of molecules that are used as a testing ground for better theories. The mean absolute error in the atomization energies (energy difference between the molecule and the individual atoms in isolation) is 3 eV in the Hartree-Fock theory. The predicted binding of the $Li_2$ clusters is a factor ten too low, and another alkali metal cluster not in the table, $Na_2$, is incorrectly predicted to be unbound. We conclude that on a practical level Hartree-Fock is not accurate enough to be useful for chemistry or for computing cluster structures.

## 1.3. *Homogeneous electron gas*

We will see next time that the density functional theory makes use of the properties of the homogeneous interacting electron gas, and it will be useful to have on hand some analytic results. There is a systematic expansion of the energy of an electron gas accurate at high density. The first two terms are contained in the Hartree-Fock theory. They are the kinetic energy of a free Fermi gas, and its exchange energy. As part of the warmup, I will now derive them.

### 1.3.1. *Free electrons*

For the free electron gas eq.(1.6) is solved without the potential energy, and in a box (a cube of side $L$) with periodic boundary conditions. Since the Hamiltonian contains only the one-particle kinetic energy operator, the equation is separable into $N$ one-particle problems. The solution are plane waves with discrete wave vectors $\mathbf{k}$ chosen to fulfill the boundary conditions $\phi(x+L,y,z)=\phi(x,y,z)$, etc.,

$$\phi_{\mathbf{k}}(\mathbf{r}) = L^{-3/2}e^{i\mathbf{kr}} \quad \text{where} \quad \mathbf{k} = \left(\frac{2\pi}{L}\right)\begin{pmatrix} n_x \\ n_y \\ n_z \end{pmatrix} \qquad n_i = 0, \pm 1, \pm 2, ... \tag{1.7}$$

According to the Pauli Principle each $\mathbf{k}$-vector can be occupied by two electrons. Thus the volume in $k$-space required for one electron is

$$V_k^0 = \frac{1}{2}\left(\frac{2\pi}{L}\right)^3 \tag{1.8}$$

and a sum over states can be replaced by an integral over $\mathbf{k}$ according to

$$\sum_{\mathbf{k}} \to 2L^3 \int \frac{d^3k}{(2\pi)^3}.$$

The kinetic energy of a particle of wave vector $\mathbf{k}$ is given by $\epsilon_{\mathbf{k}} = \hbar^2 k^2/2m$. Clearly the lowest energy state for a given number of electrons requires that the states are filled within a sphere in $k$-space, the Fermi sphere. Calling the radius of the sphere the Fermi wave number $k_F$, the number of particles can be expressed

$$N = 2L^3 \int_{k<k_F} \frac{d^3k}{(2\pi)^3} = L^3 \frac{k_F^3}{3\pi^2}$$

and the corresponding density $n$ is given by

$$n = N/L^3 = k_F^3/3\pi^2.$$

The total kinetic energy of the electron gas is found by integrating the single-electron kinetic energy over the Fermi sphere,

$$T = 2L^3 \int_{k<k_F} \frac{(\hbar k)^2}{2m} \frac{d^3k}{(2\pi)^3} = \frac{L^3\hbar^2 k_F^5}{10m\pi^2}$$

We shall later need the kinetic energy density as a function of particle density. Combining the last two equations one finds

$$t = \frac{T}{L^3} = \frac{3}{10}\frac{\hbar^2(3\pi^2)^{2/3}}{m} n^{5/3} \tag{1.9}$$

### 1.3.2. *Exchange energy*

We evaluate the interaction in the Hartree-Fock approximation to find the next term in the energy of the electron gas. I will assume that the system remains uniform in the presence of the interaction. In general this should not be taken for granted; just because the Hamiltonian has some symmetry does not imply that the solutions do also. For the uniform solution, the wave functions remain plane waves. The next problem is that the Hartree term diverges because of the long range of the Coulomb interaction. The remedy is to include an external field to compensate, adding a positive background charge density equal in magnitude to the electron charge density. This is called the *jellium model*; in it the Hartree term is canceled exactly. That leaves the exchange energy given by

$$E_x = -\frac{1}{2} 2L^6 \int_{k_F} \frac{d^3k}{(2\pi)^3} \frac{d^3k'}{(2\pi)^3} \int_L d^3r \int_L d^3r' \frac{e^2}{|\mathbf{r}-\mathbf{r}'|} \phi^*_{\mathbf{k}}(\mathbf{r}) \phi_{\mathbf{k}}(\mathbf{r}') \phi^*_{\mathbf{k}'}(\mathbf{r}') \phi_{\mathbf{k}'}(\mathbf{r}). \quad (1.10)$$

The factor of 1/2 comes from replacing the sum over orbital pairs by an unrestricted sum over orbitals. Next to it there is a single factor of 2 for spin because the two particles exchanged have to have the same spin wave function. It is easy to see how this integral will depend on the charge and on the density without going through the details of carrying it out. First, the factors of $L$ will cancel out when the expressions for the plane wave states are inserted. The dimensionful quantities left are the electron charge squared, $e^2$, and the Fermi wave number $k_F$. These have dimensions of energy-length and (length)$^{-1}$, respectively. The only quantity that can be constructed from these with dimensions of energy density is $e^2 k_F^4$. The result of carrying out the integration must be that expression multiplied by a pure number. I won't carry out the integrals here but just give the result. The exchange energy density $e_x$ as a function of number density is

$$e_x = \frac{E_x}{L^3} = -\frac{3}{4\pi} e^2 (3\pi^2)^{1/3} n^{4/3}. \quad (1.11)$$

Notice that the power of $n$ is less than in the kinetic energy term. This suggests that the expansion one is making for the energy in terms of the density is a high-density expansion. Indeed, when one evaluates the third term in the systematic many-body theory, one finds it to be small compared to the previous terms in the high density limit. But there is a surprise at that order: the expansion is not a simple Taylor series in $1/n^{1/3}$, but is nonanalytic[a]. For future reference, I will quote the full expansion up to that order:

$$e = 2.87 \frac{\hbar^2}{m} n^{5/3} - 0.74 e^2 n^{4/3} + \frac{e^4 m}{\hbar^2} (0.031 \ln(\hbar^2 n^{1/3}/me^2) - 0.062) + \ldots \quad (1.12)$$

[a]This was derived by Gell-Mann and Brueckner in 1957. [3]

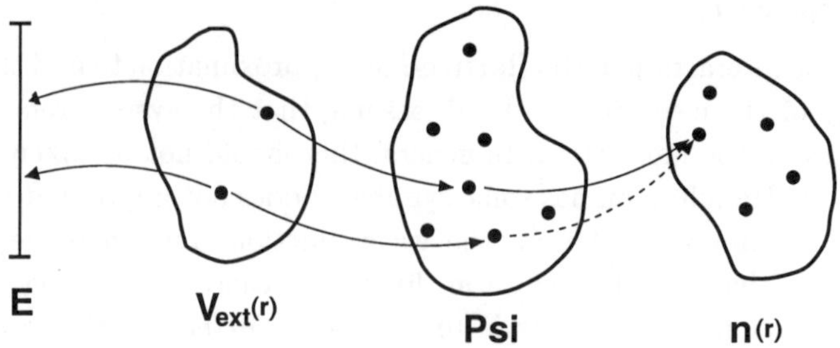


Fig. 1. Logic of the Hohenberg-Kohn theorem. The dashed mapping from a many-body wave function $\Psi$ to a density $n(r)$ is not possible, given that another mapping exists.

## 2. What is density functional theory?

The Hohenberg-Kohn theorem states that a variational functional exists for the ground state energy of the many-electron problem in which the varied quantity is the electron density. In this section I will go through the proof and then give a simple example of a candidate density functional theory.

### 2.1. *Hohenberg-Kohn theorem*

We first separate the many-electron Hamiltonian into a part $H_0$ that depends only on the electron coordinates and a part $V_{ext}$ that contains the electrons' interaction with external charges and fields. Referring back to eq. 1.1, $H_0$ contains the first two terms and $V_{ext}$ is the last term. It should not cause any confusion to use the symbol $V_{ext}$ both for the operator in the many-electron space and for the ordinary function of the coordinate. Let us call the ground-state wave function for this Hamiltonian $\Psi$. In principle we could then find the ground state energy $E = \langle\Psi|H_0 + V_{ext}|\Psi\rangle$ and the electron density in the ground state, $n(r) = \langle\Psi|\psi^\dagger(r)\psi(r)|\Psi\rangle$. These mappings are shown as the solid arrows in Fig. 1. We now ask the question of whether another Hamiltonian, differing only by the external field, could have the same ground state density. That situation would be represented by the dashed arrow in Fig. 1. This cannot happen if the ground state is nondegenerate. The proof is carried out by *reductio ad absurdum.* Suppose that a $V'_{ext}$ and a $\Psi'$ exist such that $n'(r) = n(r)$, with a ground state energy $E' = \langle\Psi'|H_0 + V'_{ext}|\Psi'\rangle$. If the ground state is nondegenerate, the other state must have a higher expectation value of the primed Hamiltonian,

$$\langle\Psi|H_0 + V'_{ext}|\Psi\rangle > E'$$

By adding and subtracting the expectation of $V_{ext}$ on the left hand side, it can be rewritten to put the inequality in the form

$$E + \int n(r)(V'_{ext} - V_{ext})d^3r > E'$$

In deriving this, we also made use of the assumption that $n = n'$. The same reasoning can be carried out starting from the expectation values of the Hamiltonian $H_0 + V'_{ext}$

in both states. That gives the inequality

$$E' + \int n(r)(V_{ext} - V'_{ext})d^3r > E$$

Adding the two inequalities gives

$$E + E' < E + E'$$

which is a contradiction. Thus, either the states are degenerate or $n \neq n'$. Since the mapping of $\Psi$ into $n$ is one-to-one, its image in $n$ is invertible. Let us call the functions of the image $n_v$ (possible densities associated with Hamiltonians of the form $H_0 + V_{ext}$), and the mapping from densities back to wave functions $\Psi[n_v]$.

We are now ready for the main theorem. The Hohenberg-Kohn theorem states that the mapping of $n_v$ to energies $E$ defined by

$$E = \langle \Psi[n_v] | H_0 + V_{ext} | \Psi[n_v] \rangle$$

is a variational principle whose minimum is the ground state energy. Given the mappings, the proof is trivial. The energy is variational in the full space of wave functions, and thus must also be variational in a restricted space that includes the minimum. Note that the derivation gives no clue on how to construct the functional, nor does it guarantee that it is smooth enough to apply the variational equations 1.4 or 1.5.

## 2.2. *A simple example: the Thomas-Fermi theory*

The simplest example of a DFT, which I will now derive, is the Thomas-Fermi model electronic systems. Looking again at the Hartree-Fock energy function, eq. (3), we see that two of the four terms are already in the desired form, i.e. depending only on the density $n(\mathbf{r})$. These are the direct electron-electron interaction and the interaction with the external potential. That leaves the kinetic energy and the exchange energy to be approximated by a function or functional of density. The Thomas-Fermi theory emerges when we ignore the exchange energy and make the simplest possible approximation for the kinetic energy. For a slowly varying density function the kinetic energy density will only depend on the number density at the same position, Taking the specific function from the Fermi gas, eq.(1.9), we arrive at the kinetic energy functional

$$T[n(\mathbf{r})] = \int \epsilon(n(\mathbf{r}))n(\mathbf{r})d^3\mathbf{r} = \int \frac{3}{10}\frac{\hbar^2(3\pi^2)^{2/3}}{m}n^{5/3}(\mathbf{r})d^3\mathbf{r} \tag{2.1}$$

The sum of the kinetic and the potential energy terms will give us the total energy within the Thomas-Fermi approximation,

$$E[n(\mathbf{r})] = \int \left[\frac{3}{10}\frac{\hbar^2(3\pi^2)^{2/3}}{m}n^{5/3}(\mathbf{r}) + \frac{e^2}{2}\int \frac{n(\mathbf{r})n(\mathbf{r}')}{|\mathbf{r}-\mathbf{r}'|}d^3\mathbf{r}' + V_{ext}n(\mathbf{r})\right]d^3\mathbf{r} \tag{2.2}$$

Note that the expression only depends on $n(r)$. In the next section we will apply the variational principle to find the Thomas-Fermi equation.

### 2.2.1. *Variational equation of Thomas-Fermi theory*

To find the ground state density function, one has to minimize the total energy functional for a fixed number of particles. This constraint is included by adding a term to the functional, $\mu \int n d^3 r$, with $\mu$ the Lagrange multiplier. The variational problem is then expressed

$$\begin{aligned} 0 &= \delta E[n(\mathbf{r})] \qquad (2.3) \\ &= \delta \int \left[ \frac{3}{10} \frac{\hbar^2 (3\pi^2)^{2/3}}{m} n^{5/3}(\mathbf{r}) + \frac{e^2}{2} \int \frac{n(\mathbf{r}) n(\mathbf{r}')}{|\mathbf{r}-\mathbf{r}'|} d^3\mathbf{r}' + V_{ext}(\mathbf{r}) n(\mathbf{r}) - \mu n(\mathbf{r}) \right] d^3\mathbf{r} \end{aligned}$$

where the variation is to be taken with respect to n($\mathbf{r}$). Eq. 1.5 then gives

$$\frac{\hbar^2}{2m} (3\pi^2)^{2/3} n^{2/3}(\mathbf{r}) + e^2 \int \frac{n(\mathbf{r}')}{|\mathbf{r}-\mathbf{r}'|} d^3\mathbf{r}' + V_{ext}(\mathbf{r}) - \mu = 0 \qquad (2.4)$$

This is an integral equation for $n(\mathbf{r})$, but it can be converted into a differential equation which is easier to solve. This is done by defining a potential

$$V_e(\mathbf{r}) = +e^2 \int \frac{n(\mathbf{r}')}{|\mathbf{r}-\mathbf{r}'|} d^3\mathbf{r}'.$$

and seeing that the potential will be a solution of the Poisson equation[b],

$$\nabla^2 V_e(r) = -4\pi e^2 n(\mathbf{r}). \qquad (2.5)$$

We first use eq. (2.4) to express $n$ as a function of $V_e$,

$$n(\mathbf{r}) = \frac{1}{3\pi^2 \hbar^3} \left(2m(-V_{ext}(\mathbf{r}) - V_e(\mathbf{r}) + \mu)\right)^{3/2} \qquad (2.6)$$

Then substitute into the right hand side of eq. (2.5) to obtain

$$\nabla^2 V_e(\mathbf{r}) = -\frac{4e^2}{3\pi\hbar^3} \left(2m(-V_{ext}(\mathbf{r}) - V_e(\mathbf{r}) + \mu)\right)^{3/2}. \qquad (2.7)$$

### 2.2.2. *Thomas-Fermi atom*

Further simplification can be obtained if particular assumptions are made for the external potential $V_{ext}$. For a spherical external field such as that of an atom, the potential functions will only depend on the radius. The potential of the atomic nucleus, $V_{ext}(r) = -Ze^2/r$, can be included into eqns.(2.6,2.7) to yield a general formulation valid for arbitrary atomic $Z$. To analyze the atomic theory, it is convenient to define a dimensionless function $\phi(r)$ according to

$$\frac{Ze^2 \phi(r)}{r} = \frac{Ze^2}{r} - V_e(r) + \mu. \qquad (2.8)$$

[b] In fact, the Poisson equation would be obtained immediately if one started from an energy function that has an explicit term for the electrostatic field energy.

Then the radial Laplacian operator reduces to a simple second derivative, and eq. (14) becomes

$$\phi''(r) = \frac{4e^3 Z^{1/2}}{3\pi\hbar^3} (2m\phi)^{3/2} r^{-1/2}. \tag{2.9}$$

The equation for the density is the same as before; in terms of the variable $\phi$ it is

$$n(r) = \frac{1}{3\pi^2\hbar^3}\left(\frac{2me^2 Z\phi}{r}\right)^{3/2}. \tag{2.10}$$

These equations can be found in textbooks as the general Thomas-Fermi equations for atoms. By taking into account that the electron potential must be finite at $r = 0$ one can fix one initial condition for $\phi$, $\phi(0) = 1$. Since it is a second-order equation, another condition must be imposed. In practice one makes a guess at the initial slope $\phi'(0)$, integrates to large $r$, sees what the solution looks like. There are three cases to be considered. The first is that $\phi$ remains positive and finite over the whole range of $r$ up to infinity. This is unphysical because eq. (2.10) would imply that the electron density was finite everywhere. The next possible is that $\phi$ goes through zero at some point $r_0$, implying (eq. (2.10) again) that $n(r_0) = 0$. We can assume that the density remains zero for larger radii. Let us integrate the density to get the total number of electrons $N$,

$$N = \int_{r_0} n(\mathbf{r}) d^3\mathbf{r}. \tag{2.11}$$

Then the potential outside of $r_0$ must be Coulombic and given by $-(Z - N)e^2/r$. Matching at $r_0$ gives the value of the chemical potential,

$$\mu = -\frac{(Z - N)e^2}{r_0}. \tag{2.12}$$

This defines the Thomas-Fermi model of an atom with $N$ electrons and atomic number $Z$. The chemical potential is negative, implying that $N < Z$.

The last case in integrating the Thomas-Fermi equation is the boundary situation where the function $\phi$ only goes to zero asymptotically at large $r$. The chemical potential is zero in this case, and the integral of the electron density is equal to the nuclear charge. This defines the Thomas-Fermi theory of the neutral atom.

In all cases the Thomas-Fermi equations require a numerical solution. Not even in the neutral-atom case it can be calculated analytically. However, eq. (2.9) has one very nice feature for that case. Namely, there is no finite $r_0$ to fix the dimension of the atom, and all of the dimensional information is contained in the coefficient on the right hand side. The equation assumes the dimensionless form $\phi''(x) = \phi^{3/2}(x)/x^{1/2}$ if we make the change of variable to $r \to (3\pi/4)^{2/3}(\hbar^2/2me^2)x$. in eq.(2.9). Thus all neutral atoms are described by the same universal function $\phi(x)$.

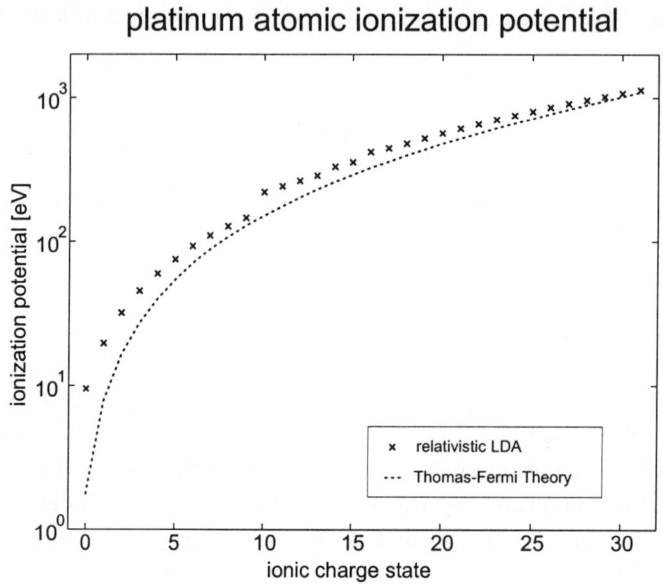


Fig. 2. Comparison of ionization potentials of platinum atoms. Dotted line is the Thomas-Fermi prediction; crosses mark the values from the Dirac-LDA theory.

### 2.2.3. *An example*

Let us see how well the Thomas-Fermi theory does for a heavy atom. I take as an example the Pt atom which has $Z = 78$. We will examine the ionization potentials starting from the neutral atom and going up to a charge $Q = Z - N = 30$. The Thomas-Fermi calculations are done in the way described in the previous section: one tries an initial condition, integrates, examines the charge, and from that refines the initial condition. For a given charge $N$ the total Thomas-Fermi energy can be computed using the energy function eq. (2.2). The ionization potential of the ion is defined as a difference in the total energies; for charge $Q$ $IP = E(N = Z-Q) - E(N = Z-Q-1)$. Of course in the Thomas-Fermi theory $N$ is a continuous parameter; we simply take the energy at the integer values. Experimentally, one only knows the ionization potentials for a few charge states. So for comparison purposes we also calculated the values using the more sophisticated Dirac-Hartree-LDA theory. The results are shown in Fig. (2). We see that the results come out quite well for high charge states, and the overall trend is reproduced. However the first ionization potential comes out poorly. Experimentally it is 9.0 eV and in the relativistic Hartree-LDA theory it is 9.5 eV, but in the Thomas-Fermi it is predicted to be only 1.8 eV.

Another failing of the Thomas-Fermi model is the complete absence of atomic shell effects. It may be seen in the Figure that there is a jump in ionization potentials between charges $Q = 9$ and $Q = 10$. This is due to the depletion of the $5d$ shell and the beginning ionization of the $5p$ shell. Shell effects such as this are in fact beyond the scope of theories that ignore the wave character of quantum mechanics. Thomas-Fermi theory fails on another fundamental property of many-electron physics, namely the existence of molecules. It can be shown (Ref. 4) that all molecules would be unstable with respect to dissociation into isolated atoms, according to the energies calculated in the Thomas-Fermi theory. Again, this points to the need for explicit wave function mechanics. From the Figure one can also see that the Thomas-Fermi systematically underpredicts the ionization potentials. This is a deficiency that can in principle be remedied by the use of a more sophisticated energy function. We shall return to this point later.

## 3. Kohn-Sham theory

We saw in the previous section that explicit quantum mechanics is needed to make a theory that would reproduce the most elementary properties of matter. Kohn and Sham proposed, in essence, to put wave mechanics into the kinetic energy functional, but retain the density variable $n(\mathbf{r})$ elsewhere. Hartree-Fock theory gives us guidance on how to proceed. We take as the primary variational functions the orbitals of the Hartree or Hartree-Fock theory, and we calculate the kinetic energy accordingly. We also use these orbitals to determine the density. The interaction terms that can be expressed directly in terms of the density we keep, and we add another term, the exchange-correlation energy $E_{xc}$ to take care of everything else. The resulting functional is

$$E_{KS}[\phi_1...\phi_N] = \sum_a^N \frac{\hbar^2}{2m}\int d^3r \nabla\phi_a^* \cdot \nabla\phi_a + \frac{1}{2}\int d^3r \int d^3r' \frac{e^2}{|\mathbf{r}-\mathbf{r}'|} n(\mathbf{r})n(\mathbf{r}') + E_{xc}[n]$$

$$+ \sum_a \int d^3r V_{ext}(\mathbf{r})|\phi_a|^2.$$

which is used together with the definition

$$n(\mathbf{r}) = \sum_a^N |\phi_a(\mathbf{r})|^2.$$

### 3.1. *Local density approximation*

Kohn and Sham proposed to treat the term $E_{xc}$ as a simple integral over an energy density $e_{ex}(\mathbf{r})$, which in term is taken as an ordinary function of the local density, i.e. $e_{ex} = f(n(\mathbf{r}))$. This is called the local density approximation (LDA). The obvious choice for the function $f$ is the corresponding energy density of the uniform electron gas. Since the kinetic energy is treated separately, the exchange-correlation energy is what is left over after taking away the first term in eq. (1.12). In fact, the idea to take the exchange energy from the electron gas was first attempted by Slater. Starting from the Fermi gas exchange energy density in eq. (1.11). Slater proposed to make a local density approximation by defining a two-body contact interaction that would have the same energy density. Taking an interaction of the $\delta$-function form $V(\mathbf{r}-\mathbf{r}') = v_0\delta(\mathbf{r}-\mathbf{r}')$, the many-body interaction energy density would be $v = 1/2v_0n^2$. Comparing with the required energy density associated with the exchange, Slater added to the Hartree Hamiltonian a two-particle contact interaction with a strength given by $v_0 = 3e^2(3/\pi)^{1/3}n^{-2/3}/2$. Then the Schrödinger equation would have an additional potential term given by

$$V_{Slater} = -\frac{3e^2}{2\pi}(3\pi^2 n_e(r))^{1/3}.$$

THIS IS WRONG. Going back to the variational principle, one sees that the one-body potential should be defined by the variation of $E_{xc}$, i.e.

$$V_{xc}(r) = \frac{\delta e_{xc}}{\delta n} = -\frac{e^2}{\pi}(3\pi^2 n_e(r))^{1/3}. \tag{3.1}$$

This is a factor 2/3 different from Slater's potential.

In practice, the LDA density functional theory takes the exchange-correlation energy from numerical calculations of the interacting electron gas. Many-body theory gives a few terms of the high-density expansion, but only numerical methods are reliable for the lower densities of interest. The classic reference is the Green's function Monte Carlo calculation of Ceperley and Alder, [5] who gave a table of energies of the homogeneous electron gas. After subtracting the kinetic energy of the free Fermi gas, the function is parameterized in a way suitable for taking the derivative needed in eq. 3.1. Such parameterizations give the "LDA" functional of density functional theory. It gives a considerable improvement over Hartree-Fock, as may be seen by the entries in Table 1. Alkali metal clusters have reasonable binding energies, and the errors in the atomization energies decrease by a factor of two. But this is still not accurate enough for chemical modeling.

### 3.2. *Spin and the local spin density approximation*

Up to now we have assumed that the electrons are all paired in the molecular orbitals: the electron density was calculated assuming two electrons occupy each spatial orbital, and the exchange-correlation energy was fit to the energy of an electron gas with equal number of up and down spin electrons. There is no reason why we can't treat up and down electrons separately, generalizing the theory to treat two (variationally) independent densities. Thus we define spin-up and spin-down densities $n_\uparrow, n_\downarrow$, and construct an exchange-correlation energy depending on both. The exchange part of the interaction is between electrons with the same spin projection, so the corresponding energy functional is a sum of two terms. Going through the same procedure we did to find the exchange interaction before, one finds the for the exchange energy density

$$e_x = -\frac{3}{4}\left(\frac{6}{\pi}\right)^{1/3} e^2(n_\uparrow^{4/3} + n_\downarrow^{4/3}) \tag{3.2}$$

The correlation energy of a spin-polarized electron gas has also been calculated and fit. The parameterization of Perdew and Zunger [6] is often used; the spin-dependent correlation energy is about half the size of the spin-dependent exchange energy, for densities in the range of those occurring for conduction electrons in metals.

Eq. 3.2 has one feature that is obviously unphysical: it is not invariant under rotations. Suppose we have a system with the electrons polarized in the $x$-direction. The electron wave function then has equal amplitudes of up and down spin, and the spin-up and spin-down densities are equal. Thus the formula would treat this incorrectly as an unpolarized system. This can be easily fixed by taking as density variables the entire spin density matrix. [7] The eigenvalues of the matrix are invariant under rotations, and the energy functional depends only on the eigenvalues. I will show an application later on.

### 3.3. *The generalized gradient approximation*

An obvious problem of the LDA is that the single-particle potential does not have the correct asymptotic behavior. The electron potential in a neutral cluster should behave as $-e^2/r$ for large separation of the electron from the cluster. But in the LDA the Coulomb potential is calculated with all the electrons and thus vanishes outside the cluster. This makes the LDA unreliable for calculating ionization potentials from the Kohn-Sham eigenvalues.

There is another argument why one should not trust the absolute numbers of the Kohn-Sham eigenvalues. Let us imagine starting with the full normalized wave function of a particle in its orbital, and take away probability a little bit at a time. Since quantum mechanics is a linear wave theory, the energy cost is the same to take a way a bit of the probability, independent of how much has been taken away already. But the self-interaction and exchange terms are nonlinear functions, and the energy in the Kohn-Sham theory does depend on how much probability is left[c]. This function behaves as $-aP^{4/3}+bP^2$. The cancellation of the self-energies comes about by having $a \approx b$. But Kohn-Sham potential comes from the derivative which is positive near $P=1$.

As an example of how this works out in a physical example, we calculated the ionization potential (IP) of the silver atom both ways. According to Hartree-Fock theory, the Ag atom has a single electron in an $s$ orbital, with an unoccupied $p$ orbital just above and a fully occupied $d$ orbital just below. Thus the IP should correspond to the energy of the $s$-orbital. Table 2 shows the experimental ionization potential (IP) compared to the Kohn-Sham eigenvalue. You see that it is predicted to be more weakly bound than is the actual IP, and the error is quite large (40%).

[c]This gives rise to a real pathology of the LDA functional, that a free electron comes out to be self-bound in its exchange potential

Table 2. Atomic properties of the Ag atom in LDA.

| | IP | First excited state |
|---|---|---|
| Experimental | 7.75 | 3.74 |
| Kohn-Sham: | | |
| eigenvalues | $\epsilon_s = 4.6$ | $\Delta\epsilon = 3.9$ |
| total energies | 8.0 | 4.1 |

What about the other method? The line labeled"total energies" was calculated as a difference between energies of the neutral atom and the singly-charged ion, i.e.

$$IP = E(\text{Ag}) - E(\text{Ag}^+). \tag{3.3}$$

The table shows that the error is only 3% when the IP is calculated this way.

Becke [8] proposed a fix to get the $-e^2/r$ asymptotic potential by adding a term to the energy functional that depends on the gradient of the density, $\nabla n_e(r)$. His proposed form works amazingly well. This is the "generalized gradient approximation", GGA. From Table 1, we see that energies can be calculated to an accuracy of tenths of an eV. Further improvements may be possible. The Kohn-Sham energy functional depends on the nonlocal quantity

$$\tau = \sum_i |\nabla\phi_i|^2$$

in the kinetic energy term. One could think of using other functional dependencies on $\tau$; an example is given in the last line of Table 1.

The functional that is most popular in quantum chemistry is called the "B3LYP". Its popularity derives in part from the fact that it is a option in a widely distributed computer program for quantum chemistry calculations. Unfortunately, the prescription is quite *ad hoc*, taking pieces from many sources include Hartree-Fock, the local spin density approximation, and the GGA. It is not even easy to find its definition without going to the original literature for the individual terms (e.g. Ref. 9).

Later on, I will show some calculations that use the GGA prescription of van Leeuwen and Baerends. [10] They write the exchange-correlation potential as

$$v_{xc}(\mathbf{r}) = -\beta n^{1/3}(\mathbf{r})\frac{x^2(\mathbf{r})}{1 + 3\beta x(\mathbf{r})\sinh^{-1}(x(\mathbf{r}))}$$

where $x = |\nabla n|/n^{4/3}$.

## 4. Numerical methods for the Kohn-Sham equation

The DFT is conceptually quite simple, but its success is in part due to the development of efficient methods to solve the Kohn-Sham equations. The algorithms may be classified by how the orbital wave function is represented. The traditional approach in quantum chemistry is to use a basis set of analytic functions centered on each atom. Gaussian basis sets are very popular because the Coulomb integrals

Table 3. Symbol definitions for quantities pertaining to the computational effort required to solve the Kohn-Sham equations of DFT.

| | Symbol | Meaning |
|---|---|---|
| $N$ | $N_\phi$ | dimension of vector representation $\phi$ |
| | $N_{\mathbf{r}}$ | real-space points |
| | $N_{\mathbf{k}}$ | reciprocal-space points |
| | $N_e$ | number of electron orbitals |
| | $N_c$ | unoccupied orbitals |
| $M$ | $M_H$ | nonzero elements in H matrix row |
| | $M_{it}$ | iterations in conjugate gradient method |
| | $M_T$ | time steps |
| | $M_\omega$ | number of frequencies |

can be evaluated exactly. Such representations have several disadvantages, however. The basis is not orthogonal, which makes it somewhat inconvenient to use. To establish convergence one has to have the flexibility to add additional functions, and this may not be easy to do, for example when one is describing weakly bound electrons. The other kind of representation is to use plane waves, i.e. a mesh in momentum space, or a mesh in coordinate space. These representations suffer from the disadvantage that the required mesh has a much finer granularity close to the nuclei than far away. It is not yet practical to deal with the full range of energy scales in the atomic problem with a uniform mesh. The way out for condensed matter theorists is to replace the nuclear potential by a much smoother pseudopotential, so that a relatively coarse grid can be used everywhere. There is much art to the construction of pseudopotentials, and I will not discuss them in detail. A popular prescription for constructing pseudopotentials was formulated in Ref. 11; a program to generate pseudopotentials this way can be download from the web site `http://bohr.inesc.pt/~jlm/pseudo.html`. Another more recent prescription together with tables of the pseudopotential parameters can be found in Ref. 12.

It is very useful to understand the computation cost of the various algorithms for solving the equations. To quantify that as much as possible, I have defined some numerical quantities in Table 3. Quantities which increase with the size of the system are denoted by the symbol $N$ with some subscript. Other quantities which may be large but are independent of the system size are denoted by the symbol $M$.

Part of solving the Kohn-Sham equations is to construct a potential $V_e$ for the electron-electron interaction, given the density $n(\mathbf{r})$. This requires solving the Poisson equation, which can be done efficiently in several ways. Very straightforward is to use the Fast Fourier Transform. This converts a function represented on a real space grid to the Fourier space, and it requires about

$$5N \log_e N$$

floating point operations to make the three-dimensional transform, with $N = N_r$ or $N_k$. Let us call the transform $\tilde{\phi}(k) = F\phi(r)$. The method achieves its efficient by representing the matrix $F$ as a product of sparse matrices $F_i$, $F = \prod_i F_i$. To solve the Poisson equation $\nabla^2\phi = -4\pi n$, one simply performs the operations in the formula

$$\phi(r) = F^{-1}\frac{4\pi}{k^2}Fn(r),$$

requiring 2 FFT's per cycle. As its name implies, the FFT is fast. But it also has some disadvantages. The full efficiency is available only for particular meshes; the largest routinely in use is $(128)^3 \approx 2 \times 10^6$ points. Also, as a finite Fourier transform, it imposes a periodicity on the system. This is no problem for the condensed matter theorists calculating properties of crystals, but for finite systems the spurious"superlattice" sometimes requires one to use large boxes that would be the case for a finite system. For ways to cope with this problem, see Ref. 13.

There is another method that is well suited to finite systems, and is also very efficient. That is the *multigrid method*, which you can find described in textbooks [14] or in a review article on numerical methods for density functional theory. [15] The method uses a finite difference formula in coordinate space to represent the Laplacian operator. You all are familiar with the three-point difference formula, $\nabla_x^2\phi\Big|_{x_0} \approx (\phi(x_0+\Delta x) + \phi(x_0-\Delta x) - 2\phi(x_0))/\Delta x^2$. Higher order approximations are more efficient in that they allow one to use a coarser mesh for a given level of accuracy. Also important to the multigrid methods the use of a sequence of coarser meshes to deal with the long-wavelength part of $\phi$. Without multigridding, the standard equation-solving techniques converge very slowly, requiring a large number of iterations. In practice, the multigrid method reduces the error in an approximate solution by a factor of two or so with each cycle of multigridding. Its speed is comparable to the FFT. [15] Given the Hartree potential, the next task is to solve the three-dimensional Schrödinger equation. With the large dimensions of the spaces required to represent the wave functions, only iterative methods are practical. These operate by refining the wave function. At the heart of any method is the operation of applying the Kohn-Sham operator $H_{KS}$ to a wave function. As a matrix multiplication, the operation requires of order $N_\phi^2$ floating point operations per wave function. Taking into account that the number of electrons grows with the size of the system, the operation scales with system size as $N_e N_\phi^3$. This would be very slow in practice. There are several algorithms that only require of the order of $N$ operations per electron.

The one used by Chelikowsky *et al.* [16] and by Yabana and Bertsch [17] takes advantage of the coordinate space representation. The potential is local or nearly local, so it only takes $N_\mathbf{r}$ operations to construct the vector $V\phi$ starting from $V$ and $\phi$ as separate $N_\mathbf{r}$-dimensional arrays. The kinetic energy is represented as a difference operator as in the multigrid method, giving a sparse matrix in the $\mathbf{r}$-representation. The operational count is thus $N_\mathbf{r} M_H$ where $M_H$ is the number of nonzero rows in the Hamiltonian. The FFT is also often used to perform the operation $H_{KS}\phi$. In what is called the split operator technique, the potential energy is evaluated in coordinate space and the kinetic energy in reciprocal space, e.g.

$$H_{KS}\phi\Big|_r = -F^{-1}\frac{k^2}{2m}F\phi(r) + V_{KS}(r)\phi(r).$$

One also has a choice of methods for the iterative refinement of the wave function. The two main methods are the imaginary time propagation and the conjugate gradient method. The imaginary time propagation simulates the operator $\exp(-\tau H)$, which when applied to a wave function, will damp the higher energy components and enhance the lower energy components. The conjugate gradient method is a workhorse for any kind of problem with many variables and equations to be satisfied simultaneously. Besides the iteration to get the wave function, there is an iteration on the potential. Unfortunately, the kinetic energy operator has too large a range of eigenvalues to make either method rapidly convergent. There is a technique to improve the convergence, called preconditioning. Here one modifies the operator to reduce the amplitude of high momentum components in determining the correction to be applied in an iteration cycle. This is most easily done in wave vector space, but it has been applied also in real space using multigrid method (see Ref. 18).

I give as an example the calculation of the ground state of a ring of 20 carbon atoms using the Yabana-Bertsch code (available at `http://www.phys.washingtone.edu/ bertsch`. The code uses the coordinate space representation and the 9-point difference formula for the kinetic energy operator. The other numerical parameters for the calculation are very simple to report. The wave functions were calculated in a cylindrical volume of 1,500 Å$^3$. The mesh spacing was 0.3 Å, giving $N_r \approx 56,000$ mesh points. The number of occupied electron orbitals is $N_e = 40$ for the $C_{20}$ cluster. Namely, there are 4 valence electrons per carbon atom, and the orbitals are doubly occupied, giving $20 \times 4/2 = 40$ altogether. The core electrons of the carbons are only included implicitly in constructing the pseudopotential.

With a 9-point difference formula, the number of nonzero elements in the operator matrix is $M_H = 25$. This does not include the computation of the nonlocal parts of the pseudopotential, but that does add significantly to the total number of operations. In the code, the conjugate gradient method is used to extract the orbital wave functions, using five calls to the $H_{KS}\phi$ operation per orbital iteration. There does not seem to be an easy way to estimate the number of iterations required. For the case here, $M_{it} = 100$ gave convergence to a meV on the occupied orbital energies. Combining these numbers, one finds of the order of $5M_{it}M_H N_r N_e \approx 30 \times 10^{+9}$ arithmetic operations. Our computer can do arithmetic operations at a pipelined speed of $0.2 \times 10^9$ s$^{-1}$, giving a time of 150 seconds. In fact about twice as much time is spent in the subroutine. This shows that memory transfer also has a significant impact on the running speed.

### 4.0.1. *Exact exchange*

It is now easy to see why one avoids making an explicit calculation of the exchange interaction. The exchange term in the Hartree-Fock equation has the following form

$$H_{ex}\phi)_{a,\mathbf{r}} = -\sum_b \int d^3r' \frac{e^2}{|\mathbf{r}-\mathbf{r}'|}\phi_a(\mathbf{r}')\phi_b^*(\mathbf{r}')\phi_b(\mathbf{r}).$$

Since there are two indices of orbitals and two position variables, the naive computational cost is proportionaly to $N_e^2 N_r^2$. This would be very time-consuming to compute, but it can somewhat ameliorated. One can use a Poisson solver to find the potential field

$$V_{ab}(r) = \int d^3r' \frac{e^2}{|\mathbf{r}-\mathbf{r}'|}\phi_a(\mathbf{r}')\phi_b^*(\mathbf{r}').$$

The computational cost here is $N_e^2 N_r M_{it}$. The second multiplication to get the exchange term is insignificant, giving an overall scaling as $N^3$. In contrast, with the LDA or GGA, the exchange potential is just a local function of position with computational cost scaling as $N_e N_\phi$. In fact, the computer speed here is actually limited by memory access rather than arithmetic.

### 4.0.2. $\mathcal{O}(N)$ *methods*

There is much discussion now about methods to solve the Kohn-Sham equation that have a computational cost that increases only linearly with the size of the system. Various approaches are discussed by Goedecker in his review article. [19] The basic issue is how to avoid the $N^2$ cost of computing $N_e$ wave functions on a basis of size $N_\phi$. There is a way around when one recognizes that the orbitals are not really the physical objects of the theory. Rather the physical quantity is the single-electron density matrix, $n(r,r')$, constructed from the orbitals. At least in insulators, this function goes to zero quickly as $|r-r'|$ becomes large. Thus, representing the density matrix requires one index $r$ to have $N_r$ values but the other index can be restricted in a way that is independent of $N$. I will just briefly mention two of the approaches that follow this line. The first is to use $n(r,r')$, suitably restricted, as the variational function. One can't just take the Kohn-Sham energy $E_{KS}[n(r,r')]$ by itself as the functional to be varied, because $n(r,r')$ must also respect the Pauli principle: not all $n(r,r')$ are compatible with an antisymmetrized many-electron wave function. This is taken care of in the orbital approach by constructing $n$ from $N_e$ orthogonal orbitals. Considering the density matrix as an operator, $\langle r|\hat{n}|r'\rangle = n(r,r')$, its characteristic properties is that it is a projection operator, with $N_e$ eigenvalues equal to one and the rest equal to zero. Now suppose we have a trial density matrix that doesn't satisfy this projector property. We can drive its eigenvalues to one or zero by using the minimization principle $\pm\mathrm{Tr}\ (3\hat{n}^2 - 2\hat{n}^3)$. We want to drive the orbitals below the Fermi energy to eigenvalue $n_i = 1$, and the orbitals above to to $n_i = 0$.

This is accomplished by the variational principle

$$\text{Tr}\ \left((H_{KS} - \mu I)(3\hat{n}^2 - 2\hat{n}^3)\right).$$

where $\mu$ is the Fermi energy.

A related $\mathcal{O}(N)$ approach is to keep an orbital representation, but restrict the domain of each orbital to some neighborhood. Again, satisfying the Pauli principle is nontrivial. Since the orthogonality must be retained at least as an approximation, constraints must be introduced into the calculational method. A method that seems to be successful [20] is as follows. First, one puts the Fermi energy into the Hamiltonian as above, $H'_{KS} = H_{KS} - \mu I$. Second, one adds a term to drive the orbitals to orthonormality. A variational principle that can do this is

$$2\sum_i \langle i|H'_{KS}|i\rangle - \sum_{i,j} \langle i|H'_{KS}|j\rangle\langle i|j\rangle.$$

The number of orbitals in the sums over $i, j$ must be larger than $N_e$. The second term obviously doesn't have any effect if the orbitals space is complete, since then the Kohn-Sham orbitals are orthogonal and the variation of that term gives zero contribution to the Kohn-Sham operator. When the orbitals have appropriately restricted domains, the overlaps with a given orbital $i$ will be zero except for a limited number of other orbitals $j$, and the number of basis amplitudes will be also be limited in the integrations. This make the method $\mathcal{O}(N)$, but does not guarantee that it convergences in a controlled way to the exact solution, or indeed that it convergences at all to the sought solution. The second method is farther along computationally, [21] but more experience is needed to assess the reliability of these approximations.

## 5. Some applications and limitations of DFT

Since I have already discussed the predictive power of the DFT for molecules and clusters, this section only mentions other applications. At the end I mention some limitations arising from the implicit treatment of correlation effects.

### 5.1. *Two examples of condensed matter*

Here I will show you the results for crystalline silicon and iron, two materials of great importance in technology. The first properties a theory should be able to reproduce the structure of the element, namely its crystallographic form and the lattice size, and its binding energy. Iron has an additional complication that it is ferromagnetic, and the theory should reproduce the magnetic moment as well.

The properties of bulk silicon are shown in Table 4, according to the LDA calculations of Ref. 22 and 23. The LDA gives the correct crystal symmetry, and one can see from the table that the lattice constant is well reproduced. The bulk modulus can be calculated by find the energy of the material as a function of assumed lattice spacing. This also comes out very well. A property that comes out much poorer is the electronic band gap. This may be defined either as the difference between electron removal and electron addition energy, or as the difference between Kohn-Sham eigenvalues for the highest occupied and the lowest unoccupied orbitals. In an infinite system these quantities are the same.

Table 4. Ground state properties of Si in the LDA, [22,23] compared to experiment.

| Property | LDA | Experimental |
|---|---|---|
| lattice constant | 5.42 | 5.42 Å |
| bulk modulus | 0.96 | 0.99 MBar |
| band gap | 0.52 | 1.17 eV |

In the case of iron, there are important differences in the predicted structures between the LDA and the GGA theories. [24] Table 5 shows these quantities for iron, both theoretically and calculated with the DFT. We see that the LDA fails the first test of the theory. Iron has a body-centered cubic structure, but the LDA has a lower energy for the hexagonal closed packed structure. On the other hand, the GGA gives the right structure and even comes very close on the lattice constant.

Table 5. Metallic Iron: an example of DFT. Theory is from Refs. 25, 26.

| Property | Exp. | LDA | GGA |
|---|---|---|---|
| Crystal symmetry | bcc | hcp | bcc |
| Lattice constant $a_0$ | 2.87 Å | -4% | -0.5% |
| Cohesive energy | 4.28 eV | +50% | +20% |
| Bulk modulus | 1.68 Mbar | +40% | -8% |
| Magnetization | 2.22 $\mu_B$ | -8% | +5% |

The DFT correctly predicts the ferromagnetism of iron if the functional includes the spin-dependence interaction of the LSDA. As one can see from the table, the net spin per atom comes out rather well for both versions of the functional. The ferromagnetism of iron is rather simple in that the spin alignment is the same on each atom and one only needs to consider densities of spin up and spin down electrons. Other ferromagnetic materials have a more complicated spin structure, with the orientation of the spin depending on position. For these systems one would need to use the generalized spin functional to have the possibility of noncollinear spin alignment. In our group in Seattle we calculated the spin alignment in small cobalt clusters and found generalized spin functional gave a predicted noncollinear alignment. [27]

## 5.2. *Vibrations*

The DFT should also be applicable to vibrational properties. Recall that the DFT is a theory of the electronic ground state for an arbitrary external potential. Thus we can displace the nuclei from the equilibrium positions and find the energy of the distorted structure. Table 5 shows the bulk modulus of iron calculated this way and compared to experiment. The error is large with the LDA but is only 8% with the GGA functional. Taking the DFT energy as a potential energy in the equations for vibrational motion, one can easily calculate the phonon dispersion relation for bulk systems. An 8% error in the force constant becomes a 4% error in the sound velocity, because of its quadratic relationship to the force constants.

In general, the DFT also works quite well for molecular vibrations. As an example I show some calculations of the infrared-active vibrations in the fullerene $C_{60}$. This is an example of historical interest, because the vibrational spectrum was one of the convincing early pieces of evidence that $C_{60}$ had icosahedral symmetry. There are $60 \cdot 3 - 6 = 174$ vibrational modes in a molecule with 60 atoms, and if there were no symmetries they could all be excited by photon absorption. However, with icosahedral symmetry, only four of the modes have nonzero dipole matrix elements. Observation of exactly four modes in the infrared absorption spectrum was one of the convincing evidences of that structure when Krätchmer and Huffman first synthesized it in bulk. [28]

Table 6. Excitation energies of $C_{60}$ infrared-active vibrations ($T_{1u}$).

| Mode | 1 | 2 | 3 | 4 |
|---|---|---|---|---|
| Experimental | 0.065 eV | 0.071 | 0.147 | 0.177 |
| LDA error | | | | |
| Ref. 29 | -2% | -2% | +2% | +10% |
| Ref. 30 | -3% | -7% | -8% | -7% |

To calculate the modes, one first computes the potential energy surface as a function of displacement of the atoms from their ground state positions. Diagonalizing the Hessian matrix (the quadratic expansion of the potential energy surface) gives the normal modes and their frequencies. The results of three different LDA calculations are shown in Table 6. The agreement with experiment is impressive, with mean absolute relative error on frequency only 4%. A more demanding test of the theory is the transition strength associated with the vibrations. The accuracy here is perhaps only a factor of two. [29] But that is a great improvement over previous theories that were completely unreliable.

### 5.3. *NMR chemical shifts*

All the applications so far only requires energies or electron densities. There are other properties, more closely related to the electron wave function, that are better described by DFT than by Hartree-Fock theory. One of these is the current induced by an external magnetic field. We add a magnetic field to the energy functional by including the gauge field $\vec{A}$ in the kinetic energy operator,

$$\int d^3r \frac{\hbar^2}{2m}\left(-\frac{\nabla}{i}\psi^\dagger - e\vec{A}\psi^\dagger\right)\cdot\left(\frac{\nabla}{i}\psi - e\vec{A}\psi\right)$$

and adding a term representing the field energy associated with $\vec{A}$,

$$\frac{1}{8\pi}\int d^3r|\nabla\times\vec{A}|^2.$$

The variation $\delta E_{KS}/\delta\phi_i^*$ then gives Kohn-Sham equations with the additional terms

$$\frac{e^2}{mc^2}|A|^2\phi_i + \frac{e}{mc}\vec{A}\cdot\nabla\phi_i.$$

The variation with respect to $\vec{A}$ gives the Maxwell equation relating the magnetic field $\vec{B} = \nabla\times\vec{A}$ to the current, where the current is defined by

$$\frac{\delta T}{\delta\vec{A}} = \frac{e}{mc^2}\vec{A}n + \frac{e}{m}\sum_i\left(\phi_i^*\nabla\phi_i - (\nabla\phi_i^*)\phi_i\right).$$

The current generates an additional magnetic field that weakly screens the external field. The fields at the nuclei are observable in nuclear magnetic resonance, and this screening gives rise to so-called chemical shifts of the NMR frequencies. Typically these are small, only hundredths of a percent, but they are easily measured and are characteristic of the chemical environment of the nucleus.

The DFT was compared with Hartree-Fock in Ref. 31. The calculations were done in an atomic orbital basis, which is far from trivial to use when dealing with gauge fields. I will just mention the results here, for the proton NMR shift in three molecules (HF, $H_2O$, $CH_4$). The proton chemical shift is of the order of $30\times10^{-6}$, and all the methods came within a few percent in the three cases. However, what is more interesting is the relative shift from one environment to another. These numbers are of the order $1\times10^{-6}$. The Hartree-Fock is found to be only accurate to a factor of two on the relative shifts, while the DFT is considerably better with only 20% errors. Interestingly, the ordinary LDA is as good or better than the GGA here.

## 6. Limitations of DFT

In some ways, the Kohn-Sham theory tries to capture the effects of correlations without building them explicitly into the wave function. There are certain systems and properties where explicit treatment of the correlations is needed, and we mention these before going on to discuss dynamics. Two well-known phenomena rely on correlations for their existence, namely superconductivity in bulk matter and the Van der Waals interaction between molecules. The van der Waals interaction is an attraction between particles that falls off as $1/r^6$ at moderately large distances. It is due to fluctuations in dipole fields associated with electronic transitions. Thus it is completely absent from a theory that represents the electron wave function by a single Slater determinant. I want to show here how one could view density functional theory in a way that would make it natural to encompass the van der Waals interaction. Let us go back to eq. (1.12), the many-body theory expansion of the energy of the electron gas. The third term gives the contribution of correlations; in perturbation theory it has the second-order expression

$$\sum_{ii'}\sum_{cc'}\langle ij|\frac{e^2}{|\mathbf{r}-\mathbf{r}'|}|cc'\rangle^2\frac{1}{\epsilon_i+\epsilon_{i'}-\epsilon_c-\epsilon_{c'}}.$$

The van der Waals interaction is obtained if one replaces the full Coulomb interaction in this expression by a dipole-dipole approximation

$$\frac{e^2}{|\mathbf{r}-\mathbf{r}'|}\approx\frac{e^2}{|\mathbf{R}-\mathbf{R}'|}+\frac{e^2(\mathbf{r}-\mathbf{R})\cdot(\mathbf{r}'-\mathbf{R}')}{|\mathbf{R}-\mathbf{R}'|^3},$$

where $\mathbf{R},\mathbf{R}'$ are the centers of the two atoms and the integrations are restricted appropriately. One way to incorporate the van der Waals in the DFT would be as follows. Separate the Coulomb interaction into two parts, one containing the long wavelength components and the other the short wavelength components. In the DFT, incorporate the short-range Coulomb interaction into a local density approximation. For the long range part, calculate the second order perturbation explicitly. In this way the full interaction would be accounted for, but the van der Waals term would arise naturally.

Correlation effects are also hard to ignore in systems with a high density of states near the Fermi level. Then there is much configuration mixing, and the basic DFT approximation taking a single configuration to generate the density is not reliable. This suggests that one use the density functional theory to find the optimal orbitals in a set of configurations, and use other methods to improve the many-particle wave function by taking the configuration mixing into account. As with the van der Waals interaction, there is a conceptual difficulty in separating the part of the interaction that is treated by the LDA functional and the part that is treated by configuration mixing. A pragmatic approach is to ignore the overcounting and take $\frac{\delta H_{KS}}{\delta n}$ as a two-particle operator that can mix configurations. One such hybrid theory is called "LDA+U". [32] The LDA is used to construct orbitals, but allows the wave to mix orbitals by an simplified interaction term on each site. That is called the "U". One finds that such an extension is essential for describing the phase diagram of metallic plutonium. [33] Another theory of this kind is called "dynamic mean field theory". It has also been applied to the metallic plutonium problem. [34]

## 7. Time-dependent density functional theory: the equations

Schrödinger proposed two equations in his original paper, the eigenvalue equation for static properties and the time-dependent equation for the dynamics. But the left-hand side of both equations was the same. The situation is the same for dynamic theories based on Hartree-Fock or DFT. The theories may be derived from the time-dependent variational principle,

$$\delta \int dt \langle\Psi|H - i\frac{\partial}{\partial t}|\Psi\rangle = 0. \tag{7.1}$$

Taking $\Psi$ to be a Slater determinant and varying with respect to $\phi_i^*(r,t)$, one obtains the time-dependent Hartree-Fock equations, first proposed by Dirac. [35] The corresponding equations for DFT are the time-dependent Kohn-Sham equations,

$$H_{KS}\phi_i = i\frac{\partial}{\partial t}\phi_i. \tag{7.2}$$

There are several properties of the TDDFT that make it very attractive as a theory of electron dynamics. The theory automatically satisfies two important conservation laws, conservation of energy and the equation of continuity. To prove energy conservation in the TDDFT, take the derivative of the energy and expand using Eq. (7.2):

$$\begin{aligned}\frac{dE_{KS}}{dt} &= \sum_j \int d^3r \left(\frac{\delta E_{KS}}{\delta\phi_j^*}\frac{\partial\phi_j^*}{\partial t} + \frac{\delta E_{KS}}{\delta\phi_j}\frac{\partial\phi_j}{\partial t}\right) \\ &= \sum_j \int d^3r \left((H_{KS}\phi_j)(iH_{KS}\phi_j^*) + (H_{KS}\phi_j^*)(-iH_{KS}\phi_j)\right) = 0\,.\end{aligned}$$

To derive the equation of continuity, we start from the expression for the density, $n(r) = \langle\Psi|\psi^\dagger(r)\psi(r)|\Psi\rangle$, taking its time derivative. If $\Psi$ were the exact wave function, the time derivative may be evaluated using the solution of the many-particle Schrödinger equation,

$$\begin{aligned}\frac{dn(r)}{dt} &= \langle\partial_t\Psi|\psi^\dagger(r)\psi(r)|\Psi\rangle\langle\Psi|\psi^\dagger(r)\psi(r)|\partial_t\Psi\rangle \\ &= i\langle(H\Psi)|\psi^\dagger(r)\psi(r)|\Psi\rangle - i\langle\Psi|\psi^\dagger(r)\psi(r)|H\Psi\rangle = i\langle\Psi|[H,\psi^\dagger(r)\psi(r)]|\Psi\rangle.\end{aligned}$$

We now assume that the Hamiltonian is local except for the kinetic energy operator $T$. Then the potential terms in the Hamiltonian commute with the density, and we only need to evaluate the kinetic energy commutator, $[T,\psi^\dagger(r)\psi(r)]$. It may be simplified using the fermionic anticommutator relations and the definition of derivatives as limits of differences. The result can be expressed in terms of the momentum density operator $\vec{p}(r)$, defined as

$$\vec{p}(r) = \frac{1}{i}\left(\psi^\dagger(r)(\nabla\psi)_r - (\nabla\psi^\dagger)_r\psi(r)\right).$$

It is also convenient to define the particle current, $\vec{j}_p = \vec{p}(r)/m$. The commutator of the density operator and the kinetic energy operator is then given by

$$[T,\psi^\dagger(r)\psi(r)] = \frac{i}{m}\nabla\cdot\vec{p}(r) = i\nabla\cdot\vec{j}_p.$$

Combining the above results and taking the operator expectation value in the wave function $\Psi$, we obtain the equation of continuity,

$$\frac{dn(r)}{dt} = -\nabla \cdot \langle \vec{j}_p(r) \rangle.$$

This derivation assumed that $\Psi$ was the exact wave function. In TDDFT, the wave function is given by Eq. (7.2), but the proof can still be carried out in the same way. The particle current will then have the expectation value

$$\langle \vec{j}_p \rangle = \frac{1}{im} \sum_j (\phi_j^* (\nabla \phi_j(r))_r - ((\nabla \phi_j)_r^* \phi_j(r)).$$

The fact the TDDFT satisfies the equation of continuity means that it will have the correct short-time behavior for the response of the system to external perturbations. This will be discussed in more detail below.

### 7.1. *Optical properties*

The most important application of the TDDFT is to the optical properties of molecules, atomic clusters, and condensed matter. In this section I will summarize for future reference a number of useful formulas for dealing with small systems. Here by small, I mean the dimension of the particle is much less than the wavelength of the photon. Then the electric dipole field dominates the interaction, and the optical response can be described with the dynamic polarizability $\alpha(\omega)$. Physically, we may come to the definition of alpha by asking the question, what is the form of an energy functional for a small finite system coupled to an external electric field? Expanding in powers of the field, we have the leading electric field-dependent terms,

$$\vec{d} \cdot \mathcal{E} + \frac{1}{2} \sum_{ij} \alpha_{ij} \mathcal{E}_i \cdot \mathcal{E}_j + \ldots$$

The first term vanishes if the system does not have a dipole moment $\vec{d}$. The coefficients $\alpha_{ij}$ define the polarizability tensor. If the system has a high symmetry or if one is content with the average over angles, the second term can be written $\frac{1}{2}\alpha \mathcal{E} \cdot \mathcal{E}$ with $\alpha = \sum_i^3 \alpha_{ii}$. The polarizability has a dimension of volume, which can be easily seen by comparing the energy expression with that for the electric field energy, $(1/8\pi) \int d^3r |\mathcal{E}|^2$.

Applying time-dependent perturbation theory for the field $x\mathcal{E}_0 \exp(-i\omega t)$, the polarizability in the $x$ direction may be calculated as

$$\alpha_{xx}(\omega) = \frac{e^2}{\hbar} \sum_a \left| \langle 0|x|a \rangle \right|^2 \left( \frac{1}{-\omega - i\eta + \omega_a} + \frac{1}{\omega + i\eta + \omega_a} \right), \tag{7.3}$$

with $a$ labeling excited states. The polarizability also has an equivalent definition as the linear coefficient of the field in the dipole moment of the system in an external electric field,

$$e\langle x\rangle = -\sum_i \alpha_{xi}\mathcal{E}_i + ... \tag{7.4}$$

The imaginary part of the dynamic polarizability is directly related to the photon absorption cross section $\sigma_{ab}$. The relation is

$$\sigma_{ab} = \frac{4\pi\omega}{c}\mathrm{Im}\,\alpha(\omega).$$

Under certain conditions the dielectric function of an extended medium is related to the polarizability of its constituents. This is the Clausius-Mossotti relation:

$$\epsilon(\omega) = \frac{1 + 8\pi\alpha n_0/3}{1 - 4\pi\alpha n_0/3}.$$

Here $n_0$ is the density of the polarizable particles. The conditions for validity of the formula are: 1) the particles interact only through the electric fields; 2) the environment of each particle is isotropic or at least has cubic symmetry.

### 7.1.1. *f-sum rule*

The cross section $\sigma_{ab}$ satisfies a very important sum rule due to Thomas, Reiche and Kuhn called the *f-sum rule*,

$$\frac{m_e c}{2\pi^2 e^2 \hbar}\int_0^\infty \sigma_{ab} dE = N_e. \tag{7.5}$$

As before, $N_e$ is the number of electrons in the particle. I will also define the dipole strength function $S(\omega)$ as

$$S(\omega) = \frac{2m_e\omega}{\pi\hbar e^2}\mathrm{Im}\alpha(\omega). \tag{7.6}$$

The strength function has units of oscillator strength per unit energy, and satisfies the sum rule $\int_0^\infty S d\omega = N_e$. It is common to express energy-integrated cross sections in units of the sum rule for a single electron. This is the quantity "$f$" in the sum rule's name. For any physical excitation, the absorption cross section is finite over some spectral region $\Delta\omega$. The *oscillator strength* $f_i$ associated with the excitation may then be defined

$$\int_{\Delta\omega} d\hbar\omega S(\omega) = f.$$

The theoretical strength for a discrete excitation $a$ is given by the formula

$$f_a = \frac{2m}{\hbar^2}| < 0|x|a > |^2 \hbar\omega_i.$$

There are many derivations of the $f$-sum rule. One derivation starts from the operator relation $[x,[H,x]] = 1/2m_e$. Another derivation that is easily applied to the TDDFT follows from the short-time behavior in an external dipole field, $V(t) =$

$ex\mathcal{E}_0\delta(t)$. The ground state wave function $\Psi_0$ is perturbed by the field; to linear order the perturbed wave function is

$$\Psi(t = 0_+) = \Psi_0 + \frac{e\mathcal{E}_0}{i}\sum_j \Psi_a\langle\Psi_a|x|\Psi_0\rangle.$$

Each eigenstate $\Psi_a$ evolves in time with the phase factor $\exp(-iE_at)$. Again to linear order in the perturbation, the dipole moment is given by

$$e\langle x\rangle = -ie^2\mathcal{E}_0\sum_a(e^{-i\omega_a t} - e^{i\omega_a t})\langle\Psi_a|x|\Psi_0\rangle^2, \qquad \text{where } \omega_a = E_a - E_0.$$

Taking the derivative with respect to time, we find the time rate of change of the dipole moment at $t = 0_+$ as

$$e\frac{d\langle x\rangle}{dt}\bigg|_{0_+} = 2e^2\mathcal{E}_0\sum_a \omega_a\langle\Psi_a|x|\Psi_0\rangle^2,$$

On the other hand, we can evaluate this derivative using the equation of continuity. Immediately after the external field is applied, the particle current is given by $\langle\vec{j}(r)\rangle_{0_+} = e\hat{x}\mathcal{E}_0 n(r)/m$. Then

$$e\frac{d\langle x\rangle}{dt}\bigg|_{0_+} = \int d^3r x\nabla\cdot\langle\vec{j}(r)\rangle_{0_+} = e\mathcal{E}_0/m\int d^3r x\frac{dn(r)}{dx} = e\mathcal{E}_0 N_e/m$$

Comparing the last two equations gives back the sum rule in the form

$$\sum_a \omega_a\langle\Psi_a|x|\Psi_0\rangle^2 = \frac{N_e}{2m}.$$

This shows the close connection between the sum rule and the short-time behavior of the system.

The first application of TDDFT was to describe the photoionization of atoms.[36] The theory has since been widely applied to clusters, molecules, and bulk matter. My own research is on applications of the TDDFT, starting with the collective excitations of atomic clusters but going on to other electronic systems and other properties. Most of the effort was in the numerical aspects of solving the equations. My work in this area was in collaboration with K. Yabana. For reference, our published work on TDDFT is in Refs. 17-43. We started using a method developed in nuclear physics, and this proved to be quite useful. In the next section I will briefly describe the leading algorithms, their advantages and disadvantages.

### 7.2. *Methods to solve the TDDFT equations*

There are many mathematical formulations of the TDDFT. Most of them assume that the solution is close to the ground state wave function, so that the time-dependent equation can be linearized. In this section we will distinguish ground state orbitals with a superscript (0), for example $\phi_i^0(\mathbf{r})$ for the $i$-th orbital. The TDDFT satisfies an important condition, that the ground state remain stable under the time evolution. It is easy to see for this case that the time-dependent orbitals evolve only by a overall phase factor, $\phi_i(\mathbf{r},t) = e^{-i\epsilon_i t}\phi_i^0(\mathbf{r})$ with $\epsilon_i$ the Kohn-Sham eigenvalue. Let us now consider a small deviation $\delta\phi_i(\mathbf{r},t)$ from this solution, expanding the time-dependent wave function as

$$\phi_i(\mathbf{r},t) = e^{-i\epsilon_i t}\left(\phi_i^0(\mathbf{r}) + \delta\phi_i(\mathbf{r},t)\right).$$

Putting the above wave function into Eq. (7.2) we have

$$H_{KS}[n_0 + \delta n + ...]\left(\phi_i^0(\mathbf{r}) + \delta\phi_i(\mathbf{r},t)\right) = \epsilon_i\phi_i^0 + \epsilon_i\delta\phi_i + i\frac{\partial}{\partial t}\delta\phi_i.$$

The transition density $\delta n$ may be obtained by a first-order expansion of $n = \sum_i |\phi_i|^2$,

$$\delta n = \sum_i(\phi_i^*\delta\phi_i + \delta\phi_i^*\phi_i).$$

Next collect all the first order terms to obtain

$$(H_{KS}^0 - \epsilon_i)\delta\phi_i + \phi_i^0\frac{\delta H_{KS}}{\delta n}\delta n = i\frac{d\delta\phi_i}{dt}$$

Here the Kohn-Sham operator $H_{KS}^0$ is defined $H_{KS} = H_{KS}[n_0]$. The functional derivative of $H_{KS}$ with respect to $n$ contains two terms, one from the direct Coulomb field and one from the exchange-correlation potential. The LDA functional derivative has the form

$$\frac{\delta H_{KS}(\mathbf{r})}{\delta n(\mathbf{r}')} = \frac{e^2}{|\mathbf{r}-\mathbf{r}'|} + \frac{dV_{xc}}{dn}\delta^3(\mathbf{r}-\mathbf{r}').$$

In the literature, $\frac{\delta H_{KS}}{\delta n}$ is often denoted by $K(\mathbf{r},\mathbf{r}')$ and in fact the identification $K = \frac{\delta H_{KS}}{\delta n}$ is called the adiabatic approximation. Once the equations are linearized, the time dependence may be found by Fourier transforming with the integral $\int dt\, e^{i\omega t}$ and solving for each frequency separately. Let us write the perturbed wave function for frequency $\omega$ as $\delta\phi_i(\mathbf{r},t) = \phi_i^+(\mathbf{r})e^{-i\omega t} + \phi_i^-(\mathbf{r})e^{i\omega t}$. It is necessary to consider both positive and negative frequencies because $\delta n$ involves both $\phi$ and its complex conjugate. The resulting equations of motion are then

$$(H_{KS}^0 - \epsilon_i)\phi_i^\pm + \frac{\delta H_{KS}}{\delta n}\delta n\phi_i^0 = \pm\omega\phi_i^\pm. \tag{7.7}$$

The transition density $\delta n$ in this equation is given by $\delta n = \sum_i(\phi_i^{0*}\phi_i^+ + \phi_i^0\phi_i^{-*})$. Eqs. (7.7) form a generalized eigenvalue problem; in general the equations have a nontrivial solution only for particular values of $\omega$.

Several choices can be made at this point to represent $\phi_i^\pm$ and solve the equations. One way to proceed is to expand the unknown functions $\phi_i^\pm$ in eigenstates of $H_{KS}^0$,

$$\phi_i^+ = \sum_c X_{ci}\phi_c^0$$

$$\phi_i^- = \sum_c Y_{ci}\phi_c^0$$

It turns out that it is only necessary to include the unoccupied orbitals in summing over $c$. Eq. 7.7 can then be expressed as the matrix eigenvalue problem

$$\begin{pmatrix} A & B \\ -B & -A \end{pmatrix} \begin{pmatrix} X \\ Y \end{pmatrix} = \omega \begin{pmatrix} X \\ Y \end{pmatrix} \tag{7.8}$$

The matrix elements $A$, $B$, are given by

$$A_{ci,c',i'} = \delta_{c,c'}\delta_{i,i'}(\epsilon_c - \epsilon_i) + \int d^3rd^3r'\psi_c^*(\mathbf{r})\psi_i(\mathbf{r})\frac{\delta H_{KS}(\mathbf{r})}{\delta n(\mathbf{r}')}\psi_{i'}^*(\mathbf{r}')\psi_{c'}(\mathbf{r}')$$

$$B_{ci,c',i'} = \int d^3rd^3r'\psi_c^*(\mathbf{r})\psi_i(\mathbf{r})\frac{\delta H_{KS}(\mathbf{r})}{\delta n(\mathbf{r}')}\psi_{i'}^*(\mathbf{r}')\psi_{c'}(\mathbf{r}')$$

The dimensionality of the matrix is the number of particle-hole combinations in the space. The normalization of the eigenvectors $(X, Y)$ is a tricky point. The matrix in Eq. (7.8) is not Hermitian, so it requires a biorthogonal basis to expand the matrix in eigenvectors. The conjugate vector to the eigenvector $(X, Y)$ is the eigenvector of the conjugate equation, which is easily seen to be $(X^*, -Y^*)$. The orthogonality and normalization conditions then become

$$\sum_{ci,c'i'} (X_{c'i'}^* X_{ci} - Y_{c'i'}^* Y_{ci}) = \delta_{ci,c'i'} \tag{7.9}$$

Another approach to solving Eq. (7.7) is to include an external field and solve the resulting inhomogeneous equations. Taking the field of the form $U(r)(e^{-i\omega t} + e^{i\omega t})$ the equations are

$$(H_{KS}^0 - (\epsilon_i \pm \omega))\phi_i^\pm + \frac{\delta H_{KS}}{\delta n}\delta n \phi_i^\pm = U\phi_i^0 \tag{7.10}$$

The direct solution of these equations is called the Sternheimer method. It can done efficiently using the conjugate gradient algorithm, if the operator on the left-hand size has a sparse matrix representation. It can also be convenient to solve them with the help of the Green's function for the operator $H_{KS}$, as is done in the linear response method described below.

### 7.2.1. *Linear response formula*

Here we seek an equation for the transition density, $\delta n$. As a first step, we formally solve eqs. (7.10) for $\phi_i^\pm$,

$$\phi_i^\pm = \frac{1}{(H_{KS}^0 - \epsilon_i \pm \omega)}(U + \delta n \frac{\delta H_{KS}}{\delta n})\phi_i^0$$

Next we insert this in the expression for the transition density, to obtain

$$\delta n = \sum_i \phi_i^0 \left( \sum_\pm \frac{1}{H_{KS}^0 - \epsilon_i \pm \omega} \right) \phi_i^0 \left( U + \frac{\delta H_{KS}}{\delta n}\delta n \right)$$

To make the equation more concrete, let us put in explicitly the coordinate dependence. The sum over the wave functions with the single-electron Green's function will be denoted $\chi_0(r,r')$,

$$\chi_0(r,r') = \sum_i \phi_i^0(r) \left( \sum_{\pm} \frac{1}{H^0_{KS} - \epsilon_i \pm \omega} \right)_{r,r'} \phi_i^0(r')$$

Then the equation has the appearance

$$\delta n(r) = \int d^3r \chi_0(r,r') \left( U(r') + \int d^3r'' \frac{\delta H_{KS}}{\delta n} \delta n(r'') \right)$$

The formal solution of the equation is

$$\delta n = (1 - \chi_0 \frac{\delta H_{KS}}{\delta n})^{-1} \chi_0 U \tag{7.11}$$

It can be solved by representing the operator $(1-\chi_0 \frac{\delta H_{KS}}{\delta n})$ as a matrix in coordinate space or in reciprocal space. The matrix is then inverted directly or by some iterative method.

### 7.3. *Dynamic polarizability*

We now describe how to compute the dynamic polarizability using the different methods to solve the TDDFT equations.

The linear response method is very easy to apply. One takes the external field of the dipole form, $U(r) = e\mathcal{E}_0 x$, calculates the response $\delta n$, and finds the resulting dipole moment:

$$\alpha_{xx}(\omega) = \frac{D}{\mathcal{E}_0} = \frac{e^2}{\mathcal{E}_0} \int \delta n(r) x d^3r$$

Because the response is linear, the field strength $\mathcal{E}_0$ can be taken with any convenient value.

The dynamic polarizability can also be calculated from the direct solution of Eq. (7.2). One starts with the ground state, and perturbs the system with an impulsive external field, $U(r,t) = -kx\delta(t)$. The orbitals acquire a phase factor due to the perturbation. Immediately following they are given by

$$\phi_i(t = 0_+) = e^{ikx} \phi_i^0$$

These are taken as the initial conditions for the time-dependent equations. Solving the equations to evolve the wave functions forward in time, one can construct the time-dependent density, $n(t)$. A time-dependent polarizability can be defined as the ratio of the dipole moment to the strength the perturbing field,

$$\alpha_{xx}(t) = \frac{e^2}{k} \int d^3r x n(t).$$

The dynamic polarizability is then given simply by the Fourier transform,

$$\alpha(\omega) = \int_0^\infty e^{-i\omega t} \alpha(t)$$

Note that all frequencies are obtained in one calculation of the time evolution. This makes this method quite efficient if the polarizability is needed over a large range of frequencies.

If one use the matrix eigenvalue method, the dynamic polarizability is given by Eq. (7.3) summing over the eigenmodes $a$ and taking the excitation energies as the eigenfrequencies $\omega_a$. The necessary dipole matrix elements $\langle 0|x|a\rangle$ are given by

$$\langle 0|x|a\rangle = \sum_{ij}\langle i|x|j\rangle(X^a_{ij} + Y^a_{ij}) \tag{7.12}$$

In principle, this method requires that all the eigenvalues and eigenvectors be found, which is certainly a disadvantage if only a limited frequency range is of interest.

### 7.4. *Dielectric function*

The methods described above can all be applied to infinite periodic systems, with some modifications. A very popular technique is to use a plane-wave basis, since it is convenient for expanding the periodic Bloch wave functions. I will show here an alternative method we developed that uses the coordinate space representation to solve the real-time equations. [42] It is easy to impose the periodic boundary conditions of the Bloch wave functions on the computed orbitals in a cubic unit cell. However, there are two subtleties. First, one cannot take the usual form for the dipole field, $V(r) = \mathcal{E}_0 x$, because it differs from cell to cell, and anyway gets large at large $r$. The other problem is that the currents can induce a surface charge on the dielectric. This charge is crucial to provide the restoring force for a plasmon at zero momentum, for example. Both these difficulties can be overcome by using a gauge field. We divide the electric field into a periodic part that can be represented with an ordinary potential $V$, and a uniform electric field arising from the gauge field $\hat{x}A(t)$:

$$\vec{\mathcal{E}} = -\vec{\nabla}V - \hat{z}\frac{dA}{dt}.$$

The equations of motion are derived from the Lagrangian of a unit cell $\Omega$: we derive the dynamic equations from the Lagrangian:

$$L = \int_\Omega d^3r\Bigg(\frac{\sum_i|\vec{\nabla}\phi_i/i - eA\hat{z}\phi_i|^2}{2m} - \frac{1}{8\pi}\vec{\nabla}V(r)\cdot\vec{\nabla}V(r) + en(r)V(r) + en_{ion}(r)V(r)$$
$$+ V_{xc}[n(r)] + V_{ion}[\rho(r,r')]\Bigg) - \frac{\Omega}{8\pi}\left(\frac{dA}{dt}\right)^2 - i\int_\Omega d^3r\sum_i \phi_i^*\frac{\partial\phi_i}{\partial t}. \tag{7.13}$$

Here the $\phi_i$ are the Bloch wave functions of the electrons, normalized so that $n(r) = \sum_i|\phi_i(r)|^2$ is the electron density. The volume of the unit cell is $\Omega$. In these formal equations, we use units with $\hbar = c = 1$.

The other pieces of the first integral are the usual terms in the Kohn-Sham energy functional. The term $en(r)V(r)$ gives the direct Coulomb interaction of the electrons, except for the surface charging. The ionic interaction is separated into a long-range part that can be associated with an ionic charge density $n_{ion}(r)$ and a short-range part $\mathcal{V}_{ion}$. The latter depends on the orbital angular momentum of the electrons in typical *ab initio* pseudopotentials. It therefore depends on the full one-electron density matrix $\rho(r,r') = \sum_i \phi_i^*(r)\phi_i(r')$. We have emphasized this point because nonlocal interactions do not respect gauge invariance. The invariance is of course restored when the density matrix is gauged. Finally, the $V_{xc}$ is the usual exchange-correlation energy density of density functional theory.

Requiring the Lagrangian action to be stationary gives equations of motion for $\phi_i$ and $A$ and the Poisson equation for $V$. The variation with respect to $\phi_i^*$ gives the Kohn-Sham equations, and varying with respect to $A$ gives its equation of motion:

$$\frac{\Omega}{4\pi}\frac{d^2A}{dt^2} - \frac{e}{m}\sum_i \langle\phi_i|\nabla_z/i|\phi_i\rangle + \frac{e^2}{m}AN_e + \frac{\delta}{\delta A}\int_\Omega \mathcal{V}_{ion}d^3r = 0 \tag{7.14}$$

where $N_e = \int_\Omega d^3rn(r)$ is the number of electrons per unit cell. The reason I have given so much detail is that we didn't arrive at the correct equations immediately. We first guessed at the form of the equations, and tried to solve them numerically. The solutions did not conserve energy, and we only found the correct equations after we had formulated the problem with the variational principle, Eq. (7.13). The lesson is that variational principles are very useful!

Let us now see how the gauge field treatment works in a simple analytically solvable model, namely the electron gas. As mentioned before, when the field $A_0$ is applied, there is no immediate response to the operator $\nabla$, since the wave function does not change instantaneously. However, in the Fermi gas, the single-particle states are eigenstates of momentum so the response remains $\langle\nabla\rangle = 0$ for all time. Putting this in Eq. (7.14), and dropping the pseudopotential term, the equation for $A$ becomes simple harmonic motion, with the solution

$$A(t) = A_0 \cos\omega_{pl}t \tag{7.15}$$

where $\omega_{pl}$ is the plasmon frequency,

$$\omega_{pl}^2 = \frac{4\pi e^2 N_e}{m\Omega} = \frac{4\pi e^2 n}{m}. \tag{7.16}$$

One sees that the derivation here is much simpler than the usual one using the Coulomb gauge. There one formulates the response in a particle-hole representation, and takes the external field to be of the form $e^{iq\cdot r}$ with $q$ finite. The dielectric function is then found by taking the $q \to 0$ limit.

## 8. TDDFT: numerical aspects

I discussed above the equations for the various methods to calculate the TDDFT, but there are also important numerical aspects that one needs to understand in order to apply the theory to large systems. In this section I will describe the computational issues for the configurational matrix method, the linear response method, and the real-time method.

### 8.1. *Configuration matrix method*

Eq. (7.8) is a nonhermitian matrix eigenvalue problem. It may be converted to Hermitian form as follows. The two equations are added and subtracted to give

$$(A+B)(X+Y)=\omega(X-Y)$$

$$(A-B)(X-Y)=\omega(X+Y)$$

Then eliminate $X+Y$ between the two equations to get

$$(A+B)(A-B)(X-Y)=\omega^2(X-Y)$$

The matrix $A-B$ is diagonal in the configuration representation with matrix elements $(A-B)_{ij,ij}=\epsilon_i-\epsilon_j$. Thus one can easily take the square root of $A-B$ and transform the last equation to

$$(A-B)^{1/2}(A+B)(A-B)^{1/2}W=\omega^2 W \tag{8.1}$$

with

$$W=(A-B)^{1/2}(X-Y)$$

Once one has $W$, the vectors $X$ and $Y$ may be obtained by the formulas $X+Y=(A-B)^{1/2}W/\omega$ and $X-Y=(A+B)(A-B)^{1/2}W/\omega^2$.

If only a few configurations are important in an excitation, this method is an obvious choice. The case of a single configuration is especially simple. There are two energies that enter, the particle-hole excitation energy $\Delta\epsilon=\epsilon_c-\epsilon_i$, and the interaction matrix element $v=\int d^3rd^3r'|\phi_i(r)\phi_c^*(r')|^2\delta H_{KS}(r)/\delta n(r')$. Putting this in eq. 8.1, one finds

$$\omega^2=\Delta\epsilon(\Delta\epsilon+2v)$$

and the amplitudes $X$, $Y$ are found to be

$$X=\sqrt{\frac{\Delta\epsilon}{\omega}}\frac{v}{\Delta\epsilon-\omega} \qquad Y=\sqrt{\frac{\Delta\epsilon}{\omega}}\frac{v}{\Delta\epsilon+\omega}.$$

If one calculates the oscillator strength with eqs. (7.3,7.12), one finds that it is independent of $v$. The interaction $v$ changes the excitation energy, but also changes the dipole matrix element in a way that preserves $f$.

The configurational matrix method is widely applied, generally with a judicious truncation of the configurational space. Besides in the many quantum chemistry programs that use this method, it is also applied in condensed matter physics using either the momentum- or the coordinate-space representation [47] to calculate the matrix elements. However the configurational matrix method requires truncation of the particle-hole space to be practical for large systems. Without truncation, the dimension $D$ of the matrices is $D=N_cN_e$. The direct solution of the eigenvalue problem requires of the order of $D^3$ arithmetic operations, thus giving an $N^6$ scaling with system size. However, iterative methods may still be useful. Two popular iterative methods are the Lanczos method and the Davidson method. These are both Krylov space methods in the terminology of computational physics. One utilizes a space of vectors generated by applying the Hamiltonian operator to some initial vector. The Lanczos method is very straightforward, just diagonalizing the Hamiltonian matrix in this space. Unfortunately, it does not work very well when there are many states at high excitation. The Davidson method uses preconditioning of the Hamiltonian matrix to suppress the overemphasis on configurations with large diagonal energies in the Krylov space. It is often used in quantum chemistry programs. All of the iterative methods are slow if one is seeking highly excited states. Namely, the iterative methods work by extracting the eigensolutions one at a time, starting from an end of the spectrum. The iterative process for each state requires orthogonalizing to the solutions already extracted. This becomes time consuming when there are many states.

With truncation of the particle-hole space, the most time-consuming task is usually the calculation of the matrix elements. This requires of the order of $N_c^2 N_e^2 N_\phi$ operations. Storage of the matrix requires a memory of size $N_c^2 N_e^2$ which can also be a limitation. These scaling factor are listed in Table 7, comparing the numerical requirements of the different methods of the computer can also be limiting.

Table 7. Size scaling of various algorithms for TDLDA–general comparison: floating point operations (FPO) and memory requirements.

| Method | FPO | Memory |
|---|---|---|
| Configuration matrix | $N_c^2 N_e^2 N_r$ | $(N_c N_e)^2$ |
| $k$-space response | $5M_\omega (N_\mathbf{k})^3$ | $N_\mathbf{k}^3$ |
| Response (Sternheimer) | $M_\omega M_{it} M_H N_e N_\mathbf{r}$ | $N_R(N_e + N_c)$ |
| Real-time | $N_e N_\mathbf{r} M_H M_T$ | $N_\mathbf{r}(N_e + 4.5)$ |

## 8.2. *Linear response method*

The linear response method works directly in the vector space representing the orbitals. For example, in a momentum space representation, the matrix $\chi_0$ has a dimension of $N_k$. This representation has been very commonly used in the last decade for calculating the dielectric properties of bulk solids. There are two computational demanding tasks to evaluate eq. (7.11). The first is to evaluate the matrix elements in $\xi$, which is done using the formula

$$\xi_0(\mathbf{k}, \mathbf{k}', \omega) = \sum_{ic} \sum_{\pm} \frac{\langle i|e^{i\mathbf{k}\cdot\mathbf{r}}|c\rangle\langle c|e^{i\mathbf{k}'\cdot\mathbf{r}}|i\rangle}{\omega - \epsilon_c + \epsilon_i + i\eta}$$

This requires $N_k^2 N_e N_c$ operations and would be the most demanding part of the computation if the space of unoccupied orbitals is unrestricted. Even with a restricted set of orbitals, one still has the matrix multiplication and inversion costs, both of which are of order $N_k^3$. This is clearly an improvement over the configuration method when one uses a complete space and direct diagonalization. In Table scaling, we have listed the computational costs assuming that the configuration space truncation is severe enough so that the matrix operations and storage are limiting. The factor $M_\omega$ is the number of frequencies one wishes to calculate.

### 8.3. *Sternheimer method*

Here one solves eq. 7.10 using the real-space representation so that sparse methods can be applied. Typically one uses the conjugate gradient method to solve the equations. The $N$-scalings are shown in Table 7. Unlike the previously discussed methods, storage is not limiting. However, the iteration process may be problematic; it is difficult to estimate the number of iterations $M_{it}$ to get a converged solution. If the frequency is well away from the eigenmodes of the system, convergence may be possible with several hundred iterations. Close to the eigenfrequencies the method does not converge at all.

### 8.4. *Real time method*

In the real time method, the linear response is computed by applying an impulsive ($\delta$-function in time) external potential to the electronic ground state, then calculating the expectation of some observable of interest as a function of time, and then Fourier transforming to get the frequency response. The real-time method is intrinsically of order $N_e N_\phi$, unlike other methods which are higher order in $N$ unless the bases are truncated in some way. The $N$-scalings are shown as the last line in Table scaling. Like the Sternheimer method, the real-time method does not require huge memory resources. Note also that all frequencies are contained in the initial perturbation and thus the entire frequency response comes out of a single time-dependent calculation. This reminds one of the modern experimental techniques in spectroscopy, which are to measure a signal in time and then Fourier transform to get a frequency or a mass spectrum. However, the time integration is done stepwise and requires a large number of steps $M_T$. I analyze this below in describing some specific algorithms to do the time integration.

We begin with the analysis of a time-independent Hamiltonian $H$. Given a state $\phi_i^0$ at time zero, the time evolution is expressed by the formula

$$\phi(t) = e^{-iHt}\phi_i^0 \tag{8.2}$$

The wave function is needed as a function of time with some interval $\Delta t$. Thus one computes for successive time steps as

$$\phi(t + \Delta t) = e^{-i\Delta t H}\phi(t).$$

If one allows oneself $M$ operations applying $H$ to a wave function, the best approximation one can make is to expand the above exponential in a $M$-order Taylor series. But if you tried this with desired values of $\Delta t$, you would not get a useful result. The problem is numerical, that any arithmetic operation introduces a truncation error. Now let us decompose the error vector into the eigenstates of $H$, calling the highest eigenvalue $|H|$. The first arithmetic operation introduces a spurious component of the highest eigenvalue, and each successive step in the expansion of the exponential multiplies that component by $\Delta t|H|/n$. For the desired $\Delta t$ this is typically much larger than one, and the spurious component is dominant after some steps. Even using a small time step, it is important to carry out an expansion of the exponential to some higher order. This method was first used in a nuclear physics context, [48] and it was found that the $4^{th}$ order expansion is especially useful. This is connected with an important requirement for the numerical representation of the exponential operator, that there not be any eigenvalues with modulus greater than one. Otherwise the vector would blow up under repeated application of the operator. To analyze the series expansion method, let us denote the $N$-order expansion by $e_N$, $e_N^{ix} = \sum_n^N \frac{(ix)^n}{n!}$. It is easy to see that $e_1$ is unsuitable:

$$|e_1^{i\Delta t|H|}|^2 = 1 + (\Delta t|H|)^2 > 1$$

On the other hand, the fourth-order expansion has the norm

$$|e_{(4)}^{ix}|^2 = 1 - \frac{8x^6}{4!2} + \frac{x^8}{4!2}. \tag{8.3}$$

This is less than one if $x = \Delta t|H| < \sqrt{8}$. In practice, the states of interest have small values of $x$ and the decrease of norm for states of large $x$ do not affect the observables. This method is the one we applied in most of our publications and is used in the code `real_time_dft.f` provided in our package of programs.

There is actually a better method, called the leapfrog, which can easily be derived from a discretized version of the variational principle, eq. (7.1). We take as variational variables the wave functions at discrete time steps, $\phi_n = \phi(r, n\Delta t)$, and approximation the time derivative as $(\phi_{n+1} - \phi_n)/i\Delta t$. The variational functional eq. (7.1) becomes

$$\Delta t \sum_n E_{KS}[\phi_n] - \sum_n \phi_n^*(\phi_{n+1} - \phi_n)/i.$$

Taking the variation with respect to $\phi_n^*$ yields the set of equations

$$\phi_{n+1} = \phi_{n-1} + 2i\Delta t H \phi_n. \tag{8.4}$$

Like the previous method, there is a limit on how large one can choose $\Delta t$. To derive the limit, first note that eq. 8.4 is invariant under the transformation $t \to t \pm n\Delta t$. Thus the time translations form a group, and the eigenstates have the time dependence $e^{i\lambda n\Delta t}$ for some value of $\lambda$ that depends on the eigenstate. Substituting this form in Eq. 8.4, we arrive at the following relation,

$$\phi_{n+1} = e^{i\lambda\Delta t}\phi_n = \phi_n e^{-i\lambda\Delta t} + 2i\epsilon\Delta t\phi_n. \tag{8.5}$$

Solving for $\epsilon$ in terms of $\lambda$ leads to

$$\sin(\lambda\Delta t) = \epsilon\Delta t. \tag{8.6}$$

from which it is clear that if $|\epsilon\Delta t| > 1$ the equation cannot be satisfied for $\lambda$ real. An imaginary part to $\lambda$ leads to an exponential growth that causes the algorithm to break down. Therefore we arrive at the condition $|H|\Delta t < 1$ for use of the leapfrog method. An example of this method in use is Ref. 49. It should be noted that it requires more care on initialization because it uses information from the previous two time steps. An error in the initialization makes an oscillation with period $2\Delta t$ and amplitude approximately equal to the magnitude of the error. However, this oscillation does not grow over time.

Besides norm conservation, the algorithm needs to conserve energy very accurately to be useful. In principle the TDDFT equations do conserve energy, but that is not guaranteed numerically. In practice, both the above algorithms produce acceptable energy conservation if $\Delta t$ is small enough to conserve the norm.

We can now easily analyze the number of time steps $M_T$ in Table 7. Let's suppose we want to get the spectrum with a resolution of $\Gamma$ on the line width. To get this resolution in the Fourier transform of the real-time response, the time interval $T$ must satisfy[d] $T > 2\pi\hbar/\Gamma$. On the other hand, from the above discussion we learned that the time step must satisfy $\Delta t < \hbar/|H_{KS}|$. In the real-space method, the highest eigenvalues are determined by the mesh spacing, $|H_{KS}| \approx 6/(m(\Delta x)^2$. Putting this together for a typical case ($\Delta x = 0.3$ Å, $\Gamma = 0.1$ eV), we find $M_T = 30,000$.

## 9. Applications of TDDFT

The main application of the linearized TDDFT is to optical and electric properties such as photon absorption cross sections, polarizabilities, and dielectric functions. I will only give a sample of the many kinds of systems that the method can be applied to, discussing metal clusters, various carbon systems, and two condensed materials.

### 9.1. *Simple metal clusters*

I shall show the results of numerical calculations of the TDDFT on excitations in clusters of alkali metals and of silver. These elements have a single valence electron in an $s$-shell, giving them a relatively simple electronic structure. It is always helpful to have simple models to compare with, and for these systems a simple qualitatively understanding is possible with the surface plasmon model. Here one assumes that the electrons behave classically and are free within the system, and compute the dynamics generated by the direct Coulomb field. This problem can be solved exactly for spherical systems. The frequency of the surface plasmon excitation is given by

$$\omega_{sp}^2 = \frac{4\pi e^2 n}{3m}, \tag{9.1}$$

[d]This is the full width at half maximum when the Fourier transform is filtered with the function $1 - 3(t/T)^2 + 2(t/T)^3$.

where $n$ is the density of conduction electrons[e]. Because all the valence electrons move coherently in the model, the predicted oscillator strength $f$ is equal to the number of atoms in the metal particle. A more sophisticated model is the jellium model, which has the quantum mechanics of the electron wave functions, but replaces the ionic potentials of the real materials by a uniform positively charge background. Table 8 shows a sample of the predicted excitation energy of the strong transitions and comparison to experiment.

These metal clusters do show a highly collective transition, confirming the existence of the surface plasmon mode. However, its frequency is lower than the surface plasmon formula predicts. For example, for sodium clusters the formula gives $\omega_{sp} \approx 3.4$ eV but the observed surface plasmon is lower by about 20% for clusters in the size range of several tens to hundreds of atoms. One possibility to explain the red shift would be with a lower density of atoms in the cluster than in the bulk, but that is not supported by DFT calculations of the cluster structure. Another mechanism to explain the red shift is the spillout of the electrons, a quantum effect that the orbitals extend farther than the boundaries of the background charge distribution. This mechanism can be analyzed quantitatively by calculating the TDDFT response in the jellium model. There is a computer code, called JellyRpa, that I wrote and distributed [50] to calculate the response of spherical jellium, making no other approximations on the dynamics.

Indeed there is an effect of the spillout. The peak is red shifted from the surface plasmon formula by 10%, about half of what is needed to explain the empirical position. Another possibility is that the local density approximation might not be accurate enough. In particular, we will see shortly that the local approximation to exchange can produce serious errors in infinite systems. This question was addressed in finite clusters by Madjet *et al.*, [52] who examined and compared the different treatments of exchange. They found that the LDA exchange was quite satisfactory for sodium clusters, and could not be the reason for the discrepancy.

Lastly, the jellium approximation might be inadequate; the full TDDFT of course includes a realistic treatment of the ionic potentials. Using pseudopotentials for the sodium ions, one obtains somewhat different results depending on the prescription used for the pseudopotential. The simplest realistic prescription gives very similar results to the jellium model. However, another prescription, [53] produces an additional red shift, bringing the surface plasmon close to the observed position. In the end, it is rather disquieting that a seemingly small change in the ionic potential would have a very noticeable effect on the absorption spectrum.

Lithium clusters show an even larger red shift than sodium: the surface plasmon formula predicts 4.6 eV, while the observed peak in the absorption spectrum is lower by 35%. [54] From the Table, one sees that the TDDFT reproduces the observed position with its shift quite well. Here it is much easier to understand how the shift arises from the ionic potential. The ionic potential is different for $s$ and $p$ waves

[e]In applying this formula, we will use the value of the density of atoms in the bulk solid.

because there is a core $s$ orbital that is excluded from the valence wave function but no corresponding excluded $p$ state. Thus, the ionic potential is effectively more attractive for $p$ orbitals. This makes it easier to excite the electrons from the ground state, and lowers the excited state energies. The experimental peak is rather broader than in sodium and has an asymmetric shape. These features are also reproduced by the TDDFT. [17]

Table 8. Surface plasmon in simple metal clusters

| Element | Cluster size | Surface Plasmon | TDDFT | Exp. |
|---|---|---|---|---|
| Na | 8 | 3.4 eV | 2.7 eV | 2.5 eV [51] |
| Li | 139 | 4.6 | 2.9 [17] | 2.9 |
| Ag | 8 | 5.4 | 3.2,3.9 [40] | 3.2,3.9 |

There is also a large red shift of the surface plasmon for silver clusters. Here the formula gives 5.4 eV, but the observed peak in the absorption much lower, closer to 3.5 eV. For the $Ag_8$ clusters, shown in Table 8, the excitation is split into two peaks. Both are reproduced in the TDDFT assuming a certain geometry for the cluster (see Ref. 40 for details). The origin of the red shift is not the same as in the lithium. Silver has a filled $d$ shell just below the valence $s$ shell which is rather easy to polarize, effectively screening the charge of the valence electrons. This physics is contained in the TDDFT, provided the $d$ electrons are treated explicitly. One can even separate out their influence in doing the calculation. Effectively, the charges are screened by a factor of two.

I finally want to mention the strength of the surface plasmon. In sodium clusters, the strength is close to the theoretical maximum of $f$ equal to the number of $s$ electrons. In lithium, there is some reduction that can be motivated by putting an effective mass $m^* > m$ into the surface plasmon formula. In silver, the theoretical reduction factor is 1/4 corresponding to the effective charge screening, $e^* = e/2$. However, while theory and experiment are in accord for sodium and lithium, the experimental reduction of strength for silver is not as large as predicted. Probably one should wait for additional experiments before drawing a conclusion here.

### 9.2. *Carbon structures*

There is a rich variety of conjugated carbon systems, ranging from small clusters and molecules to fullerenes, nanotubes, and graphite. I use the word *conjugated* to mean that the orbitals can be classified as $\pi$-type or $\sigma$-type and that the $\pi$ orbitals are at the Fermi surface[f]. A very simple theory, the Hückel model, provides the same kind of guidance for these systems as the jellium model does for the alkali metal clusters. The $\pi$ orbitals on different atoms couple rather weakly, and the

[f]The classification requires also that the carbons lie on a plane or a surface that can be considered locally planar. The $\sigma$ and $\pi$ orbitals have even or odd reflection symmetry, respectively.

Hückel models the wave function by the amplitude of the $\pi$ orbital on each carbon, constructing the eigenstates from the simple hopping Hamiltonian $H = -\beta \sum_{ij} c_i^\dagger c_j$. Here the sum $i, j$ runs over pairs of adjacent carbon atoms. The hopping parameter $\beta$ depends on distance $d$ between carbon atoms when that varies; see Ref. 41 for a parameterization.

Table 9. $\pi - \pi$ transitions in conjugated carbon systems

| System | Number of carbons | Hückel model | TDDFT | Exp. |
|---|---|---|---|---|
| chain | 9 | 1.7 eV | 4.6 eV | 4.2 eV |
| polyene | 10 | 2.1 | 3.2 | 3.4 |
| benzene | 6 | 5.2 | 6.9 | 6.9 |
| fullerene | 60 | 2 | 3.4-6.6 | 3.8-6 |

A sampling of the results for strong transitions is shown in Table 9. The first entry is the carbon chain $C_9$, taken from Ref. 37. The Hückel model predicts that the lowest particle-hole excitation is at 1.7 eV, and the number agrees well with the difference of the Kohn-Sham eigenvalues for the orbitals just above and below the Fermi energy. However, the time-varying potential in the equations of the TDDFT increase the frequency of the oscillation, pushing the state up to 4.6 eV. From the Table it may be seen that this is close to the observed position. The behavior is similar in the polyenes (molecules of structure $CH_2(CH)_nCH_2$): the particle-hole gap in the Kohn-Sham spectrum is well reproduced by the Hückel model, but the interaction pushes the collective excitation up in frequency. The energies for the polyene $C_{10}H_{12}$ are shown in the Table, taken from Ref. 41.

The optical response of benzene is very well known experimentally, making it a good test of the theory. Besides the numbers in the Table, Fig. 3 of the absorption experiment shows an excellent agreement between theory and experiment. The sharp peak at 6.9 eV is the strong $\pi$-$\pi$ transition; its predicted strength $f = 1.1$ is only 25% off from the measured value. As one goes higher in energy, there are many more states that can be excited, and the real-time TDDFT gives a prediction for the entire spectrum. The broad bump centered at 18 eV is due to the more tightly bound $\sigma$ electrons. Its center position and overall width is correctly described by the theory. Of course the TDDFT does not describe all the details of the spectrum. Between the two main peaks one can see some tiny structure in the experimental spectrum that reminds one of tuft of grass. These are the so-called Rydberg states, having a very loosely bound electron in the Coulomb field of the ion. These states cannot be described in the LDA because of its incorrect asymptotic potential field.

### 9.3. *Diamond*

I will take diamond as an example of the application of the TDDFT to a bulk insulator. Fig. 4 shows the imaginary part of its dielectric function. The strong absorption centered at 12 eV is fairly well reproduced, but the band gap, seen by the energy where the function becomes nonzero, is too low by a couple of eV.

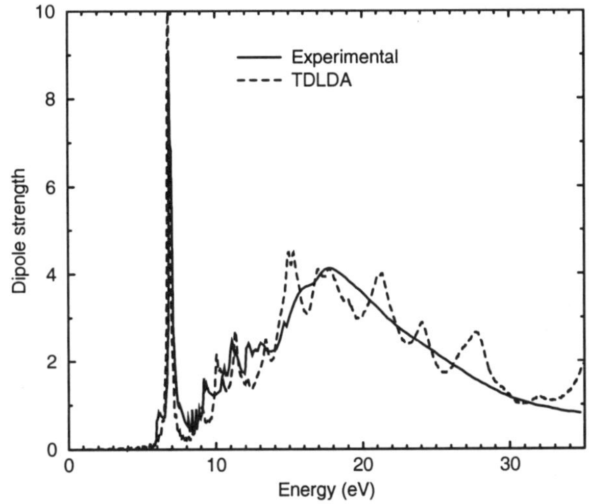


Fig. 3. Comparison of experimental and theoretical absorption spectrum in benzene, from Ref. 41.

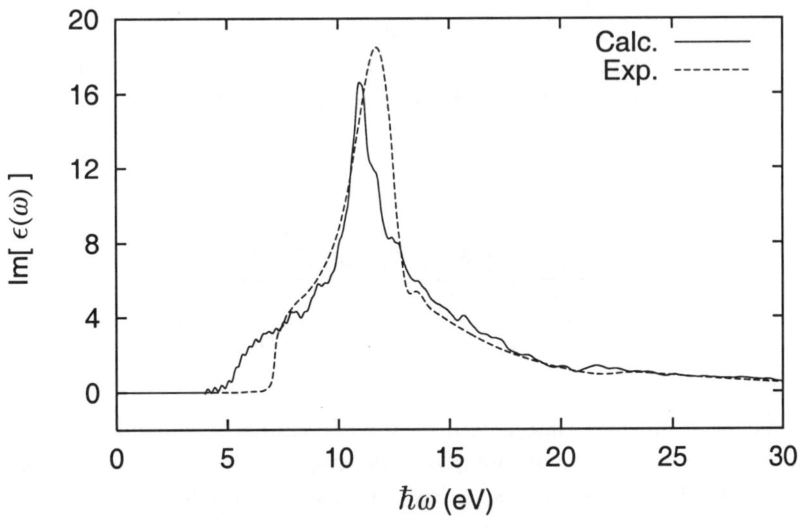


Fig. 4. Imaginary parts of the dielectric function $\epsilon(\omega)$ for diamond, calculated by TDDFT [45] and compared with experiment. [55]

Nevertheless the error here is compensated elsewhere in some way: if one uses this result to calculate the dielectric constant (real part of the dielectric function at zero frequency), one finds excellent agreement with experiment. See Ref. 45 for further details.

### 9.4. *Other applications*

In this section we briefly mention three other applications of of TDDFT: electron-vibration coupling, optical rotatory power, and nonlinear excitation.

We have seen that the DFT works very well for vibrational properties, and the TDDFT is a useful theory for the excitations. What about the coupling between them? The best way to see if it applies to the couplings is to try it out. Fortunately this is very simple theory to treat the couplings called the reflection approximation. It requires the probability distribution of the nuclear coordinates (arising from zero point motion or thermal excitation or both), and the response is calculated by averaging over that distribution. This approximation can be derived by a semiclassical treatment of the coupling, using a short-time expansion of the response of the system. [56] This joins well to the TDDFT, which we saw could also be viewed as a short-time expansion. I won't go into the details here, but just show some of the results for benzene. [46] That molecule makes a good test case because essentially all of its properties are known. Besides the strong $\pi$-$\pi$ transition at 6.9 eV that we saw earlier, there are $\pi - \pi$ excitations at lower energy that are forbidden by symmetry to be excited by dipole photons. However, the zero-point vibrational motion breaks the symmetry and allows the states to be weakly excited. Table 10 shows the oscillator strength for the three excitation in the $\pi - \pi$ manifold. The strength range over three orders of magnitude, and theory reproduces the numbers to 35% or so. This is really a remarkable achievement of the TDDFT.

Table 10. Oscillator strength of the excitations of benzene in the $\pi$-$\pi$ manifold. See 46 for details.

| $f/10^{-3}$ | TDDFT | CASSCF | Exp. |
|---|---|---|---|
| $^1B_{2u}$ | 1.6 | 0.5 | 1.3 |
| $^1B_{1u}$ | 59 | 75 | 90 |
| $^1E_{1u}$ | 1100 | | 900-950 |

Molecules or clusters that lack a center of symmetry will interact differently with left- and right-circularly polarized photons. In the language of multipoles, such molecules have a mixed parity and the electric dipole and magnetic dipole operators are coherent. The interference terms that are responsible for the optical rotatory power have the form

$$R_n = -\frac{e^2\hbar}{2mc}\langle\Phi_0|\sum_i \vec{r}_i|\Phi_n\rangle \cdot \langle\Phi_n|\sum_i \vec{r} \times \vec{\nabla}|\Phi_0\rangle. \tag{9.2}$$

where the first matrix element is the ordinary electric dipole and the second is the magnetic dipole. This quantity can be extracted with the real-time response to an impulsive dipole field. The difference from what we did before is that the dipole moment that is measured as a function of time is not the electric but the magnetic. For details see our article, Ref. 39. We tried it on a simple ten-atom molecule and on the chiral fullerene $C_{76}$. We found that the theory give the correct order of magnitude of the chiral properties, but on a quantitative level was much less reliable than when calculating oscillator strengths. For the lowest transitions in the ten-atom molecular, the $R_n$ values had the correct sign but were a factor of three off in magnitude. I should mention that the calculations were done with the LDA functionals, and it might be that the GGA could give quite different results.

All the previous applications have used the linear response of the system. The nonlinear response is described by hyperpolarizability coefficients, [57] extending the power series eq. (7.4) to second and higher terms in the electric field,

$$e\langle x\rangle = -\sum_j \alpha_{ij}\mathcal{E}_j + \frac{1}{2}\sum_{jk}\beta_{ijk}\mathcal{E}_k\mathcal{E}_j + \frac{1}{6}\sum_{jkl}\gamma_{ijkl}\mathcal{E}_k\mathcal{E}_j\mathcal{E}_l + \ldots$$

The second-order nonlinear polarizability $\gamma_{ijk}$ is only observable if the molecule can be oriented. It is important in technology for devices that can control optics by external electric fields. For a simple case, the water molecule, the TDDFT give a result that is about a factor of two different from the experimental. Unfortunately, the derived numbers depend on the choice of energy functional. There are also situations where the systems are exposed to extremely large fields. One of these is with excitation of clusters or molecules by slow-moving, highly charged ions. Such processes have been calculated in Ref. 58, 59. Another application of strong-field TDDFT is the excitation of clusters by intense laser fields. Calculations have also been made in this area, [60] but so far one has not made detailed enough calculations to model some of the most interesting experimental observations.

### 9.5. *Limitations*

In the diamond spectrum one saw that the theoretical excitations started at a lower energy than experiment. This is a generic problem of the TDDFT in infinite systems, that the band gap is too small. The band-gap problem originates in the LDA treatment of exchange. To see the origin of the problem, one needs only to look in more detail at the exchange contribution to the electron energies in the Fermi gas. The Fock energy of an electron of momentum $k$ is given by

$$-\sum_i \langle ki|\frac{e^2}{|r-r'|}|ik\rangle = -\frac{e^2k_F}{\pi}\left(1+\frac{k_F^2-k^2}{2kk_F}\log\left|\frac{k+k_F}{k_F-k}\right|\right). \tag{9.3}$$

From this equation one sees that the exchange potential of an electron in a Fermi gas has a weak logarithmic singularity at the Fermi surface. Particle-hole excitations across the Fermi surface have a higher energy for a given momentum difference than the quadratic kinetic energy functional. Because the singularity is weak, it doesn't show up for small systems. Clearly, any local approximation will miss the singularity. An effective, but computationally costly method to overcome this is to calculate the electron energies from the many-body perturbation theory. [61] In the GW approximation, the electron-self energy is calculated including exact exchange. The GW theory gives an enormous improvement to the band gap. But the computational demands of the theory has so far restricted its application to relatively simple systems.

The GW approximation fixes the energies of single-electron excitations, but the electron-hole interaction can also be important as well. In a large system with a band gap, the electron and hole are attracted to each other by their Coulomb interaction, and the bound states are called excitons. This part of the interaction is treated by a zero-range approximation in the density functional theory, so it has no possibility to produce weakly bound hydrogenic states. This is fixed up with an approved treatment of the exchange called the Bethe-Salpeter equation. This is even more computational intensive than the GW, but has achieved the best success so far in describing the optical properties in insulators near the band gap.

## Acknowledgments

We would like to thank T. Fennel for help in preparing the lecture notes, J.-I. Iwata for preparing the computer programs and J. Giansiricusa for additional help on the distribution package. GB acknowledges discussions with C. Guet, J. Rehr, and E. Krotscheck. GB's research mentioned in these lectures was supported by the Department of Energy under Grant FG06-90ER-40561.

## References

1. J. P. Perdew, K. Burke, and M. Ernzerhof, Phys. Rev. Lett. **77**, 3865 (1996).
2. J. P. Perdew, S. Kurth, A. Zupan, and P. Blaha, Phys. Rev. Lett. **82**, 2544 (1999).
3. M. Gell–Mann and K. A. Brueckner, Phys. Rev. **106**, 364 (1957).
4. E. Teller, Rev. Mod. Phys. **34**, 627 (1962).
5. D. M. Ceperley and B. J. Alder, Phys. Rev. Lett. **45**, 566 (1980).
6. J. P. Perdew and A. Zunger, Phys. Rev. B **23**, 5048 (1981)
7. U. Barth and L. Hedin, J. Phys. C **5**, 1629 (1972).
8. A. D. Becke, Phys. Rev. A **38**, 3098 (1988).
9. P. J. Stevens, *et al.,* J. Phys. Chem. **98**, 11623 (1994).
10. R. van Leeuwen and E. J. Baerends, Phys. Rev. A **49**, 2421 (1994).
11. N. Troullier and J. L. Martins, Phys. Rev. B **43**, 1993 (1991).
12. C. Hartwigen, S. Goedecker, and J. Hutter, Phys. Rev. B **58** 3641 (1998).
13. A. Castro, A. Rubio and M. J. Stott, preprint arXiv:physics/0012024.
14. W. H. Press, *et al., Numerical Recipes. The Art of Scientific Computing*, Second Edition, (Cambridge University Press, Cambridge, 1992), Sect. 19.6.
15. T. L. Beck, Rev. Mod. Phys. **72**, 1041 (2000).
16. J. Chelikowsky, *et al.,* Phys. Rev. **50** 11355 (1994).
17. K. Yabana and G. F. Bertsch, Phys. Rev. B **54**, 4484 (1996).
18. J.-L. Fattebert and J. Berholc, Phys. Rev. B **62**, 1713 (2000).

19. S. Goedecker, Reviews of Modern Physics **71**, 1085 (1999).
20. J. Kim, F. Mauri, and G. Galli, Phys. Rev. B **52**, 1640 (1995).
21. P. Ordejón, Phys. Stat. Sol. B **217**, 335 (2000).
22. S. Baroni and R. Resta, Phys. Rev. B **33**, 7017 (1986).
23. M. Hybertson and S. Louie, Phys. Rev. B **34**, 5390 (1986).
24. Asada and Tera, Phys. Rev. B **46**, 13599 (1992).
25. J. Cho and M. Scheffler, Phys. Rev. B **53**, 10685 (1996).
26. E. G. Moroni, *et al.,* Phys. Rev. B **56**, 15629 (1997).
27. C. Kohl and G. F. Bertsch, Phys. Rev. B **60**, 4205 (1999).
28. W. Krätschmer, *et al.,* Nature **347**, 354 (1990)
29. G. F. Bertsch, A. Smith, and K. Yabana, Phys. Rev. B **52**, 7876 (1995).
30. J. Kohanoff, *et al.,* Phys. Rev. B **46**, 4371 (1992).
31. A. M. Lee, N. C. Handy, and S. M. Colwell, J. Chem. Phys. **103**, 10095 (1995).
32. V. Anisimov, *et al.,* Phys. Rev. B **48**, 16929 (1993).
33. J. Bouchet, B. Siberchicot, F. Jollet, A. Pasturel, J. Phys. Condensed Matter **12**, 1723 (2000).
34. S. Savrasov, G. Kotliar, and E. Abrahams, Nature. **410**, 793 (2001).
35. P. Dirac, Proc. Cambridge Phil. Soc. **26**, 376 (1930).
36. A. Zangwill and P. Soven, Phys. Rev. A **21**, 1561 (1980).
37. K. Yabana and G. F. Bertsch, Zeit. Phys. D **42**, 219 (1997).
38. K. Yabana and G. F. Bertsch, Phys. Rev. A **58**, 2604 (1998).
39. K. Yabana and G. F. Bertsch, Phys. Rev. A **60**, 1271 (1999).
40. K. Yabana and G. F. Bertsch, Phys. Rev. A **60**, 3809 (1999).
41. K. Yabana and G. F. Bertsch, Int. J. Quant. Chem. **75**, 55 (1999).
42. G.F. Bertsch, J-I. Iwata, A. Rubio, and K. Yabana, Phys. Rev. B **62**, 7998 (2000)
43. J-I Iwata, K. Yabana and G. F. Bertsch, J. Chem. Phys. **115**, 8773 (2001).
44. K. Yabana, G. F. Bertsch, and A. Rubio, preprint arXiv: phys/0003090.
45. G. F. Bertsch, J. I. Iwata, A. Rubio, and K. Yabana, Phys. Rev. B **62**, 7998 (2000).
46. G. F. Bertsch, A. Schnell, and K. Yabana, J. Chem. Phys. **115**, 4051 (2001).
47. I. Vasiliev, S. Ogut, and J. Chelikowsky, Phys. Rev. Lett. **82**, 1919 (1999).
48. H. Flocard, S. Koonin, and M. Weiss, Phys. Rev. C **17**, 1682 (1978).
49. J. J. Rehr and R. Alben, Phys. Rev. B **16**, 2400 (1977).
50. G. Bertsch, Comp. Phys. Comm. **60**, 247 (1990). The program JELLYRPA may be downloaded from the author's Web site, www.phys.washington.edu/˜bertsch.
51. K. Selby, *et al.,* Phys. Rev. B **43**, 4565 (1991).
52. M. Madjet, C. Guet and W. R. Johnson, Phys. Rev. A **51**, 1327 (1995).
53. S. G. Louie, S. Froyen, and M. L. Cohen, Phys. Rev. B **26**, 1738 (1982).
54. C. Brechignac, *et al.,* Phys. Rev. Lett. **70**, 2036 (1993).
55. *CRC Hanbook of Chemistry and Physics* (CRC Press, Boca Raton, 77th ed. 1996). p. 12-133.
56. E. Heller, J. Chem Phys. **68**, 2066 (1978).
57. P. N. Butcher and D. Cotter, *Elements of Nonlinear Optics,* (Cambridge University Press, Cambridge, 1990), eq. 4.83.
58. K. Yabana, T. Tazawa, P. Bozek, and Y. Abe, Phys. Rev. A **57**, R3165 (1998).
59. R. Nagano, K. Yabana, T. Tazawa, and Y. Abe, Phys. Rev. A **62**, 062721 (2000).
60. C. A. Ullrich, P. G. Reinhard, and E. Suraud, J. Phys. B **31**, 1871 (1998).
61. L. Hedin, J. Phys. Condens. Matter **11**, R489 (1999).

# CHAPTER 2

# MICROSCOPIC DESCRIPTION OF QUANTUM LIQUIDS

Artur Polls
*Departament de Estructura i Constituents de la Matèria,*
*Facultat de Física, Diagonal 647, Barcelona 08028, Spain*
*E-mail:artur@ecm.ub.es*

Ferran Mazzanti
*Departament d'Electrònica, Engineria i Arquitectura La Salle,*
*Pg. Bonanova 8, Barcelona 08022, Spain*
*E-mail:mazzanti@salleurl.edu*

A pedagogical introduction to the macroscopic description of helium liquids is presented. Particular attention is paid to the methodological aspects in the framework of the variational theory. After a short introduction to the physics of helium liquids, it is shown how a properly chosen trial wave function leads to a correct description of the ground state of an infinite and homogeneous boson system. The extension of the method allows also describing the main trends of the spectrum of elementary excitations. Furthermore, the problem of one $^3$He impurity on a background of $^4$He atoms is also analyzed and used to introduce perturbation theory in the framework of the Correlated Basis Function theory. The formalism is afterwards extended to treat Fermi systems. Finally, the excitation spectrum in a Fermi liquid and general aspects of the dynamic structure function are also discussed.

## 1. Introduction

There is no doubt that the liquefaction of helium by Kammerlingh Onnes (1908) [1] signaled the birth of modern low–temperature physics. Since then, atomic helium fluids became an endless source of physical motivations for both theorists and experimentalists and have played a crucial role in the evolution of low temperature physics during the last century.

After the hydrogen, helium is the second most abundant element in the universe. However, it is much rarer on earth. Helium is the lightest element of the noble gases and has two stable isotopes: $^4$He and $^3$He, the former being a 99.99986% of the total population. The atomic structure of the He atom is very simple, having a couple of electrons in the $1s$ orbital in a spin singlet state around a nucleus, composed of two protons and two neutrons in the case of the $^4$He, and two protons and one neutron for $^3$He. As the number of fermionic constituents of the $^4$He atom is even, its total spin is integer and therefore it behaves as a boson. On the other hand, the number of fermionic constituents in $^3$He is odd, its total spin is $1/2$ and it behaves as a fermion.

The different statistical character of the two helium isotopes produces dramatic differences in their physical properties. So far, a huge amount of experimental information on helium liquids has already been accumulated, including precise knowledge of the equation of state and most of its thermodynamic properties. [2,3,4] In addition, the dynamics and the excitation spectrum have also been systematically investigated by inelastic neutron scattering. [5] One of the most remarkable facts is the phase transition that takes place in liquid $^4$He at 2.17 K at saturated vapor pressure. When the liquid is cooled below this temperature, its viscosity drops to almost zero and it is able to flow freely through very narrow capillary tubes in such a way that the velocity becomes independent of the pressure gradient along the channel, as it was first shown by Kapitza in 1938. [6] Under these conditions, liquid $^4$He not only shows this *super* flow property but also becomes a surprisingly good heat conductor. One of the most spectacular effects where both properties are manifested is the so called fountain effect, discovered by Allan and Jones in 1938. [7] Superfluidity in liquid $^3$He was discovered much later in 1972 [8] by Osheroff, Lee and Richardson. In this case the transition occurs at 2 mK, a temperature three orders of magnitude smaller than the transition temperature of pure $^4$He.

Nowadays, the activity concerning helium liquids is continuously pushed further to explore new situations. An example is the physical realization of almost two–dimensional (films) [9] and one–dimensional (nanotubes) [10] systems, and the consequent possibility to study the dependence on dimensionality and the effects of confining geometries on the quantum behavior of those liquids.

For theoreticians, and this will be our point of view, helium liquids can be considered as excellent laboratories to study the quantum many–body problem and to test and develop many–body theories. This is quite relevant because, in some sense, nearly all branches of physics deal with many–body physics at the most microscopic level of understanding.

In the present case, the energy and length scales relevant to describe helium liquids define the atoms as the microscopic constituents of our system. Besides, the interaction among the atoms is simple, depends only on the distance, and is well known. In addition, the quantum behavior has macroscopic manifestations and the effects of having different statistics can be investigated.

Actually, the fact that helium remains liquid at zero temperature is a consequence of the large zero point motion of the atoms in the fluid and can be considered as a macroscopic quantum effect. Moreover the superfluid behavior is a clear macroscopic manifestation of its quantum nature.

Helium liquids can be considered in the bulk limit but also in surfaces. Besides and as mentioned above, almost two–dimensional and one–dimensional realizations are also possible. In addition, droplets of $^4$He or $^3$He allow to study the dependence of properties on the number of particles. Actually helium clusters is another very interesting field which has evolved very quickly, and the present experimental capabilities make possible to investigate problems as the minimum number of atoms in a cluster required to have superfluidity. [11] Also the existence of the $^3$He-$^4$He mixtures permits the study of different types of excitations and the interplay of the correlations and statistics on the excitation spectrum.

In spite of the enormous progress in the last two decades [12] a complete and satisfactory explanation of some experimental facts has not yet been achieved. In this line, it is worth recalling that a full microscopic description of superfluidity in liquid $^3$He is still missing, or that a fully consistent microscopic theory describing the physics of strongly correlated Fermi systems at finite temperature is still lacking.

Quantum mechanics is the natural tool to describe helium liquids. Indeed, helium liquids at low temperature are quantum fluids, which differ from classical fluids in the sense that they can not be described in the framework of classical statistical physics. In a quantum fluid, the thermal de Broglie wavelength is of the order of the mean interparticle distance, and therefore there can be large overlaps between the wave functions of different atoms. In this situation quantum statistics has important consequences and one expects to find large differences between liquid $^4$He and liquid $^3$He at low temperatures.

In this chapter, an introduction to the microscopic description of helium systems at zero temperature is presented. Efforts are concentrated on the description of the ground state, although the spectrum of elementary excitations and the role played by the statistics are also analyzed. Methodological aspects are overemphasized, as the techniques explained can be used in general to describe strongly correlated systems, *i.e.*, systems which can not be properly described by mean field theories and that require the full machinery of the Quantum Many Body Methods (QMBM) to understand their behavior.

There are several review papers which can serve as good references complementary to these lecture notes. First to mention are the contributions of S. Fantoni, A. Fabrocini and E. Krotscheck to the proceedings of the previous Summer School on *Microscopic Quantum Many–Body Theories and Their Applications* that was held in Valencia in 1997. [12] Also interesting are the review articles by S. Rosati, [13] J. G. Zabolotzky, [14] C. Campbell, [15] J.W. Clark, [16] and S. Fantoni and Rosati. [17] Finally it is mandatory to single out the book of E. Feenberg, *Theory of Quantum Liquids,* [18] which contains the most complete monography on the variational theory applied to Quantum Liquids and that has enlightened most of the developments of the theory during the last thirty years.

## 1.1. *General properties*

In order to get familiar with the energy and length scales associated to liquid helium, some of its general properties will be first reviewed. The helium atom has a very stable configuration that makes it chemically inert. Besides, its lowest excitation

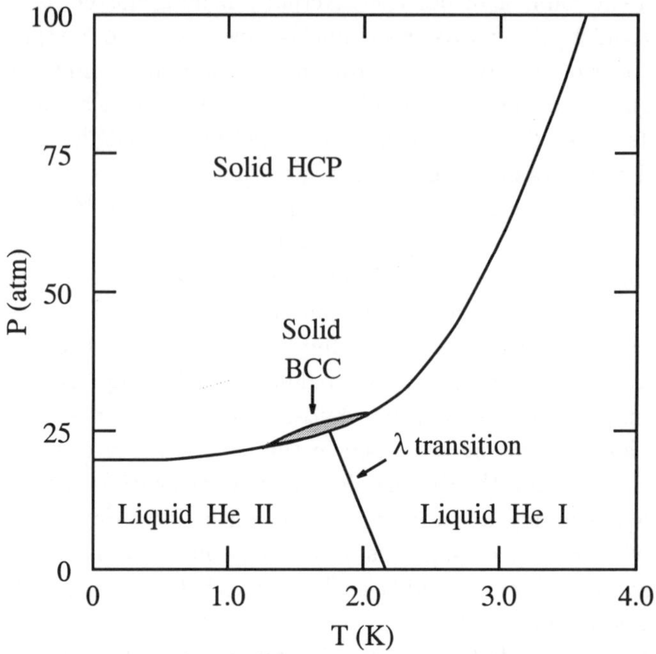


Fig. 1. Schematic phase diagram of liquid $^4$He in the PT plane.

energy is around 20 eV and the ionization energy is 24.56 eV. These energies are quite high compared with the energy scales involve in the description of the liquid, which are of the order of few Kelvins (K) (notice that 1 eV equals 11604 K). For the same reasons, the polarizability of the He atom is very small, and as a consequence, the van der Waals interaction when two atoms are close together is rather small and translates into a weak attraction that nevertheless turns into a strong repulsion when the atoms overlap. The situation is such that one can forget about the internal structure of the atoms and consider them as the elementary constituents of the quantum liquid interacting through a two–body potential. Actually this interaction is so weak that two $^4$He atoms are very weakly bound, while two $^3$He atoms are not bound at all. Notice also that the interaction between the atoms is due to the polarization of the electronic cloud, and as consequence the interaction between $^4$He-$^4$He, $^3$He-$^3$He, and $^4$He-$^3$He pairs can be considered to be the same. Keeping this in mind, it is easy to understand that, being the $^4$He-$^4$He system very weakly bound, the $^3$He-$^3$He and $^4$He-$^3$He systems, which have lighter mass but the same interaction, are not bound.

As it has been previously mentioned, both $^4$He and $^3$He remain liquid at zero temperature. It is commonly thought that cooling down a substance reduces the average kinetic energy of its atoms. If the temperature is decreased low enough, the atoms lose their mobility and get confined to fixed positions: at this point the substance solidifies. However, in this qualitative argument one is forgetting the

Table 1. Characteristic properties of helium liquids.

| | $^4$He | $^3$He |
|---|---|---|
| $\hbar^2/2m$ | 6.02 K Å$^2$ | 8.03 K Å$^2$ |
| $\rho_0$ | 0.365 $\sigma^{-3}$ | 0.274 $\sigma^{-3}$ |
| $e_0$ | -7.17 K | -2.47 K |
| $\langle U \rangle$ | $\sim$ -21 K | $\sim$ -15 K |
| $\langle T \rangle$ | $\sim$ 14 K | $\sim$ 12 K |
| $1/\rho_0$ | $\sim$ 46 Å$^3$ | $\sim$ 61 Å$^3$ |
| $V_{\text{atom}} * \rho_0$ | 0.2 | 0.15 |
| $v_{\text{sound}}$ | 238 m/s | 183 m/s |
| $P_{\text{sol}}$ | 25 atm | 34 atm |

uncertainty principle, which tells us that if the atoms are confined to a small region in space, then their momentum is not fixed, and so they acquire a non–zero average kinetic energy, the so called *zero point motion.* One way to estimate this energy is to associate the volume occupied by each particle, $1/\rho$, to a spherical box, and to approximate the kinetic energy to the ground state energy of a particle moving in this spherical box $t \sim \hbar^2\pi^2/(2mR^2)$, $R$ being its radius. At the saturation density of liquid $^4$He, $\rho_0 = 0.0218$ Atoms/Å$^3$, one finds a volume of 46 Å$^3$, with $R = 2.22$ Å and $E \sim 12.3\,$K. This kinetic energy is large enough to compensate the interaction between the atoms that tries to locate them in fixed equilibrium positions, and that is the reason why helium remains liquid at $T = 0$. This is a macroscopic manifestation of quantum mechanics. The other noble gases are heavier and easier to polarize, and become solid at zero temperature. The phase diagram of liquid $^4$He in the PT plane at low temperature is shown in Fig. 1. The line marked as $\lambda$–transition separates the normal (liquid He I) from the superfluid (Liquid He II) phases.

Several characteristic properties of $^4$He and $^3$He are reported in Table 1. Some of these properties can be qualitatively understood taking into account the different masses, the different statistics, and the fact that $^4$He and $^3$He have the same interaction. The lighter mass of $^3$He atoms translates into a larger mobility and therefore into a smaller saturation density and smaller binding energy per particle. Notice also that the binding energy results from a strong cancellation between the potential and the kinetic energies. Actually, the value of the kinetic energy for liquid $^4$He is a direct measure of the relevance of correlations. In the absence of interactions, the ground state has all the particles in the zero momentum state. When the interaction is turned on, particles are promoted out of the zero momentum state and the system ends up with a kinetic energy of about 14 K. However, there still remains a macroscopically large number of particles in the zero momentum state, that is measured by the condensate fraction value, $n_0 = N_0/N \neq 0$. At saturation density, the condensate fraction of liquid $^4$He is of the order of 8%.

The volume per particle occupied by liquid $^4$He is smaller than that of $^3$He. Therefore, it is easier to compress liquid $^3$He, and as a consequence the speed of sound in liquid $^3$He is smaller than in liquid $^4$He when they are compared at the corresponding saturation densities. When the interactions are turned off, the $^3$He system is described by means of a Slater determinant of plane waves filling the Fermi sphere or free Fermi sea. The radius of the Fermi sphere is determined by the Fermi momentum which, at saturation density of liquid $^3$He, corresponds to $k_F = 0.78$ Å$^{-1}$. The average kinetic energy per particle associated to the free Fermi sea is $e_F = 3\hbar^2 k_F^2/10m$ which is, at the same density, $e_F \sim 3K$. The Fermi momentum defines the Fermi surface, a very important concept that is meaningful even for a normal Fermi liquid when the interactions are turned on.

Another quantity of interest is the binding energy of one $^3$He impurity in liquid $^4$He, $\mu = -2.78$ K at the $^4$He saturation density. Notice that a $^3$He impurity is more bound in liquid $^4$He than the binding energy of liquid $^3$He. This indicates that it is energetically more favorable to solve $^3$He in $^4$He than to have separate phases. As a consequence we have stable liquid mixtures of both isotopes, with a maximum solubility of $\sim 6\%$.

It is also worth noticing that the ratio of the volume occupied by each atom ($V_{atom} = 4\pi R^3/3$ with R$\sim$ 1.3 Å) to the volume it has at its disposal ($V = 1/\rho_0$) is approximately 0.2 for pure $^4$He, thus indicating that the system is highly packed and that therefore the influence of correlations is strong. This same ratio for *nuclear matter* (an homogeneous system of nucleons that simulates matter at the center of a nucleus) would be 0.05. In this sense, liquid helium is more correlated than nuclear matter. From this point of view, convergence of perturbative calculations will be harder to achieve in liquid helium than in nuclear matter. However, the many–body problem in nuclear matter gets much more involved due to the complexity and uncertainties of the nucleon–nucleon interaction.

Finally, it is interesting to notice that in spite of the different scales in energy and length, from Kelvins to MeVs and from Angstroms to Fermis, the physics of both systems, liquid $^3$He and nuclear matter, is similar. Also from the methodological point of view, and this is one of the strongest points of the many–body theory, one can use the same techniques to study them.

## 2. Microscopic description

The microscopic description of a quantum many–body system can be performed once the Hamiltonian, written in terms of the masses of the atoms and the interaction, is established

$$H = \sum_{i=1}^{N} \frac{p_i^2}{2m} + \sum_{i<j}^{N} V(r_{ij}) \, . \tag{2.1}$$

In the following, only the description of infinite, homogeneous systems at $T = 0$ and in the thermodynamic limit ($N \to \infty$, $\Omega \to \infty$ but $\rho = N/\Omega \to cte$) will be considered.

As the Hamiltonian is written in first quantization, the indistinguishability of particles is accounted in the symmetry of the operators entering in $H$. However, the Hamiltonian does not distinguish if the identical particles are bosons or fermions, and therefore this information should be added by hand to the wave function. As usual, the latter has to be even for a system of bosons and odd for a system of fermions under the exchange of any pair of particle coordinates.

As it has already been said, and due to the fact that the electronic structures of the $^4$He and $^3$He atoms are equal, the interactions between the different pairs of atoms, $^4$He-$^4$He, $^3$He-$^3$He, and $^4$He-$^3$He can be considered to be identical. The interaction depends only on the interparticle distance and a simple representation is the Lennard–Jones potential

$$V(r) = 4\epsilon \left[ \left(\frac{\sigma}{r}\right)^{12} - \left(\frac{\sigma}{r}\right)^{6} \right] , \tag{2.2}$$

where $\epsilon = 10.22$ K defines the depth of the potential and $\sigma = 2.556$ Å the length scale. Nowadays, the more accurate Aziz (HFDHF2) potential [19]

$$V_{Aziz}(r) = \epsilon \left[ Ae^{-\alpha x} - F(x) \left( \frac{C_6}{x^6} + \frac{C_8}{x^8} + \frac{C_{10}}{x^{10}} \right) \right] , \tag{2.3}$$

where

$$F(x) = \begin{cases} \exp\left[ -\left(\frac{D}{x} - 1\right)^2 \right] & x < D \\ 1 & x \geq D \end{cases} \tag{2.4}$$

and

$$x = \frac{r}{r_m} \tag{2.5}$$

with parameters

$$\begin{aligned} \epsilon &= 10.8K & C_6 &= 1.3732412 \\ r_m &= 2.9673\,\text{Å} & C_8 &= 0.4253785 \\ D &= 1.241314 & C_{10} &= 0.1781 \\ \alpha &= 13.353384 & A &= 5.448504 \cdot 10^5 \ , \end{aligned} \tag{2.6}$$

and its revised version HFD-B(HE), [20] are used in realistic calculations. The He-He potential is characterized by a strong short range repulsion (such that in a first approximation, the atoms can be considered as hard spheres of diameter $\sim 2.6$ Å) and a weak attraction at medium and large distances. The Lennard–Jones and the Aziz potential are compared in Fig. 2. Despite the apparent similarities, $V_{Aziz}(r)$ appreciably differs from $V_{LJ}(r)$ at short distances where in particular the former does not diverge. In this sense, $V_{Aziz}(r)$ has a softer core.

To solve the many–body Schrödinger equation is not an easy task and only in very few exceptional cases one is able to find an exact analytic solution. In addition, the presence of the hard core in the interaction makes the application of standard perturbative methods difficult. This is because the matrix elements entering in the

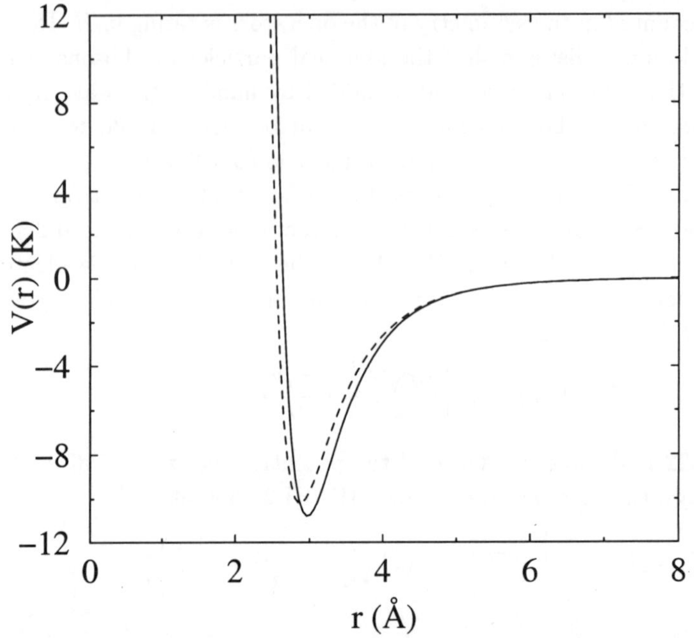


Fig. 2. Comparison between the Lennard–Jones (dashed–line) and the Aziz (solid line) potentials.

perturbative series would be too large, and also due to the fact that these are very dense systems. An efficient way to handle the short range correlations induced by the core of the potential is to embed them, from the very beginning, in a trial wave function $\Psi_T$ describing the system with $N$ atoms

$$\Psi_T(1,\ldots,N) = F(1,\ldots,N)\phi(1,\ldots,N) . \tag{2.7}$$

The correlation operator factor $F$, which is symmetrical with respect to the exchange of particles, incorporates to the wave function in a direct but approximate way the dynamic correlations induced by the interaction between the atoms. $\phi(1,\ldots,N)$ is the wave function corresponding to the free gas, which describes the system when the interactions are turned off and keeps the correct symmetry of the total wave function. For the ground state, one usually places all particles in the zero momentum state and so $\phi(1,\ldots,N) = \Omega^{-N/2}$. For fermions, the model function $\phi(1,\ldots,N)$ is chosen as the Slater determinant associated to the free Fermi sea

$$\phi(1,\ldots,N) = det \mid \psi_{\alpha_i}(j) \mid . \tag{2.8}$$

The single–particle wave functions $\psi_{\alpha_i}(j)$, where the subscript refers to the state, are taken as plane waves

$$\psi_{\alpha_i}(j) = \frac{1}{\Omega^{1/2}} e^{i\mathbf{k}_{\alpha_i}\cdot\mathbf{r}_j}\sigma(j) , \tag{2.9}$$

$\sigma(j)$ representing the spin contribution. In the $^3$He case, $\sigma(j)$ refers to the third component of the spin and so it can be either $\mid +\rangle$ or $\mid -\rangle$. In this way, each momentum state has a degeneracy 2. These plane waves satisfy the boundary conditions in a large cube of volume $\Omega$ which becomes infinite when the thermodynamic limit is considered. The allowed momenta fill a Fermi sphere of radius $k_F = (6\pi^2\rho/\nu)^{1/3}$, where $\nu = 2$ is the spin degeneracy. Once a trial wave function is defined, the variational principle guarantees that the expectation value of the Hamiltonian

$$\frac{\langle\Psi_T \mid H \mid \Psi_T\rangle}{\langle\Psi_T \mid \Psi_T\rangle} = E_T, \tag{2.10}$$

is an upper bound to the ground state energy $E_0$. Obviously, for the method to be efficient, the trial wave function must yield a *good* representation of the real many–body ground state wave function. Simple as it may look, the evaluation of the expectation value is by no means an easy task, and very sophisticated algorithms requiring large computer capabilities have been devised during the last years. [12]

In this way, the ingredients for a variational calculation are the Hamiltonian and the wave function. In addition, one needs a powerful and efficient machinery to evaluate the expectation values. At this point, one should appeal to the physical intuition in order to choose an appropriate trial wave function describing the system under consideration. A good starting point is to use a Jastrow correlation operator [21]

$$F(1,...,N) = F_2(1,...,N) = \prod_{i<j} f_2(r_{ij}) \ , \tag{2.11}$$

$f_2(r)$ being a properly chosen two–body correlation function that should be zero or almost zero when the interparticle distance is smaller than the range of the repulsive part of the potential that characterizes the interaction between the atoms, it should go to unity at large distances manifesting the absence of correlations when particles are separated far away from each other.

One way to proceed is to use a correlation function with an analytical expression depending on several free parameters, whose values are chosen to minimize the total energy. In this way, one determines the best function within the functional space defined by the variation of these parameters. Another alternative that will be discussed later on is to solve the Euler–Lagrange equations associated to the problem

$$\frac{\delta}{\delta f_2}\frac{\langle\Psi_T \mid H \mid \Psi_T\rangle}{\langle\Psi_T \mid \Psi_T\rangle} = 0 \tag{2.12}$$

which yields the optimum two–body correlation factor This is relevant if one seeks to recover the right asymptotic behaviors of the different quantities involved in the calculation. However and as commented above, before discussing the determination of the optimum two–body correlation factor, one must find a good and efficient way to evaluate the expectation value of the Hamiltonian. One possibility is to use Monte Carlo methods, based on the Metropolis algorithm, to evaluate the multidimensional integrals appearing in the expressions of $E_T$, and this particular application is known as variational Monte Carlo method (VMC). Monte Carlo methods are extensively discussed in the lectures by Prof. Senatore in this same volume. One important thing to keep in mind is that, apart from finite size effects associated to the fact that the Monte Carlo calculation takes place in a finite box with a finite number of particles, Monte Carlo provides an exact result for that given trial wave function and can be used as an accuracy check for the calculations described here. Another alternative, which is actually the one to be used here, is to perform a cluster expansion of the multidimensional integrals involved in the evaluation of the expectation value of the energy and to try a systematic summation of the different terms. Before developing this program, it is useful to introduce first the concept of distribution function, which appear in a natural way in the evaluation of the expectation value of a many–body Hamiltonian.

The calculation of the expectation value of the potential energy per particle can be carried out as follows. The indistinguishability of particles is reflected in the potential operator, which is symmetric under the exchange of coordinates, and in the wave function, which is symmetric for bosons and antisymmetric for fermions. Therefore, any pair of particles contributes the same to the potential energy, which then becomes

$$\frac{1}{N}\langle V\rangle \equiv \frac{1}{N}\frac{\langle \Psi_T \mid \sum_{i<j} V(r_{ij}) \mid \Psi_T\rangle}{\langle \Psi_T \mid \Psi_T\rangle} = \frac{N(N-1)}{2}\frac{1}{N}\frac{\langle \Psi_T \mid V(r_{12}) \mid \Psi_T\rangle}{\langle \Psi_T \mid \Psi_T\rangle} . \quad (2.13)$$

After that, one can carry out the integrals over coordinates $\mathbf{r}_3, \ldots, \mathbf{r}_N$ and write

$$\frac{1}{N}\langle V\rangle = \frac{1}{N}\frac{1}{2}\rho^2 \int d\mathbf{r}_1 d\mathbf{r}_2 V(r_{12}) \frac{N(N-1)}{\rho^2}\frac{\int d\mathbf{R}_{12}\Psi_T^*(1,\ldots,N)\Psi_T(1,\ldots,N)}{\int d\mathbf{R} \mid \Psi_T(1,\ldots,N)\mid^2} , \quad (2.14)$$

where the innermost term coincides with the definition of the two–body radial distribution function

$$g(r) = \frac{N(N-1)}{\rho^2}\frac{\int \mid \Psi_T(1,\ldots,N)\mid^2 d\mathbf{r}_3 \cdots d\mathbf{r}_N}{\int \mid \Psi_T(1,\ldots,N)\mid^2 d\mathbf{r}_1 \cdots d\mathbf{r}_N} . \quad (2.15)$$

In the previous expressions, $d\mathbf{R}$ and $d\mathbf{R}_{12}$ are shortcut notations for $d\mathbf{r}_1\, d\mathbf{r}_2 \ldots d\mathbf{r}_N$ and $d\mathbf{r}_3\, d\mathbf{r}_4 \ldots d\mathbf{r}_N$, respectively.

For uniform and homogeneous systems, $g(r)$ depends only on the magnitude of $\mathbf{r}$ and is proportional to the probability of finding two particles separated by a given distance $r$, normalized such that $g(r) \to 1$ when $r \to \infty$. Therefore, changing from variables $(\mathbf{r}_1, \mathbf{r}_2)$ to $(\mathbf{r}_1, \mathbf{r} = \mathbf{r}_2 - \mathbf{r}_1)$, and performing the trivial integration over $\mathbf{r}_1$, one finds

$$\frac{1}{N}\langle V\rangle = \frac{1}{2}\rho \int d\mathbf{r}\, V(r) g(r) , \quad (2.16)$$

and thus the calculation of the potential energy is translated into the calculation of the distribution function $g(r)$.

Similar things happen with the evaluation of the kinetic energy. In order to illustrate how it can be calculated, the simple case of a bosonic system described through a Jastrow wave function will be considered. Again and due to the indistinguishability of particles reflected in the symmetry of the wave function, all terms contribute the same and one has

$$\frac{1}{N}\langle T\rangle = -\frac{1}{N}\sum_{i=1}^{N}\frac{\hbar^2}{2m}\frac{\langle\Psi_T\mid\sum\nabla_i^2\mid\Psi_T\rangle}{\langle\Psi_T\mid\Psi_T\rangle} \equiv -\frac{\hbar^2}{2m}\frac{\langle\Psi_T\mid\nabla_1^2\mid\Psi_T\rangle}{\langle\Psi_T\mid\Psi_T\rangle} . \tag{2.17}$$

Although not necessary, it is convenient in the next step to take advantage of the anti–hermitian character of the gradient operator. This can be easily done performing an integration by parts which leads to the so–called Jackson–Feenberg identity [22]

$$\int d\mathbf{R}\,\Psi^*\nabla_1^2\Psi \equiv \frac{1}{4}\int d\mathbf{R}\left[\Psi^*(\nabla_1^2\Psi)+(\nabla_1^2\Psi^*)\Psi-2(\nabla_1\Psi^*)(\nabla_1\Psi)\right] , \tag{2.18}$$

which for real wave functions, as is the case of the ground state of a boson liquid as will be shown later, reduces to

$$\frac{1}{2}\int d\mathbf{R}\left[(\nabla_1^2\Psi)\Psi-(\nabla_1\Psi)(\nabla_1\Psi)\right] . \tag{2.19}$$

In the case considered here, $\Psi(1,\ldots,N)=F(1,\ldots,N)=\prod_{i<j}f_2(r_{ij})$, and the action of the $\nabla$ and $\nabla^2$ operators is straightforward

$$\nabla_1F(1,\ldots,N)=\sum_{i=2}^{N}\frac{\nabla_1f_2(r_{1i})}{f_2(r_{1i})}F(1,\ldots,N) , \tag{2.20}$$

and

$$\begin{aligned}\nabla_1^2F(1,\ldots,N)=&\left[\sum_{i=2}^{N}\frac{(\nabla_1^2f_2(r_{1i}))f_2(r_{1i})-(\nabla_1f_2(r_{1i}))^2}{f_2^2(r_{1i})}\right.\\&\left.+\left(\sum_{i=2}^{N}\frac{\nabla_1f_2(r_{1i})}{f_2(r_{1i})}\right)\left(\sum_{j=2}^{N}\frac{\nabla_1f_2(r_{1j})}{f_2(r_{1j})}\right)\right]F(1,\ldots,N) .\end{aligned} \tag{2.21}$$

Notice that the piece in the second row of the previous equation contains three body terms, whose evaluation requires from the knowledge of a three–body distribution function $g(\mathbf{r}_1,\mathbf{r}_2,\mathbf{r}_3)$. However, it is easy to see that this contribution is exactly canceled by the $(\nabla_1\Psi)(\nabla_1\Psi)$ term of Eq. (2.19). Keeping this cancellation in mind, one can write

$$\begin{aligned}\frac{1}{N}\langle T\rangle_{JF}&=-\frac{1}{N}\frac{\hbar^2}{2m}\frac{1}{2}N\int d\mathbf{R}_{12}\frac{F(1,\ldots,N)^2}{\langle\Psi\mid\Psi\rangle}\sum_{i=2}^{N}\frac{(\nabla_1^2f_2(r_{1i}))f_2(r_{1i})-(\nabla f_2(r_{1i}))^2}{f_2^2(r_{1i})}\\&=-(N-1)\frac{\hbar^2}{2m}\frac{1}{2}\int d\mathbf{R}_{12}\frac{F(1,\ldots,N)^2}{\langle\Psi\mid\Psi\rangle}\frac{(\nabla_1^2f_2(r_{12}))f_2(r_{12})-(\nabla f_2(r_{12}))^2}{f_2^2(r_{12})}\end{aligned} \tag{2.22}$$

Now taking into account that

$$\frac{(\nabla_1^2f_2(r_{12}))f_2(r_{12})-(\nabla f_2(r_{12}))^2}{f_2^2(r_{12})}=\frac{2f_2'}{rf_2}+\frac{f_2''}{f_2}-\frac{(f_2')^2}{f_2^2}=\triangle_{r_{12}}\ln f_2(r_{12}) , \tag{2.23}$$

identifying the expression of the distribution function, and performing the same change of variables done in the evaluation of the potential energy, one finally finds

$$\frac{1}{N}\langle T\rangle_{JF} = \frac{1}{2}\rho\frac{\hbar^2}{2m}\int d\mathbf{r} g(r)(-\triangle \ln f_2(r)) \ . \tag{2.24}$$

Notice that even if $T$ is a one–body operator, the presence of two–body correlations in the wave function makes its expectation value depend on the two–body radial distribution function $g(r)$.

Collecting results (2.16) and (2.24), one finally arrives at the expression of the total energy of a boson fluid described by a Jastrow wave function

$$\frac{1}{N}\frac{\langle \Psi_T \mid H \mid \Psi_T\rangle}{\langle \Psi_T \mid \Psi_T\rangle} = \frac{1}{2}\rho\int d\mathbf{r} g(r)\left[V(r) - \frac{\hbar^2}{2m}\triangle \ln f(r)\right] \ . \tag{2.25}$$

The angular integration can be trivially performed, and therefore if $g(r)$ is known, the evaluation of the energy reduces to a one dimensional quadrature. Thus the whole problem has been translated into the evaluation of the two–body radial distribution function.

Obviously there are ways other than using the Jackson–Feenberg identity to evaluate the kinetic energy per particle. These yield the so called Pandharipande–Bethe [23] or Clark–Westhaus [24] expressions. In contrast, the Jackson–Feenberg method is more convenient since all these alternatives require from the knowledge of the three–body distribution function $g(\mathbf{r}_1, \mathbf{r}_2, \mathbf{r}_3)$ which is harder to calculate.

A rather simple $f_2(r)$ which has been extensively used in the study of $^4$He and $^3$He liquids is the so called McMillan or Schiff–Verlet correlation factor [25,26]

$$f_2(r) = \exp\left[-\frac{1}{2}\left(\frac{b\sigma}{r}\right)^5\right] \ . \tag{2.26}$$

This function approximates well the short range behavior of the optimal correlation factor, and approaches quickly to unity. Besides, it is basically zero inside the core of the potential. The energy is minimized with respect to $b$, which is taken as a variational parameter.

Before concentrating on the calculation of the distribution function, it is useful to discuss a couple of general properties of the groundstate wave function of a bosonic system, mainly because they can be easily shown by using the variational principle, and also because they serve as a guide to write a general *ansatz* for the variational many–body wave function. In a boson system, the latter presents two important properties: i) It has no nodes, that is, it does not change sign and so can be chosen real and positive; ii) is not degenerate. The proof given here has been extracted from a review article on quantum liquids by C. Campbell, [27] and begins by writing the ground state wave function of the many–body boson systems in a general exponential form

$$\Psi_0(\mathbf{r}_1, \ldots, \mathbf{r}_N) = \exp\left[\frac{1}{2}\chi(\mathbf{r}_1, \ldots, \mathbf{r}_N)\right] \ , \tag{2.27}$$

where $\chi$ is a complex function

$$\chi(\mathbf{r}_1,\ldots,\mathbf{r}_N) = \chi_R(\mathbf{r}_1,\ldots,\mathbf{r}_N) + \imath\chi_I(\mathbf{r}_1,\ldots,\mathbf{r}_N) \tag{2.28}$$

with $\chi_R$ and $\chi_I$ real and symmetric under the exchange of particle coordinates. Note that this form is general enough to represent the ground state, allowing for the presence of nodes in $\Psi_0$.

The first thing to realize is that the expectation value of the potential energy does not depends on $\chi_I$

$$\begin{aligned} \langle V\rangle &= \frac{\langle\Psi_0 \mid \sum_{i<j}^N V(r_{ij}) \mid \Psi_0\rangle}{\langle\Psi_0 \mid \Psi_0\rangle} \\ &= \frac{\int d\mathbf{r}_1...d\mathbf{r}_N \sum_{i<j}^N V(r_{ij}) \exp[\chi_R(\mathbf{r}_1,...,\mathbf{r}_N)]}{\int d\mathbf{r}_1...d\mathbf{r}_N \exp[\chi_R(\mathbf{r}_1...,\mathbf{r}_N)]} \end{aligned} \tag{2.29}$$

while, on the other hand, and using the Jackson–Feenberg identity , the kinetic energy can be written as

$$\begin{aligned} \langle T\rangle &= \frac{\langle\Psi_0 \mid \frac{-\hbar^2}{2m}\sum_{i=1}^N \nabla_i^2 \mid \Psi_0\rangle}{\langle\Psi_0 \mid \Psi_0\rangle} \\ &= \frac{\int d\mathbf{r}_1...d\mathbf{r}_N\, e^{\chi_R}\left[-\frac{\hbar^2}{8m}\sum_{i=1}^N(\nabla_i^2\chi_R) + \frac{\hbar^2}{8m}\sum_{i=1}^N(\nabla_i\chi_I)^2\right]}{\int d\mathbf{r}_1...d\mathbf{r}_N \exp[\chi_R]} . \end{aligned} \tag{2.30}$$

The kinetic energy splits in two pieces, the second one depending on $\chi_I$ and being positive definite. As a consequence, one can always decrease the energy by setting $\chi_I = 0$. But if $\chi_I = 0$, then it is clear that the wave function has no nodes, is real and does not change sign. The fact that $\Psi_0$ is non–degenerate is a consequence of the absence of nodes. If there was another eigenstate with the same energy, it would be possible to take it orthogonal to $\Psi_0$. However, in order to be orthogonal to $\Psi_0$, this new eigenstate would have to show at least one node, and therefore $\chi_I(\mathbf{r}_1,...,\mathbf{r}_N) \neq 0$. But then it would be possible to lower once again the energy of that state by choosing $\chi_I = 0$, contradicting the hypothesis that $E_0$ was the lowest possible energy. This can not happen and therefore $\Psi_0$ is non–degenerate.

In conclusion, the ground state wave function of a Bose system can be written in the form

$$\Psi_0(\mathbf{r}_1,\ldots,\mathbf{r}_N) = \exp\left[\frac{1}{2}\chi_R(\mathbf{r}_1,\ldots,\mathbf{r}_N)\right] , \tag{2.31}$$

where the function $\chi_R$ can be decomposed into $n$–body functions $u_n$, with $1 \leq n \leq N$

$$\chi_R(\mathbf{r}_1,\ldots,\mathbf{r}_N) = \sum_{n=1}^N \sum_{i_1,\ldots,i_n} u_n(\mathbf{r}_{i_1},\ldots,\mathbf{r}_{i_n}) , \tag{2.32}$$

with the summation on $i_1 \ldots i_n$ running over all possible choices of $n$ indices from the set $(1, \ldots, N)$. More explicitly

$$\Psi_0(\mathbf{r}_1, \ldots, \mathbf{r}_N) = \exp\left[\frac{1}{2}\sum_{i<j}^{N} u_2(r_{ij}) + \frac{1}{2}\sum_{i<j<k} u_3(\mathbf{r}_i, \mathbf{r}_j, \mathbf{r}_k) + \cdots\right] \tag{2.33}$$

which is known as the Feenberg *ansatz*. An equivalent way to write this wave function would be

$$\Psi_0(\mathbf{r}_1, \ldots, \mathbf{r}_N) = \prod_{i<j} f_2(r_{ij}) \prod_{i<j<k} f_3(\mathbf{r}_i, \mathbf{r}_j, \mathbf{r}_k) \ldots . \tag{2.34}$$

with $f_n = e^{u_n/2}$.

In this way, the Jastrow wave function containing only two body correlations

$$\Psi_J(\mathbf{r}_1, \ldots, \mathbf{r}_N) = \prod_{i<j} f_2(r_{ij}) \tag{2.35}$$

can be considered as the first step towards a systematic approach to the exact groundstate wave function of a boson fluid.

The previous expressions serve to identify the general structure of the wave function, but do not provide enough information to determine the correlation functions entering. One knows from basic grounds, however, that there are general requirements they should fulfill

- They should be real functions in order to yield a real total wave function.
- They must be symmetric under the exchange of particle coordinates.
- They must be invariant under translation of the center of mass of the particles involved in the correlation

$$\mathbf{R} = \frac{1}{n}\sum_{i=1}^{n} \mathbf{r}_i \, . \tag{2.36}$$

- They have to be invariant under rigid rotations around the center of mass **R**.
- They have to be invariant under simultaneous inversion of all the coordinates.
- They should fulfill the cluster property, which means that $u_n(\mathbf{r}_1, \ldots, \mathbf{r}_N)$ has to vanish whenever any one of the coordinates becomes arbitrarily large.

Particularly, the cluster condition implies that the decomposition of $\chi_R$ given in Eq. (2.33) is unique.

All those conditions must be also fulfilled when the correlation operator is expressed in terms of $F(1, \ldots, N)$. In fact, if any subset, $i_1, \ldots i_p$, of particles is separated from the rest, $F(1, \ldots, N)$ decomposes in a product of two factors

$$F(1, \ldots, N) = F_p(i_1, \ldots, i_p) F_{N-p}(i_{p+1}, \ldots, i_N) \, . \tag{2.37}$$

Finally, notice that the cluster property particularly implies that

$$f^{(2)}(r \to \infty) \to 1 \,, \tag{2.38}$$

and that all the previous properties apply equally well in the case when one uses an approximate wave function of the Jastrow type with two–body correlation factors only.

## 3. Hypernetted–chain equations

It has been shown in the previous section that the energy per particle of an homogeneous bosonic system can be easily calculated once the two–body radial distribution function $g(r)$ is known. In this sense, $g(r)$ is the key quantity required in most variational calculations. The main goal of this section is to learn how it can be calculated. For a Jastrow wave function, $g(r)$ becomes

$$g(r) = \frac{N(N-1)}{\rho^2} \frac{\int \prod_{i<j} f_2^2(r_{ij})\, d\mathbf{R}_{12}}{\int \prod_{i<j} f_2^2(r_{ij})\, d\mathbf{R}} \,. \tag{3.1}$$

In order to calculate $g(r)$ it is useful to introduce the *cluster* function $h(r) = f_2^2(r) - 1$. Because of the healing property of $f_2(r)$ that makes it approach unity when the distance increases, the function $h(r)$ goes quickly to zero and provides a plausible expansion parameter. In fact, both the numerator and the denominator of Eq. (3.1) can be expanded in powers of the function $h(r)$

$$g(r) = \frac{N(N-1)\Omega^{-N}}{\rho^2 \Omega^{-N}} \frac{\int d\mathbf{R}_{12}\, (1 + \sum_{i<j} h(r_{ij}) + \sum_{i<j,k<l} h(r_{ij})h(r_{kl}) + ....)}{\int d\mathbf{R}\, (1 + \sum_{i<j} h(r_{ij}) + \sum_{i<j,k<l} h(r_{ij})h(r_{kl}) + ....)} \,, \tag{3.2}$$

where a factor $\Omega^{-N}$ corresponding to the normalization of the non–interacting wave function has been added both in the numerator and the denominator. The resulting contributions become multidimensional integrals involving one or more $h(r)$ factors, which are referred to as *cluster terms.* Doing so leads to an expression that apparently looks much worse than the original one, but that is easier to deal with as it will be shown in the following. Two important facts allowing for enormous simplifications should be taken into account: a) many integrals give the same result, and b) there are substantial cancellations between infinite sets of terms. The best way to understand how to take advantage of these facts is, as in many branches of modern physics, by introducing a diagrammatic notation.

Cluster diagrams are built with dashed lines representing functions $h(r_{ij})$ connecting particles $i$ and $j$, and points or circles denoting the particles themselves. In order to get used to this diagrammatic notation, Fig. 3 shows some of the diagrammatic contributions resulting from the cluster expansion of the numerator. Particles 1 and 2, which are not integrated, appear as open circles in the diagrams. On the other hand, internal points corresponding to integrated coordinates are represented by solid circles. Many different terms, as for example $\int d\mathbf{r}_3 h(r_{13})h(r_{23})$ and $\int d\mathbf{r}_7 h(r_{17})h(r_{27})$, yield the same value since internal points are nothing but dummy

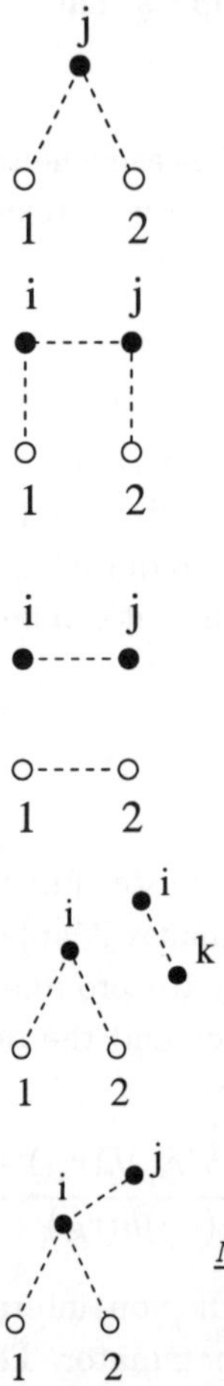


$$\frac{N-2}{\Omega}\int d\mathbf{r}_j h(r_{1j})h(r_{2j}) \to \rho \int d\mathbf{r}_3 h(r_{13})h(r_{23})$$

$$\frac{(N-2)(N-3)}{\Omega^2}\int d\mathbf{r}_i d\mathbf{r}_j\, h(r_{1i})h(r_{2j})h(r_{ij})$$
$$\to \rho^2 \int d\mathbf{r}_3 d\mathbf{r}_4\, h(r_{13})h(r_{24})h(r_{34})$$

$$h(r_{12})\frac{(N-2)(N-3)}{2\Omega}\int d\mathbf{r}_{ij} h(r_{ij})$$

$$\frac{N-2}{\Omega}\int d\mathbf{r}_i h(r_{1i})h(r_{i2})\frac{(N-3)(N-4)}{2\Omega}\int d\mathbf{r}\, h(r)$$

$$\frac{N-2}{\Omega}\int d\mathbf{r}_i h(r_{1i})h(r_{i2})\frac{N-3}{\Omega}\int d\mathbf{r}\, h(r)$$

Fig. 3. Several cluster diagrams appearing in the numerator of Eq. (3.2).

integration variables. They are represented by the same diagrammatic structure but with different labeling of the internal points. In this way, it is clear that disregarding the labeling, a given diagram corresponds to many different cluster integrals. In the diagrammatic formalism, all these terms are represented by a single unlabeled diagram multiplied by a properly chosen weighting factor that takes this fact into account.

Weighting factors are built from combinations of three different quantities: first, terms of the form $(N-\alpha)$ which have to do with the different ways in which internal points can be drawn from a set of $N-2$ particles (corresponding to the total number of particles $N$ minus the two external points that are fixed); second, factors $\Omega$ resulting from the cancellation of the global $1/\Omega^N$ term (Eq. (3.2)) and the integral over the coordinates not appearing in the diagram; and third, overall symmetry factors associated to the fact that sometimes permutations of the labels corresponding to internal points in a given diagram leads to exactly the same labeled diagram. All this is summarized in Fig. 3, where it is also indicated how in the thermodynamic limit factors of the form $(N-\alpha)/\Omega$ turn into densities $\rho$.

As an example, the third diagram has a function $h(r_{12})$ multiplying because neither coordinate 1 nor 2 are being integrated. It also contains a factor $(N-2)(N-3)$ corresponding to choosing two points ($i$ and $j$ in the diagram) from $N-2$ possible choices. There is also an overall $\Omega$ dividing, that results from the cancellation of the $1/\Omega^N$, and a factor $\Omega^{N-4}$ from the integration over all the remaining particle coordinates (different from 1, 2, $i$ and $j$), a factor $\Omega$ from the change of variables $\int d\mathbf{r}_1 d\mathbf{r}_2 \to \int d\mathbf{R} d\mathbf{r} = \Omega \int d\mathbf{r}$ with $\mathbf{R} = (\mathbf{r}_1 + \mathbf{r}_2)/2$ and $\mathbf{r} = \mathbf{r}_1 - \mathbf{r}_2$ and a factor $\Omega^2$ coming from the $\rho^2$ in the definition of $g(r)$. And finally, a symmetry factor $1/2$ associated to the fact that exchanging the labels $i$ and $j$ leads to exactly the same diagram. All other diagrams in the figure can be analyzed in the same way.

Furthermore and as mentioned above, there are strong cancellations between diagrams in the numerator and the denominator of $g(r)$. These cancellations can be identified once the different cluster integrals are properly classified. Since integrals are represented by diagrams, the problem of classifying the cluster integrals is directly mapped into the problem of classifying diagrams.

The standard classification of diagrams is illustrated in Fig. 4. Reduced to its fundamental topological essence, diagrams may be linked and unlinked. Unlinked diagrams contain separated pieces not connected to each other by correlation lines. In contrast, linked diagrams form one single piece and therefore one can always find at least one path of lines connecting any pair of points in the diagram. Examples of linked and unlinked diagrams are depicted in the first two rows of Fig. 4.

Linked diagrams can be classified as *reducible* and *irreducible*. Reducible diagrams are those with at least one *articulation point*. An articulation point is a circle separating the diagram in two or more different parts, such that any path of lines joining one particle in one part to another particle in another part passes always through it. Furthermore, all the external points in a reducible diagram remain in the same part. Examples of reducible diagrams are also shown in Fig. 4, pointing with an arrow the articulation points. Due to the translational invariance of $h(r)$, the integral expression associated with a reducible diagram can always be factorized into independent pieces. Despite the apparent simplicity of the classification, it is still difficult to decide which diagrams are relevant in the evaluation of $g(r)$. Fortunately, there is a useful theorem that simplifies the situation a lot. This theorem states that one needs to consider only the irreducible diagrams of the numerator. All unlinked and reducible diagrams cancel against the denominator of Eq. (3.2) up to order 1/N. A somewhat elaborated proof of the theorem, as an special application of the more general fermionic case, has been given by Fantoni and Rosati. [29] On the other hand, this problem is completely equivalent to the calculation of the radial

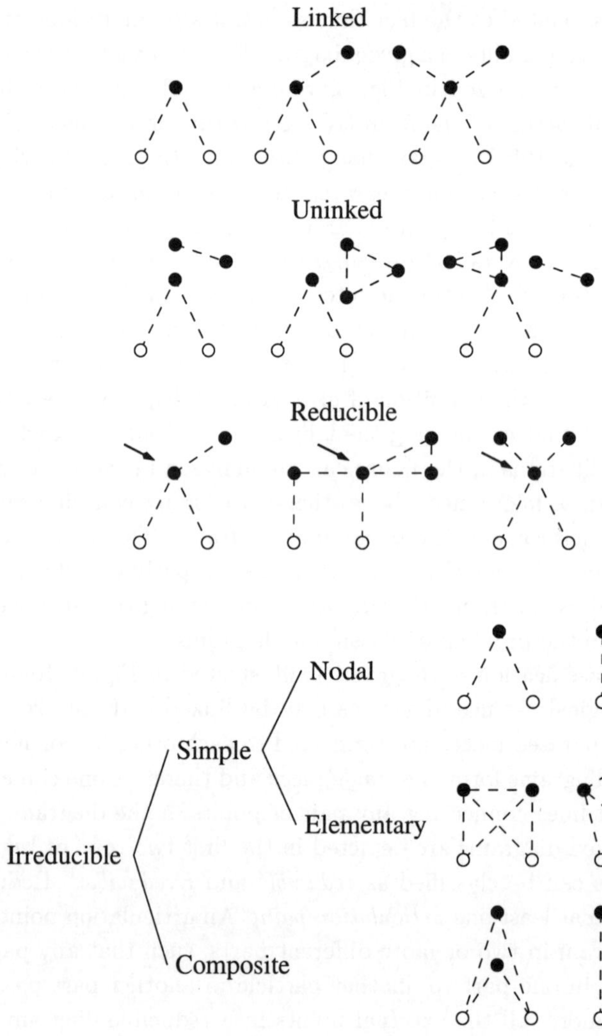


Fig. 4. Classification of cluster diagrams.

distribution function of a classical fluid at finite temperature

$$g(r) = \frac{N(N-1)}{\rho^2} \frac{\int d\mathbf{R}_{12} \exp\left[-\sum V(r_{ij})/kT\right]}{\int d\mathbf{R} \exp\left[-\sum V(r_{ij})/kT\right]}, \tag{3.3}$$

where $k$ is Boltzmann's constant. The parallelism is established by identifying, $f_2(r_{ij})$ with $\exp[-V(r_{ij})/2kT]$. Therefore, one can use the experience accumulated in this field, [30] for which a diagrammatic notation was also introduced, and derive a Hypernetted–Chain (HNC) equation in a similar way as Leeuwen, Groeneveld and de Boer did for classical systems. [31]

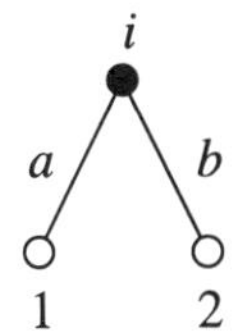


$$\rho\,(a(r_{1i}) \mid b(r_{i2})) = \rho \int d\mathbf{r}_i a(r_{1i}) b(r_{i2})$$

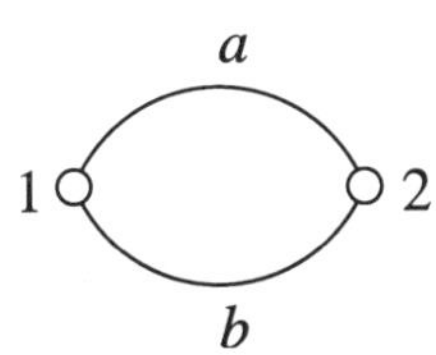


$$a(r_{12}) b(r_{12})$$

Fig. 5. Mathematical operations associated to the construction of nodal and composite diagrams.

In any case, one still has to face the problem of adding up the contribution of all the irreducible diagrams in the numerator. To this end, one needs to dive a little bit more into diagrammatic concepts and introduce new definitions. Irreducible diagrams can be classified as either *simple.* or *composite* . A simple diagram is a diagram containing only one $1-2$ *subdiagram*, while a composite diagram contains two or more $1-2$ *subdiagrams.* An $i-j$ *subdiagram* is a part of a diagram connected to the rest of the diagram through points $i$ and $j$ only. In particular, the first two diagrams in Fig. 3 are examples of diagrams with only one $1-2$ subdiagram. Besides, simple diagrams can be divided in *nodal* and *elementary.* A *nodal* diagram is a diagram that has at least one *node.* A *node* is a point of the diagram, such that all paths of lines going from one external point to the other passes through it. An elementary diagram is a simple diagram that is non–nodal. In Fig. 4 there is a scheme showing the complete classification of diagrams. The final diagrammatic rules are very simple: diagrams are built with correlation lines and points. External points, representing particle coordinates 1 and 2, are depicted as open circles, while internal points (solid circles) imply an integration over the corresponding coordinates times a factor $\rho$. The sum of all topologically distinct irreducible diagrams defines the radial distribution function $g(r)$.

There are two mathematical operations associated to the construction of diagrams (see Fig. 5): the convolution and the algebraic products. The convolution product of two generic functions $a(r_{ik})$ and $b(r_{kj})$ representing diagrammatic links between points $i$ and $k$ and $k$ and $j$ respectively

$$\rho(a(r_{ik}) \mid b(r_{kj})) \equiv \rho \int d\mathbf{r}_k a(r_{ik}) b(r_{kj}) \ , \tag{3.4}$$

generates a nodal $i-j$ subdiagram that has point $k$ as a nodal point. This is easy to understand since once the integral is performed, point $k$ becomes and internal point that is the only link connecting the external point $i$ in $a(r_{ik})$ to the external point $j$ in $b(r_{kj})$, thus being a node. On the other hand, the algebraic product of two functions $a(r_{ij})$ and $b(r_{ij})$, which can be taken as two $i-j$ subdiagrams, generates a composite $i-j$ subdiagram.

At this point one is ready to derive the HNC equation in its simplest form, adding the contribution of all diagrams except those containing elementary pieces, which will be discussed later. The first row of Fig. 6 shows the diagrammatic contents of the chain equation at the first level of iteration. Let $N(r)$ be the sum of all the nodal diagrams, and take only, for the sake of simplicity, the subset of nodal diagrams resulting from the first iteration of the chain equations, as illustrated in the upper part of Fig. 6. If now one performs the convolution product of $N(r)$ and $h(r)$, which is the basic brick used to construct the chain, one gets all the diagrams of $N(r)$ except the first one which is the convolution product of two $h(r)$ function, as it is shown in the lower part of Fig. 6. The chosen set of diagrams satisfy therefore the equation

$$\rho(h(r_{1j}) \mid N(r_{j2})) = N(r_{12}) - \rho(h(r_{1j}) \mid h(r_{j2})) \ . \tag{3.5}$$

The function $N(r)$ built in this way can then be used to construct composite diagrams. Composite diagrams have two or more $1-2$ subdiagrams, and so their sum has to be built from terms of the form $N^2(r), N^3(r), \ldots$ producing composite diagrams with $2, 3$ or more $1-2$ subdiagrams. However, these contributions must include appropriate symmetry factors in order to avoid undesired overcountings. As for example, $N^2(r)$ is the product of a sum of nodal diagrams with itself, thus producing diagonal terms squared and cross terms multiplied by 2. In the diagrammatic formalism, diagonal terms squared have to be divided by 2, which is the right symmetry factor correcting the fact that exchanging all particles in one of the subdiagrams with all particles in the other leads to exactly the same composite diagram. On the other hand, the factor 2 in the cross terms has to be removed as the previous situation does not happen here. Both facts are properly taken into account

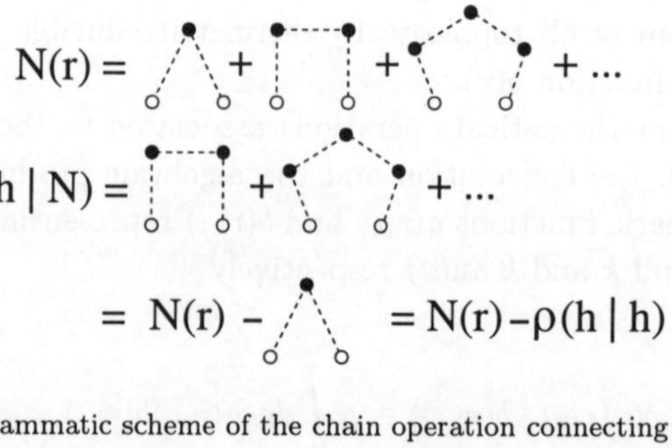


Fig. 6. Diagrammatic scheme of the chain operation connecting diagrams.

by replacing $N^2(r)$ with $N^2(r)/2!$. Similar analysis leads to the conclusion that composite diagrams containing three $1-2$ subdiagrams have to be built from the combination $N^3(r)/3!$, and so on. The total sum of composite diagrams formed in this way becomes $N^2(r)/2!+N^3(r)/3!+N^4(r)/4!+\cdots$ which in the thermodynamic limit coincides with $\exp[N(r)] - N(r) - 1$. This is schematically shown in Fig. 7, where the diagrammatic contents of the exponential is explicitly depicted. Finally, it should be noticed that one can still build new composite diagrams by multiplying $\exp[N(r)]$ with $h(r)$, thus producing the same composite diagrams but with an extra correlation line connecting external points 1 and 2. Since both $\exp[N(r)]$ and $h(r)\exp[N(r)]$ are possible, and $1 + h(r) \equiv f_2^2(r)$, one finally finds the total sum of composite diagrams $X(r)$ to be

$$X(r) = f_2^2(r)\exp(N(r)) - N(r) - 1 \,. \tag{3.6}$$

$X(r)$ and $N(r)$ obtained through Eqs. (3.5) and (3.6) are the result of the first iteration in the HNC scheme. One can then use this $X(r)$ back as the building block of the chain and obtain $N(r)$ in the second iteration

$$\rho(X(r_{1j}) \mid N(r_{j2})) = N(r_{12}) - \rho(X(r_{1j}) \mid X(r_{j2})) \,, \tag{3.7}$$

and use this new nodal function in Eq. (3.6) to obtain $X(r)$ in the second iteration. This process must be repeated until convergence is reached. At that point, $X(r)$ and $N(r)$ simultaneously satisfy Eqs. (3.5) and (3.6), which constitute the set of HNC equations.

This scheme is commonly known as HNC/0, and its solution yields the sum of all diagrams but those containing elementary contributions. Elementary diagrams can be incorporated to the HNC process by including them in the definition of $X(r)$

$$X(r) = f_2^2(r)\exp\left[N(r) + E(r)\right] - N(r) - 1 \,, \tag{3.8}$$

although they can not be treated in an exact way because there are not simple relations connecting them to $N(r)$ or $X(r)$ in a closed form.

In the end, the distribution function is expressed as the sum of all nodal, composite and elementary diagrams

$$g(r) = 1 + X(r) + N(r) = f_2^2(r)\exp[N(r) + E(r)] \,. \tag{3.9}$$

The HNC equations (Eq. (3.7) and (3.8)) form a non linear set of integral equations, whose solution suggests in a natural way the iterative process discussed above that is illustrated in the scheme shown in Fig. 8. Notice also that while doing the iterations, one can take advantage of the known properties of the Fourier Transform (FT), which turns convolutions into algebraic products and thus helps for solving at each step the nodal contributions from Eq. (3.7)

$$\tilde{N}(k) = \frac{\tilde{X}^2(k)}{1 - \tilde{X}(k)} \,, \tag{3.10}$$

where it is customary to define the FT in the HNC context as

$$\tilde{a}(k) = \rho \int d\mathbf{r}\, e^{i\mathbf{k}\cdot\mathbf{r}} a(r) \ , \tag{3.11}$$

including a density factor $\rho$.

According to the scheme in the figure, in the iterative process one usually calculates $X(r)$ (sum of composite and elementary diagrams) in $r$ space and the nodal function in $k$ space. In order to do so, good FT subroutines able to accurately treat correlations inside the core of the potential are required. Alternatively, the whole process can be carried out in position space, although convolutions may be harder

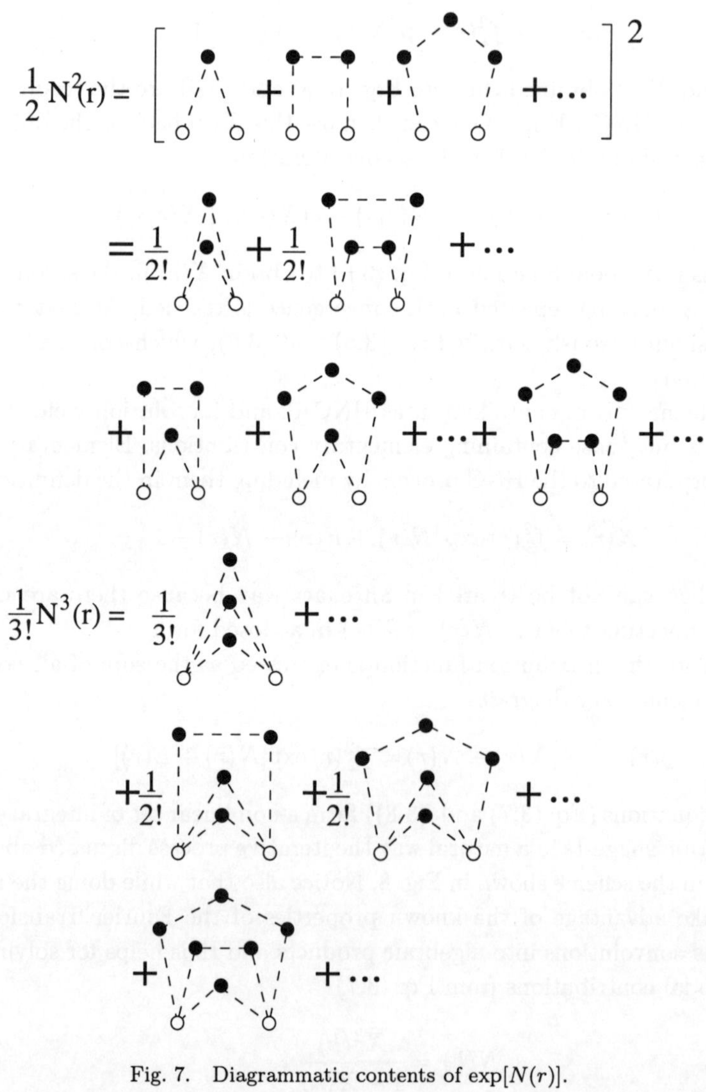

Fig. 7. Diagrammatic contents of $\exp[N(r)]$.

to deal with. Usually, ten to twenty iterations are enough to reach convergence, at least for correlation factors with no long range order. Since each iteration is built from the results obtained in the previous one, the number of diagrams included at each steps grows exponentially.

One function which is experimentally accessible is the static structure function, defined as the Fourier transform of the radial distribution function

$$S(k) \equiv 1 + \rho \int d\mathbf{r}\, e^{i\mathbf{k}\cdot\mathbf{r}}(g(r) - 1) = 1 + \tilde{X}(k) + \tilde{N}(k) \ . \tag{3.12}$$

Taking into account the HNC equation, the nodal and the composite diagrams can be expressed in terms of the static structure function as

$$\tilde{N}(k) = \frac{(S(k) - 1)^2}{S(k)} \tag{3.13}$$

and

$$\tilde{X}(k) = \frac{S(k) - 1}{S(k)} \ . \tag{3.14}$$

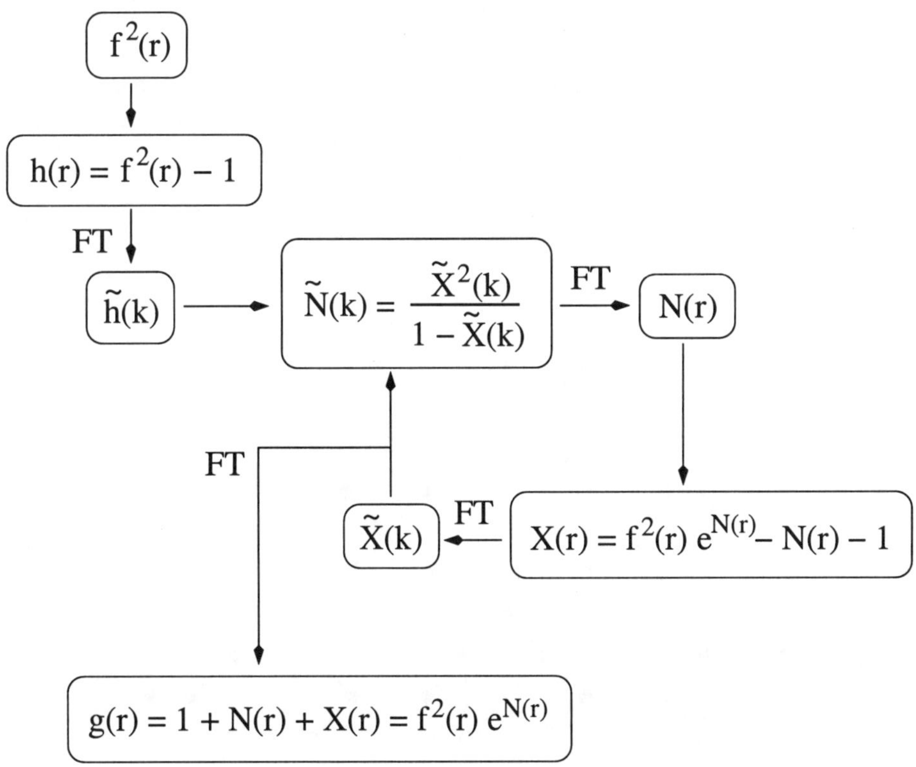


Fig. 8. Iterative scheme to solve the HNC equations.

These last two equations are valid independently of whether the elementary diagrams are included or not.

Before showing numerical results, it may be worth pushing the formalism a little further and discuss how to introduce the contribution of elementary diagrams and how to improve the trial wave function by including triplet correlations. The simplest elementary diagram, shown in Fig. 9, has four points and a symmetry factor 1/2. Actually, there is no way to sum up the contribution of all the elementary diagrams in a closed form, and so one has to rely on suitable approximations. It is customary to classify the HNC scheme according to the type of elementary diagrams included. For instance and as previously mentioned, the HNC/0 scheme disregards any contribution from elementary diagrams, while HNC/4 means that only the four points elementary diagram $E_4$ is implemented. Stated as it is, the HNC/4 scheme would only include one elementary diagram out of the infinite number one can draw by adding more and more internal points. This is not the case if one replaces the $h(r)$ bonds in $E_4$ by dressed lines $g(r) - 1$. Since $g(r)$ already contains an infinite sum of diagrams, the dressed $E_4$ built in this way becomes a massive sum of elementary contributions, thus yielding a *better* approximation to the exact result. In this way, the dressed $E_4$ has to be recalculated at each iteration in the HNC scheme until convergence is reached. Alternatively, there has been a lot of work done in the HNC/5 scheme where a selected subset of all the possible five points elementary diagrams is included.

Other approaches to sum elementary contributions try to exploit the numerically observed cancellation between the elementary and the composite diagrams. This is the idea behind the Percus–Yevick approximation, [32] in which it is assumed that

$$\exp[N_{PY}(r) + E_{PY}(r)] \sim 1 + N_{PY}(r) , \tag{3.15}$$

and therefore

$$E_{PY}(r) = \ln(1 + N_{PY}(r)) - N_{PY}(r) . \tag{3.16}$$

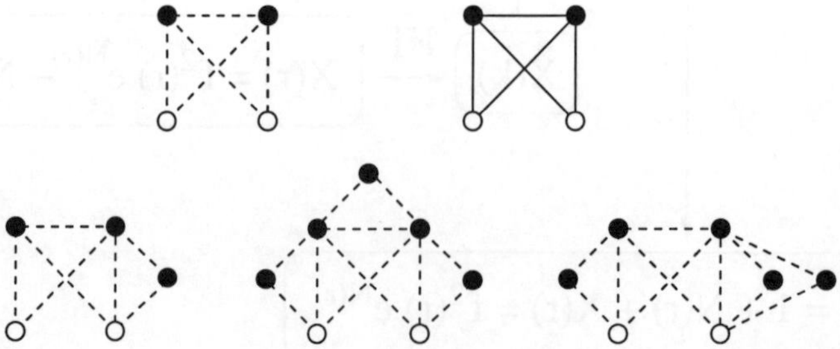

Fig. 9. Four–points elementary diagrams and other elementary diagrams with the same basic structure. Dashed and solid lines represent factors $h(r)$ and $g(r) - 1$, respectively.

In the Percus–Yevick scheme, all definitions are the same except the function X(r), which reads

$$X_{PY}(r) = f_2^2(r)(1 + N_{PY}(r)) - N_{PY}(r) - 1 \; . \tag{3.17}$$

A simple way to calculate their contribution is based on the empirical observation that, at least in the pure $^4$He case, the sum of the elementary diagrams show approximately the same spatial dependence as the four–points elementary diagram $E_4$. This fact suggests as a working recipe that the sum of all elementary diagrams may be well approximated by a scaling factor times $E_4$. The scaling factor can be determined in different ways. A reasonable strategy is to adjust it by equating the kinetic energies calculated with the Jackson–Feenberg and the Pandharipande–Bethe expressions, which certainly coincide when all the diagrams are taken into account. [33] A second choice is to impose the resulting energy to be equal to the experimental energy of $^4$He at saturation density. Both ways give similar results in the end.

A possible improvement of the trial wave function is to introduce triplet or three–body correlations

$$\Psi_{JT}(\mathbf{r}_1, \ldots, \mathbf{r}_N) = \prod_{1=i<j}^{N} f_2(r_{ij}) \prod_{1=i<j<k}^{N} f_3(r_{ij}, r_{ik}, r_{jk}) \; , \tag{3.18}$$

which, among others, may present the following parameterization [34]

$$f(r_{ij}, r_{ik}, r_{jk}) = exp\left[ -\frac{\lambda}{2} \sum_{cyc} \xi(r_{ij}) \xi(r_{ik}) \hat{\mathbf{r}}_{ij} \hat{\mathbf{r}}_{ik} \right] \; , \tag{3.19}$$

where the summation runs over all cyclic permutations of the indices $i, j$ and $k$.

The introduction of triplet correlations in the wave function modifies the expression of the energy, whose evaluation now requires the knowledge of the three–body distribution function $g_3$

$$\begin{aligned} \frac{E}{N} &= \frac{1}{2}\rho \int d\mathbf{r}\, g(r) \left[ V(r) - \frac{\hbar^2}{2m} \nabla^2 \ln f_2(r) \right] \\ &\quad - \frac{\hbar^2}{8m}\rho^2 \int d\mathbf{r}_{12} d\mathbf{r}_{13}\, g(r_{12}, r_{13}, r_{23}) \nabla_1^2 \ln f_3(r_{12}, r_{13}, r_{23}) \; , \end{aligned} \tag{3.20}$$

with

$$g(r_{12}, r_{13}, r_{23}) = \frac{N(N-1)(N-2)}{\rho^3} \frac{\int d\mathbf{R}_{123}\, \Psi^*(\mathbf{r}_1, \mathbf{r}_2, \ldots, \mathbf{r}_N) \Psi(\mathbf{r}_1, \mathbf{r}_2, \ldots, \mathbf{r}_N)}{\int d\mathbf{R}\, |\Psi(\mathbf{r}_1, \mathbf{r}_2, \ldots, \mathbf{r}_N)|^2} \tag{3.21}$$

independent of the variational form of the trial wave function.

The incorporation of the triplet correlations to the two–body radial function is rather straightforward. One performs a cluster expansion in terms of the function

$$\chi(r_{12}, r_{13}, r_{23}) = f_3^2(r_{12}, r_{13}, r_{23}) - 1 \tag{3.22}$$

which is diagrammatically represented by a triangle whose vertices correspond to particle coordinates 1, 2 and 3, see Fig. 10.

As seen from the wave function, a cluster expansion can now be performed in terms of the two independent functions $h(r)$ and $\chi(r_{ij}, r_{ik}, r_{jk})$. In this way, one can think about the addition of triplet correlations as the contribution of new terms that have to be added on top of the previous expansion, with cross terms mixing the final contributions.

For instance, the equation defining nodal diagrams remains unaltered, since nodes can only be built by linking independent subdiagrams through one external point, as shown in Eq. (3.7). Despite that, the nodal function generated in this way certainly contains additional contributions which have to be included in $X(r)$.

At the HNC/0 level, $X(r)$ is the sum of composite diagrams only. Composite diagrams are made from two or more $1-2$ subdiagrams. Each of these subdiagrams have two external points and one or more internal points that can be connected by $h(r)$ and/or $\chi(r_{12}, r_{1k}, r_{2k})$ bonds. In this way, subdiagrams can now be classified according to the kind of correlations that reach the external points. Points 1 and 2 can be directly linked or not by a triplet correlation. If not, the subdiagram may be nodal or elementary as before, but keeping in mind that these diagrams may also contain triplet correlations connecting inner parts. However, a new type of diagram is found when the external points are directly linked by a triplet correlation. Actually, one such piece has one internal point $k$ (that is, the third point different from 1 and 2 in the triplet), and all kind of subdiagrams $1-k$ and $2-k$ *dressing* the sides of the triangle. Since these sides can be dressed with *all* kind of diagrams, they should be represented by complete $g(r)$ lines. As a consequence, the relevant

$f_3^2(r_{12}, r_{13}, r_{23}) - 1$

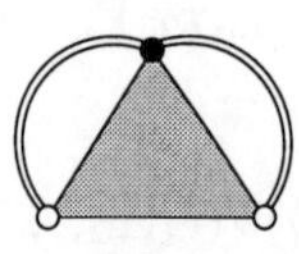

$C(r_{12}) = \rho \int d\mathbf{r}_3 \left[f_3^2(r_{12}, r_{13}, r_{23}) - 1\right] g(r_{13}) g(r_{23})$

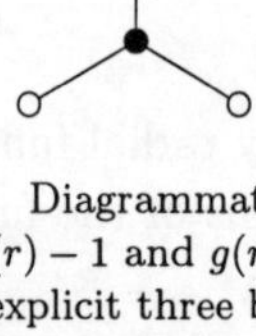

$A_4(r_{12}, r_{13}, r_{23}) = \rho \int d\mathbf{r}_4 \, (g(r_{14}) - 1) \, (g(r_{24}) - 1) \, (g(r_{34}) - 1)$

Fig. 10. Diagrammatic representation of triplet correlations. Single and double lines represent factors $g(r) - 1$ and $g(r)$, respectively. The lowest diagram refers to the four–points Abe diagram, without explicit three body correlations.

triplet function becomes

$$C(r_{12}) = \rho \int d\mathbf{r}_3 \left[ f_3^2(r_{12}, r_{13}, r_{23}) - 1 \right] g(r_{13}) g(r_{23}) , \tag{3.23}$$

which is then incorporated into $X(r)$ in the usual form to produce composite diagrams with the appropriate symmetry factors

$$X(r) = f_2^2(r) \exp\left[ N(r) + E(r) + C(r) \right] - N(r) - 1 . \tag{3.24}$$

As seen from Eq. (3.20), the inclusion of triplet correlations leads to an expression of the energy per particle containing the three–body distribution function. Consequently, a variational model of $g(r_{12}, r_{13}, r_{23})$ is required. Proceeding as before, a cluster expansion of this quantity can be easily carried out, but now taking into account that there are three external points corresponding to particle coordinates 1, 2 and 3. This quantity can be built from two different contributions, mainly those linking pairs of points and those linking the three external points all together. Since there is no restriction on the kind of $1-2$, $1-3$ and $2-3$ subdiagrams, each of these pieces can be replaced by complete $g(r)$ factors. In this way, one component contributing to the three–body distribution function is the product $g(r_{12})g(r_{13})g(r_{23})$. Actually, the substitution $g(r_{12}, r_{13}, r_{23}) = g(r_{12})g(r_{13})g(r_{23})$ is commonly used and becomes the so called *Superposition Approximation.* However, a complete description of $g(r_{12}, r_{13}, r_{23})$ contains the second piece which links all three external points. One direct way to link these points is through a bare $\chi(r_{12}, r_{13}, r_{23})$ factor, but this is not the only one. There is still the possibility to form $1-2-3$ subdiagrams with appropriate combinations of one or several internal points, and $h(r)$ and $\chi(r_{ij}, r_{ik}, r_{jk})$ links. As before, contributions of this kind may contain one or more $1-2-3$ subdiagrams, and so their total sum must include the necessary symmetry factors. Putting all this together one finds the final expression of the three–body distribution function

$$g(r_{12}, r_{13}, r_{23}) = f_3^2(r_{12}, r_{13}, r_{23}) g(r_{12}) g(r_{13}) g(r_{23}) \exp\left[ A(r_{12}, r_{13}, r_{23}) \right] , \tag{3.25}$$

where the function $A(r_{12}, r_{13}, r_{23})$ takes care of the so called Abe diagrams. [35] The four–points Abe diagram without considering explicit triplet correlations is shown in Fig. 10. In this case the triplet correlations are indirectly taken into account in the two–body bonds $g(r) - 1$ used to construct the diagram

$$A_4(r_{12}, r_{13}, r_{23}) = \rho \int d\mathbf{r}_4 \, (g(r_{14}) - 1)(g(r_{24}) - 1)(g(r_{34}) - 1) . \tag{3.26}$$

Reducing this expression to the case of two–body correlations is as easy as taking $f_3(r_{12}, r_{13}, r_{23}) \equiv 1$.

### 3.1. *Results for liquid $^4$He*

A characteristic feature of the energy per particle as a function of density for a liquid is the presence of a minimum, which defines the saturation density ($\rho_0$) and

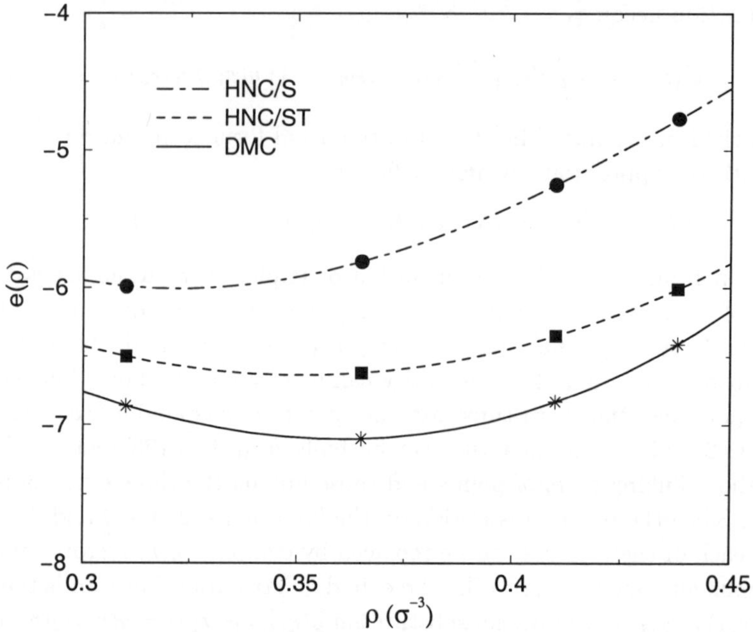


Fig. 11. Energy per particle of liquid $^4$He as a function of density in several approximations. HNC/S and HNC/ST correspond to Jastrow and Jastrow plus triplet correlations using the scaling approximation. DMC stands for *Diffusion Monte Carlo*, which can be taken as the exact solution.

corresponds to zero pressure. At this density it would be not necessary to exert pressure on the liquid to maintain it in a recipient. In Fig. 11, one can see the energy per particle of liquid $^4$He calculated in several approximations and using the Aziz potential. To the left of the saturation density the pressure is negative and the liquid enters a metastable zone, to the right of $\rho_0$ the pressure is positive.

As a starting point, HNC/0 results for the old Lennard–Jones potential and a Jastrow wave function built from a Schiff–Verlet two–body correlation factor are presented and discussed. At the experimental saturation density ($\rho_0^{exp} = 0.365\sigma^{-3}$), the variational parameter $b = 1.17$ (see Eq. (2.26)) defines the minimum energy $e_{HNC/0}(\rho_0^{exp}) = -4.44$ K, to be compared with the Monte Carlo result $e_{VMC}(\rho_0^{exp}) = -5.73$ K for the same trial wave function. The kinetic and potential energies per particle are $t_{HNC/0} = 14.7$ K and $v_{HNC/0} = -19.2$ K, respectively. Notice that one gets an upper bound to $e_{exp}(\rho_0^{exp}) = -7.17$ K although this was not guaranteed because the contribution from elementary diagrams has been entirely neglected.

To have an idea of the shape of the different functions entering in the HNC process, $f_2^2(r)-1$ (dot–dashed line), the sum of composite diagrams $X(r)$ (lower dashed line), the nodal function $N(r)$ (solid line) and the first chain of nodal diagrams built

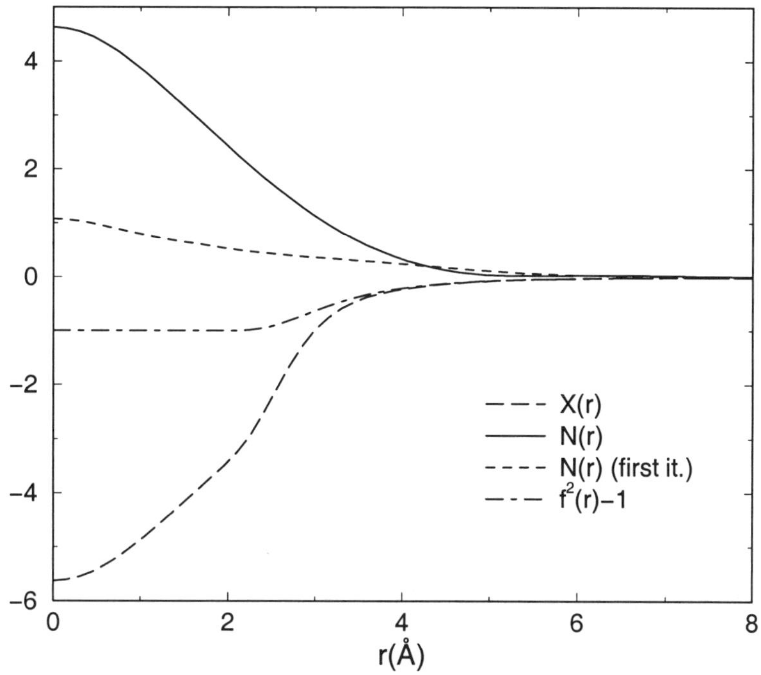


Fig. 12. Composite diagrams $X(r)$ (lower dashed line), $h(r) = f^2(r) - 1$ (dot–dashed line), nodal diagrams $N(r)$ (solid line) and the sum of nodal diagrams at the first iteration (upper dashed line) for a Macmillan correlation factor with $b = 1.17$ and at the experimental saturation density $\rho = 0.365\sigma^{-3}$.

with $f_2^2(r) - 1$ (upper dashed line) are shown in Fig. 12. As seen from the picture, $N(r)$ in the first iteration and the complete $N(r)$ function are quite different, thus stressing the need to incorporate all other nodal diagrams resulting from the whole set of HNC iterations.

The square of the correlation function (dot–dashed line), the HNC/0 radial distribution function (solid line) and the variational Monte Carlo $g(r)$ corresponding to the same correlation function (dashed line) are shown in Fig. 13. Notice that VMC can be considered as the *exact* result for that given trial wave function. Both the two–body correlation factor and the radial distribution function are practically zero inside the repulsive core of the potential. Also remarkable is the presence of a maximum around $r \sim 3.2$ Å, not appearing in $f^{(2)}(r)$, which indicates the existence of a characteristic distance at which the probability of finding another particle becomes maximal. The difference between the Monte Carlo and the HNC/0 calculations measures the contribution of the elementary diagrams.

The incorporation of the elementary diagrams, to the HNC process, using the scaling method, yields $e_{HNC/S}(\rho_0^{exp}) = -5.71$ K, in fairly good agreement with the VMC result.

Getting closer to the experimental numbers requires improving the trial wave function. This can be achieved by solving the corresponding Euler–Lagrange equations (2.12) to find the optimal two–body correlation factor. It turns out, as will be seen in the next section, that the optimization of the correlation function is very important if one seeks to recover the large–$r$ asymptotic behavior of the distribution function, but that otherwise does not influence very much on the binding energy, which is basically dominated by the short–range structure of the radial distribution function $g(r)$. The next step is to introduce three–body correlations, which can be of the form proposed in Eq. (3.19).

$$f_3(r_{ij}, r_{ik}, r_{jk}) = \exp\left[-\frac{1}{2}\lambda \sum_{cyc} \xi(r_{ij})\xi(r_{ik})\hat{\mathbf{r}}_{ij}\hat{\mathbf{r}}_{ik}\right] . \qquad (3.27)$$

Variational Monte Carlo results of the early 80's [34] were performed using the following analytic form for the $\xi(r)$ function entering in $f_3(r_{ij}, r_{ik}, r_{jk})$

$$\xi_{MC}(r) = \begin{cases} r \exp\left[-\left(\frac{r-r_t}{\omega}\right)^2\right]\left(\frac{r-r_B}{r_B}\right)^3 & r < r_B \\ 0 & r \geq r_B \ , \end{cases} \qquad (3.28)$$

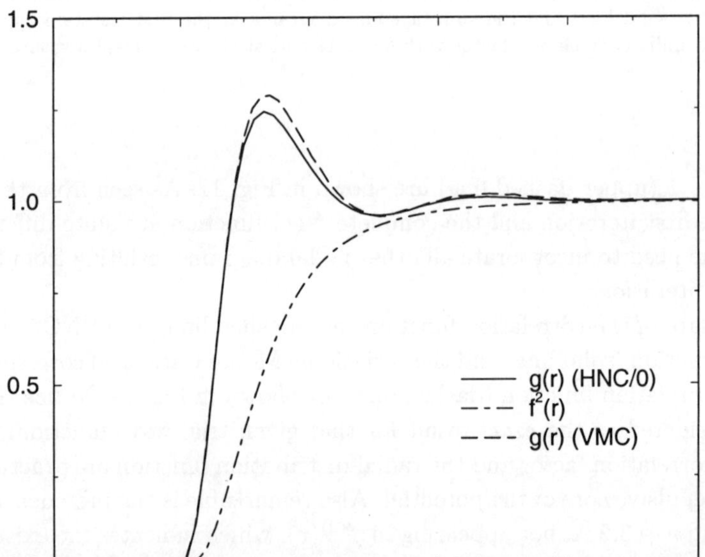


Fig. 13. The square of the Schiff–Verlet two–body correlation factor with $b = 1.17$ (dot–dashed line), the two–body distribution function calculated in the HNC/0 approximation (solid line) and the VMC result (dashed line) for $\rho = 0.365\sigma^{-3}$.

Table 2. Saturation density and energy per particle of liquid $^4$He in different approximations using the Lennard–Jones potential. See the text for further explanations.

| | $\rho$ ($\sigma^{-3}$) | $E/N$ (K) |
|---|---|---|
| HNC/0 | 0.291 | -5.12 |
| HNC/S | 0.330 | -5.82 |
| HNC/ST | 0.364 | -6.55 |
| VMC (J+T) | 0.360 | -6.53 |
| DMC | 0.365 | -6.85 |
| Exp | 0.365 | -7.17 |

were $\lambda$, $\omega$ and $r_t$ are treated as variational parameters. Using the Schiff–Verlet correlation function with b=1.17 and this triplet correlation with the values of $\lambda$, $\omega$ and $r_t$ given in Ref. 34 ($\lambda = -8\sigma^{-2}$, $r_t = 0.82\sigma$, $\omega = 0.5\sigma$ and $r_B = [(108/\rho)^{1/3}]/2$) one gets $e_{VMC} = -6.53$ K. On the other hand, the HNC procedure including triplet correlations and a suitable generalization of the scaling method, known as HNC/ST, yields $e_{HNC/ST}(\rho_0^{exp}) = -6.54$ K, [36] in very good agreement with the Monte Carlo result. Three–body correlations are important and account for a good part of the difference between the Jastrow and the experimental results. Obviously the best three–body correlation factor is the one resulting from the solution of the corresponding optimization equation. [38,37] However, this procedure is technically quite involved and will not be discussed here. In any case, the inclusion of three–body correlations improves the results and gives confidence to the technology employed, in such a way that one can face the calculation of the equation of state in the full range of densities where the system remains liquid.

The energies per particle for the Lennard–Jones potential, calculated at the saturation density of each approximation, are reported in Table 2. The difference between the Diffusion Monte Carlo (DMC) result and the HNC/ST or the VMC results with triplet correlations should be assigned to the optimization of the two– and the three–body correlations, and also to the lack of higher order correlation factors that are only taken into account in the DMC solution. On the other hand, differences between the DMC and the experimental results indicates some deficiencies in the interaction $V(r)$ and the possible lack of a three–body interatomic potential.

In fact, using the more realistic Aziz potential and a quasi–optimal (QO) two–body correlation factor allowing for sort of a long–range behavior

$$f_{QO} = exp\left[-\frac{1}{2}\left(\frac{b\sigma}{r}\right)^5\right]\left[A + B\exp\left(-\frac{(r-D)^2}{\tau r^4}\right)\right] \tag{3.29}$$

one gets, at the experimental saturation density, an energy per particle $e_{HNC/S}(\rho_0^{exp}) = -5.81$ K with the optimized parameters $b = 1.18$, $A = 0.85$, $D = 3.80$ Å, and $\tau = 0.043$ Å$^{-2}$. The introduction of triplet correlations of slightly different form [36] compared to $\xi_{MC}(r)$ above

$$\xi(r) = r \exp\left[-\left(\frac{r - r_t}{\omega}\right)^2\right] \tag{3.30}$$

with $\lambda = -1.08\sigma^{-2}$, $r_t = 0.80\sigma$ and $\omega = 0.41\sigma$, produces an energy per particle at the experimental saturation density $e_{HNC/ST} = -6.62$K (see Fig. 11). Finally, it is also worth mentioning that the DMC results with the Aziz (HFDHE2) and the AzizII potentials give saturation densities $\rho_o = 0.3647\sigma^{-3}$ and $\rho_0 = 0.3606\sigma^{-3}$, respectively, with energies $e_0^{Aziz} = -7.103$ K and $e_0^{Aziz\ II} = -7.277$ K. [39]

Once the $T = 0$ equation of state $e(\rho)$ is known, one can study the behavior of the first and second derivatives of the energy, which are related to the pressure, the chemical potential and the isothermal compressibility

$$P(\rho) = -\left(\frac{\partial E}{\partial \Omega}\right)_N = \rho^2 \frac{de(\rho)}{d\rho} , \tag{3.31}$$

$$\mu(\rho) = \left(\frac{\partial E}{\partial N}\right)_\Omega = e(\rho) + \frac{P(\rho)}{\rho} \tag{3.32}$$

$$\kappa(\rho) = \frac{1}{\rho}\left(\frac{\partial \rho}{\partial P}\right)_N , \tag{3.33}$$

the latter being also related to the sound velocity through the expression

$$c(\rho) = \frac{1}{\sqrt{m\kappa\rho}} . \tag{3.34}$$

Actually, the best way to calculate these quantities from the variational point of view would be to carry out a diagrammatic expansion of each one and to add the resulting series in much the same way it is done with the total energy. However, it is much easier and common to parametrize the equation of state with a suitable function containing few free parameters, adjust them to a previously calculated set of $e(\rho)$ points, and calculate $P$, $\mu$ and $\kappa$ from the fit. A functional form that has been frequently used is

$$e(\rho) = e(\rho_0) + b_1\left(\frac{\rho - \rho_0}{\rho_0}\right)^2 + b_2\left(\frac{\rho - \rho_0}{\rho_0}\right)^3 , \tag{3.35}$$

with no linear term due to the fact that the pressure is zero at the saturation density. When fitted to the experimental values corresponding to pure $^4$He, $e_0 = -7.17K$ and $\rho_0 = 0.365\sigma^{-3}$, $b_1 = 13.65$ and $b_2 = 7.67$. Another quite efficient function which reproduces very well the $T = 0$ equation of state in all the liquid regime and up to the solidification point is [40]

$$e(\rho) = B\rho + C\rho^{1+\gamma} \tag{3.36}$$

with $B = 444.4$ KÅ$^3$, $C = 5.2277 \times 10^6$ KÅ$^{3(1+\gamma)}$ and $\gamma = 2.8$.

The HNC/ST energies shown in Fig. 11 yield reasonable sound velocities at the experimental saturation density, $c(\rho_0^{exp}) = 222$ m/s, to be compared with the experimental value $c_{exp} = 238$ m/s, and a compressibility $\kappa = 0.014$ atm$^{-1}$ in front of $\kappa_{exp} = 0.012$ atm$^{-1}$.

## 4. Minimization of the energy: Optimal two–body correlation functions

Instead of minimizing the energy by adjusting few free parameters in a two–body correlation factor of a given functional form, one can try a more ambitious project and obtain the best $f_2(r) \equiv f(r)$ corresponding to the problem in use. This is done performing a functional variation of the expectation value of the Hamiltonian with respect to the (yet unknown) two–body correlation factor, and demanding the resulting expression to be stationary around the solution. This is a very instructive exercise and it is worth giving the derivation in some detail. For the sake of simplicity, the derivation will be restricted to a Jastrow wave function for which the Jackson–Feenberg identity leads to the energy

$$e = \frac{1}{2}\rho \int d\mathbf{r}\, g(r) \left[ v(r) - \frac{\hbar^2}{2m} \triangle \ln f(r) \right] . \tag{4.1}$$

As seen from this expression, the energy per particle can be taken as a functional of $g(r)$ and $f(r)$ alone. In addition, $g(r)$ and $f(r)$ are related through the HNC equations, and therefore the following functional dependences hold

$$e = e[f, g] \tag{4.2}$$

$$g = g[f] . \tag{4.3}$$

One way to proceed would be to eliminate $g$ from both equations and to perform the functional derivative with respect to $f$. However, it turns out to be more convenient to eliminate $f$ and do the variation with respect to $g$. In order to make it as simple as possible, only the HNC/0 approximation will be considered. In this scheme, one can easily express the two–body correlation factor $f$ as a function of $g$

$$\ln f(r) = \frac{1}{2} \left[ \ln g(r) - \frac{1}{(2\pi)^3 \rho} \int \frac{(S(k)-1)^2}{S(k)} e^{i\mathbf{k}\cdot\mathbf{r}}\, d\mathbf{k} \right] , \tag{4.4}$$

where use has been made of the fact that the sum of nodal diagrams can be expressed in terms of the static structure function in the form

$$\tilde{N}(k) = \frac{(S(k)-1)^2}{S(k)} . \tag{4.5}$$

Inserting Eq. (4.4) into (4.1), the energy per particle becomes a function of $g(r)$ alone

$$e = \frac{\rho}{2} \int d\mathbf{r}\, g(r) v(r) - \frac{\hbar^2}{8m} \frac{1}{(2\pi)^3 \rho} \int d\mathbf{k}\, k^2 \frac{(S(k)-1)^3}{S(k)} - \frac{\rho}{2} \frac{\hbar^2}{4m} \int d\mathbf{r}\, g(r) \triangle \ln g(r) , \tag{4.6}$$

since $S(k)$ is nothing but the FT of the two–body radial distribution function.

At this point, it is convenient to introduce a new function $G(r)$, defined as $G^2(r) = g(r)$, as in this way the resulting expression is slightly easier to deal with.

The optimization condition with respect to $g(r)$ is equivalent to the optimization with respect to $G(r)$, which then becomes

$$0 = \frac{\delta e(\rho)}{\delta G} = \frac{\delta}{\delta G}\left[\frac{\rho}{2}\int d\mathbf{r}\, G^2(r)v(r) - \frac{\hbar^2}{8m}\frac{1}{(2\pi)^3\rho}\int d\mathbf{k}\, k^2 \frac{(S(k)-1)^3}{S(k)} + \frac{\hbar^2}{2m}\rho\int d\mathbf{r}\,(\nabla G(r))^2\right] . \quad (4.7)$$

The variation of the term containing the static structure function is easy to carry out, leading to

$$\delta\left[-\frac{\hbar^2}{8m}\frac{1}{(2\pi)^3\rho}\int d\mathbf{k}\, k^2\frac{(S-1)^3}{S}\right] = -\frac{\hbar^2}{8m}\frac{1}{(2\pi)^3\rho}\int d\mathbf{k}\, k^2 \frac{\delta}{\delta S}\left(\frac{(S-1)^3}{S}\right)\delta S$$
$$= -\frac{\hbar^2}{8m}\frac{1}{(2\pi)^3\rho}\int d\mathbf{k}\, k^2\left(\frac{S-1}{S}\right)^2 (2S+1)2\,\widetilde{G\delta G} . \quad (4.8)$$

Now one can use the Parseval identity

$$\rho\int d\mathbf{r} A(r)B(r) = \frac{1}{(2\pi)^3\rho}\int d\mathbf{k}\,\tilde{A}(k)\tilde{B}(k) \quad (4.9)$$

to write the previous equation in coordinate space

$$\frac{1}{(2\pi)^3\rho}\int d\mathbf{k}\left(-\frac{\hbar^2k^2}{4m}(2S+1)\left(1-\frac{1}{S}\right)^2\right)\widetilde{G\delta G} = \rho\int d\mathbf{r}\ \omega_0(r)G\delta G , \quad (4.10)$$

where $w_0(r)$ is the Fourier transform of

$$\tilde{\omega}(k) = -\frac{1}{2}\frac{\hbar^2k^2}{2m}(2S(k)+1)\left(1-\frac{1}{S(k)}\right)^2 . \quad (4.11)$$

Finally, taking into account the variation of the last term in Eq. (4.6)

$$\delta\left[\frac{\hbar^2}{2m}\rho\int d\mathbf{r}\,(\nabla G(r))^2\right] = -\frac{\hbar^2}{m}\rho\int d\mathbf{r}\,(\Delta G(r))\delta G(r) \quad (4.12)$$

and the variation of the potential energy term, one arrives at the equation

$$\frac{\delta}{\delta G}(e(\rho)) = 0 \implies -\frac{\hbar^2}{m}\Delta G(r) + [v(r)+\omega_0(r)]\,G(r) = 0 , \quad (4.13)$$

which looks like a Schrödinger–equation for $G(r)$, where the influence of the medium is contained in the so called *induced interaction* $w_0(r)$.

A completely equivalent way to express the optimization condition, useful to study the behavior of the the distribution function in momentum space, is obtained by doing the variations with respect to $S(k)$ instead of $g$ or $G$ [37]

$$\frac{\delta e(\rho)}{\delta S(k)} = 0 \implies S(k) = \frac{1}{\sqrt{1+\frac{2}{t(k)}\tilde{V}_{ph}(k)}} \quad (4.14)$$

where $t(k) = \hbar^2 k^2/2m$ and

$$V_{ph}(r) = g(r)v(r) + \frac{\hbar^2}{m} \mid \nabla \sqrt{g(r)} \mid^2 + [g(r) - 1] \, \omega_0(r) \tag{4.15}$$

is the *particle-hole effective interaction.* This formulation is especially suitable for finding the optimal $g(r)$ through an iterative procedure, which does not require at any moment on the knowledge of the optimal correlation function $f(r)$. Starting from a reasonable first guess for $g(r)$ and the corresponding $S(k)$, one calculates the new $\tilde{\omega}_0(k)$ and $V_{ph}(r)$. After a FT one gets the new $S(k)$ (Eq. (4.14)) which is then used back to define the new $g(r)$. The process is then repeated until convergence is reached.

This formulation allows also to establish stability conditions which give physical insight into the Euler–Lagrange equations. For instance, to have an acceptable solution, the term under the square root in Eq. (4.14) must be positive, which implies in turn that $\tilde{V}_{ph}(0^+) > 0$. The violation of the stability condition can either indicate that the wave function at hand does not describe correctly the real physical situation or signal the onset of a phase transition. [41]

Another possibility to find the optimization condition is to perform the variations in Eq. (4.1) with respect to $\ln f(r)$. Following this route one arrives at

$$\frac{\delta e(\rho)}{\delta \ln f(r)} = 0 \quad \Longrightarrow \quad g'(r) - \frac{\hbar^2}{4m} \triangle g(r) = 0 \,, \tag{4.16}$$

where $g'(r)$ is the distribution function in which the special link $v_{JF}(r) f^2(r)$, with

$$v_{JF}(r) = v(r) - \frac{\hbar^2}{2m} \ln \triangle f(r) \,, \tag{4.17}$$

appears in the external side of the diagrams joining the points 1 and 2, or in all internal sides, but only once in each diagram.

The HNC equations for this distribution function can be obtained using the same kind of arguments employed in the derivation of the equations for $g(r)$. The new function $g'(r)$ is expressed as the sum of nodal and non–nodal diagrams

$$g'(r) = N'(r) + X'(r) \,, \tag{4.18}$$

while the HNC equation in momentum space reads

$$\tilde{N}'(k) = \tilde{X}'(k)(S^2(k) - 1) \,. \tag{4.19}$$

On the other hand, the sum of non–nodal diagrams becomes

$$X'(r) = v_{JF}(r) g(r) + N'(r)(g(r) - 1) \,, \tag{4.20}$$

which leads to the final result

$$g'(r) = (v_{JF}(r) + N'(r)) g(r) \,. \tag{4.21}$$

Finally, it is interesting to notice that, in momentum space, the condition expressed by Eq. (4.16) reads

$$\tilde{g}'(k) + \frac{\hbar^2 k^2}{4m} (S(k) - 1) = 0 \,, \tag{4.22}$$

which is the same condition encountered in the paired phonon analysis (PPA). [18,37,42] This condition can also be recovered by imposing

$$\langle \Psi \mid \rho_{\mathbf{k}} \rho_{-\mathbf{k}} (H - E_0) \mid \Psi \rangle = 0 \tag{4.23}$$

where $\rho_{\mathbf{k}}$ represents the density fluctuation operator

$$\rho_{\mathbf{k}} = \sum_j e^{i\mathbf{k}\cdot\mathbf{r}_j} \ . \tag{4.24}$$

Notice that the condition in Eq. (4.23) is trivially satisfied for the true ground-state wave function. However, this condition also holds for a Jastrow wave function built from the optimal two–body correlation factor. The derivation of Eq. (4.23), is quite interesting and illustrates different technical aspects of the variational procedure.

Let us consider a generic trial wave function

$$\Psi_T(\mathbf{r}_1, \mathbf{r}_2, ..., \mathbf{r}_N) = \exp\left[\frac{1}{2}\sum_{i<j} u_2(r_{ij}) + \frac{1}{2}\sum_{i<j<k} u_3(r_{ij}, r_{ik}, r_{jk}) + ...\right] , \tag{4.25}$$

and suppose that the energy can be lowered using a different choice of $u_2(r_{ij})$, i.e, by using the improved wave function

$$\Psi(\mathbf{r}_1, \mathbf{r}_2, \ldots, \mathbf{r}_N) = \Psi_T(\mathbf{r}_1, \ldots, \mathbf{r}_N) \exp\left[\frac{1}{2}\sum_{i<j} \Delta u(r_{ij})\right] \tag{4.26}$$

requiring that

$$\frac{\delta}{\delta(\Delta u(r))} \frac{\langle \Psi \mid H \mid \Psi \rangle}{\langle \Psi \mid \Psi \rangle} = 0 \ . \tag{4.27}$$

A simpler expression of the variational condition is obtained assuming the system lays inside a big box of volume $\Omega$ and introducing the discrete Fourier transform of $\Delta u(r)$

$$\Delta u(r) = \frac{1}{N}\sum_k c_{\mathbf{k}} e^{i\mathbf{k}\cdot\mathbf{r}} , \tag{4.28}$$

$c_{\mathbf{k}}$ being the Fourier coefficients. Taking into account that $u(r)$ is real, $c_{\mathbf{k}}$ equals $c_{-\mathbf{k}}$ and one can express the correction to the wave function in terms of the Fourier coefficients as

$$\frac{1}{2}\sum_{i\neq j} \Delta u(r) = \frac{1}{2}\frac{1}{N}\sum_{\mathbf{k}} c_{\mathbf{k}} \sum_{i\neq j} e^{i\mathbf{k}\cdot(\mathbf{r}_i - \mathbf{r}_j)} , \tag{4.29}$$

which can also be written in terms of the density fluctuation operator in the form

$$\sum_{i<j} \Delta u(r_{ij}) = \frac{1}{2}\sum_{\mathbf{k}} c_{\mathbf{k}} \left(\frac{\rho_{\mathbf{k}}\rho_{-\mathbf{k}}}{N} - 1\right) . \tag{4.30}$$

Using this notation, the improved wave function becomes

$$\Psi(\mathbf{r}_1,\ldots,\mathbf{r}_N) = \Psi_T(\mathbf{r}_1,\ldots,\mathbf{r}_N)\prod_{\mathbf{k}}\exp\left[-\frac{c_{\mathbf{k}}}{4}\right]\prod_{\mathbf{k}}\exp\left[\frac{1}{4}\left(\frac{\rho_{\mathbf{k}}\rho_{-\mathbf{k}}}{N}\right)c_{\mathbf{k}}\right], \quad (4.31)$$

and the variational condition, expressed in terms of these Fourier coefficients, reads

$$\frac{\delta}{\delta c_{\mathbf{k}}}\frac{\langle\Psi(c_{\mathbf{k}}) \mid H \mid \Psi(c_{\mathbf{k}})\rangle}{\langle\Psi(c_{\mathbf{k}}) \mid \Psi(c_{\mathbf{k}})\rangle} = 0\,. \quad (4.32)$$

To impose this condition, one makes the variation of one of the coefficients $c_{\mathbf{k}}$ and calculates the variation of the energy keeping only the linear terms in $\delta c_{\mathbf{k}}$. The change in the wave function when only one coefficient is varied is given by

$$\Psi(c_{\mathbf{k}}+\delta c_{\mathbf{k}}) = \Psi(c_{\mathbf{k}})\left[1-\frac{\delta c_{\mathbf{k}}}{4}+\frac{1}{4}\frac{\rho_{\mathbf{k}}\rho_{-\mathbf{k}}}{N}\delta c_{\mathbf{k}}+\ldots\right], \quad (4.33)$$

while the expectation value becomes

$$\begin{aligned} E(c_{\mathbf{k}}+\delta c_{\mathbf{k}}) &= \frac{\langle\Psi(c_{\mathbf{k}}+\delta c_{\mathbf{k}}) \mid H \mid \Psi(c_{\mathbf{k}}+\delta c_{\mathbf{k}}\rangle}{\langle\Psi(c_{\mathbf{k}}+\delta c_{\mathbf{k}}) \mid \Psi(c_{\mathbf{k}}+\delta c_{\mathbf{k}})\rangle} \\ &= \frac{E+\delta c_{\mathbf{k}}\left[-\frac{1}{2}E+\frac{1}{2}\langle\Psi(c_{\mathbf{k}}) \mid \frac{\rho_{\mathbf{k}}\rho_{-\mathbf{k}}}{N}H \mid \Psi(c_{\mathbf{k}})\rangle+\ldots\right]}{1+\delta c_{\mathbf{k}}\left[-\frac{1}{2}+\frac{1}{2}\frac{\rho_{\mathbf{k}}\rho_{-\mathbf{k}}}{N}\right]+\ldots} \\ &= \left[E+\delta c_{\mathbf{k}}\left(-\frac{1}{2}E+\frac{1}{2}\langle\Psi(c_{\mathbf{k}}) \mid \frac{\rho_{\mathbf{k}}\rho_{-\mathbf{k}}}{N}H \mid \Psi(c_{\mathbf{k}}\rangle\right)+\ldots\right] \\ &\quad\times\left[1-\delta c_{\mathbf{k}}\left(-\frac{1}{2}+\frac{1}{2}\frac{\rho_{\mathbf{k}}\rho_{-\mathbf{k}}}{N}\right)+\ldots\right]. \end{aligned} \quad (4.34)$$

Collecting the linear terms in $\delta c_{\mathbf{k}}$ one finally finds

$$E(c_{\mathbf{k}}+\delta c_{\mathbf{k}}) = E+\delta c_{\mathbf{k}}\frac{1}{2N}\left[\langle\Psi(c_{\mathbf{k}}) \mid \rho_{\mathbf{k}}\rho_{-\mathbf{k}}(H-E) \mid \Psi(c_{\mathbf{k}})\rangle\right]+(\delta c_{\mathbf{k}})^2\ldots \quad (4.35)$$

and therefore the minimization equation becomes

$$\frac{\delta}{\delta c_{\mathbf{k}}}\frac{\langle\Psi \mid H \mid \Psi\rangle}{\langle\Psi \mid \Psi\rangle} = 0 \implies \langle\Psi \mid \rho_{\mathbf{k}}\rho_{-\mathbf{k}}(H-E) \mid \Psi\rangle = 0 \quad (4.36)$$

which is nothing but the condition discussed above.

### 4.1. *Asymptotic behavior*

The analysis of the optimal equations, or equivalently the PPA procedure, shows that the static structure function increases linearly with $k$ at low momenta when the potential is of shorter range than the Coulomb interaction. This $k \to 0$ behavior translates into a long range in the radial distribution function $g(r)$, which must approach unity as 1 minus a suitable constant times $r^{-4}$.

Instead of deriving the linear behavior of $S(k)$ when $k \to 0$ by looking at the optimization equation, which is otherwise a well known fact from very general sum rules arguments, [18] one can *assume* it and analyze the consequences it has on the radial distribution function and the two–body correlation factor. In this way, the starting point becomes the condition

$$\lim_{k\to 0} S(k) = \frac{\hbar k}{2mc} , \tag{4.37}$$

where $c$ is the sound velocity. This behavior can only be accurately reproduced by a long ranged $f(r)$ which yields in turn a long ranged $g(r)$. In fact, a short ranged $f(r)$ of the Schiff–Verlet type (Eq. (2.26)) yields a finite value of $\lim_{k\to 0} S(k)$ that is different from zero.

In the forthcoming analysis, two different $k \to 0$ behaviors will be of interest: $\alpha k$ and $\alpha / k$ where $\alpha$ is a constant. To figure out what these conditions imply in coordinate representation, it is useful to recall the asymptotic expansion

$$\int_0^\infty F(k)\sin(kr)dk = \frac{F(0)}{r} - \frac{F''(0)}{r^3} + \frac{F''''(0)}{r^5} .... \tag{4.38}$$

valid when both $F(k)$ and its derivatives are well behaved at the origin.

Due to the radial dependence of the functions $A(r)$ considered so far, the Fourier transform takes the following form

$$A(r) = \frac{1}{2\pi^2 \rho r}\int dk\, \tilde{A}(k) k \sin(kr) , \tag{4.39}$$

and therefore, applied to the radial distribution function

$$g(r) = 1 + \frac{1}{2\pi^2 \rho r}\int_0^\infty (S(k)-1)\sin(kr)kdk \tag{4.40}$$

one finds, taking into account the previous asymptotic expansion and the linear dependence of $S(k)$ near the origin, the following large–$r$ behavior

$$g(r\to\infty) \to 1 - \frac{\hbar}{2\pi^2 \rho m c}\frac{1}{r^4} . \tag{4.41}$$

The analysis can be easily extended to study the functions $\tilde{N}(k)$ and $\tilde{X}(k)$. This can be done recalling the HNC relations in Eqs. (3.13), (3.14) which express $\tilde{N}(k)$ and $\tilde{X}(k)$ in terms of $S(k)$, leading to

$$\tilde{X}(k) = 1 - \frac{1}{S(k)} \quad \Longrightarrow \quad \lim_{k\to 0}\tilde{X}(k) = 1 - \frac{2mc}{\hbar}\frac{1}{k} \tag{4.42}$$

$$\tilde{N}(k) = \frac{\tilde{X}^2(k)}{1-\tilde{X}(k)} \quad \Longrightarrow \quad \lim_{k\to 0} = -2 + \frac{2mc}{\hbar}\frac{1}{k} + \frac{\hbar}{2mc}k. \tag{4.43}$$

Notice that it is necessary to keep the $k^0$ and $k^1$ terms in the previous expressions in order to recover the proper linear behavior of $S(k)$ when expressed in the form $S(k) = 1 + \tilde{X}(k) + \tilde{N}(k)$. The low $k$ behaviors are translated after the FT into the corresponding long–range behaviors in configuration space.

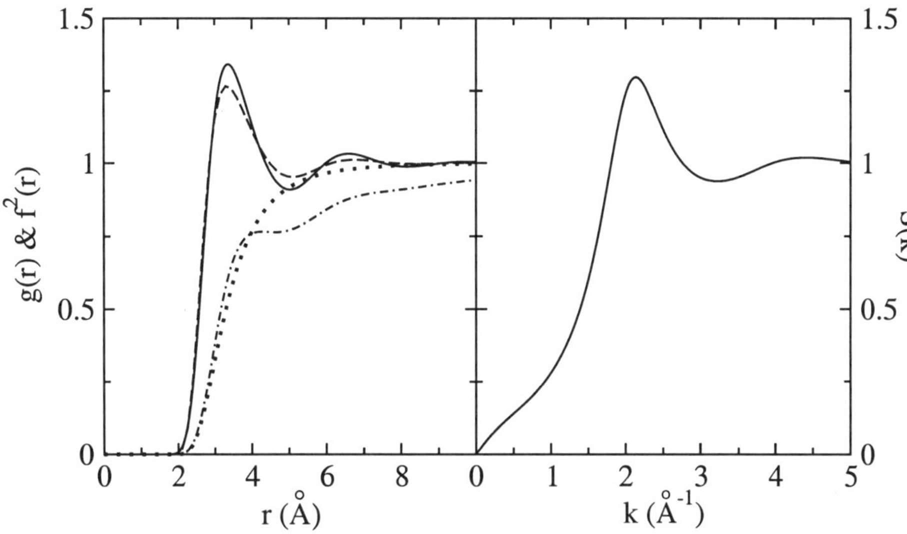


Fig. 14. HNC/0 Results for $^4$He at the experimental saturation density. Left panel: Optimal $f^2(r)$ and $g(r)$ (dot–dashed and solid lines) compared with $f^2(r)$ and $g(r)$ from a Schiff–Verlet two–body correlation factor (dotted and dashed lines). Right panel: optimal $S(k)$.

At the Jastrow level, it is also easy to see that a two–body correlation factor satisfying the condition

$$f^2(r \to \infty) - 1 \ \to \ -\frac{mc}{\pi^2 \hbar \rho}\frac{1}{r^2} \tag{4.44}$$

leads to the right $r \to \infty$ behavior of $g(r)$ already in the HNC/0 approximation.

Let us close this section recalling that the effect of the optimization on the energy is rather small, for instance at the experimental saturation density, and working with the Aziz potential one gets $e_{HNC/0}(\rho^{exp}) = -4.10$ correlation function with $b = 1.20$ while the optimize $f(r)$ provides $e_{opti}(\rho^{exp}) = -4.33$ in Fig. 14, where the distribution function is also shown. Besides the long range behavior, the optimization induces a little bit more of structure at an intermediate range. The static structure function at the same density is shown in Fig. 14, where one can appreciate the linear behavior at low $k$.

## 5. Low excited states

Up to this point the analysis has been centered exclusively on the study of the groundstate. However, the power of the developed variational techniques goes far beyond that and can be used for instance to analyze the spectrum of elementary excitations of a strongly correlated system such as pure $^4$He. In this section, a short introduction on how this is done and what the low energy excitation spectrum of a strongly correlated Bose system looks like is presented and discussed. More details will be presented in Saarela's lectures in this volume. [43]

Excited states can be accessed in essentially two ways: by increasing the temperature, and by exciting the system with an external probe as is done in neutron scattering. At finite temperature, the system populates some of the excited states and the elementary excitation spectrum is required if one seeks to understand the thermodynamic properties of the system. As en example, a quantity that directly depends on the density of excited states and therefore on the excitation energies is the specific heat. On the other hand, in a scattering event the probe transfers momentum and energy to the system, and the cross section for fixed momentum transfer as a function of $E$ shows marked peaks at the excitation energies.

Landau [44] stated many years ago that some of the thermodynamic properties of liquid $^4$He as the specific heat can be easily understood if the excitation spectrum is characterized by both a linear dispersion relation near the origin

$$\epsilon(k \approx 0) = \hbar k c \ , \tag{5.1}$$

where $c$ is the sound velocity, and a region characterized by a quadratic minimum at a finite momentum $k_0$, known as the roton region

$$\epsilon(k \approx k_0) = \Delta + \frac{\hbar^2 (k-k_0)^2}{2\mu} \ . \tag{5.2}$$

The value of the parameters that best fit the measured low temperature heat capacity are

$$\frac{\Delta}{k_B} = 8.6K, \quad k_0 = 1.9\,\text{Å}^{-1}, \quad \mu = 0.16\, m_{^4\text{He}} \ . \tag{5.3}$$

In a completely different way, and using a variational *ansatz*, Feynman [45] showed years later that, in the long wavelength limit, the wave function of the lowest excited states corresponding to a momentum $\mathbf{k}$ takes the form

$$| \Psi_{\mathbf{k}} \rangle = \rho_{\mathbf{k}} \mid \Psi_0 \rangle \ , \tag{5.4}$$

where $\Psi_0$ is assumed to be the *exact* ground state. The excitation energy associated with this trial wave function becomes then an upper bound to the exact excitation energy.

Taking into account that

$$\rho^\dagger_{\mathbf{k}} \rho_{\mathbf{k}} = \sum_i e^{-i\mathbf{k}\cdot\mathbf{r}_i} \sum_j e^{i\mathbf{k}\cdot\mathbf{r}_j} = N + \sum_{i\neq j} e^{i\mathbf{k}\cdot(\mathbf{r}_j - \mathbf{r}_i)} \ , \tag{5.5}$$

the normalization constant becomes

$$\frac{\langle \Psi_{\mathbf{k}} \mid \Psi_{\mathbf{k}} \rangle}{\langle \Psi_0 \mid \Psi_0 \rangle} = N S(k) \ , \tag{5.6}$$

and the excitation energy reads

$$\epsilon(k) = \frac{\langle \Psi_{\mathbf{k}} \mid H - E_0 \mid \Psi_{\mathbf{k}} \rangle}{\langle \Psi_{\mathbf{k}} \mid \Psi_{\mathbf{k}} \rangle} = \frac{\langle \Psi_0 \mid \rho^\dagger_{\mathbf{k}} \, (H - E_0) \, \rho_{\mathbf{k}} \mid \Psi_0 \rangle}{\langle \Psi_0 \mid \rho^\dagger_{\mathbf{k}} \rho_{\mathbf{k}} \mid \Psi_0 \rangle} \tag{5.7}$$

which, after shifting the position of the rightmost density fluctuation operator, leads to

$$\epsilon(k) = \frac{\langle \Psi_0 \mid \rho^\dagger_{\mathbf{k}} \rho_{\mathbf{k}} \, (H - E_0) \mid \Psi_0 \rangle}{\langle \Psi_0 \mid \rho^\dagger_{\mathbf{k}} \rho_{\mathbf{k}} \mid \Psi_0 \rangle} + \frac{\langle \Psi_0 \mid \rho^\dagger_{\mathbf{k}} \, [(H - E_0), \rho_{\mathbf{k}}] \mid \Psi_0 \rangle}{\langle \Psi_0 \mid \rho^\dagger_{\mathbf{k}} \rho_{\mathbf{k}} \mid \Psi_0 \rangle} \, . \tag{5.8}$$

The first term of the previous equation is zero when $\mid \psi_0 \rangle$ is the exact groundstate but also when it is a variational model fulfilling the two–body optimization condition of Eq. (4.23).

As for the second term, the commutator of $\rho_{\mathbf{k}}$ with $E_0$ is trivially zero and, if the potential is momentum independent, the commutator with the potential operator also cancels. In this way one is left with the kinetic energy contribution

$$\epsilon(k) = \frac{\langle \Psi_0 \mid \rho^\dagger_{\mathbf{k}} [-\frac{\hbar^2}{2m} \sum_i \nabla^2_i, \rho_{\mathbf{k}}] \mid \Psi_0 \rangle}{\langle \Psi_{\mathbf{k}} \mid \Psi_{\mathbf{k}} \rangle} \tag{5.9}$$

which can be easily simplified in the following way: taking into account the anti–hermitian character of the $\nabla$ operator, and the fact that the groundstate of a Bose system is always real, one has

$$\int d\mathbf{R} \, \Psi_0 f(r) (\nabla_1 \Psi_0) = -\frac{1}{2} \int d\mathbf{R} \, \Psi_0 (\nabla_1 f) \Psi_0 \tag{5.10}$$

which, together with the linear behavior of commutators, allows one to write

$$\epsilon(k) = -\frac{\hbar^2}{2m} \sum_{i=1}^{N} - \frac{\langle \Psi_0 \mid (\nabla_i \rho^\dagger_{\mathbf{k}})(\nabla_i \rho_{\mathbf{k}}) \mid \Psi_0 \rangle}{\langle \Psi_0 \mid \Psi_0 \rangle} \left( \frac{\langle \Psi_{\mathbf{k}} \mid \Psi_{\mathbf{k}} \rangle}{\langle \Psi_0 \mid \Psi_0 \rangle} \right)^{-1} \tag{5.11}$$

thus leading to

$$\epsilon(k) = \frac{\hbar^2 k^2}{2m} \sum_i^N \left( \frac{\langle \Psi_{\mathbf{k}} \mid \Psi_{\mathbf{k}} \rangle}{\langle \Psi_0 \mid \Psi_0 \rangle} \right)^{-1} = \frac{\hbar^2 k^2}{2m S(k)} \, . \tag{5.12}$$

Recalling the low $k$ behavior of $S(k)$, one finally recovers

$$\epsilon(k)_{k \to 0} = \hbar k c \tag{5.13}$$

which is the typical linear dispersion relation of phonon excitations. As one can see in Fig. 15 where the calculated $\epsilon(k)$ is compared with the experimental results, the variational approximation with the Feynman *ansatz* reproduces the experimental dispersion law at low momenta up to $k \sim 0.6\text{Å}^{-1}$.

As it has been mentioned above, the excitation spectrum of the system can be explored by means of inelastic neutron scattering. The setup of a typical scattering experiment is shown in Fig. 16. A beam of neutrons is emitted from the source, and falls on the sample after it is collimated to match the desired energy and momentum. Neutron detectors are placed at a large distance surrounding the sample. The scattering angle is directly related to the energy $E$ and the momentum $\mathbf{q}$ transferred from the probe to the sample. An important aspect to bear in mind is the fact that the range of energies and momenta of the neutrons used in the experiment are precisely those matching the energies and momenta characteristic of the

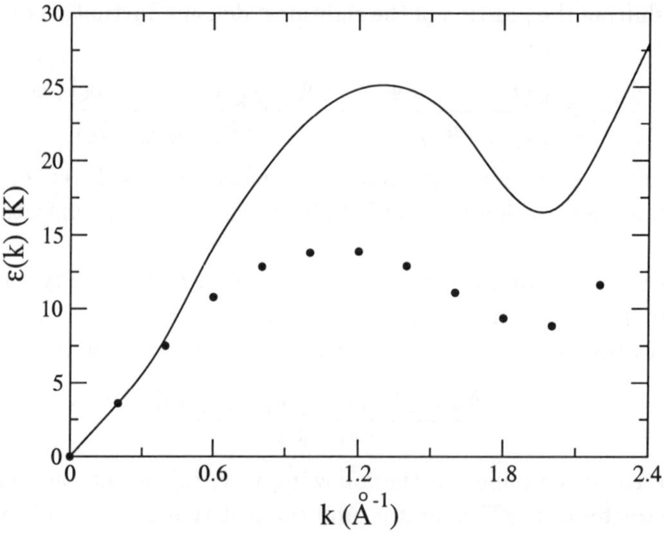


Fig. 15. Experimental excitation spectrum of liquid $^4$He (circles) compared to the Feynman approximation (solid line).

elementary excitations and collective modes of a typical condensed matter system. In addition, the neutrons only interact with the nucleus of the atoms in the sample, and the nuclear interaction is really short–ranged compared with the wave length of the incident beam (of the order of few Å). This fact implies that the scattering basically happens in $s$–wave and that the effective potential the neutron sees can be well characterized by a Fermi pseudo potential

$$V(r) = \frac{2\pi\hbar^2}{m} b \sum_{j=1}^{N} \delta(\mathbf{r} - \mathbf{r}_j) \ , \tag{5.14}$$

where $b$ is the scattering length. On the other hand, the use of Fermi' "Golden Rule" allows to write the cross–section in the form

$$\frac{d^2\sigma}{dE d\Omega} = \frac{k_1}{k_0} b^2 S(q, E) \ , \tag{5.15}$$

where $S(q, E)$ is the dynamic structure function

$$S(q, E) = \frac{1}{N} \sum_n \mid \langle n \mid \rho_{\mathbf{q}} \mid 0 \rangle \mid^2 \delta(E - (E_n - E_0)) \ , \tag{5.16}$$

$\rho_{\mathbf{q}}$ being the density fluctuation operator. The measurement performed at each detector can be associated with well defined values of the energy and momentum transferred, and the number of counts registered in each detector gives precise information about the intensity of the scattering. This intensity is characterized by the double differential cross section defined above, which becomes the product of a

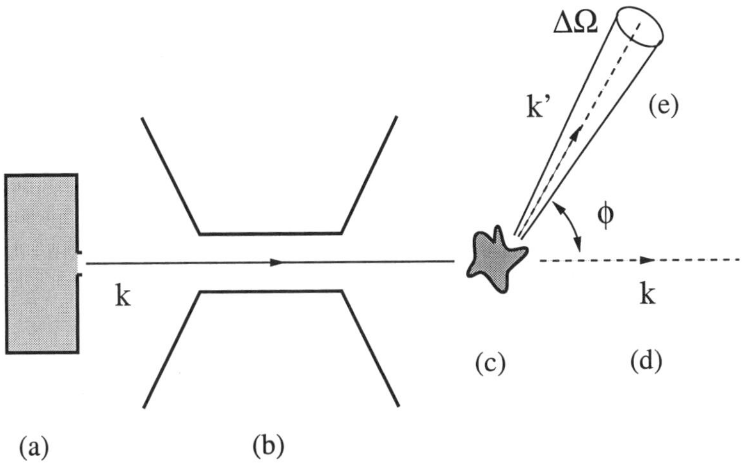


Fig. 16. Experimental setup of a typical neutron scattering experiment. (a) neutron source, (b) momentum selector, (c) sample, (d) transmitted neutrons, (e) neutrons with momentum **k'** scattered in a solid angle $\Delta\Omega$.

kinematic term containing the initial and the final momentum of the neutron, the scattering length that defines the interaction, and the new function, $S(q, E)$, which incorporates all the information one can extract in the experiment corresponding to the system under study. The dynamic structure function for a given excitation operator collects the contribution of all excited states which are accessible with that excitation operator, and that are compatible with the external energy and momentum being transferred to the system. The cross section then presents marked peaks at these energies. [58]

As an example, one can calculate $S(q, E)$ in the one–phonon approximation, where $\rho_{\mathbf{q}} \mid 0\rangle$ are the only allowed excited states. This is a very crude approximation, but since this *ansatz* gives a good description of the spectrum at very low energies, one can expect the resulting $S(q, E)$ to be accurate enough to describe the response in this regime.

The set of normalized excited states considered are then

$$\mid n\rangle \equiv \frac{\mid \Psi_{\mathbf{k}}\rangle}{\langle \Psi_{\mathbf{k}} \mid \Psi_{\mathbf{k}}\rangle^{1/2}} = \frac{\rho_{\mathbf{k}} \mid \Psi_0\rangle}{N^{1/2} S(k)^{1/2}} , \tag{5.17}$$

and the required matrix elements become

$$\langle n \mid \rho_{\mathbf{q}} \mid 0\rangle = \delta_{\mathbf{k},\mathbf{q}} N^{1/2} (S(q))^{1/2} , \tag{5.18}$$

which lead to the expression

$$S(q, E) = \frac{1}{N} \sum_{\mathbf{k}} \delta_{\mathbf{kq}} N S(q) \delta \left( E - \frac{\hbar^2 k^2}{2m S(k)} \right) . \tag{5.19}$$

Now, one can use the $\delta_{\mathbf{kq}}$ to perform the summation over $\mathbf{k}$ to finally find

$$S(q,E) = S(q)\delta\left(E - \frac{\hbar q^2}{2mS(q)}\right) . \tag{5.20}$$

In this approximation, $S(q,E)$ becomes a $\delta$ function centered at the excitation energy with a strength given by the static structure function $S(q)$. There is only one excited state taking all the strength of the excitation operator because in this approximation there is only one state able to take the whole momentum and energy transferred to the system.

Furthermore, one can also perform an analysis in terms of the energy weighted sum rules satisfied by $S(q,E)$, which is a common tool used to study the dynamic structure function of generic systems. This is done by evaluating the different energy moments of $S(q,E)$ and inferring its behavior from them. Sum rules of the response are defined as

$$m_k(q) = \int_0^\infty dE\, E^k S(q,E) , \tag{5.21}$$

which, due to the energy integration, do not depend on the detailed structure of the excited states and reduce to ground state expectation values of commutators of the Hamiltonian and the density fluctuation operator.

Here only the first two moments $m_0$ and $m_1$ will be considered. The idea is to first carry out the energy integration and to use the closure property when appropriate. The $m_0$ moment is extremely simple and becomes

$$\begin{aligned}
m_0(q) &= \int dE\, S(q,E) \\
&= \frac{1}{N}\int dE \sum_n |\langle n \mid \rho_{\mathbf{q}} \mid 0\rangle|^2\, \delta(E-(E_n-E_0)) \\
&= \frac{1}{N}\sum_n |\langle n \mid \rho_{\mathbf{q}} \mid 0\rangle|^2 = \frac{1}{N}\sum_n \langle 0 \mid \rho_{\mathbf{q}}^\dagger \mid n\rangle\langle n \mid \rho_{\mathbf{q}} \mid 0\rangle \\
&= \frac{1}{N}\langle 0 \mid \rho_{\mathbf{q}}^\dagger \rho_{\mathbf{q}} \mid 0\rangle = S(q) .
\end{aligned} \tag{5.22}$$

The evaluation of $m_1(q)$ is also straightforward

$$\begin{aligned}
m_1 &= \int_0^\infty dE\, S(q,E)\, E \\
&= \frac{1}{N}\sum_n (E_n - E_0)\langle 0 \mid \rho_{\mathbf{q}}^\dagger \mid n\rangle\langle n \mid \rho_{\mathbf{q}} \mid 0\rangle \\
&= \frac{1}{N}\sum_n \langle 0 \mid \rho_{\mathbf{q}}^\dagger \mid n\rangle\langle n \mid [H,\rho_{\mathbf{q}}] \mid 0\rangle \\
&= \frac{1}{2}\frac{1}{N}\langle 0 \mid [\rho_{\mathbf{q}}^\dagger, [H,\rho_{\mathbf{q}}]] \mid 0\rangle
\end{aligned} \tag{5.23}$$

where in the last step use has been made of the time reversal invariance of the system.[46]

Only the kinetic operator from $H$ contributes to the innermost commutator, and after little algebra one finds the final result

$$m_1(q) = \frac{\hbar^2 q^2}{2m} \,, \tag{5.24}$$

thus showing that the first moment of the response, also known as the *f-sum rule*, depends on the details of the system only through the mass of its constituents. Notice that this result has been obtained on the basis that the system is initially in its groundstate, and that it is infinite, homogeneous and invariant under time reversal transformations. These conditions are quite common and therefore the range of validity of this sum rule is quite wide. Of course, the fact that $m_1(q)$ equals $\hbar^2 q^2/2m$ is very interesting from the theoretical point of view as it imposes a severe model independent constraint on any realistic calculation of the response.

At this point it is straightforward to check that $S(q,E)$ calculated in the one–phonon approach, $S(q,E) = S(q)\delta(E - \hbar^2 q^2/2mS(q))$, fulfills these two sum rules. Of course, this fact does not mean that this is the exact response. Actually, one can reverse the question and ask what should the form of $S(q,E)$ be if the sum rule is exhausted by only one state. In that case, one can write

$$S(q,E) = Z(q)\delta(E - E_q) \tag{5.25}$$

and require the two first sum rules to be fulfilled

$$m_o(q) = \int_0^\infty S(q,E)\, dE = S(q) \equiv Z(q) \tag{5.26}$$

and

$$m_1(q) = \int_0^\infty S(q,E) E\, dE = \frac{\hbar^2 q^2}{2m} \equiv Z(q) E_q \,, \tag{5.27}$$

so that one can recover $E_q = m_1/m_0$ to get $E_q = \hbar^2 q^2/2mS(q)$. This shows that the one–phonon approximation is the exact solution in this case when only one state is populated.

## 6. A $^3$He impurity in liquid $^4$He

As an extension of the concepts developed so far, the problem of one $^3$He impurity in liquid $^4$He is discussed in this section. In addition, the excitation spectrum of the impurity, which will serve to introduce perturbation theory in the framework of correlated basis function, is also considered. In any case, the theoretical study of one impurity in liquid $^4$He is very helpful in understanding the properties of $^3$He-$^4$He mixtures. In fact, the liquid $^3$He-$^4$He mixture is stable at zero temperature, with a maximum solubility of the $^3$He component at zero pressure of $\sim 6\%$. [55] In the excitation spectrum of the mixture coexist two types of excitations, the lowest energy one corresponds to a particle–hole band associated to $^3$He quasi–particles. $^3$He quasi–particles have a single particle spectrum characterized by an effective mass, $m_I^*$, mainly due to the interaction with $^4$He atoms. A little higher in energy lies the collective phonon–roton branch, associated to $^4$He. Both branches have been clearly separated in recent neutron scattering experiments [56] and in this way, it has been possible to identify the impurity excitation spectrum. The experiments show a significant deviation from the quadratic Landau–Pomeranchuk (LP) spectrum, [57] $\epsilon_{LP}(q) = \hbar^2 q^2/2m_I^*$. The measured spectrum is well described by a modified LP (MLP) spectrum

$$\epsilon_{MLP}(q) = \frac{\hbar^2 q^2}{2m_I^*} \frac{1}{1+\gamma q^2} , \tag{6.1}$$

with $m_I^* = 2.2\, m_3$ and $\gamma = 0.13\,\text{Å}^2$.

Since the $^3$He concentration $x_3$, in the mixture is small, it is justified to study the $x_3 \to 0$ limit or the impurity problem. The analysis of the impurity behavior not only provides clues to understand the mixture itself but also gives useful information on pure liquid $^4$He. As will be seen, the impurity can be used as a theoretical probe to examine the kinetic and potential energies of the bulk system. [59]

As it has been mentioned in the introduction, the chemical potential of the $^3$He impurity at $^4$He saturation density and zero temperature is $\mu_I = -2.78\,\text{K}$, to be compared with that of liquid $^3$He, $\mu_3 = -2.47\,\text{K}$ at its own saturation density. The microscopic description of the impurity problem requires setting up a Hamiltonian

$$H = -\sum_{i=1}^{N_4} \frac{\hbar^2}{2m}\nabla_4^2 + \sum_{i<j}^{N_4} V(\mid \mathbf{r}_i^{(4)} - \mathbf{r}_j^{(4)} \mid) - \frac{\hbar^2}{2m_I}\nabla_I^2 + \sum_{i}^{N_4} V(\mid \mathbf{r}^{(I)} - \mathbf{r}_i^{(4)} \mid) \tag{6.2}$$

and a trial wave function for the ground state of $N_4$ $^4$He atoms plus one $^3$He impurity. In order to keep the formalism simple, a two–body correlation factor (which is the simplest extension of the wave function of pure $^4$He) is proposed

$$\Psi_T(I,\ldots,N_4) = \prod_{i=1,N_4} f_2^{(I,4)}(r_{Ii}) \prod_{m>l=1,N_4} f_2^{(4,4)}(r_{lm}) \,. \tag{6.3}$$

The extension to include three–body correlations is rather straightforward and do not bring new insights to the purpose of this section.

The variation in the energy of the system when one impurity is added at constant volume is measured by the chemical potential

$$\mu_I = \frac{\langle \Psi_T(I;N_4) \mid H(I,N_4) \mid \Psi_T(I;N_4)\rangle}{\langle \Psi_T(I;N_4) \mid \Psi_T(I;N_4)\rangle} - \frac{\langle \Psi_T(N_4) \mid H(N_4) \mid \Psi_T(N_4)\rangle}{\langle \Psi_T(N_4) \mid \Psi_T(N_4)\rangle} \,. \tag{6.4}$$

The quantities to be subtracted are of order $N_4$, while the result is of order unity. Therefore, one must explicitly take into account the cancellations between these large quantities to have a good estimate of $\mu_I$.

These cancellations may be effectively dealt with by writing the expectation value of $H(I,N_4)$ as $E_0(I,N_4) = E_4 + \mu_I$, where $E_4$ is the energy of the medium (proportional to $N_4$) and $\mu_I$ is the chemical potential of the $^3$He impurity. Then, the background energy, $E_4$ cancels in the difference of Eq. (6.4). The background energy, can be obtained within the Jackson–Feenberg prescription (Eq. (2.25)), while the chemical potential of the impurity is decomposed in two pieces, $\mu_I = \mu_{int} + \mu_{re}$. The interaction term reads

$$\mu_{int} = \rho_4 \int d\mathbf{r}\, g^{(I,4)}(r) \left[ V(r) - \frac{\hbar^2}{4\tilde{m}} \nabla^2 \ln f^{(I,4)}(r) \right] , \tag{6.5}$$

$\tilde{m}$ being the reduced mass $\tilde{m} = m_I m_4/(m_I + m_4)$, while the rearrangement term becomes

$$\mu_{re} = \frac{1}{2}\rho_4^2 \int d\mathbf{r}\, g_I^{(4,4)}(r) \left[ V(r) - \frac{\hbar^2}{4m_4} \nabla^2 \ln f^{(4,4)}(r) \right] . \tag{6.6}$$

The $^4$He-$^4$He radial distribution function has two types of diagrammatic contributions, those with the impurity as an internal point, $g_I^{(4,4)}(r)$ and those without the impurity ($g^{(4,4)}(r)$). In this way, the distribution functions are defined by

$$g^{(4,4)}(r_{12}) + \frac{1}{\Omega} g_I^{(4,4)}(r_{12}) = \frac{N_4(N_4-1)}{\rho_4^2} \frac{\int d\mathbf{r}_I d\mathbf{r}_3 \ldots \mid \Psi(I;N_4) \mid^2}{\int d\mathbf{r}_I d\mathbf{R} \mid \Psi(I;N_4) \mid^2}, \tag{6.7}$$

and

$$g^{(I,4)}(r_{I1}) = \frac{\Omega N_4}{\rho_4} \frac{\int d\mathbf{r}_2 d\mathbf{r}_3 \ldots \mid \Psi(I;N_4) \mid^2}{\int d\mathbf{r}_I d\mathbf{R} \mid \Psi(I;N_4) \mid^2}. \tag{6.8}$$

Once more, the problem of calculating the expectation value has been translated into the evaluation of the distribution functions. These functions can be expressed in terms of nodal and non–nodal contributions as it was done for pure liquid $^4$He

$$g^{(\alpha,4)}(r) = 1 + N^{(\alpha,4)}(r) + X^{(\alpha,4)}(r) , \tag{6.9}$$

with $\alpha = I, 4$. The HNC equations are

$$N^{(\alpha,4)}(r) = \rho_4 (X^{(\alpha,4)} \mid X^{(4,4)} + N^{(4,4)}) , \tag{6.10}$$

while the sum of composite diagrams $X^{(\alpha,4)}$ are, in the HNC/0 approximation given by

$$X^{(\alpha,4)}(r) = (f^{(\alpha,4)})^2 \exp[N^{(\alpha,4)}(r)] - 1 - N^{(\alpha,4)}(r) . \tag{6.11}$$

The distribution function associated to the renormalization of the medium, which has the impurity inside, can be expressed as

$$g_I^{(4,4)} = g^{(4,4)}(r) \left[ N_{I'}^{(4,4)}(r) + N_I^{(4,4)}(r) \right] , \tag{6.12}$$

where

$$N_{I'}^{(4,4)}(r) = (g^{(4,I)} - 1 \mid g^{(I,4)} - 1) \tag{6.13}$$

is the sum of the nodal diagrams having one $^3$He impurity on a node, and

$$N_I^{(4,4)}(r) = 2\rho_4 (g^{(4,4)} - 1 \mid X_I^{(4,4)}) + \rho_4 (g^{(4,4)} - 1 \mid \rho_4 (X_I^{(4,4)} \mid g^{(4,4)} - 1)) \tag{6.14}$$

is the the sum of nodal diagrams having one $^3$He inside that is not located in a nodal point. The composite diagrams $X_I^{(4,4)}(r)$ are given by

$$X_I^{(4,4)}(r) = g_I^{(4,4)}(r) - N_I^{(4,4)}(r) - N_{I'}^{(4,4)}(r) . \tag{6.15}$$

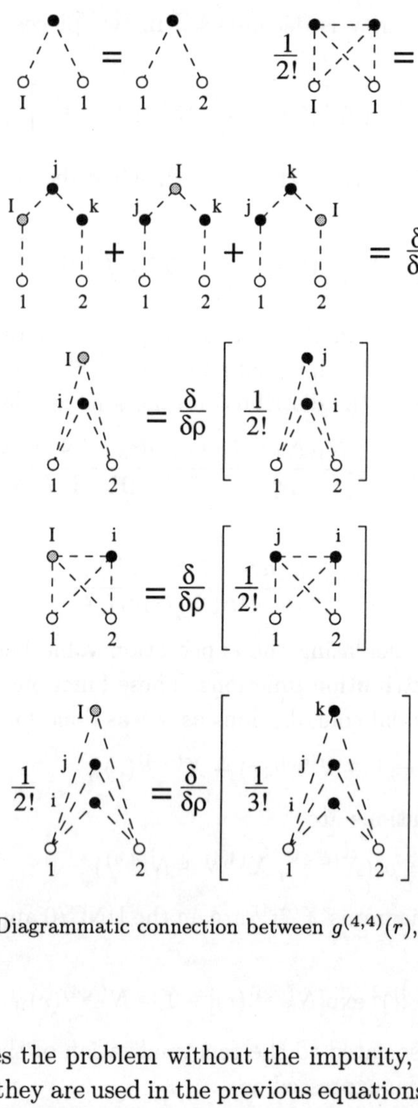

Fig. 17. Diagrammatic connection between $g^{(4,4)}(r)$, $g^{(I,4)}(r)$ and $g_I^{(4,4)}(r)$.

Usually one solves the problem without the impurity, and once $N^{(4,4)}$, $X^{(4,4)}$ and $g^{(4,4)}$ are known, they are used in the previous equations to find the other quantities. In this way, the convergence of the iterative process is rather fast.

Based on the fact that the interaction between the different pairs of atoms is the same and that the masses of the bare atoms are close to each other, the impurity problem has been often analyzed within the average correlation approximation (ACA). [60] The ACA is obtained by taking all dynamical correlation functions identical. In this case, $g^{(I,4)}(r) = g^{(4,4)}(r)$ and

$$g_I^{(4,4)}(r_{12}) = \frac{\partial g^{(4,4)}(r_{12})}{\partial \rho_4} , \tag{6.16}$$

as it can be seen from Eq. (6.7). The diagrammatic contents of these relations is illustrated in Fig. 17. As one can see from the figure, the symmetry factors associated to the different diagrams, are properly obtained by doing the derivative with respect to the density. Taking into account these diagrammatic relations one can express $\mu_{int}^{ACA}$ as

$$\mu_{int}^{ACA} = 2e(\rho_4) + \left(\frac{m_4}{m_I} - 1\right) t^{(4)}(\rho_4) \ , \tag{6.17}$$

where $e(\rho_4)$ and $t^{(4)}(\rho_4)$ are the total and the kinetic $^4$He energies per particle, respectively. On the other hand, $\mu_{re}$ can be related to the thermodynamical pressure $P(\rho_4)$ through

$$\mu_{re}^{ACA} = \frac{P(\rho_4)}{\rho_4} - e(\rho_4) \ . \tag{6.18}$$

In this way, the chemical potential is expressed as

$$\mu_I^{ACA} = e(\rho_4) + \frac{P(\rho_4)}{\rho_4} + \left(\frac{m_4}{m_I} - 1\right) t^{(4)}(\rho_4) = \mu_4(\rho_4) + \left(\frac{m_4}{m_I} - 1\right) t^{(4)}(\rho_4). \tag{6.19}$$

The chemical potential of the impurity in the ACA equals the chemical potential of pure $^4$He corrected by a kinetic energy term properly scaled to take into account the different mass of the impurity. Actually this formula can be easily generalized to wave functions containing $n$–body correlations and coincides with the expression previously derived by Baym. [61]

A recent diffusion Monte Carlo calculation [62], has given $t_4 = 14.32$ K and $\mu_4 = -7.27$ K at the $^4$He saturation density. Employing these values, one gets $\mu_3^{ACA} = -2.58$ K, to be compared with the diffusion Monte Carlo value $\mu_3^{DMC} = -2.79$ K in excellent agreement with the experimental value $\mu_3^{exp} = -2.785$.

### 6.1. *The $^3$He impurity as a probe in liquid $^4$He*

The difference of expectation values in Eq. (6.4) shows that $\mu_I$ is not necessarily an upper bound to the chemical potential, *i.e.*, the difference of two upper bounds is not always an upper bound. However, if one assumes that the trial wave function is the exact wave function of liquid $^4$He, then $\mu_I^{ACA}$ provides an exact upper bound to $\mu_I^{exp}$

$$\mu_I^{exp}(\rho_4) \leq \mu_4^{exp} + \left(\frac{m_4}{m_I} - 1\right) t(\rho_4) \ . \tag{6.20}$$

Since $\mu_4^{exp}(\rho_4)$ and $\mu_I^{exp}$ are experimentally known, the previous inequality can be used to establish a model independent lower bound (LB) to the kinetic energy of liquid $^4$He [59]

$$t(\rho_4) \geq t_{LB}(\rho_4) \equiv \left[\frac{m_I}{m_4 - m_I}\right] [\mu_I^{exp}(\rho_4) - \mu_4^{exp}(\rho_4)] \ . \tag{6.21}$$

At the same time, an upper bound (UB) to the potential energy per particle can be determined by

$$v(\rho_4) \leq v_{UB}(\rho_4) \equiv e_{exp}(\rho_4) - t_{LB}(\rho_4) \ , \tag{6.22}$$

where $e^{exp}(\rho_4)$ is the $^4$He energy per particle. At the experimental saturation density, $\mu^{exp}(\rho_0) = -7.17$ K and $\mu_I^{exp}(\rho_0) = -2.78$ K, therefore $t_{LB}(\rho_0) = 13.4$ K and $v_{UB}(\rho_o) = -20.6$ K. These bounds are basically respected by all the theoretical calculations performed with the Aziz potential. Also the extraction of the kinetic energy from recent deep inelastic neutron scattering data is consistent with the lower–bound. [28]

### 6.2. *The excitation spectrum of the $^3$He impurity*

To calculate the excitation spectrum of the impurity, one should first choose a trial wave function describing an excited state of momentum $\mathbf{q}$. As in the pure $^4$He case, the simplest *ansatz* would be

$$\Psi_{\mathbf{q}}(I, N_4) = \rho_I(\mathbf{q})\Psi_0(I, N_4) \ , \tag{6.23}$$

where $\rho_I(\mathbf{q}) = e^{i\mathbf{q}\cdot\mathbf{r}_I}$ is the impurity excitation operator. The excitation energy associated to this wave function is

$$\begin{aligned}\epsilon_0(q) &= \frac{\langle \Psi_{\mathbf{q}}(I, N_4) \mid H(I, N_4) - E_0(I; N_4) \mid \Psi_{\mathbf{q}}(I; N_4)\rangle}{\langle \Psi_{\mathbf{q}}(I; N_4 \mid \Psi_{\mathbf{q}}(I; N_4)\rangle} \\ &= \frac{\langle \Psi_0(I, N_4) \mid \rho_I(\mathbf{q})^\dagger (H(I, N_4) - E_0(I; N_4))\rho_I(\mathbf{q}) \mid \Psi_0(I; N_4)\rangle}{\langle \Psi_{\mathbf{q}}(I; N_4 \mid \Psi_{\mathbf{q}}(I; N_4)\rangle} \ .\end{aligned} \tag{6.24}$$

In this case, the evaluation of the norm of $\Psi_{\mathbf{q}}(I; N_4)$ is trivial and leads to

$$\epsilon_0(q) = \frac{\langle \Psi_0(I, N_4) \mid \rho_I(\mathbf{q})^\dagger \left[H(I, N_4), \rho_I(\mathbf{q})\right] \mid \Psi_0(I, N_4)\rangle}{\langle \Psi_0(I; N_4) \mid \Psi_0(I; N_4)\rangle} \ . \tag{6.25}$$

Taking into account that

$$[H(I, N_4), \rho_I(\mathbf{q})] = \rho_I(\mathbf{q})\frac{\hbar^2 q^2}{2m_I} \ , \tag{6.26}$$

one finally obtains

$$\epsilon(q) = \frac{\hbar^2 q^2}{2m_I} \ . \tag{6.27}$$

The excitation energy is therefore the dispersion relation corresponding to the free particle. In this way, the mass associated to this parabolic dispersion relation equals the bare mass of the impurity, and therefore the effective mass

$$\frac{m_I^*}{m_I} = \left[\frac{m_I}{\hbar^2}\frac{1}{q}\frac{\partial \epsilon(q)}{\partial q}\right]^{-1} \tag{6.28}$$

is 1. A better *ansatz*, which takes into account the backflow of $^4$He atoms around the $^3$He impurity is provided by

$$\Psi_{\mathbf{q}}^{BF}(I; N_4) = \rho_I(\mathbf{q})F_{BF}(\mathbf{q}; I, N_4)\Psi_0(I; N_4) \ , \tag{6.29}$$

with

$$F_{BF}(\mathbf{q}; I; N_4) = \prod_{i=1,N_4} e^{i\mathbf{q}\cdot(\mathbf{r}_i - \mathbf{r}_I)\eta(r_{Ii})} \ . \tag{6.30}$$

Backflow correlations do not change the binding energy of the impurity and only affect the excitation spectrum, which, however, remains parabolic

$$\epsilon_{BF}(q) = \frac{\hbar q^2}{2m_I}\left[1 + a_1 + a_2 + a_3\right] , \tag{6.31}$$

where

$$a_1 = \rho_4 \int d\mathbf{r}\, g^{(I,4)}(r) \left(2\eta(r) + \frac{2}{3} r\eta'(r)\right)$$

$$a_2 = \frac{m_I}{\tilde{m}} \rho_4 \int d\mathbf{r}\, g^{(I,4)}(r) \left(\eta(r)^2 + \frac{1}{3}\left[r^2(\eta'(r))^2 + 2\eta(r) r\eta'(r)\right]\right)$$

$$a_3 = \rho_4^2 \int d\mathbf{r}_{I1} d\mathbf{r}_{I2}\, g_3^{(I,1,2)} \Bigg(\eta(r_{I1})\eta(r_{I2}) + \frac{1}{3}\left[r_{I1}\eta'(r_{I1})\eta'(r_{I2}) r_{I2} (\hat{\mathbf{r}}_{I1}\cdot\hat{\mathbf{r}}_{I2})^2 + 2\eta(r_{I1})\eta'(r_{I2}) r_{I2}\right]\Bigg) , \tag{6.32}$$

whit $g_3^{(I,1,2)}$ an impurity–$^4$He-$^4$He three–body distribution function.

A good choice for the variational function $\eta(r_{I1})$ has been found to be

$$\eta(r) = A_0 exp\left(-\left[\frac{(r - r_0)}{\omega_0}\right]^2\right) , \tag{6.33}$$

where the parameters $r_0$, $\omega_0$ and $A_0$ are varied to find the minimum. With this variational *ansatz*, one gets a $q$–independent effective mass $m^*/m_3 \sim 1.7$. A full functional variation with respect to $\eta(r)$ [47] does not change this result.

The three–body distribution function $g_3^{(I,1,2)}$ entering in the evaluation of $a_3$ is defined by

$$g_3^{(I,1,2)}(\mathbf{r}_{31}, \mathbf{r}_{32}) = \frac{N_4(N_4-1)\Omega}{\rho^2} \frac{\int d\mathbf{r}_3 ... d\mathbf{r}_{N_4} \mid \Psi_0(I, N_4)(\mathbf{r}_1, ....\mathbf{r}_{N_4}, \mathbf{r}_I) \mid^2}{\int d\mathbf{r}_1 ... d\mathbf{r}_{N_4} d\mathbf{r}_I \mid \Psi_0(I, N_4)(\mathbf{r}_1, ....\mathbf{r}_{N_4}, \mathbf{r}_I) \mid^2} . \tag{6.34}$$

Depending on the approximation used to calculate the three–body distribution function, the change in the effective mass is of the order of 3 %. The same is true if triplet correlations are included in the variational wave function. [51]

### 6.3. *Correlated perturbative approach*

A systematic way to incorporate effects beyond the two–body backflow is provided by the correlated basis function (CBF) perturbation theory.

The underlying strategy consists in incorporating the non–perturbative correlations directly into the basis functions and then develop a perturbative expansion in this basis. Since the correlated states are a good approximation to the eigenstates of the Hamiltonian, the first orders of the perturbation series are usually enough to get reliable and realistic results. At its simple level where only one correlated state in this basis is taken, this reduces to ordinary variational theory. In the impurity case, the idea is to allow the total momentum $\mathbf{q}$ to be shared between the impurity and the phonons in the medium. The correlated basis is then classified according to the number of Feynman phonons

$$\Psi_{\mathbf{q};\mathbf{q}_1..\mathbf{q}_n}(I;N_4) = \rho_I(\mathbf{q}-\mathbf{q}_1-..-\mathbf{q}_n)\rho_4(\mathbf{q}_1)...\rho_4(\mathbf{q}_n)\Psi_0(I,N_4). \tag{6.35}$$

The multiphonon states will mix in the perturbative process with the unperturbed state $\Psi_{\mathbf{q}} = \rho_I(\mathbf{q})\Psi_0(I;N_4)$. The states (6.35) are not orthogonal to each other, and therefore they must either be properly orthonormalized following a Lödwin–like procedure, [48] or the perturbative expansion must be carried on in a non orthogonal basis. [50]

Necessary ingredients to perform a perturbative expansion are the unperturbed and the interaction Hamiltonians, which here are defined via their matrix elements

$$H_{0,ij} = \delta_{ij}\frac{\langle\Psi_i \mid H \mid \Psi_j\rangle}{\langle\Psi_i \mid \Psi_i\rangle} = \delta_{ij}E_i^v \ , \tag{6.36}$$

$$H_{I,ij} = (1-\delta_{ij})\frac{\langle\Psi_i \mid H - E_q \mid \Psi_j\rangle}{(\langle\Psi_i \mid \Psi_i\rangle\langle\Psi_j \mid \Psi_j\rangle)^{1/2}} = (1-\delta_{ij})(H_{ij} - E_q N_{ij}) \ , \tag{6.37}$$

where $E_q = E_0(I;N_4) + \epsilon_0(q) + \delta\epsilon(q)$ is the eigenvalue of $H$ for the state with momentum $\mathbf{q}$. Notice the presence of the energy that one tries to determine together with the overlap between the different states in the definition of $H_I$ due to the non–orthogonality of the basis states. By definition, the unperturbed Hamiltonian is diagonal, and the eigenvalues are just the variational estimates of the energy for the basis states. On the other hand, the diagonal matrix elements of $H_I$ are zero by construction, and therefore there are no first order perturbative corrections.

The Brillouin–Wigner series for the perturbative correction $\Delta E_q$ to $E_q^v$, appropriate for non–orthogonal states is given by [50]

$$\Delta E_q = \sum_{j\neq q}\frac{(H_{qj} - E_q N_{qj})(H_{jq} - E_q N_{jq})}{E_q - E_j^v} + \dots \ , \tag{6.38}$$

where $E_q$ is the perturbed energy and the overlap matrix elements, $N_{qj} = \langle\Psi_q \mid \Psi_j\rangle$, take care of the non–orthogonality. The series is first expanded around the correction to the energy of the medium, $\Delta E_0 = E_4 - E_4^v$, and then resummed in such a way to cancel all the terms diverging in the $N_4 \to \infty$ limit coming from both the cluster expansion of the matrix elements and the perturbative expansion itself. This procedure has been devised to describe the ground state energy of an infinite Fermi system [49] and then generalized to the case of the impurity. [51]

Some of the difficulties related to the divergences can be avoided by working in the orthogonalized scheme. In fact, one can used the Gram–Schmidt procedure to orthogonalize the $n$–phonon states to states with a lower number of phonons. Then the overlaps between those orthogonalized states would be zero and the expression for the perturbative correction (6.38) simplifies a lot. For instance, the orthogonalized one–phonon (OP) state reads

$$\mid \mathbf{q};\mathbf{q}_1\rangle = \frac{\mid\Psi_{\mathbf{q};\mathbf{q}_1}\rangle - \mid\Psi_{\mathbf{q}}\rangle\langle\Psi_{\mathbf{q}} \mid \Psi_{\mathbf{q};\mathbf{q}_1}\rangle}{\langle\Psi_{\mathbf{q};\mathbf{q}_1} \mid \Psi_{\mathbf{q};\mathbf{q}_1}\rangle^{1/2}}, \tag{6.39}$$

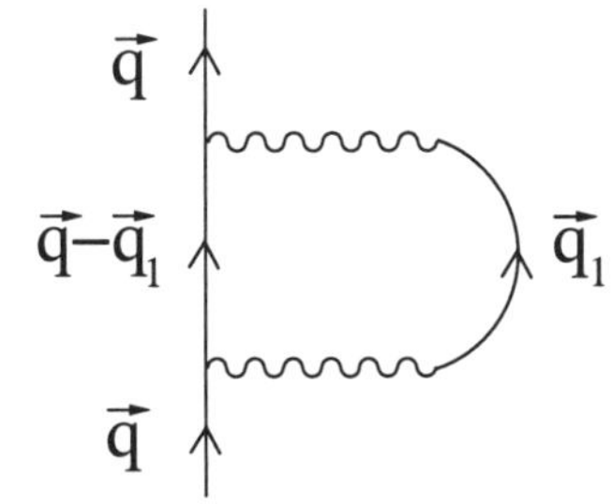


Fig. 18. Diagram corresponding to second order in the one–phonon state.

while the two phonon state, $\Psi_{\mathbf{q};\mathbf{q}_1\mathbf{q}_2}$, may be orthogonalized in a similar way to $\Psi_{\mathbf{q}}$, $\Psi_{\mathbf{q};\mathbf{q}_1+\mathbf{q}_2}$ , $\Psi_{\mathbf{q};\mathbf{q}_1}$ and $\Psi_{\mathbf{q};\mathbf{q}_2}$. The orthogonalization makes the convergence of the series faster. One can introduce a diagrammatic notation to represent the different perturbative corrections. Fig. 18, shows the diagram corresponding to the second order in the one–phonon space. Notice that there are two levels of diagrams, the perturbative diagrams like the one shown in Fig. 18, and the HNC type diagrams used to evaluate the matrix elements of the Hamiltonian between the different states.

The second order perturbative correction in the one–phonon subspace (diagram shown in Fig. 18) is

$$\Delta\epsilon_{OP}(q) = \sum_{\mathbf{q}_1} \frac{\mid \langle \mathbf{q} \mid H \mid \mathbf{q};\mathbf{q}_1\rangle \mid^2}{\epsilon(q) - \epsilon_0(\mid \mathbf{q}-\mathbf{q}_1 \mid) - \omega_F(q_1)} , \tag{6.40}$$

where $\epsilon(q) = \epsilon_0(q) + \Delta\epsilon_{OP}(q)$. The matrix elements needed to evaluate the previous expression are associated to the normalization of the states

$$\langle \Psi_{\mathbf{q};\mathbf{q}_1} \mid \Psi_{\mathbf{q};\mathbf{q}_1}\rangle = \frac{\langle \Psi_0(I;N_4) \mid \rho_4(\mathbf{q})^\dagger \rho_4(\mathbf{q}) \mid \Psi_0(I;N_4)\rangle}{\langle \Psi_0(I;N_4) \mid \Psi_0(I;N_4)\rangle} = N_4 S(q) + \mathcal{O}(1) \ , \tag{6.41}$$

and to the overlaps

$$\frac{\langle \Psi_{\mathbf{q}} \mid \Psi_{\mathbf{q};\mathbf{q}_1}\rangle}{\langle \Psi_0(I;N_4) \mid \Psi_0(I;N_4)\rangle} = \rho \int d\mathbf{r}_{1I}\, e^{i\mathbf{q}_1\cdot(\mathbf{r}_1-\mathbf{r}_I)} g^{(4,I)}(r_{1I}) \equiv S^I(q_1) \ . \tag{6.42}$$

The matrix elements of the Hamiltonian, can be easily evaluated by assuming that either $\mid \Psi_0\rangle$ is the true ground state of the system of $N_4+1$ particles or that the two– and the three–body correlations are solutions of the corresponding Euler equations. Finally, the matrix elements between the orthonormal states have the following expressions

$$\langle \mathbf{q} \mid H(I,N_4) \mid \mathbf{q}\rangle = E_o(I,N_4) + \epsilon_0(q) \tag{6.43}$$

$$\langle \mathbf{q};\mathbf{q}_1 \mid H(I,N_4) \mid \mathbf{q};\mathbf{q}_1\rangle = E_0(I;N_4) + \epsilon_0(\mid \mathbf{q}-\mathbf{q}_1 \mid) + \omega_F(q_1) \ . \tag{6.44}$$

The diagonal matrix elements yield the ground state energy of the system of $N_4+1$ particles added to the sum of the energy associated to the excitations present in the state, *i.e.*, the plane wave for the impurity and the Feynman phonon for the medium. The matrix element between the one–phonon state and the state without phonons, $\mid \mathbf{q}\rangle$ is given by

$$\langle \mathbf{q} \mid H \mid \mathbf{q}; \mathbf{q}_1 \rangle = -[N_4 S(q_1)]^{-1/2} \frac{\hbar^2}{2m_I} \mathbf{q} \cdot \mathbf{q}_1 S_I(q_1) \ . \tag{6.45}$$

In Brillouin–Wigner perturbation theory, the correction itself depends on $\epsilon(q)$ and the series must be self–consistently summed. The sum over momenta in Eq. (6.40) is converted into an integral, and thus

$$\Delta\epsilon_{OP}(q) = \frac{\Omega}{(2\pi)^3} \left(\frac{\hbar^2}{2m_I}\right)^2 \int d^3q_1 \frac{1}{N_4 S(q_1)} \frac{[S^I(q_1)\mathbf{q}\cdot\mathbf{q}_1]^2}{\epsilon(q) - \epsilon_0(\mid \mathbf{q} - \mathbf{q}_1 \mid) - \omega_F(q_1)} \ . \tag{6.46}$$

For $q = 0$, the effective mass is given by a simple quadrature

$$\frac{m_I^*}{m_I} = \left[1 - \frac{1}{4\pi^2 \rho_4} \frac{\hbar^2}{2m_I} \frac{2}{3} \int_0^\infty dq \frac{q^2 S^I(q)^2}{\frac{S(q)\hbar^2}{2m_I} + \frac{\hbar^2}{2m_4}}\right]^{-1} \ . \tag{6.47}$$

The spectrum obtained by taking only the OP states is very close to the Landau–Pomeranchuk one with an effective mass similar to that given by the variational calculation with backflow correlations. Actually, it has been pointed out that the second order perturbative expansion with OP states builds two–body backflow correlations into the wave function. In this approach, one finds $m_I^*(OP) = 1.8m_3$, in good agreement with the variational estimate.

The matrix elements involving two–independent phonons, whose lengthy expressions can be found in Ref. 51, involve two– and three–body structure functions, *i.e.*, the Fourier transforms of the two– and the three–body distribution functions.

The results are rather sensitive to the approximations used to evaluate the three–body structure functions. A calculation including both one– and two–phonon states and evaluating the four–body correction to the three–body structure functions (the Abe terms [35]) and the one–phonon re–scattering terms provides $m_I^* = 2.2m_I$ at saturation density, which is in good agreement with experimental results and with a recent diffusion Monte Carlo estimation. [62]

Recent CBF calculations reproduce the full spectrum at $q > 0$, adjust very well to the MLP spectrum (Eq. (6.1)) [63] and are in good agreement with other calculations using time dependent correlations [64] and shadow wave functions. [65]

## 7. Variational description of Fermi systems

The variational methods developed so far allow for the correct description of the ground state of infinite and homogeneous interacting Bose systems like pure $^4$He. However, other systems of interest as liquid $^3$He or nuclear matter obey Fermi statistics. Besides dynamical correlations, Fermi systems are affected by statistical correlations which differ from those found in a Bose fluid. In this way, therefore, the proper description of an interacting Fermi liquid requires an alternative treatment. The main goal of this section is to introduce the reader to this new subject.

Systems of identical fermions are described by antisymmetric wave functions. In the spirit of the variational theory, an efficient wave function for an infinite and homogeneous strongly correlated Fermi system could be of the form

$$\Psi_T(1, ..., N) = \prod_{i<j} f_2(r_{ij}) \phi(1, ..., N) \ , \tag{7.1}$$

that is, a Jastrow factor, similar to that used for bosons, times a model wave function, which is the Slater determinant associated to the free Fermi sea. This model wave function describes the system in the absence of interactions and incorporates the antisymmetry to the trial wavefunction

$$\phi(1, ..., N) = det \mid \psi_{\alpha_i}(j) \mid . \tag{7.2}$$

As stated in Eq. (2.9), the single particle wave functions $\Psi_{\alpha_i}(j)$, are taken as plane waves

$$\psi_{\alpha_i}(j) = \frac{1}{\Omega^{1/2}} e^{i\mathbf{k}_{\alpha_i} \cdot \mathbf{r}_j} \sigma(j) \ , \tag{7.3}$$

$\sigma(j)$ being the third component of the spin. These plane waves satisfy periodic boundary conditions in a large cube of volume $\Omega$, which becomes infinite in the thermodynamic limit. The allowed momenta fill a Fermi sphere of radius $k_F = (6\pi^2 \rho/\nu)^{1/3}$ with $\nu$ the spin degeneracy. In the $^3$He case, $\nu = 2$. The energy per particle of the free system, for which $\phi(1, .., N)$ is an eigenfunction, is the kinetic energy associated to the free Fermi sea

$$\frac{1}{N} \frac{\langle \phi \mid -\sum_i \frac{\hbar^2}{2m} \nabla_i^2 \mid \phi \rangle}{\langle \phi \mid \phi \rangle} = \frac{\nu \frac{\Omega}{(2\pi)^3} \int_{k<k_F} d^3k \frac{\hbar^2 k^2}{2m}}{\nu \frac{\Omega}{(2\pi)^3} \int_{k<k_F} d^3k} = \frac{3}{5} \frac{\hbar^2 k_F^2}{2m} \ , \tag{7.4}$$

where the sum over momentum states $\sum_k$ has been substituted by an integral over the Fermi sphere $\Omega/(2\pi)^3 \int_{k<k_F} d\mathbf{k}$.

As in the Bose case, a little bit of algebra shows that the calculation of the energy associated to this trial wave function requires the previous knowledge of the two–body radial distribution function which, for the wave function in Eq. (7.1), becomes

$$g(r) = \frac{N(N-1)}{\rho^2} \frac{\int d\mathbf{R}_{12} \, \phi^*(1, \ldots, N) \prod_{i<j} f_2(r_{ij}) \prod_{k<l} f_2(r_{kl}) \phi(1, ..., N)}{\int d\mathbf{R} \mid \Psi_T(1, \ldots, N) \mid^2} \tag{7.5}$$

where a trace over spin or spin–isospin variables has to be understood.

In order to perform the cluster decomposition of $g(r)$, one can proceed as usual and introduce the function $h(r) = f_2^2(r) - 1$ upon which a cluster expansion is performed

$$\mid \Psi_T \mid^2 = \left(1 + \sum h_{ij} + \sum h_{ij} h_{kl} + \ldots\right) \mid \phi \mid^2 \ . \tag{7.6}$$

Although this expression looks similar to the one found in the Bose case, one still has to learn how to manipulate the square of the Slater determinant. As it will be seen later, one can integrate over all internal coordinates that are not touched by a dynamical correlation in a given cluster. At the end, one needs to introduce new graphical elements in order to have a diagrammatical representation of the square of the Slater determinant.

Before performing this integration, it is illustrative to evaluate the expectation value of the potential operator on the free Fermi sea. For a two–body potential depending only on the distance between the particles

$$\frac{\langle V \rangle}{N} = \frac{1}{N} \frac{\langle \phi \mid \sum_{i<j} v(r_{ij}) \mid \phi \rangle}{\langle \phi \mid \phi \rangle}$$
$$= \frac{1}{2}\frac{1}{N} \sum_{\alpha\beta} \langle \psi_\alpha(1)\psi_\beta(2) \mid v(r_{12}) \mid \psi_\alpha(1)\psi_\beta(2) - \psi_\beta(1)\psi_\alpha(2) \rangle \ , \tag{7.7}$$

and thus becomes the sum over all possible states two particles can occupy. Performing the sum over spin and momenta, one gets

$$\frac{\langle V \rangle}{N} = \frac{1}{2}\frac{1}{N}\frac{\Omega^2}{(2\pi)^6} \int_0^{k_F} d\mathbf{k} \int_0^{k_F} d\mathbf{k}' \sum_{s_1 s_2} \langle \mathbf{k}s_1, \mathbf{k}'s_2 \mid v(r) \mid \mathbf{k}s_1, \mathbf{k}'s_2 - \mathbf{k}'s_2, \mathbf{k}s_1 \rangle \ , \tag{7.8}$$

while the evaluation of the two–body matrix elements for the plane wave single particle wave function yields

$$\frac{\langle V \rangle}{N} = \frac{1}{2}\frac{1}{N}\frac{\Omega^2}{(2\pi)^6} \int_0^{k_F} d\mathbf{k} \int_0^{k_F} d\mathbf{k}' \int d\mathbf{r}_1 \int d\mathbf{r}_2 v(r_{12})$$
$$\times \frac{1}{\Omega^2} \left[ \sum_{s_1 s_2} 1 - \left( \sum_{s_1 s_2} \delta_{s_1 s_2} \right) e^{i(\mathbf{k}-\mathbf{k}')\cdot(\mathbf{r}_2 - \mathbf{r}_1)} \right] \ . \tag{7.9}$$

The sum over the spin variables is equivalent to the calculation of the trace of the identity and the exchange spin operator $P_\sigma = (1 + \sigma_1 \sigma_2)/2$

$$\sum_{s_1 s_2} 1 = T_r(I) = \nu^2$$
$$\sum_{s_1 s_2} \langle s_1 s_2 \mid s_2 s_1 \rangle = \sum_{s_1 s_2} \delta_{s_1 s_2} = T_r(P_\sigma) = \nu \ . \tag{7.10}$$

After that, one can perform the momentum integration. The first integral yields simply the volume of the Fermi sphere, $\int_0^{k_F} d\mathbf{k} = 4/3\pi k_F^3 = (2\pi)^3 \rho/\nu$. The evaluation of the second integral gives

$$\int_{k<k_F} d\mathbf{k}\, e^{i\mathbf{k}\cdot\mathbf{r}} = \frac{(2\pi)^3 \rho}{\nu} \ell(rk_F) \ , \tag{7.11}$$

where $\ell(rk_F)$ is the *Slater function*, defined by

$$\ell(rk_F) = \frac{3 j_1(k_F r)}{k_F r} = \frac{3}{k_F r} \left( \frac{\sin(k_F r)}{(k_F r)^2} - \frac{\cos(k_F r)}{k_F r} \right) \ . \tag{7.12}$$

Taking into account these results and the cancellations of the different factors, one arrives at

$$\frac{\langle V \rangle}{N} = \frac{1}{2}\frac{1}{N} \rho^2 \int d\mathbf{r}_1 d\mathbf{r}_2 \ v(r) \left( 1 - \frac{1}{\nu} \ell^2(rk_F) \right) \ , \tag{7.13}$$

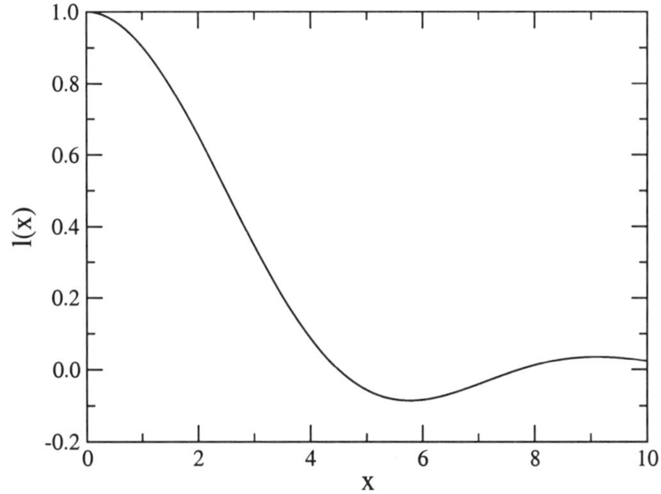


Fig. 19. The Slater function $l(x)$.

and changing to relative coordinates

$$\frac{\langle V\rangle}{N} = \frac{1}{2}\int d\mathbf{r}\; v(r)\left(1-\frac{1}{\nu}\ell^2(rk_F)\right) \equiv \frac{1}{2}\rho\int d\mathbf{r}\; v(r)g^F(r)\;, \tag{7.14}$$

which allows to identify the two–body radial distribution function of the free Fermi gas

$$g^F(r) = 1-\frac{\ell^2(rk_F)}{\nu}\;. \tag{7.15}$$

The function $\ell(x)$ is represented in Fig. (19). As $\ell(0) = 1$, $g(0) = 1 - 1/\nu$, which means that if $\nu = 1$, $g(0) = 0$ according to the Pauli principle. If $\nu = 2$, $g(r) = 1/2$ because two of the four possible two–particle spin states are forbidden (both particles with spin up or down), and therefore the probability to find two particles in the same position is 1/2. Even if there are no dynamical correlations, $g(r)$ presents a hole at the origin due to the statistical correlations reflected in the antisymmetry of the wave function.

There are several properties of the function $l(x)$ which are easy to derive and worth to be reported

$$\int d\mathbf{r}\; \ell(rk_F) = \int d\mathbf{r}\; \ell^2(rk_F) = \frac{\nu}{\rho} \tag{7.16}$$

$$\int d\mathbf{y}\, \ell(yk_F)\ell(\mid \mathbf{y}-\mathbf{z}\mid k_F) = \frac{\nu}{\rho}\ell(zk_F)\;, \tag{7.17}$$

while for the Fourier transforms

$$\rho\int d\mathbf{r}\, \ell(rk_F)\, e^{i\mathbf{k}\cdot\mathbf{r}} = \nu\theta(k_F-k) \tag{7.18}$$

$$\rho \int d\mathbf{r}\, \ell^2(rk_F)\, e^{i\mathbf{k}\cdot\mathbf{r}} = \nu\theta(2k_F - k)\frac{(k^3 - 12kk_F^2 + 16k_F^3)}{16k_F^3} . \tag{7.19}$$

Taking into account the last relations, the structure function of the free Fermi gas becomes

$$S^F(q) = 1 + \rho \int d\mathbf{r}\, (g^F(r) - 1)\, e^{i\mathbf{k}\cdot\mathbf{r}} = 1 - \theta(2k_F - q)\frac{q^3 - 12qk_F + 16k_F^3}{16k_F^3} . \tag{7.20}$$

Moreover, $S^F(q)$ shows a linear low–$q$ behavior which translates into a $g^F(r)$ with the appropriate long range.

At this point one has enough information to face the calculation of $\phi^*\phi$. The first thing to realize is that the product $\phi^*\phi$ can also be written as a determinant

$$\phi^*\phi = det \mid \rho(i,j) \mid \equiv \Delta_N(1,\ldots,N) , \tag{7.21}$$

whose elements are given by

$$\begin{aligned} \rho(i,j) &= \left(\sum_{p=1}^{\nu} s_p^\dagger(i)s_p(j)\right)\frac{1}{\Omega}\sum_k e^{-i\mathbf{k}\cdot\mathbf{r}_i}e^{i\mathbf{k}\cdot\mathbf{r}_j} \\ &= \left(\sum_{p=1}^{\nu} s_p^\dagger(i)s_p(j)\right)\frac{\rho}{\nu}\ell(\mid \mathbf{r}_j - \mathbf{r}_i \mid k_F) , \end{aligned} \tag{7.22}$$

and satisfying the property

$$\int d\mathbf{r}_j\, \rho(i,j)\rho(j,k) = \rho(i,k) , \tag{7.23}$$

where a spin summation is also implied.

Now, one can define new $p$–particle sub–determinants

$$\Delta_p(1,\ldots,p) = det \mid \rho(i,j) \mid , \tag{7.24}$$

fulfilling

$$\int \Delta_{p+1}(1,\ldots,p+1)\, d\mathbf{r}_{p+1} = (N-p)\Delta_p(1,\ldots,p) . \tag{7.25}$$

This last relation is crucial to integrate over the particles that are not reached by any dynamical correlation in a given cluster term. Notice in particular that

$$\int \Delta_N(1,\ldots,N)\, d\mathbf{r}_2 ... d\mathbf{r}_N = 1\cdot 2\cdot ...(N-1)\rho(1,1) , \tag{7.26}$$

where

$$\Delta(1,1) = \rho(1,1) = \rho , \tag{7.27}$$

and therefore

$$\int \Delta_N(1,\ldots,N) d\mathbf{r}_1 ... d\mathbf{r}_N = (N-1)!\frac{N}{\Omega}\Omega = N! . \tag{7.28}$$

Using these properties one can easily derive the two–body radial distribution function of the free Fermi sea

$$g_2^F(r) = \frac{N(N-1)}{\rho^2 N!} \int d\mathbf{R}_{12}\, \phi^* \phi = \frac{N(N-1)}{N! \rho^2} (N-2)! \Delta_2(1,2) \,, \tag{7.29}$$

where the factor $N!$ in the denominator comes from the normalization of $\phi$. Expanding the determinant $\Delta_2$ and taking into account that $\rho(i,i) = \rho$, one finally finds

$$g_2^F(r) = \frac{1}{\rho^2} \left[\rho(1,1)\rho(2,2) - \rho(1,2)\rho(2,1)\right] = 1 - \frac{\ell^2(rk_F)}{\nu} \,, \tag{7.30}$$

where in the last step use has been made of the spin summation

$$\sum_p^\nu \sum_{p'}^\nu s_p^\dagger(1) s_p(2) s_{p'}^\dagger(2) s_{p'}(1) = \nu \,. \tag{7.31}$$

As a further example, one can also calculate the three–body distribution function

$$g_3^F(1,2,3) = \frac{N(N-1)(N-2)}{N!\rho^3} \int d\mathbf{R}_{123}\, \phi^* \phi = \frac{N(N-1)(N-2)}{N!\rho^3} (N-3)! \Delta_3 \,, \tag{7.32}$$

where the expansion of $\Delta_3(1,2,3)$ leads to

$$\begin{aligned} g_3^F(1,2,3) = 1 &- \frac{1}{\nu}\ell^2(r_{12}k_F) - \frac{1}{\nu}\ell^2(r_{13}k_F) - \frac{1}{\nu}\ell^2(r_{23}k_F) \\ &+ \frac{2}{\nu^2}\ell(r_{12}k_F)\ell(r_{13}k_F)\ell(r_{23}k_F) \,. \end{aligned} \tag{7.33}$$

In expanding one of those sub–determinants, one should take into account all possible permutations which in turn are decomposed in products of cyclic permutations with no common elements. These permutations are represented by loops and products of separate loops of $-\frac{\ell}{\nu}$ functions. Each loop brings a factor $-2\nu$, the only exception being the two–particle loop, which carries a factor $-\nu$. These functions can be depicted by means of oriented lines, each one representing a $-\frac{\ell}{\nu}$ function. In Fig. 20 one can see the diagrammatic representation of the two– and the three–body distribution function of the free Fermi sea.

Once the diagrammatic contents of Fermi statistics is understood, one can go back to the original problem and face the calculation of the energy of an interacting Fermi liquid. The appropriate expression of $g(r)$ that will be used to perform the cluster expansion reads

$$g(r_{12}) = \frac{N(N-1)}{N!\rho^2} \frac{\int d\mathbf{R}_{12}\, \Delta_N F^2}{\frac{1}{N!}\int d\mathbf{R}\, \Delta_N F^2} \,, \tag{7.34}$$

where the $N!$ in the numerator and the denominator have been added in order to keep $\phi(1,...,N)$ normalized.

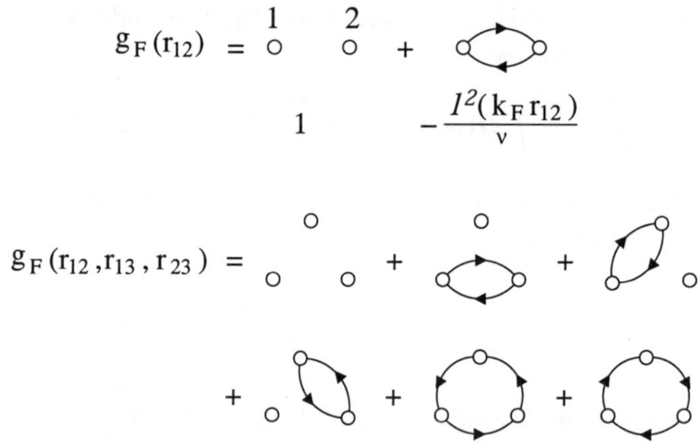


Fig. 20. Diagrammatic representation of the two– and three–body distribution functions of the free Fermi gas.

It was shown by Fantoni and Rosati [29] that the cluster expansion of Eq. (7.34) is linked and irreducible, and that therefore only contributions from the numerator survive. In this way

$$F^2(1,\ldots,N) = f_2^2(r_{12})\left(1+\sum_{i\neq 1,2} X^{(3)}(12;i) + \sum_{i<j\neq 1,2} X^{(4)}(12;ij)+\ldots\right), \quad (7.35)$$

where $X^{(p)}$ correlates particles 1 and 2 with $p-2$ particles in the background. As $\Delta_N$ is symmetric under the exchange of coordinates, all terms of $X^{(p)}$ which differ only in the labels of their arguments can be relabeled and summed up. A typical term in the expansion looks like

$$\frac{N(N-1)}{\rho^2}\frac{1}{N!}f_2^2(r_{12})\int d\mathbf{R}_{12}\Delta_N\frac{(N-2)!}{(p-2)!(N-p)!}X^{(p)}(12;3...p) \quad (7.36)$$

and then, integrating over the $A-p$ coordinates not affected by $X^{(p)}(12;3...p)$, one gets $(N-p)!\Delta_p$ with $\Delta_p = \rho^p g_p^F(1,...,p)$, and the previous expression reduces to

$$f_2^2(r_{12})\frac{1}{(p-2)!}\rho^{p-2}\int d\mathbf{r}_3...d\mathbf{r}_p X^{(p)}(12;3...p)g_p^F(1,...,p)\,. \quad (7.37)$$

Now one can permute the $p-2$ internal particles, to get $(p-2)!$ contributions which Will cancel the $(p-2)!$ in the denominator. Finally, a symmetry factor corrects for those permutations that give rise to exactly the same diagram. The suitable expression of the numerator for a cluster expansion then becomes

$$f_2^2(r_{12})g_2^F(r_{12}) + f_2^2(r_{12})\sum_{p\geq 3}\rho^{p-2}\int d\mathbf{r}_3...d\mathbf{r}_p X^{(p)}(12;3,...,p)g_p^F(1,...,p)\,. \quad (7.38)$$

In the end, $g(r)$ becomes the sum of all irreducible diagrams resulting from this expression.

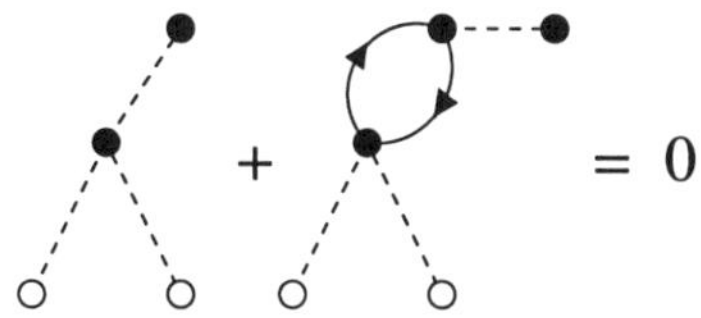

Fig. 21. Diagrammatic representation of the cancellation in Eq. (7.39).

### 7.1. *Diagrammatic rules and Fermi hypernetted chain equations*

The diagrams representing the different cluster contributions are constructed respecting simple rules. They are built with points, dashed lines and oriented lines. External points are represented by open circles, while internal points are represented by solid circles and imply an integration over the corresponding coordinates times a factor $\rho$.

There are two types of lines, dashed and oriented solid. Dashed lines account for the dynamical correlations while oriented lines take care of the statistical correlations. A dashed line connecting two points $i$ and $j$ gives a factor $h(r_{ij}) = f_2^2(r_{ij}) - 1$. An oriented solid line connecting two points $i$ and $j$ brings a factor $-\ell(r_{ij}k_F)/\nu$.

Every internal point should be the extremity of at least one dynamical line. Exchange lines form closed loops with no common points. Each closed loop carries a factor $-2\nu$, except for the two particle loop which has factor $-\nu$.

The expansion is linked since all the disconnected diagrams exactly cancel with the denominator. This fact is translated into an additional diagrammatical rule: each internal point, besides being the extremity of at least one dynamical line, should be connected to the external points by at least one path of correlation (dynamical and statistical) lines.

Moreover, the expansion is irreducible, [29] and therefore only the irreducible diagrams of the numerator contribute. The cancellation of diagrams in the numerator and the denominators is now *exact*, in contrast to what happens in the Bose case where it is only up to order $1/N$. In Fig. 21, a characteristic example of the cancellations that lead to the irreducible of the expansion is shown. The explicit expression for this cancellation is

$$\rho \int d\mathbf{r}_3\, h(r_{13})h(r_{23})\rho \int d\mathbf{r}\, h(r)$$
$$+\rho \int d\mathbf{r}_3\, h(r_{13})h(r_{23}) \left[\rho \int d\mathbf{r}' \left(-\frac{\ell^2(r'k_F)}{\nu}\right)\right] \rho \int d\mathbf{r}\, h(r) = 0 \qquad (7.39)$$

because the integration inside the square brackets gives exactly $-1$ (see Eq. (7.16)). The proof can be easily extended to any articulation point affected only by dynamical correlations.

The concepts of $i - j$–subdiagram, composite, nodal and non–nodal sub–diagrams are the same as in the Bose case. In addition, sub–diagrams must be classified according to the type of correlations reaching points $i-j$. In Fig. 22 the different classes of $i-j$–sub–diagrams are represented, which can be of the following types: $dd$ (*dynamical–dynamical*) both points $i$ and $j$ are not affected by statistical lines, $de$ (*dynamical–exchange*) point $i$ is not affected by statistical lines, while point $j$ is a common extremity of two statistical lines, superimposed or not with dynamical lines, $ee$ (*exchange–exchange*) both $i$ and $j$ are extremities of two statistical lines, superimposed or not with dynamical lines, and $cc$ (*cyclic–cyclic*) both points $i$ and $j$ are touched by only one statistical line superimposed or not with dynamical lines.

At this point, it is convenient to introduce the different types of nodal and non–nodal diagrams used in the expansion. $N_{yz}$ is the sum of all nodal diagrams of the type specified by the subscripts $yz$ ($y,z=e,d$ or $c$). Moreover, $X_{yz}$ represents the sum of all non–nodal diagrams of the $yz$–type. The functions $N_{yz}$ may be constructed by making chain connections of the various $X_{yz}$. The diagrammatic rules allow only for the following connections: $X_{dd}$ can be connected with $X_{dd}, X_{ee}$ and $X_{de}$; $X_{de}$ with $X_{dd}$ and $X_{de}$, while $X_{ee}$ can be connected with $X_{dd}$ and $X_{de}$. Finally, $X_{cc}$ can be joined to $X_{cc}$ and $-\ell/\nu$. Obviously, chain connections between two $N_{yz}$ functions are not allowed, because they would generate more than once the same diagram. However, it is clear that connecting $N_{dd}$ and $X_{dd}$ reproduces part of $N_{dd}$.

The generalization to fermions of the chain connections, taking into account the above diagrammatic rules leads to the following set of non–linear coupled integral equations

$$
\begin{aligned}
N_{dd}(r_{12}) &= \rho(X_{dd}(r_{13})+X_{de}(r_{13}) \mid N_{dd}(r_{32})+X_{dd}(r_{32})) \\
&\quad +\rho(X_{dd}(r_{13}) \mid N_{ed}(r_{32})+X_{ed}(r_{32})) \\
N_{de}(R_{12}) &= \rho\,(X_{dd}(r_{13})+X_{de}(r_{13}) \mid N_{de}(r_{32})+X_{de}(r_{32})) \\
&\quad +\rho\,(X_{dd}(r_{13}) \mid N_{ee}(r_{32})+X_{ee}(r_{32})) \\
N_{ee}(r_{12}) &= \rho\,(X_{ed}(r_{13}) \mid N_{ee}(r_{32})+X_{ee}(r_{32})) \\
&\quad +\rho\,(X_{ed}(r_{13})+X_{ee}(r_{13}) \mid N_{de}(r_{32})+X_{de}(r_{32}))
\end{aligned}
\tag{7.40}
$$

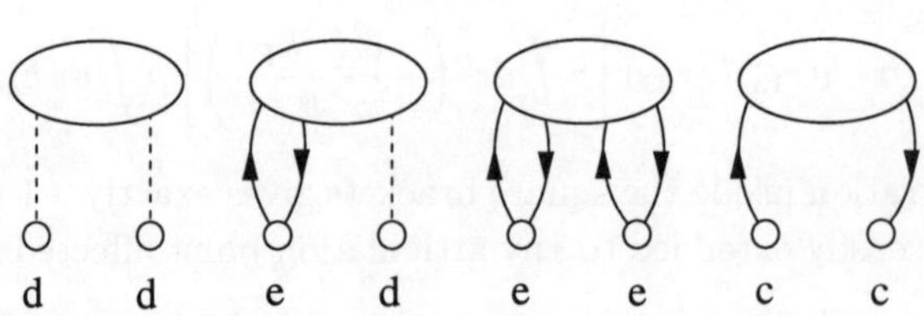


Fig. 22. Classification of subdiagrams according the type of correlations reaching the external points.

For $N_{cc}$, the situation is more delicate. $N_{cc}$ corresponds to the sum of all nodal diagrams constructed with $X_{cc}$ and $-\ell/\nu$. In those chains, however, $-\ell/\nu$ can appear superimposed or not with dynamical lines. In addition, the value of the chain does not change if the order of the components in the chain is altered, as can be seen by writing the corresponding diagram in momentum space. Therefore all factors $-\ell/\nu$ can be brought together. Then using Eq. (7.17), all the factors $-\ell/\nu$, which appear in a given chain with no dynamical lines superimposed, can collapse in only one or none $-\ell/\nu$ factors. This statement is compatible with the following equation

$$N_{cc}(r_{12}) = \rho \left( X_{cc}(r_{13}) \mid N_{cc}(r_{32}) + X_{cc}(r_{32}) - \frac{\ell(r_{32}k_F)}{\nu} \right) . \tag{7.41}$$

The equations for the non–nodal diagrams that complete the system of FHNC equations are

$$\begin{aligned}
X_{dd}(r) &= F(r) - N_{dd}(r) - 1 \\
X_{de}(r) &= F(r)\left\{N_{de}(r) + E_{de}(r)\right\} - N_{de}(r) \\
X_{ee}(r) &= F(r)\left\{N_{ee}(r) + E_{ee}(r) + [N_{de}(r) + E_{de}(r)]^2 - \nu[N_{cc}(r) + E_{cc}(r)]^2\right. \\
&\qquad \left. +2\ell(rk_F)[N_{cc}(r) + E_{cc}(r)] - \frac{1}{\nu}\ell^2(rk_F)\right] - N_{ee}(r) \\
X_{cc}(r) &= F(r)\left[N_{cc}(r) + E_{cc}(r) - \frac{1}{\nu}\ell(rk_F) - N_{cc}(r)\right] ,
\end{aligned} \tag{7.42}$$

where $E_{yz}(r)$ stands for the sum of all elementary diagrams, which are also classified according to their external legs, and

$$F(r) = f_2^2(r) \exp\left[N_{dd}(r) + E_{dd}(r)\right] . \tag{7.43}$$

As seen from the above equations, the proper treatment of the statistics in the Fermi case leads to a system of four coupled non–linear integral equations which would allow for the exact evaluation of the distribution function if the contribution from the elementary diagrams was known.

In the end, the radial distribution function is given by

$$\begin{aligned}
g(r) &= f_2^2(r) \exp\left[N_{dd}(r) + E_{dd}(r)\right] \\
&\quad \times \left[1 - \frac{1}{\nu}\ell^2(rk_F) + N_{ee}(r) + E_{ee}(r) + [N_{de}(r) + E_{de}(r)]^2\right. \\
&\quad +2\left[N_{de}(r) + E_{de}(r)\right]\nu\left[N_{cc}(r) + E_{cc}(r)\right]^2 \\
&\quad \left. +2\ell(rk_F)\left[N_{cc}(r) + E_{cc}(r)\right]\right]
\end{aligned} \tag{7.44}$$

which has the correct weighting factors imposed by the diagrammatic rules. As in the Bose case, it is possible to get a more compact expression introducing the non–nodal diagrams $X_{yz}(r)$, thus producing

$$g(r) = 1 + X_{dd}(r) + N_{dd}(r) + X_{ee}(r) + N_{ee}(r) + 2X_{de}(r) + 2N_{de}(r). \qquad (7.45)$$

An alternative but equivalent way to write $g(r)$ is obtained by defining a new function $L(r) = \ell(rk_F) - \nu\,[N_{cc}(r) + E_{cc}(r)]$, in terms of which

$$g(r) = f_2^2(r)\exp\left[N_{dd}(r) + E_{dd}(r)\right] \times\left[-\frac{1}{\nu}L^2(r) + N_{ee}(r) + E_{ee}(r) + [1 + N_{de}(r) + E_{de}(r)]^2\right]. \qquad (7.46)$$

The static structure function, $S(k) = 1 + \widetilde{g-1}$ can then be expressed as

$$S(k) = 1 + \tilde{X}_{dd}(k) + \tilde{N}_{dd}(k) + \tilde{X}_{ee}(k) + \tilde{N}_{ee}(k) + 2\tilde{X}_{de}(k) + 2\tilde{N}_{de}(k)\ , \qquad (7.47)$$

or equivalently

$$S(k) = \frac{1 + \tilde{X}_{ee}(k)}{\left[1 - \tilde{X}_{de}\right]^2 - \left[1 + \tilde{X}_{ee}(k)\right]\tilde{X}_{dd}(k)}\ . \qquad (7.48)$$

It is important to notice that not all approximations guarantee that the static structure functions vanishes in the $k \to 0^+$ limit, which is the behavior of the structure function of the exact ground state. In order to accomplish this condition, it is necessary to take into account other effects, called Fermi cancellations, occurring among diagrams with a given number of dynamical correlations and exchange correlations involving elementary diagrams with different number of particles. This is precisely the scheme suggested by Krotscheck and Ristig, [66,67,68] who proposed a different set of FHNC equations which take the Fermi cancellation into account. This scheme is very useful to study $S(k)$ at low $k$ and to establish the optimization conditions for fermions. [69]

### 7.2. *Results for* $^3$*He*

The first thing to notice when discussing the results for pure $^3$He is the fact that while the potential energy has the same expression as in the Bose case, see Eq. (2.16), the kinetic energy presents additional contributions originating from the acation of the Laplacian to the Slater determinant. The kinetic energy

$$\frac{\langle T\rangle}{N} = -\frac{\hbar^2}{2m}\frac{1}{N}\sum_{i=1}^{N}\frac{\int d\mathbf{R}\,\phi^*(1,\ldots,N)F(1,\ldots,N)\nabla_i^2F(1,\ldots,N)\phi(1,\ldots,N)}{\langle\Psi_T\mid\Psi_T\rangle} \qquad (7.49)$$

becomes, using the Jackson–Feenberg identity

$$\frac{\langle T\rangle}{N} = -\frac{\hbar^2}{2m}\frac{1}{N}\sum_{i=1}^{N}\frac{\int d\mathbf{R}\left[\phi^*F^2\nabla_i^2\phi - \frac{1}{4}F^2\nabla_i^2(\phi^*\phi) + \frac{1}{2}\phi^*\left(F(\nabla_i^2F) - (\vec{\nabla}_iF)^2\right)\phi\right]}{\langle\Psi_T\mid\Psi_T\rangle}\ . \qquad (7.50)$$

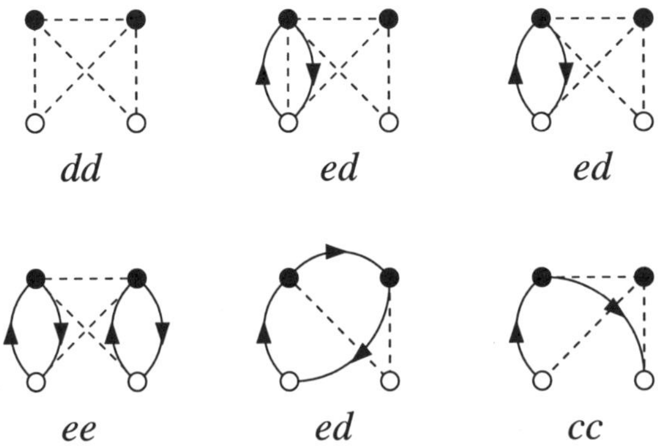


Fig. 23. Classification of four point elementary diagrams.

The first term on the right hand side gives rise to the average energy of the free Fermi gas, while the third term with the derivatives acting on the dynamical correlations produces a term formally identical to that found in the Bose case. Finally, the application of the Laplacian on the square of the model wave function produces terms which are not trivial to evaluate. The final result becomes

$$\frac{\langle T\rangle}{N} = \frac{3}{5}\frac{\hbar^2 k_F^2}{2m} - \frac{\hbar^2}{4m}\rho \int d\mathbf{r}\, g(r)\nabla^2 \ln f(r) - \frac{\hbar^2}{2m}\int d\mathbf{R}\, \frac{F^2(-\frac{1}{4}\nabla_1^2 \mid \phi \mid^2)}{\langle \Psi \mid \Psi\rangle} \quad (7.51)$$

The last term in Eq. (7.50) can be decomposed in two pieces, $t_2^s$ and $t_3^s$, with

$$t_2^s = -\frac{\hbar^2}{8m}\rho \int d\mathbf{r}\, f_2^2(r) \exp\left[N_{dd} + E_{dd}\right]\left(\frac{1}{\nu}\nabla^2 \ell^2(rk_F) - 2N_{cc}(r)\nabla^2 \ell(rk_F)\right) \quad (7.52)$$

and $t_3^s$ a the three–body contribution that is rather small and often disregarded. [52]

Elementary diagrams are harder to deal with. In the Bose case there is only one four–point elementary diagram, while in this case and due to the different graphical elements entering in the diagrams there are many more elementary diagrams which are also classified according to the type of links reaching points 1 and 2. Fig. 23 shows different types of four–point elementary diagrams.

Several strategies have been proposed to sum up the contribution of elementary diagrams. In close analogy with the Bose case, an extension of the *scaling* approximation has been devised. [53] In this approximation, only the *dd* elementary diagrams $E_{dd}$ are scaled while all the others $E_{yz}$ with $yz \neq dd$ are taken equal to zero.

A calculation of the energy per particle at $\rho_0 = 0.277\sigma^{-3}$ using a Schiff–Verlet correlation function (Eq. (2.26)) with $b = 1.15$, the Aziz potential and the scaling method to evaluate the elementary diagrams, yields $-1.08$ K, [53] in excellent agreement with a variational Monte Carlo [54] calculation ($-1.08$ K) for the same correlation function at the same density, but still far away from the experimental result $-2.47$ K.

One can follow the same route as in the Bose case to improve over this result. The first thing to do would be to find the optimal two–body correlation factor. However and compared to the Bose case, this problem is much harder to solve. [37,69] An alternative is to use the optimal correlation function for the underlying Bose system (a system of bosons at the same density but with the $^3$He mass), and to use this *optimal* correlation function in the full FHNC machinery. This leads to an energy per particle at $\rho_o^{exp}$ of $-1.28$K. Once again, the inclusion of three–body correlation improves the result, which now becomes $-1.83$ K, to be compared with $e^{exp}(\rho_0^{exp}) = -2.47$ K. It is also interesting to mention that the energy of polarized $^3$He (that is, $^3$He with all the spins pointing in the same direction), calculated either with a wave function including Jastrow correlations or even two– and three–body correlations, lies below the energy of the unpolarized system. If these numbers were right, polarized $^3$He would be the stable phase found in nature, which is not true. One can further improve the wave function by including other type of correlations in order to reverse this unphysical situation. Introducing momentum dependent backflow correlations similar to those used in the study of the impurity, one finally gets $e = -2.36$K for the normal phase and $e = -2.09$K for the fully polarized phase, making the unpolarized phase stable. The underlying physical reason is that the effect of backflow correlations on the polarized phase is rather small. This is due to the fact that as a consequence of the Pauli principle, there are no pairs of particles with zero relative angular momentum in the polarized phase, while the effect of the backflow correlations is larger in this partial wave.

Recent calculations based on Diffusion Monte Carlo methods reproduce very well the equation of state of liquid $^3$He in the full range of densities and provide a polarized phase that is less bound than the normal phase. [71] They also predict a good parameterization of the equation of state at zero temperature, which allow for the calculation of other quantities such as the chemical potential or the speed of sound. Using the parameterization in Eq. (3.35), the appropriate value of the parameters are $e(\rho_0) = -2.464$ K, $\rho_0 = 0.274\,\sigma^{-3}$, $B = 6.21$ K and $C = 3.49$ K.

Finally, it is also worth mentioning that the idea of using optimal correlations has also been pushed further, paying particular attention to the long–range behavior of the different functions describing the system. Quite recently, E. Krotscheck has reviewed the situation for pure $^3$He considering the optimization of two– and three–body correlations. [69]

### 7.3. *Excited states and the dynamic structure function*

In order to illustrate the different nature of the excited states in comparison with the Bose case, the excitation spectrum and the dynamic structure function associated to the free Fermi system is briefly discussed. [70]

The dynamic structure function, as in the Bose case reads

$$S(q,E) = \frac{1}{N} \sum_n \mid \langle n \mid \rho_q^\dagger \mid 0 \rangle \mid^2 \delta(E - (E_n - E_0)), \tag{7.53}$$

where $| \ 0\rangle$ and $| \ n\rangle$ are the ground state and excited states of the the free Fermi gas, which are Slater determinants of plane waves. The sum over $n$ extends over the whole set of excited states that can be connected to the ground state through the density fluctuation operator. The latter takes the following form

$$\hat{\rho}_{\mathbf{q}} = \sum_{\mathbf{k},\mathbf{l}} \langle \mathbf{k} \mid \exp\left[\imath \mathbf{q} \cdot \mathbf{r}\right] \mid \mathbf{l} \rangle a^{\dagger}_{\mathbf{k}} a_{\mathbf{l}} \tag{7.54}$$

where $a_{\mathbf{k}}$ and $a^{\dagger}_{\mathbf{k}}$ are the annihilation and creation operators associated to the plane wave basis. Momentum conservation allows simplifying this expression which, for an infinite and homogeneous system, becomes

$$\hat{\rho}_{\mathbf{q}} = \sum_{\mathbf{l}} a^{\dagger}_{\mathbf{l}+\mathbf{q}} a_{\mathbf{l}} \ . \tag{7.55}$$

The density fluctuation operator creates particle–hole excitations of total momentum $\mathbf{q}$. Therefore, only the 1–particle–1–hole configurations contribute to $S(q, E)$, and thus

$$S(q,E)=\frac{\nu}{(2\pi)^3\rho}\int d\mathbf{k}\Theta(k_F-k)\Theta(\mid \mathbf{k}+\mathbf{q}\mid -k_F)\delta\left[E-\left(\frac{\hbar^2(\mathbf{k}+\mathbf{q})^2}{2m}-\frac{\hbar^2k^2}{2m}\right)\right] . \tag{7.56}$$

The energy conservation condition expressed by the $\delta$ function allows for excitations inside a band of energies (shown in Fig. 24) satisfying

$$0 \leq E \leq \frac{\hbar^2 q^2}{2m} + \frac{\hbar^2 q k_F}{m} \quad q < 2k_F \ , \tag{7.57}$$

and

$$-\frac{\hbar^2 q k_F}{m} + \frac{\hbar^2 q^2}{2m} \leq E \leq \frac{\hbar^2 q^2}{2m} + \frac{\hbar^2 q k_F}{m} \quad q > 2k_F \ . \tag{7.58}$$

In this case, $S(q, E)$ counts the number of particle–hole configurations compatible with a given momentum and energy transferred by the fluctuation to the system. It is quite easy to explicitly evaluate $S(q, E)$ for $q > 2k_F$, where $\Theta(\mid \mathbf{k}+\mathbf{q} \mid)$ is always 1. Therefore

$$S(q,E) = \frac{\nu}{(2\pi)^3\rho}\int d\mathbf{k}\, \Theta(k_F - k)\delta\left[E - \left(\frac{\hbar^2(\mathbf{k}+\mathbf{q})^2}{2m} + \frac{\hbar^2k^2}{2m}\right)\right] \ , \tag{7.59}$$

which can be easily integrated to produce

$$S(q,E) = \frac{\nu}{4\pi^2\rho}\frac{m}{q}\int_{|Y|}^{k_F} kdk = \frac{\nu}{4\pi^2\rho}\frac{m}{\hbar^2 q}\left(\frac{k_F^2}{2} - \frac{1}{2}\left(\frac{mE}{\hbar^2 q} - \frac{q}{2}\right)^2\right) \tag{7.60}$$

with $Y = mE/q - q/2$ the West scaling variable. [72] $S(q, E)$ for $q = 3$ $k_F$ is shown in Fig. 24 in dimensionless units. It is worth to notice that, for $q > 2k_F$, $S(q, E)$ can be brought to a form that scales only on $Y$

$$J(Y) = \frac{\hbar^2 q}{m} S(q,E) \ . \tag{7.61}$$

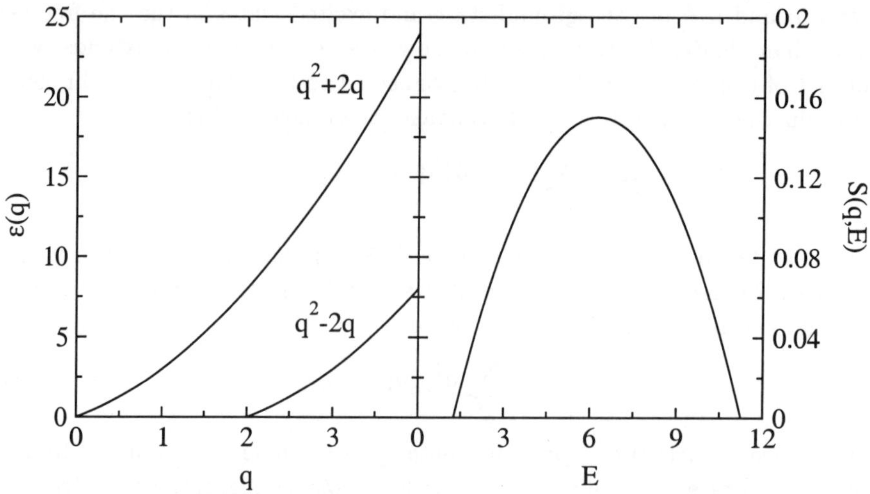


Fig. 24. The left panel shows the particle–hole band and the right one shows $S(q, E)$ for $q/k_F$=2.5. All quantities are dimensionless.

As in the Bose case, one can also perform an analysis of $S(q, E)$ in terms of energy weighted sum rules. For $q > 2k_F$ one finds $m_0(q) = S(q) = 1$ and $m_1(q) = \hbar^2 q^2/2m$. However, in this case $S(q, E)$ does not peak at a single energy, and therefore one can not recover the full $S(q, E)$ from the knowledge of $m_0$ and $m_1$.

In a more realistic interacting system, besides a similar particle–hole band, there are residual interactions between the particle–hole configurations that could generate collective states of higher energy. [46]

## 8. Summary

In this lecture, a pedagogical introduction to the microscopic description of quantum fluids, mainly focusing on liquid $^4$He and liquid $^3$He, has been presented. Helium liquids are characterized by an interatomic potential with a very strong repulsion at short distances, and therefore it is very inefficient to use traditional perturbative methods for calculating the energy. On the other hand, the variational method with a trial wave function of the Jastrow type incorporating the correlations induced by the potential from the very beginning turns out to be very efficient, and more sophisticated wave functions including two– and three–body correlation can accurately reproduce the equation of state at zero temperature.

Particular attention has been devoted to the optimization of the two–body correlation factor in the Bose case, and the consequences for the long range behavior of the two–body radial distribution function have been analyzed. The low energy excitation of liquid $^4$He has been also studied in the spirit of the variational theory, using the Feynman *ansatz* as a trial wave function.

As a first extension of the formalism, the impurity problem of one $^3$He atom in bulk $^4$He has been considered. The chemical potential of the impurity has been analyzed, and the experimental values for this same quantity have proved to be suitable to extract model independent information on the energetics of liquid $^4$He. Finally, the study of the excitation spectrum of the impurity has allowed to introduce perturbation theory in the framework of correlated basis functions. As the correlations originated by the short range repulsion of the potential are already included in the wave function, it is expected that perturbation theory using this correlated basis will not require higher order perturbative terms to produce reliable results. As a counterpart the evaluation of the matrix elements can be rather difficult.

Variational methods suitable for the description of Fermi systems have also been introduced, focusing particularly on the calculation of the energy and the evaluation of the two–body radial distribution function in the framework of the Fermi hypernetted chain equations. Moreover as a way to illustrate the different nature of elementary excitations in a Fermi liquid compared to the Bose case, the spectrum of elementary excitations of the free Fermi gas has been analyzed, showing that it is defined in an interval of energies characterized by a the particle–hole band.

Other extensions and further developments of the theory are discussed in this volume. However there are a couple of things that are worth pointing out as a guide for future research: the need of a finite temperature formalism (specially for fermions and mixtures), the improvement of the correlated basis perturbation theory for fermions, and the development of hybrid methods combining semi–analytical techniques and the Monte Carlo algorithms.

Finally let us conclude this lecture recalling that the experimental developments in the study of inhomogeneous systems, low–dimensionality systems and the new world of clusters should point to the future needs of theoretical developments in the field.

## 9. Acknowledgments

The authors are very indebted to A. Fabrocini and J. Boronat for long standing collaboration on the subjects discussed in this work. A. Polls is also very grateful to R. Guardiola, his thesis advisor, who initiated him to the secrets of the HNC and scientific computing, and also to S. Rosati and S. Fantoni that with lots of generosity provided him with invaluable help at the early stages of the FHNC development. F.Mazzanti wants also to acknowledge E.Krotscheck for stimulating discussions on most of the topics developed here, and for his kind support over the last years. This work has profited from a lot of stimulating discussions with numerous colleagues, specially C.E. Campbell, J. W. Clark, E. Krotscheck, and M. Saarela. Financial support from DGICYT (Spain) grant No. PB98–1247, and from the program 1998SGR00011 from the *Generalitat de Catalunya* are also acknowledged.

## References


1. H. Kammerlingh Onnes, Proc. R. Acad. Amsterdam **11**, 168 (1908).
2. J. Wilks, *The properties of Liquid and Solid Helium* (Clarendon Press, Oxford, 1967).
3. *The Physics of Liquid and Solid Helium*, edit. by K. H. Bennemann and J. B. Ketterson (John Wiley & Sons, New York, 1976).
4. J. Wilks and D. S. Betts, *An Introduction to Liquid Helium* (Clarendon Press, Oxford, 1987).
5. H. R. Glyde, *Excitations in Liquid and Solid Helium* (Clarendon Press, Oxford, 1994).
6. P. L. Kapitza, Nature (London) **141**, 74 (1938).
7. J. F. Allen and H. Jones, Nature (London) **141**, 75 (1938).
8. D. D. Osheroff, R.C. Richardson, and D. M. Lee, Phys. Rev. Lett. **28**, 1098 (1972).
9. *Excitations in Two–Dimensional and Three–Dimensional Quantum Fluids*, Vol. 257 of NATO *Advanced Study Institute, Series B: Physics,* A. F. G. Wyatt and H. J. Lauter, eds. (Plenum, New York, 1991).
10. W. Teizer, R. B. Hallock, E. Dujardin, and T. W. Ebbesen, Phys. Rev. Lett. **82**, 5305 (1999).
11. S. Grebenev, J. P. Toennies, and A. Vilesov, Science **279**, 2083 (1998).
12. *Microscopic Quantum Many–Body Theories and Their Applications*, J. Navarro and A. Polls eds., Lecture Note in Physics, Vol. 510 (Springer-Verlag, Heidelberg, 1998).
13. S. Rosati, in *International School of Physics Enrico Fermi, Course LXXIX*, edited by A. Molinari, 1981 (North-Holland, Amsterdam), 73.
14. J. G. Zabolitzky, in *Advances in Nuclear Physics* , Vol. 12, edited by J. W. Negele and E. Vogt (Plenum, New York, 1980).
15. C. E. Campbell, in *Progress in Liquid Physics*, edited by C.A. Croxton, Ch. 6 (Wiley, New York, 1977).
16. J. W. Clark, in *Progress in Particle and Nuclear Physics*, Vol. 2, edited by D. H. Wilkinson, p.89 (Pergamon, Oxford, 1979).
17. S. Rosati and S. Fantoni, in *The Many–Body Problem: Jastrow Correlations Versus Brueckner Theory*, (Lecture Notes in Physics, Vol. 138), edited by R. Guardiola and J. Ros, p. 1 (Springer, Berlin, 1981).
18. E. Feenberg, *Theory of Quantum Fluids* (Academic Press, New York, 1969).
19. R. A. Aziz, V. P. S. Nain, J. S. Caley, W. L. Taylor, and G. T. McConville, J. Chem. Phys. **70**, 4330 (1979).
20. R. A. Aziz, F. R. W. McCourt, and C. C. K. Wong, Mol. Phys. **61**, 1487 (1987).
21. R. Jastrow, Phys. Rev. **98**, 1479 (1955).
22. H. W. Jackson and E. Feenberg, Ann. of Phys. **15**, 266 (1961).
23. V. R. Pandharipande and H. A. Bethe, Phys. Rev. C **7**, 1312 (1973).
24. J. W. Clark and P. Westhaus, Phys. Rev. **149**, 990 (1966).
25. W. L. McMillan, Phys. Rev. **138**, 442 (1965).
26. D. Schiff and L. Verlet, Phys. Rev. **160**, 208 (1967).
27. C. E. Campbell, in *Progress in Liquid Physics*, C. A. Croxton, ed., Ch. 6 (Wiley, New York, 1977).
28. R. T. Azuah, W. G. Stirling, H. R. Glyde, M. Bonisegni, P. E. Sokol, and S. M. Bennington, Phys. Rev. B **56**, 14620 (1997).
29. S. Fantoni and S. Rosati, Nuovo Cimento **20A**, 179 (1974).
30. J. P. Hansen and I. R. McDonald, *Theory of Simple Fluids* (Academic Press, New York, 1976).
31. J. M. J. van Leeuwen, J. Groeneveld, and J. de Boer, Physica **25**, 792 (1959).
32. J. K. Percus and G. J. Yevick, Phys. Rev. **110**, 1 (1958).
33. Q. N. Usmani, B. Friedman, and V. R. Pandharipande, Phys. Rev. B **25**, 4502 (1982).

34. K. Schmidt, M. H. Kalos, M. A. Lee, and G. V. Chester, Phys. Rev. Lett. **45**, 573 (1981).
35. R. Abe, Orig. Theor. Phys. **21**, 421 (1959).
36. Q. N. Usmani, S. Fantoni, and V. R. Pandharipande, Phys. Rev. B **26**, 6123 (1982).
37. E. Krotscheck, Phys. Rev. B **33**, 3158 (1986).
38. C. C. Chang and C. E. Campbell, Phys. Rev. B **15**, 4238 (1977).
39. J. Boronat and J. Casulleras, Phys. Rev. B **49**, 8920 (1994).
40. S. Stringari and J. Treiner, Phys. Rev. B **36**, 8369 (1987).
41. L. Castillejo, A. D. Jackson, B. K. Jennings, and R. A. Smith, Phys. Rev. B **20**, 3631 (1979).
42. C. E. Campbell and E. Feenberg, Phys. Rev. **188**, 346 (1969).
43. M. Saarela, *Elementary excitations and dynamic structure of quantum fluids*, this volume.
44. L. Landau, Phys. Rev. **60**, 356 (1941). L. Landau, J. Phys. U.S.S.R. **5**, 71, (1941).
45. R. P. Feynman, Phys. Rev. **94**, 262 (1954).
46. D. Pines and P. Nozières, *The Theory of Quantum Liquids*, Vol. I (Addison Wesley, 1989).
47. J. C. Owen, Phys. Rev. B **23**, 5815 (1981).
48. S. Fantoni and V. R. Pandharipande, Phys. Rev. C **37**, 1697 (1988).
49. S. Fantoni, Phys. Rev. B **29**, 2544 (1984).
50. P. M. Morse and H. Feshbach, *Methods of Theoretical Physics* (McGraw Hill, New York, 1953).
51. A. Fabrocini, S. Fantoni, S. Rosati, and A. Polls, Phys. Rev. B **33**, 6057 (1986).
52. S. Fantoni and S. Rosati, Phys. Lett. B **84**, 33 (1979).
53. E. Manousakis, S. Fantoni, V. R. Pandharipande, and Q. N. Usmani, Phys. Rev. B **28**, 3770 (1983).
54. K. Schmidt, M. A. Lee, M. H. Kalos, and G. V. Chester, Phys. Rev. Lett. **47**, 807 (1981).
55. C. Ebner and D. O. Edwards, Phys. Rep. **2**, 77, (1970).
56. B. Fåk, K. Guckelsberger, M. Krofer, R. Scherm and A. J. Dianoux, Phys. Rev. B **41**, 8732 (1990).
57. L. D. Landau and I. M. Khalatnikov, Zh. Eksp. Teor. Fiz. **2**, 637 (1948).
58. F. Mazzanti, J. Boronat, A. Polls, Phys. Rev. B **53**, 5561 (1996).
59. J. Boronat, A. Fabrocini, and A. Polls, Phys. Rev. B **39**, 2700 (1989).
60. J. Boronat, A. Fabrocini, and A. Polls, J. Low Temp. Phys. **74**, 347 (1989).
61. G. Baym, Phys. Rev. Lett. **17**, 952 (1966).
62. J. Boronat and J. Casulleras, Phys. Rev. B **59**, 8844 (1999).
63. A. Fabrocini and A. Polls, Phys. Rev. B **58**, 5209 (1998).
64. E. Krotscheck, J. Paaso, M. Saarela, K. Schökhuber, and R. Zillich, Phys. Rev. B **58**, 12282 (1998).
65. D. E. Galli, G. L. Masserini and L. Reatto, Phys. Rev. B **60**, 3476 (1999).
66. E. Krotscheck and M. L. Ristig, Nucl. Phys. A **242**, 389 (1975).
67. E. Krotscheck, Phys. Lett. A **54**, 123 (1975).
68. E. Krotscheck, Phys. Rev. A **15**, 397 (1977).
69. E. Krotscheck, J. Low Temp. Phys. **119**, 103 (2000).
70. F. Mazzanti, A. Polls and J. Boronat, Phys. Lett. **A** 220, 251 (1996).
71. J. Casulleras and J. Boronat, Phys. Rev. Lett. **84**, 3121 (2000).
72. J. B. West, Phys. Rep. **18C**, 263 (1975).

34. [illegible] Schmidt, M. E. Kuroda, M. A. Lee, and [illegible], Chem. Phys. Lett. 49, 473 (19[illegible]).
35. K. [illegible], Prog. Theor. Phys. 41, 427 (1969).
36. [illegible] Usmani, [illegible] Pandit and V. R. Pandharipande, Phys. Rev. B 36, 5128 (1987).
37. [illegible] Feng, Phys. Rev. B 13, [illegible] (1976).
38. C. C. [illegible] and C. E. Campbell, Phys. Rev. B 15, 4238 (1977).
39. [illegible] Bishop and J. Ganley, Phys. Rev. B 40, 5530 (1989).
40. S. [illegible] and [illegible], Phys. Rev. B [illegible] (1982).
41. [illegible]
42. [illegible], Phys. Rev. [illegible] (19[illegible]).
43. [illegible]
44. [illegible] Jordan, Phys. Rev. 85, 370 (1952); [illegible] (1962).
45. F. P. Rymer, Phys. Rev. 94, 20 (1954).
46. R. [illegible] and [illegible], The Theory of [illegible] (19[illegible]).
47. [illegible], Phys. Rev. B 23, 5613 (1981).
48. [illegible] and V. R. Pandharipande, Phys. Rev. B 27, 5973 (1983).
49. [illegible], Phys. Rev. B 28, [illegible] (1983).
50. [illegible] and [illegible], Methods of Theoretical Physics [illegible] (1953).
51. A. Robinson, [illegible] and A. [illegible], Phys. Rev. B [illegible], 5657 (1990).
52. [illegible] and S. Rosati, Phys. Lett. B 67, [illegible] (1977).
53. [illegible], [illegible] Pandit, V. R. Pandharipande and [illegible], Phys. Rev. B 38, 2710 (1988).
54. [illegible], Phys. Rev. Lett. 47, 607 (1981).
55. [illegible] (1970).
56. [illegible], Phys. Rev. [illegible], 2782 (1990).
57. [illegible], J. Low Temp. Phys. 20, [illegible] (19[illegible]).
58. [illegible], Phys. Rev. B 28, 5601 (1983).
59. [illegible] and [illegible], Phys. Rev. B 33, 8700 (1986).
60. [illegible] and [illegible], J. Low Temp. Phys. 71, 347 (1988).
61. [illegible], Phys. Rev. Lett. 17, 552 (1966).
62. [illegible] and [illegible], Phys. Rev. B 36, 5350 (1987).
63. [illegible] and [illegible], Phys. Rev. C [illegible], 5500 (1996).
64. [illegible], Phys. [illegible] 64, 10226 (1996).
65. [illegible] and L. [illegible], Phys. Rev. B [illegible] (1989).
66. [illegible] and M. [illegible], Proc. Phys. [illegible] (1976).
67. [illegible] (1975).
68. [illegible], Phys. Rev. A 15, 387 (1977).
69. [illegible], Phys. [illegible] 118, [illegible] (2000).
70. [illegible] and [illegible], [illegible] (2000).
71. [illegible] and [illegible], Phys. Rev. [illegible] (2000).
72. [illegible] (1976).

## CHAPTER 3

# THE COUPLED CLUSTER METHOD AND ITS APPLICATIONS

J. Navarro
*IFIC (Centre Mixt CSIC - Universitat de València)*
*Apdo. 22085, E-46.071 Valencia, Spain,*
*E-mail: navarro@ific.uv.es*

R. Guardiola, I. Moliner
*Dept. Física Atómica y Nuclear, Facultat de Física*
*Dr. Moliner 50, E-46.100 Burjassot, Spain,*
*E-mail: Rafael.Guardiola@uv.es, Inaki.Moliner@uv.es*

The purpose of these lectures is to present a pedagogical and illustrative, rather than exhaustive, review on the Coupled Cluster Method and some of its applications. The basic formalism to calculate the ground state energy of a general many-body system is presented, and a brief outline of its extensions to analyze other observables is also given. The main SUB(n) or CCn hierarchy of approximations is discussed, and some applications to systems interacting through a Coulomb potential are shown. The problems posed by the presence of a repulsive hard-core in the potential are briefly considered, and the HCSUB(n) approximation is formulated. To deal with finite systems such as nuclei or helium droplets, a translationally invariant approximation is formulated. This is first considered for a bosonic system, both in configuration and coordinate representation, and it is then generalized to the fermionic case. Results for the $^4$He and $^{16}$O atomic nuclei are presented, including only pair correlations in the wave function, and using semirealistic interactions of V4 form, and the V6 part of several realistic interactions. It is shown that the problems related to the existence of a very strong repulsion at short distances may be overcome by using a mixed description which incorporates scalar Jastrow correlations. This hybrid J-TICCn description, alternative to the HCSUB(n) approximation, is applied to the study of finite helium clusters, including up to three-body correlations in a Configuration-Interaction scheme. Detailed results are presented and compared with stochastic results for clusters of isotope $^4$He, both for ground and excited states. Ground state properties of $^3$He droplets are also discussed. To complement the pedagogical purpose of the lectures, some exercises are proposed to help the readers deepen their understanding of the Coupled Cluster approach to the study of many-body systems.

## 1. Introduction

The Coupled Cluster method (CCM) or exp(S) formalism is a general method to solve the many-body problem. It was invented over forty years ago by Coester and Kümmel [1,2] to deal with doubly magic atomic nuclei. Their idea was to develop a non-perturbative method to calculate the ground-state energy of a nucleus using realistic nucleon-nucleon interactions. The method was rediscovered nearly ten years later by Čížek and Paldus, [3–6] who introduced it into quantum chemistry showing that it is able to eliminate the size-extensivity problem. They performed the first applications of CCM to calculate the ground-state structure of atoms and molecules with a considerable success: the CCM is nowadays recognized as being one of the most effective and accurate schemes for atoms and small molecules. The success of CCM in quantum chemistry produced a revival of the method in nuclear physics, and Kümmel and his group at Bochum started a systematic program [7–9] to apply the method to atomic nuclei. They succeeded in calculating the ground state energy of such nuclei as $^4$He, $^{16}$O and $^{40}$Ca, using realistic NN interactions and the then available three-body interactions. Although the CCM formalism was initially developed to calculate the energy of the ground-state of a closed-shell nucleus, further developments made possible to deal with excited states, the dynamical response to external perturbations or expectation values of arbitrary operators, for example. The CCM is a general microscopic formulation of the many-body problem, which works for both bosonic and fermionic systems regardless the type and range of the interaction, and it has proved to be a useful tool in a wide variety of fields besides quantum chemistry and nuclear physics. We refer the reader to several comprehensive reviews of the method, both within physics [10,11] and chemistry. [12,13]

The purpose of these lectures is to present a pedagogical and illustrative, rather than exhaustive, review. Thus the guiding line will be the coupled-cluster (CC) structure of the wave function to calculate the ground state energy of a system of interacting particles. We shall compare systematically the results obtained with those provided by other techniques, aiming to stress the virtues and flaws of the approximations made. The present lectures aims at complementing the above mentioned reviews, as well as the pedagogical material presented at the previous summer school by R.F. Bishop. [14] Some recent applications to such finite systems as light nuclei and helium drops are presented.

The next Section presents the basic ingredients of the formalism, namely the general CC form of the wave function and the CC equations. The Configuration Interaction method (CIM) will also be discussed in order to facilitate the comprehension of the advantages related to the peculiar exponential form of the wave function. The reference state is the starting point to construct the wave function, and possible choices will be analyzed. Extensions of the method will be briefly outlined, in particular the CC parameterization of the bra state. For practical purposes some approximations are to be considered, and this is the subject of Section 3. In particular, the so-called SUB(2) approximation, which is a description of the wave function containing only pair correlations, will be analyzed from a diagrammatic point of view to investigate the physical contents of this approximation. The problems related to the existence of an infinite core in the potential will be shown, as well as its corresponding hard-core approximation. Section 4 is devoted to applications to finite nuclei considering an increasing complexity of the nucleon-nucleon interaction used, from a purely scalar potential to more complicated forms with state-dependent terms. Only pair correlations will be considered into the wave function description, but translational invariance will be imposed from the very beginning. The equations will be deduced for bosons first, both in configuration and coordinate representation, and afterwards, they will generalized to the fermionic case. Results for the nuclei $^4$He and $^{16}$O, employing semirealistic and realistic interactions of the V4 and V6 forms respectively, will be given and analyzed. The use of realistic interactions will show immediately the flaws of this approximation. One possible remedy is the use of a correlated reference state, which will be presented at an exploratory level for finite nuclei. Such an exploratory study in nuclei will be then exported to helium drops, the subject of Section 5. Two- and three-body correlations will be included, together with Jastrow correlations. Results for finite drops formed by either $^4$He or $^3$He will be given and compared to the results obtained with other techniques.

## 2. The Coupled Cluster formalism

The basic idea of the CCM is to decompose the wave function of a quantum many-body system in terms of excitations of clusters containing a finite number of particles. The amplitudes of these excitations provide a measure of the $n$-body correlations. In this section we shall show how the true wave function is represented within the CCM scheme as an exponentiated correlation operator, acting on a reference state. This CCM structure is compared with the apparently simpler wave function provided by the Configuration Interaction Method (CIM), and the advantages of the $\exp(S)$ form are stressed. The Schrödinger equation is transformed into a set of coupled non-linear equations for the unknown amplitudes. The Section is terminated by a discussion on the possible choices for the reference state, and the extensions of the method to compute other observables besides the ground-state energy.

### 2.1. *The exponential form of the wave function*

We shall give here some simple arguments [15] leading to an exponential form of the wave function. To be specific we shall consider system of $N$ fermions, for which we wish to obtain the general structure of its ground-state wave function $|\Psi\rangle$. Clearly, an important component of $|\Psi\rangle$ is the Slater determinant of occupied states

$$|\Phi\rangle = a^+_{\nu_N} \dots a^+_{\nu_1} |0\rangle \quad , \tag{2.1}$$

where $|0\rangle$ refers to the vacuum, and the set of fermionic creation operators $\{a^+_{\nu_i}\}$ refers to the occupied single-particle states defining the Fermi sea of the system. For the sake of simplicity we shall assume that the lowest energy state is represented by a single Slater determinant, *i. e.* the fermions are closing major shells. Here and henceforth, the labels $\nu, \mu, \lambda$ indicate occupied states in the reference state $|\Phi\rangle$; the labels $\sigma, \rho, \tau$ indicate unoccupied states, and the labels $\alpha, \beta$ refer to both.

Let us assume for the moment that our system is an atomic nucleus, and that the reference state $|\Phi\rangle$ can be determined by a self-consistent variational Hartree-Fock calculation using an effective interaction. The underlying physical picture is well-known: each nucleon moves independently in the mean potential produced by the averaged interaction with the other nucleons. However, even if $|\Phi\rangle$ is an important part of the true wave function $|\Psi\rangle$, it is clear that one given nucleon cannot move independently of the other nucleons. We shall now construct systematically the components of $|\Psi\rangle$ describing these dynamical correlations.

As a first step, we may imagine that the mutual interaction of two particles can lift them out of the Fermi sea. This process can be represented by means of an operator

$$S_2 = \frac{1}{(2!)^2} \sum_{\rho_1\rho_2\nu_1\nu_2} \langle \rho_1\rho_2|S_2|\nu_1\nu_2\rangle_A \; a^+_{\rho_1} a^+_{\rho_2} a_{\nu_2} a_{\nu_1} \quad , \tag{2.2}$$

where $\langle \rho_1\rho_2|S_2|\nu_1\nu_2\rangle_A$ is a *c*-number amplitude, and the subindex $A$ means anti-symmetrization with respect to the exchange of indices $\nu_1$ and $\nu_2$. The action of this operator on $|\Phi\rangle$ produces a component $S_2|\Phi\rangle$ of the wave function in which a pair of particles is above the Fermi sea while the remaining fermions are in their previous states. It may also occur that $m$ pairs do this independently, and this process will be described by applying $m$-times the operator $S_2$ to $|\Phi\rangle$, including a statistical weight $1/m!$ to avoid an overcounting of the same pair. We are thus led to a contribution

$$\frac{1}{m!} {S_2}^m |\Phi\rangle \tag{2.3}$$

to the wave function. In this component, there are $2m$ particles out of the Fermi sea and $2m$ holes in the Fermi sea. The total amplitude for these processes of independent-pair excitations can be written as

$$\sum_{m=0}^{\infty} \frac{1}{m!} {S_2}^m |\Phi\rangle \equiv \mathrm{e}^{S_2} |\Phi\rangle \quad . \tag{2.4}$$

Obviously, this series is finite for a system with $N$ particles because the contribution of the terms with $m > N/2$ is null; it is nevertheless convenient to write the series as an exponential. One may also imagine that three particles are simultaneously excited. This process can be described by a contribution $S_3|\Phi\rangle$ to the wavefunction

$$S_3 = \frac{1}{(3!)^2} \sum_{\rho_1\rho_2\rho_3\nu_1\nu_2\nu_3} \langle \rho_1\rho_2\rho_3|S_3|\nu_1\nu_2\nu_3\rangle_A \; a^+_{\rho_1} a^+_{\rho_2} a^+_{\rho_3} a_{\nu_3} a_{\nu_2} a_{\nu_1} \quad , \tag{2.5}$$

and the simultaneous excitation of $p$ independent triplets leads to a contribution

$$\frac{1}{p!} {S_3}^p |\Phi\rangle \tag{2.6}$$

to the wavefunction. Moreover, we can imagine the excitation of $m$ pairs and $p$ triplets, all independently from each other, which would be described by the amplitude

$$\frac{1}{p!m!} {S_3}^p {S_2}^m |\Phi\rangle \quad . \tag{2.7}$$

Since these pair and triplet excitations are independent processes, the operators $S_2$ and $S_3$ commute, and summing over all these pair and triplet independent excitations we find

$$\mathrm{e}^{S_2+S_3}|\Phi\rangle \quad . \tag{2.8}$$

Proceeding systematically in the same way with the excitation of clusters of four, five, ... $N$ particles we arrive to

$$\mathrm{e}^{S_2+S_3+\ldots+S_N}|\Phi\rangle \quad . \tag{2.9}$$

Finally, it could also happen that in the interaction of any cluster of particles, only one nucleon is excited out of the Fermi sea. Such a process can be described by means of an operator $S_1$ which acts on $|\Phi\rangle$ to produce a single p-h pair in the Fermi sea, and the repeated action of this operator will contribute to the wavefunction with $\mathrm{e}^{S_1}|\Phi\rangle$. Analogously, processes where the interaction of a cluster of $n$ particles produce the excitation of a number $m < n$ of particles are considered by the terms referred to operator $S_m$.

In conclusion, we find the following general expression for the wavefunction

$$|\Psi\rangle = \mathrm{e}^{S}|\Phi\rangle \quad , \tag{2.10}$$

where the operator $S$ is given by

$$S = \sum_{n=1}^{N} S_n \quad , \tag{2.11}$$

and contains a sum of one-, two-, three-, ..., N-particle operators with the following structure

$$S_n = \frac{1}{(n!)^2} \sum_{\rho_1\ldots\rho_n\nu_1\ldots\nu_n} \langle\rho_1\ldots\rho_n|S_n|\nu_1\ldots\nu_n\rangle_A a^+_{\rho_1}\ldots a^+_{\rho_n} a_{\nu_n}\ldots a_{\nu_1} \quad . \tag{2.12}$$

This $S_n$ operator creates $n$ holes in the Fermi sea and $n$ particles above the Fermi sea. A convenient factor $1/(n!)^2$ is included, but it should be changed if dealing with bosons. This structure of the wave function is a natural decomposition in terms of excitations of clusters, where each independent excitation has been considered with its appropriate multiplicity.

It is instructive to write down formally the wave function in coordinate space. [7] The single-particle wave functions will be denoted by $\langle x|\nu\rangle$, where $x$ stands for the coordinate and intrinsic degrees of freedom of the particle. The $n$p-$n$h amplitudes will be written as

$$\langle x_1x_2\ldots x_n|S_n|\nu_1\nu_2\ldots\nu_n\rangle = \sum_{\rho_1\ldots\rho_n} \langle x_1|\rho_1\rangle\ldots\langle x_n|\rho_n\rangle\langle\rho_1\ldots\rho_n|S_n|\nu_1\ldots\nu_n\rangle \quad . \tag{2.13}$$

These amplitudes are antisymmetric both in the particle and state labels, $x_i$ and $\nu_i$. Assuming that $S_1 = 0$, a choice which will be justified later, the wave function may be written as

$$\begin{aligned}\langle x_1 \dots x_n|\Psi\rangle = \langle x_1 \dots x_n|\Phi\rangle &+ \mathcal{A}\left[\langle x_1x_2|S_2|\nu_1\nu_2\rangle\langle x_3|\nu_3\rangle \dots \langle x_n|\nu_n\rangle\right] \\ &+\mathcal{A}\left[\langle x_1x_2x_3|S_3|\nu_1\nu_2\nu_3\rangle\langle x_4|\nu_4\rangle \dots \langle x_n|\nu_n\rangle\right] + \dots \\ &+\mathcal{A}\left[\langle x_1x_2|S_2|\nu_1\nu_2\rangle\langle x_3x_4|S_2|\nu_3\nu_4\rangle\langle x_5|\nu_5\rangle \dots \langle x_n|\nu_n\rangle\right] \\ &+ \dots \quad , \end{aligned} \tag{2.14}$$

where the symbol $\mathcal{A}$ means antisymmetrization with respect to particle and single-particle state labels. This result leads to a unique definition of correlations as those parts of the wave function which cannot be described in terms of single-particle wave functions. Note that each term of this series is a Slater determinant, and it is well known that the set of Slater determinants built up from a complete set of single-particle states is a complete set of many-particle states. Therefore the exponential representation may be regarded as an expansion of the true wave function in a complete orthonormal basis, and everything is exact up to now.

### 2.2. *The Configuration Interaction Method (CIM)*

The use of an exponential structure for the wave function might seem an artificial complication, and we shall try now to clarify its interest and meaning. To this respect it is worth recalling that the exponential form appeared for the first time in the fifties in connection with the Gell-Mann and Low theorem in quantum field theory, and also in connection with the Goldstone linked-cluster theorem in time-independent perturbation theory.

An apparently simpler alternative to obtain the $np$-$nh$ content of the full wave function is to consider directly its projection onto a generic $np$-$nh$ state

$$a^+_{\rho_1} \dots a^+_{\rho_n} a_{\nu_n} \dots a_{\nu_1}|\Phi\rangle \quad , \tag{2.15}$$

and write the wave function as a linear combination of all possible $np$-$nh$ states

$$|\Psi\rangle = \left(1 + \sum_{n=1}^{N} F_n\right)|\Phi\rangle \quad . \tag{2.16}$$

The operator $F_n$ is given by

$$F_n = \frac{1}{(n!)^2} \sum_{\rho_1 \dots \rho_n \nu_1 \dots \nu_n} \langle \rho_1 \dots \rho_n|F_n|\nu_1 \dots \nu_n\rangle a^+_{\rho_1} \dots a^+_{\rho_n} a_{\nu_n} \dots a_{\nu_1} \quad , \tag{2.17}$$

where $\langle \rho_1 \dots \rho_n|F_n|\nu_1 \dots \nu_n\rangle$ represents the probability amplitude for the excitation of fermions from states $(\nu_1 \dots \nu_n)$ in $|\Phi\rangle$ to states $(\rho_1 \dots \rho_n)$ out of the reference state, with all other particles in the occupied states $(\nu_{n+1} \dots \nu_N)$. This is known as the Configuration Interaction Method (CIM), widely used in quantum chemistry. Basically, it is a generalized shell model where, for practical purposes, one chooses some subspace of the full many-body Hilbert space characterized by a number $N$ of different configurations. Both the amplitudes and the eigenenergies are obtained by solving a secular problem, *i. e.* diagonalizing the $N \times N$ Hamiltonian matrix. One clear advantage of CIM is that at each level of approximation, characterized by the dimension $N$ of the subspace, rigorous variational upper bounds are obtained for the eigenvalues. Moreover, uniform convergence is guaranteed as $N$ increases, although the convergence rate depends on the specific interaction. If no approximation is made the two descriptions, CCM and CIM, coincide because both are formally exact. However, some truncations are to be made in the wave function expansion for practical purposes, and the equivalence is lost. For example, we have already mentioned that CIM results are always variational at any level of truncation. This characteristic is not shared by CCM, as it will be clear in the next section, but there are other significant differences between CCM and CIM, which we shall consider now.

The $S_n$ and $F_n$ operators are obviously related to each other since both are describing the $n$p-$n$h content of the true wave function with respect to the same reference state. It is easy to establish that

$$
\begin{aligned}
F_1 &= S_1 \quad , \\
F_2 &= S_2 + \frac{1}{2}S_1^2 \quad , \\
F_3 &= S_3 + S_2 S_1 + \frac{1}{6}S_1^3 \quad , \\
F_4 &= S_4 + S_3 S_1 + \frac{1}{2}S_2^2 + \frac{1}{2}S_2 S_1^2 + \frac{1}{24}S_1^4 \quad , \ldots
\end{aligned}
\tag{2.18}
$$

Whereas the $S_n$ operators are linked by construction, we can see in these expressions that the $F_n$ operators contain a mixture of both linked and unlinked operators. For instance even if we assume $S_1 = 0$, as it is usual, we see in $F_4$ a genuine linked operator $S_4$ and an unlinked operator $S_2^2$. It is instructive to consider the $n$-particle subsystem amplitude

$$
\langle x_1 \ldots x_n | \Psi_n | \nu_1 \ldots \nu_n \rangle_A \equiv \langle \Phi | a^+_{\nu_1} \ldots a^+_{\nu_n} a(x_1) \ldots a(x_n) | \Psi \rangle \quad , \tag{2.19}
$$

which is the amplitude for $(N-n)$ particles being in occupied states $(\nu_1 \ldots \nu_n)$ of the reference state $|\Phi\rangle$, whereas the remaining $1 \ldots n$ particles are allowed to move to positions $x_1 \ldots x_n$. These amplitudes can also be written in terms of $S_n$ (or $F_n$) amplitudes

$$
\begin{aligned}
\langle x_1 | \Psi_1 | \nu_1 \rangle &= \langle x_1 | \nu_1 \rangle + \langle x_1 | S_1 | \nu_1 \rangle \\
\langle x_1 x_2 | \Psi_2 | \nu_1 \nu_2 \rangle_A &= \{ \langle x_1 | \Psi_1 | \nu_1 \rangle \langle x_2 | \Psi_1 | \nu_2 \rangle \}_A + \langle x_1 x_2 | S_2 | \nu_1 \nu_2 \rangle_A \\
\langle x_1 x_2 x_3 | \Psi_3 | \nu_1 \nu_2 \nu_3 \rangle_A &= \{ \langle x_1 | \Psi_1 | \nu_1 \rangle \langle x_2 | \Psi_1 | \nu_2 \rangle \langle x_3 | \Psi_1 | \nu_3 \rangle \}_A \\
&\quad + S_{123} \{ \langle x_1 x_2 | S_2 | \nu_1 \nu_2 \rangle \langle x_3 | \Psi_1 | \nu_3 \rangle \}_A \\
&\quad + \langle x_1 x_2 x_3 | S_3 | \nu_1 \nu_2 \nu_3 \rangle_A \quad ,
\end{aligned}
\tag{2.20}
$$

etc, where the subscript $A$ indicates antisymmetrization of the ket states ($\nu_i$ indices), and the symbol $S_{123}$ generates the sum of all cyclic permutations of the labels $i$.

Consider the case with $n = 2$. These equations show that the amplitude $\langle x_1x_2|S_2|\nu_1\nu_2\rangle_A$ is that part of the two-body amplitude $\langle x_1x_2|\Psi_2|\nu_1\nu_2\rangle_A$ which cannot be described by one-body amplitudes $\langle x_1|\Psi_1|\nu_1\rangle$, and this is the reason why they can be properly called two-body correlation amplitudes. A similar interpretation does not exist for amplitudes of operators $F_n$. For the case $n = 2$ the correspondent expression is

$$\begin{aligned}\langle x_1x_2|\Psi_2|\nu_1\nu_2\rangle_A = &\langle x_1x_2|\nu_1\nu_2\rangle_A + S_{12}\{\langle x_1|\nu_1\rangle\langle\rho_2|\Psi_1|\nu_2\rangle\rangle\}_A \\ &+\langle x_1x_2|F_2|\nu_1\nu_2\rangle_A \quad .\end{aligned} \tag{2.21}$$

In general, the amplitude of the operator $S_n$ is that part of the amplitude of $\Psi_n$ which cannot be described in terms of one-, two-, ..., $n$-body amplitudes. The amplitudes of $S_n$ describe thus correlations occurring within an $n$-body cluster, and consequently they seem to be of much more physical interest than the amplitudes of $F_n$.

We shall refer now to the so-called size-extensivity problem. Suppose we separate the system spatially into two subsystems $A$ and $B$, with $N_A$ and $N_B$ particles respectively. We assume that as the separation increases ($r_{AB} \to \infty$) the two subsystems cease to interact with each other, so that the Hamiltonian becomes the sum of two pieces, $H_A$ and $H_B$, mutually commuting

$$H \longrightarrow H_A + H_B \quad , \quad [H_A, H_B] = 0 \quad , \tag{2.22}$$

and the energy becomes additively separable

$$E \longrightarrow E_A + E_B \quad , \tag{2.23}$$

while the wave function becomes multiplicatively separable

$$|\Psi\rangle \longrightarrow |\Psi^{(A)}\rangle \otimes |\Psi^{(B)}\rangle \quad . \tag{2.24}$$

A system is said to be size-extensive when it presents the above separability properties, in which case the energy has the (correct) linear dependence with the number of constituents. In the limit of infinite distance between the two subsystems $A$ and $B$ the wave functions factorize as

$$\begin{aligned}&\text{CIM} \to (1 + F^{(A)})(1 + F^{(B)})|\Phi^{(A)}\rangle|\Phi^{(B)}\rangle \\ &\text{CCM} \to \exp(S^{(A)})\exp(S^{(B)})|\Phi^{(A)}\rangle|\Phi^{(B)}\rangle \quad .\end{aligned} \tag{2.25}$$

Suppose now that we have made the usual approximation of neglecting cluster excitation operators involving more than $n$ particles (which will be explained in detail in the next section). In this approximation it is clear that

$$\begin{aligned}&(1 + F_1 + F_2 + \ldots F_n)|\Phi\rangle \neq \\ &(1 + F_1^{(A)} + F_2^{(A)} + \ldots F_n^{(A)})(1 + F_1^{(B)} + F_2^{(B)} + \ldots F_n^{(B)})|\Phi^{(A)}\rangle|\Phi^{(B)}\rangle \quad ,\end{aligned} \tag{2.26}$$

*i. e.* the initial correlation operator for the whole system is not equal to the product of the separate correlation operators for the subsystems. The separated parameterization in the second line would require up to $2n$ excitations, going beyond the assumed approximation. On the other hand, the separability relation is guaranteed by the exponential parameterization independently of the truncations made in the correlation operators.

In conclusion, the apparent artificiality of the exponential form results in a precise physical meaning of the clusters, which are described by linked operators, thus satisfying the Goldstone theorem. Moreover, this form guarantees the size-extensivity property no matter how the cluster correlation operator $S$ is truncated.

### 2.3. *The Coupled Cluster equations*

The next step now is to determine the amplitudes $S_n$ or $\Psi_n$, by means of which the exact ground state energy can be obtained. These amplitudes are to be determined by solving the Schrödinger equation

$$He^S|\Phi\rangle = E\,e^S|\Phi\rangle. \tag{2.27}$$

As the intermediate normalization condition $\langle\Phi|\Psi\rangle = 1$ holds, the ground state energy of the system may be written as

$$E = \langle\Phi|H|\Psi\rangle = \langle\Phi|He^S|\Phi\rangle \quad . \tag{2.28}$$

It is easy to see that for a system of nucleons interacting through a two-body interaction, only the amplitudes of $\Psi_1$ and $\Psi_2$ are necessary to determine the ground state energy

$$E = \sum_{\nu}\langle\nu|T\Psi_1|\nu\rangle + \frac{1}{2}\sum_{\nu\nu'}\langle\nu\nu'|V\Psi_2|\nu\nu'\rangle_A \quad , \tag{2.29}$$

and complete sets of states may be freely inserted to obtain the explicit expression for the energy

$$E = \sum_{\nu\alpha}\langle\nu|T|\alpha\rangle\langle\alpha|\Psi_1|\nu\rangle + \frac{1}{2}\sum_{\nu\nu'}\sum_{\alpha\alpha'}\langle\nu\nu'|V|\alpha\alpha'\rangle\langle\alpha\alpha'|\Psi_2|\nu\nu'\rangle_A \quad . \tag{2.30}$$

A set of equations for the corresponding amplitudes can be obtained as follows. Eq. (2.28) results from the projection of the Schrödinger equation onto the bra state $\langle\Phi|$, which contains 0p-0h. One can also project the Schrödinger equation onto the series of bra states

$$\begin{gathered}\langle\Phi|a^+_{\nu_1}a_{\rho_1}\\ \langle\Phi|a^+_{\nu_1}a^+_{\nu_2}a_{\rho_2}a_{\rho_1}\\ \langle\Phi|a^+_{\nu_1}a^+_{\nu_2}a^+_{\nu_3}a_{\rho_3}a_{\rho_2}a_{\rho_1}\\ \dots\ ,\end{gathered}$$

representing the projection onto 1p-1h, 2p-2h, 3p-3h, etc. It is convenient to introduce now a simplified notation in order to write down compact equations. The cluster operator $S$ will be written as

$$S = \sum_{I\neq 0} S_I C_I^+ \quad , \tag{2.31}$$

where $C_I^+$ stands for a characteristic particle-hole operator $a^+_{\rho_1} \dots a^+_{\rho_I} a_{\nu_I} \dots a_{\nu_1}$. The equations for the correlation amplitudes will thus be obtained projecting the Schrödinger equation onto a generic particle-hole excitation represented by the bra state $\langle\Phi|C_I$, and one obtains the set of coupled equations

$$\langle\Phi|C_I H \mathrm{e}^S|\Phi\rangle = E\,\langle\Phi|C_I \mathrm{e}^S|\Phi\rangle \quad , \quad I \neq 0 \quad , \tag{2.32}$$

to determine the amplitudes $\mathcal{S}_I$. The energy $E$ appears explicitly in these equations and may be troublesome for extended systems, since it is a macroscopic quantity. In fact, there are cancellations between unlinked terms in both sides of these equations, so that the problem is only apparent. However it is worth treating this point with more detail. As an alternative way to obtain equations for the amplitudes $\mathcal{S}_I$ we can pre-multiply the Schrödinger equation by $\mathrm{e}^{-S}$ to get the equivalent equation

$$\mathrm{e}^{-S} H \mathrm{e}^S|\Phi\rangle = E\,|\Phi\rangle \quad . \tag{2.33}$$

The projection onto the 0p-0h state $\langle\Phi|$ gives

$$E = \langle\Phi|\mathrm{e}^{-S} H \mathrm{e}^S|\Phi\rangle \equiv \langle\Phi|H \mathrm{e}^S|\Phi\rangle \quad . \tag{2.34}$$

To obtain the last expression we have used $\langle\Phi|\mathrm{e}^{-S} = \langle\Phi|$, which is true because $S$ contains only creation operators above the Fermi sea and gives a null result when acting on the bra reference state. We have thus obtained the same expression for the energy as before. Projecting now onto the bra state $\langle\Phi|C_I$ we get the following set of equations

$$\langle\Phi|C_I \mathrm{e}^{-S} H \mathrm{e}^S|\Phi\rangle = 0 \quad , \quad I \neq 0 \quad , \tag{2.35}$$

where no macroscopic quantity like the energy appears.

***EXERCISE 1.*** *Probe that both sets of CCM equations (2.32) and (2.35) are equivalent.*
*Hint: Use the identities*

$$1 = \mathrm{e}^S \mathrm{e}^{-S} = \sum_J \mathrm{e}^S C_J^\dagger |\Phi\rangle\langle\Phi| C_J \mathrm{e}^{-S} \quad .$$

*The first one holds without need of defining an inverse operator for $S$ or $\mathrm{e}^S$. In the second one, the decomposition of the unity in terms of the complete set of states defined by $C_J$ has been used, and the sum includes thus the term $J = 0$.*

Another reason to use this second set of equations for the amplitudes $\mathcal{S}$ is related to the fact that the operator $\mathrm{e}^{-S} H \mathrm{e}^S$ can be written in terms of a series of nested commutators

$$\mathrm{e}^{-S} H \mathrm{e}^S = H + [H,S] + \frac{1}{2!}[[H,S],S] + \frac{1}{3!}[[[H,S],S],S] + \dots \tag{2.36}$$

For a general operator $S$ this is an infinite series. However, the operator $S$ is formed by creation operators $C_I^+$ which commute with each other, $[C_I^+, C_J^+] = 0$. This implies that the only non-vanishing terms in the series come from the contractions between the Hamiltonian and $S$, and consequently the series is finite. In particular, for a two-body Hamiltonian the series terminates at fourth order in $S$. Such non-vanishing components form the so-called elementary links, and the operator $\mathrm{e}^{-S}H\mathrm{e}^{S}$ is a fully linked operator. This property holds even when approximations to $S$ are made.

Assuming a two-body Hamiltonian, the projection onto the bra state $\langle\Phi|C_I$ can be expressed in a compact form

$$\begin{aligned}&\langle\Phi|C_I H|\Phi\rangle + \sum_J \mathcal{S}_J\langle\Phi|C_I[H,C_J]|\Phi\rangle \\ &+\frac{1}{2!}\sum_{JK}\mathcal{S}_J\mathcal{S}_K\langle\Phi|C_I[[H,C_J],C_K]|\Phi\rangle \\ &+\frac{1}{3!}\sum_{JKL}\mathcal{S}_J\mathcal{S}_K\mathcal{S}_L\langle\Phi|C_I[[[H,C_J],C_K],C_L]|\Phi\rangle \\ &+\frac{1}{4!}\sum_{JKLM}\mathcal{S}_J\mathcal{S}_K\mathcal{S}_L\mathcal{S}_M\langle\Phi|C_I[[[[H,C_J],C_K],C_L],C_M]|\Phi\rangle = 0 \quad .\end{aligned} \tag{2.37}$$

This expression corresponds to a fourth-order coupled set of non-linear equations in the amplitudes $\mathcal{S}_I$ order.

## 2.4. *The reference state*

In order to describe many-body correlations we always need a reference state with respect to which the correlations are defined. It is worth keeping in mind that in the exact CCM formulation the choice of the reference state is somehow irrelevant, as it is just the starting point to span the full Hilbert space of the many-body system. However, approximations are to be made for practical applications, so that one only considers a Hilbert subspace and the choice of the reference state can be of paramount importance to obtain converged and reliable ground-state energies. For many purposes the CCM reference state $|\Phi\rangle$ may be considered as a generalized vacuum state, so that the states in the full many-body Hilbert space can be expressed in terms of many-body creation operators acting on $|\Phi\rangle$. In some cases the choice of $|\Phi\rangle$ may be determined by simple physical ideas, but it is important to realize that this choice may be not unique.

For a system of $N$ fermions, such as finite nuclei or drops formed by $^3$He atoms, we have already chosen the reference state as a Slater determinant of single-particle states taken from some complete single-particle basis $\{|\alpha_i\rangle\}$. These single-particle states are those of lowest energy in the case of the ground state. The single-fermion creation and destruction operators obey the usual anti-commutation relations

$$\begin{aligned}\{a_\alpha, a_\beta\} = \{a_\alpha^+, a_\beta^+\} = 0 \\ \{a_\alpha, a_\beta^+\} = \delta_{\alpha,\beta}\end{aligned} \tag{2.38}$$

and the single-particle states are obtained by the action of the operators on the vacuum state $|0\rangle$

$$|\alpha\rangle = a^+_\alpha |0\rangle \quad . \tag{2.39}$$

The reference state may be written in the second quantization form as

$$|\Phi\rangle = \prod_{i=1}^{N} a^+_{\nu_i} |0\rangle \quad , \tag{2.40}$$

and the question now is how to obtain the single-particle states. In the case of atomic systems the usual procedure is to solve the self-consistent Hartree-Fock equations. In other cases, as finite nuclei, one may also solve the Hartree-Fock equations using some effective interaction, which may or may not be consistent with the two-body interaction. Later on we shall simply use harmonic oscillator single-particle wave functions. In the case of a homogeneous system containing $N$ fermions in a box of volume $\Omega$, we are usually interested in the thermodynamic limit, where both $N$ and $\Omega$ tend to infinity, but particle density $\rho = N/\Omega$ remains constant. In such a case it is more convenient to use plane waves as single-particle states

$$|\alpha\rangle \rightarrow |\mathbf{k}, i\rangle \quad , \tag{2.41}$$

where $i$ labels the internal quantum numbers. Occupied states $|\nu\rangle$ have wave numbers such that $k < k_F$, where $k_F$ is the Fermi wave number. The choice of a Slater determinant for the reference state of a fermion system may seem obvious. However, it is worth noting that other choices may be more convenient depending on the type of physics one is interested in. Let us only mention the BCS state, which is also an Slater determinant but formed from quasi-particle states. These quasi-particle states are linear combinations of the previous particle and hole states.

We are going now to consider a system of bosons. An obvious choice for the reference state to be used in the description of the ground state is the Bose condensate, in which all $N$ particles condense into the lowest-energy single-particle state, denoted by $\alpha \equiv 0$. The normalized reference state is thus written as

$$|\Phi\rangle = \frac{1}{\sqrt{N!}} \left(b^+_0\right)^N |0\rangle \quad . \tag{2.42}$$

The single-particle wave functions are described by

$$|\alpha\rangle = b^+_\alpha |0\rangle \quad , \tag{2.43}$$

and the single-boson creation and annihilation operators satisfy the bosonic commutation relations

$$\begin{aligned} [b_\alpha, b_\beta] = [b^+_\alpha, b^+_\beta] = 0 \\ [b_\alpha, b^+_\beta] = \delta_{\alpha,\beta} \quad . \end{aligned} \tag{2.44}$$

We refer the reader to Ref. 14 for a discussion on useful choices of the reference state for other physical systems. Note that up to now we have considered that the reference state is an uncorrelated state, but it may be useful in some cases to incorporate also correlations to the reference state. In Sections 4 and 5 we shall employ such a correlated reference state, including Jastrow correlations. Obviously, for practical purposes and also to keep on the CCM spirit, these correlations have to be determined in advance, either by a previous variational calculation or by other physical considerations, in such a way that all the effort to determine correlations will be focused in the use of CCM.

### 2.5. *The Bra Ground state*

Up to this point we have only considered the ket ground state $|\Psi\rangle$. The ground state energy is extracted from the Schrödinger equation projecting upon $n$p-$n$h excited states. Once the correlation operator $S$ has been approximated, the CCM equations are of finite order in the amplitudes $\mathcal{S}_I$. We could calculate the ground-state energy as an expectation value

$$E = \frac{\langle\Psi|H|\Psi\rangle}{\langle\Psi|\Psi\rangle} = \frac{\langle\Phi|\mathrm{e}^{S^+}H\mathrm{e}^{S}|\Phi\rangle}{\langle\Phi|\mathrm{e}^{S^+}\mathrm{e}^{S}|\Phi\rangle} \quad , \tag{2.45}$$

but the resulting expression is in general of infinite order in the amplitudes $\mathcal{S}_I^+$ and $\mathcal{S}_I$, regardless how the operator $S$ is truncated. This is also true for the ground-state expectation value of any arbitrary operator and it has led to an *independent* CCM parameterization of the bra and the ket states.

In the so-called normal CCM (NCCM) the bra ground state is constructed from the bra reference state as

$$\langle\hat{\Psi}| = \langle\Phi|\hat{\mathcal{S}}\mathrm{e}^{-S} \quad , \tag{2.46}$$

with the new $\hat{\mathcal{S}}$ operator given by

$$\hat{\mathcal{S}} = 1 + \sum_{I\neq 0} \hat{\mathcal{S}}_I C_I \quad . \tag{2.47}$$

This parameterization preserves the explicit normalization

$$\langle\hat{\Psi}|\Psi\rangle = \langle\Phi|\Psi\rangle = \langle\Phi|\Phi\rangle = 1 \quad , \tag{2.48}$$

but if the set $\{I\}$ is truncated the exact hermiticity will be generally lost. The full set of *independent* amplitudes $\{\hat{\mathcal{S}}_I, \mathcal{S}_I\}$ provides a complete parameterization of the ground state. The expectation value of an arbitrary operator $Q$ may be written as

$$\langle Q\rangle = \langle\hat{\Psi}|Q|\Psi\rangle = \langle\Phi|\hat{\mathcal{S}}\mathrm{e}^{-S}Q\mathrm{e}^{S}|\Phi\rangle = q(\hat{\mathcal{S}}_I, \mathcal{S}_I) \quad , \tag{2.49}$$

which is fully linked even though the operator $\hat{\mathcal{S}}$ contains unlinked pieces.

The NCCM equations for $\hat{\mathcal{S}}$ are obtained from the Schrödinger equation written as $\langle\hat{\Psi}|H = \langle\hat{\Psi}|E$, and projecting onto states $C_I^+|\Phi\rangle$ to get

$$\langle\Phi|\hat{\mathcal{S}}\left(\mathrm{e}^{-S}H\mathrm{e}^{S} - E\right)C_I^+|\Phi\rangle = 0 \quad , \quad I \neq 0 \quad . \tag{2.50}$$

The ground state energy corresponds to the case $I = 0$, and is given by

$$E = \langle\Phi|\hat{S}e^{-S}He^{S}|\Phi\rangle = \frac{\langle\Psi|H|\Psi\rangle}{\langle\Psi|\Psi\rangle} = \frac{\langle\Phi|e^{S^+}He^{S}|\Phi\rangle}{\langle\Phi|e^{S+}e^{S}|\Phi\rangle} \quad . \tag{2.51}$$

The problem is solved in two steps. First the usual CCM equations are solved to obtain the amplitudes $\mathcal{S}$ and the ground state energy. Afterwards the equation for the amplitudes $\hat{\mathcal{S}}$ is solved. The interesting point with this formulation is that this bra parameterization is derivable from the Hellmann-Feynman theorem, and as a consequence the expectation value of an arbitrary operator may be calculated diagrammatically, from the same set of diagrams as for the energy, replacing the potential lines with operator lines.

The extended CCM (ECCM) is another formulation of CCM, which goes beyond the NCCM in the sense that it avoids the presence of unlinked pieces. We shall not going to discuss these extensions, and we refer the reader once again to Ref. 14 for a more detailed discussion on these points.

## 3. Approximation schemes

The equations presented up to now are exact, and some type of approximations have to be made to solve them in practical cases. A great advantage of the CCM is that the exponentiated form of the wave function has an easy physical interpretation. This readily lends itself to approximation schemes which are motivated by the physics of the problem rather than the mathematical convenience. One may argue, for instance, that for relatively low density systems it is not likely that more than a few particles are lifted simultaneously out of the Fermi sea. This results in the so-called SUB($n$) or CCn approximation hierarchy, in which all clusters with more than $n$ particles are neglected and the remaining $n$ coupled equations are to be solved. Note that disregarding the correlation operators $S_m$ with $m > n$ still leaves us with a rich wave function

$$|\Psi\rangle = e^{S_1+S_2+\ldots+S_n}|\Phi\rangle \quad , \tag{3.1}$$

where few-body effects, high-excitations, collective effects, etc, are taken into account. In this Section we shall discuss in some detail this approximation at the SUB(2) level. We shall also consider the case of an interaction having a hard-core at short distances. The approximation SUB(2) has to be adapted to this situation, resulting in the so-called HCSUB(n) approximation.

### 3.1. *The SUB(n) or CCn approximation*

Let us start by explicitly writing the two first projections following Eq. (2.32). The 1p-1h projection generates the equation:

$$\langle\alpha_1|T\Psi_1|\nu_1\rangle + \sum_{\nu}\langle\alpha_1\nu|V\Psi_2|\nu_1\nu\rangle_A + \sum_{\nu}\langle\alpha_1\nu|T(2)S_2|\nu_1\nu\rangle_A$$

$$+\frac{1}{2}\sum_{\nu,\nu'}\langle\alpha_1\nu\nu'|V(23)\chi_3(1;23)|\nu_1\nu\nu'\rangle_A$$
$$=\sum_{\nu}h_{\nu\nu_1}\langle\alpha_1|\Psi_1|\nu\rangle \quad , \tag{3.2}$$

where we have defined

$$\chi_3(1;23)=S_2(13)\Psi_1(2)+S_2(12)\Psi_1(3)+S_3(123) \quad . \tag{3.3}$$

We have used the convention that the integers in parentheses after a particular operator refer that operator to those quantum labels in the associated bra or ket in the corresponding numerical positions (counting from the left). This rule has to be used in connection with the insertion of unit operators, as in Eq. (2.30), to derive explicit expressions. The quantity $h_{\nu_1\nu_2}$ appearing in these equations is defined as

$$h_{\nu_1\nu_2}=\langle\nu_1|T\Psi_1|\nu_2\rangle+\sum_{\nu'}\langle\nu_1\nu'|V\Psi_2|\nu_2\nu'\rangle_A \quad . \tag{3.4}$$

Using the same conventions, the 2p-2h projection generates the equation

$$\begin{aligned}
&\langle\rho_1\rho_2|[T(1)+T(2)]S_2|\nu_1\nu_2\rangle_A\\
&+\langle\rho_1\rho_2|V\Psi_2|\nu_1\nu_2\rangle_A+\frac{1}{2}\sum_{\nu\nu'}\langle\rho_1\rho_2|S_2|\nu\nu'\rangle_A\langle\nu\nu'|V\Psi_2|\nu_1\nu_2\rangle_A\\
&+\sum_{\nu}\langle\rho_1\rho_2\nu|T(3)S_3|\nu_1\nu_2\nu\rangle_A\\
&+\sum_{\nu}\langle\rho_1\rho_2\nu|[V(13)\chi_3(2;13)+V(23)\chi_3(1;23)]|\nu_1\nu_2\nu\rangle_A\\
&+\frac{1}{2}\sum_{\nu\nu'}\langle\rho_1\rho_2\nu\nu'|V(34)\chi_4(12;34)|\nu_1\nu_2\nu\nu'\rangle_A\\
&=\sum_{\nu}(h_{\nu\nu_1}\langle\rho_1\rho_2|S_2|\nu\nu_2\rangle_A+h_{\nu\nu_2}\langle\rho_1\rho_2|S_2|\nu_1\nu\rangle_A \quad ,
\end{aligned} \tag{3.5}$$

where we have defined

$$\begin{aligned}
\chi_4(12;34) &= S_2(13)S_2(24)+S_2(14)S_2(23)\\
&+S_3(123)\Psi_1(4)+S_3(124)\Psi_1(3)+S_4(1234) \quad .
\end{aligned} \tag{3.6}$$

Without need of writing down the remaining equations for $n>2$, one may easily realize that the equation resulting from the $n$p-$n$h projection contains all the $S_m$ amplitudes up to $m=n+2$. The Schrödinger equation is converted into a set of $N$ non-linear coupled equations for the amplitudes, and for the moment not too much progress has been obtained for practical purposes.

Let us consider now the SUB(1) approximation. To determine the amplitudes of $S_1$ we have to solve Eq. (3.2) with $S_2=0$, $S_3=0$,

$$\langle\alpha_1|T\Psi_1|\nu_1\rangle+\sum_{\nu}\langle\alpha_1\nu|V\Psi_1\Psi_1|\nu_1\nu\rangle_A=\sum_{\nu}h_{\nu\nu_1}\langle\alpha_1|\psi_1|\nu\rangle \quad , \tag{3.7}$$

which is the Hartree-Fock equation with a self-consistent potential given by

$$\langle\alpha_1|U\Psi_1|\nu_1\rangle = \langle\alpha_1\nu|V\Psi_1\Psi_1|\nu_1\nu\rangle_A \quad , \tag{3.8}$$

and this should not be a surprise. The structure of the wave function in the SUB(1) approximation is simply

$$|\Psi\rangle = \mathrm{e}^{S_1}|\Phi\rangle \quad , \tag{3.9}$$

with

$$S_1 = \sum_{\rho,\nu}\langle\rho|S_1|\nu\rangle a_\rho^+ a_\nu \quad . \tag{3.10}$$

The Thouless theorem ensures that $|\Psi\rangle$ is nothing more than a general Slater determinant non-orthogonal to the reference state $|\Phi\rangle$. This can justify the use of $S_1 = 0$ in specific systems, using either a Hartree-Fock wave function as the reference state or a differently obtained reference state, for instance the one constructed from the harmonic oscillator, as we shall show in Sections 4 and 5. As we have mentioned before, an important test is that the results should be ideally independent of the reference state.

Let us assume $S_1 = 0$, and consider the SUB(2) approximation. The equation for amplitudes of $S_2$ is given by Eq. (3.5) with $S_3 = S_4 = 0$. We obtain

$$\begin{aligned}
&-\langle\rho_1\rho_2|[T(1)+T(2)]S_2|\nu_1\nu_2\rangle_A \\
&+\sum_\nu(h_{\nu\nu_1}\langle\rho_1\rho_2|S_2|\nu\nu_2\rangle_A + h_{\nu\nu_2}\langle\rho_1\rho_2|S_2|\nu_1\nu\rangle_A \\
&= +\langle\rho_1\rho_2|V\Psi_2|\nu_1\nu_2\rangle_A + \frac{1}{2}\sum_{\nu\nu'}\langle\rho_1\rho_2|S_2|\nu\nu'\rangle_A\langle\nu\nu'|V\Psi_2|\nu_1\nu_2\rangle_A \\
&+\sum_\nu\langle\rho_1\rho_2\nu|[V(13)(S_2(12)+S_2(23))+V(23)(S_2(12)+S_2(13))]|\nu_1\nu_2\nu\rangle_A \\
&+\frac{1}{2}\sum_{\nu\nu'}\langle\rho_1\rho_2\nu\nu'|V(34)(S_2(13)S_2(24)+S_2(14)S_2(23))|\nu_1\nu_2\nu\nu'\rangle_A \quad .
\end{aligned} \tag{3.11}$$

The equation seems to be very complicated when one inserts unit operators as sums over a complete set of states to obtain the explicit and workable expressions. But we should keep in mind that this equation contains all two-body effects, and it is exact apart from having neglected higher order correlations. However, the underlying structure of the equation is rather simple and it has a clear physical interpretation. It is interesting to have a different look into this equation and represent it in a diagrammatic form. All the terms of this equation containing interactions are represented diagrammatically in Fig. 1. In the diagrams, particle and hole states (*i. e.* unoccupied and occupied single-particle states in the reference state) are represented by lines with arrows pointing upwards and downwards respectively. The diagrams are thus time-ordered in the sense of Goldstone perturbation theory. The interaction is represented by a dashed line, and the amplitude $S_2$ by an oval. We do not enter into the rules for signs and factors associated to each diagram, because

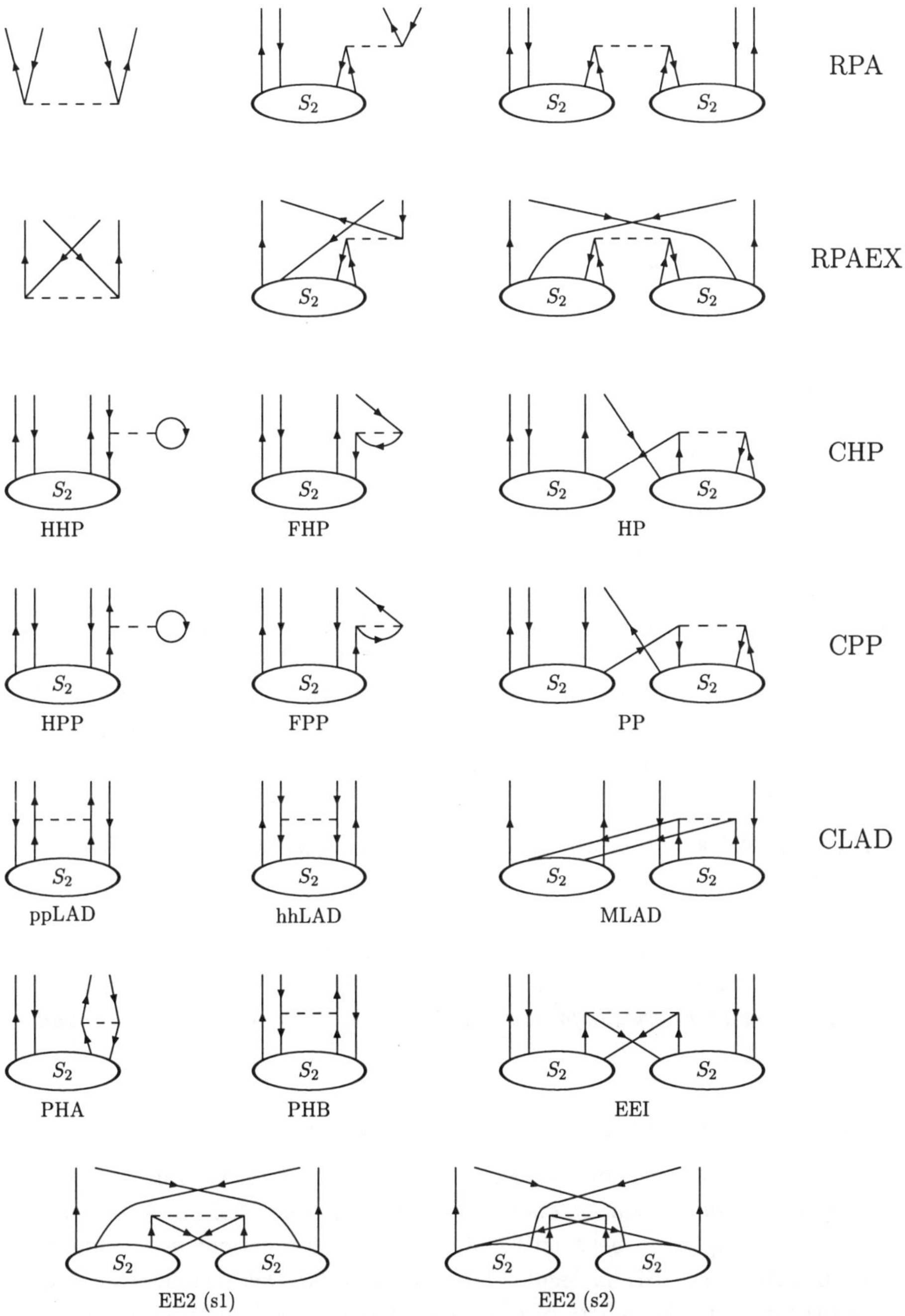


Fig. 1. Diagrammatic expression of terms containing the potential in the SUB(2) equations.

we only wish to make explicit the physical insight of the different contributions to this equation. To be precise, we may consider that this SUB(2) equation is used to describe an uniform system of fermions with spin one half, electrons for instance. The form of the equation can be cast in the following way: put on the left hand side the amplitude $S_2$ multiplied by a factor depending on the relevant moments, which comes from the kinetic energy contribution, and put on the right hand side all the terms containing the two-body interaction. The latter are represented in Fig. 1.

We can easily identify the diagrams associated to the ring approximation (labelled RPA) and their counterpart including the exchange diagrams (labelled RPAE). It seems clear that they generate the full set of ring or bubble diagrams, hence the choice of name for these diagrams. The terms labelled CHP, which have an insertion into one of the two hole lines of the amplitude $S_2$, generate the complete-hole-potential terms. We distinguish the Hartree-hole-potential (HHP), the Fock-hole-potential (FHP) and the hole-potential (HP) terms. Analogously, the terms labelled CPP correspond to the complete-particle-potential insertion, and they are referred to as the Hartree-particle-potential (HPP), the Fock-particle-potential (FPP) and the particle-potential (PP). The complete sum of ladder diagrams for two-particle and two-hole scattering in the medium are generated by the diagrams labelled CLAD, with particle-particle ladder (ppLAD), hole-hole ladder (hhLAD) and a mixed ladder (MLAD) term. Next we have the two kinds of particle-hole ladder (PHA, PHB) terms. Finally, the terms EE1 and EE2 are extra exchange terms required by Fermi statistics. In conclusion, in the SUB(2) approximation, which contains all two-body effects and is exact apart from having neglected higher order correlations, we are summing an enormous class of diagrams in terms of perturbation theory.

***EXERCISE 2.*** *Calculate the ground-state energy, in successive SUB(n) approximations, of the one-dimensional harmonic oscillator plus a quartic term, described by the Hamiltonian* $H = \frac{1}{2}\left(p^2 + x^2\right) + \lambda x^4$. *You are asked to: i) choose a reference state, ii) construct the SUB(n) ansatz for the ground state wave function, iii) derive the equations for the amplitudes, iv) solve them numerically for a given value of* $\lambda$, *and v) calculate the ground state energy.*

### 3.2. *Applications to Coulomb interacting systems*

The CCM has become one of the most effective schemes to perform accurate calculations of electron correlation effects in atoms and small molecules. In 1972, the first *ab initio* comparison between CCM and CI wave functions was made in the particular case of the BH molecule. [6] Interestingly, it was demonstrated that a large contribution to the energy comes from genuine three-body excitations, which cannot be decomposed in terms of one- and two-body excitation components. In this Section we shall illustrate with a few selected examples the power of CCM to deal with systems interacting through a Coulomb potential. To learn more details

Table 1. Total CI energies (in a.u.) of several molecules at equilibrium ($R_e$) and in stretched geometries ($1.5R_e$ and $2R_e$). Also indicated are the errors (in millihartree) relative to CI obtained in succesive SUB(n) approximations.

| Molecule | Geometry | FCI | SUB(2) | SUB(3) | SUB(4) |
|---|---|---|---|---|---|
| BH | $R_e$ | -25.227627 | 1.79 | 0.068 | 0.001 |
| | $1.5R_e$ | -25.175976 | 2.64 | 0.026 | 0.000 |
| | $2.0R_e$ | -25.127350 | 5.05 | -0.091 | 0.001 |
| HF | $R_e$ | -100.250969 | 3.01 | 0.266 | 0.018 |
| | $1.5R_e$ | -100.160395 | 5.10 | 0.646 | 0.041 |
| | $2.0R_e$ | -100.081107 | 10.18 | 1.125 | 0.062 |
| $H_2O$ | $R_e$ | -76.256624 | 4.12 | 0.531 | 0.023 |
| | $1.5R_e$ | -76.071405 | 10.16 | 1.784 | 0.139 |
| | $2.0R_e$ | -75.952269 | 21.40 | -2.472 | -0.015 |

about the accomplishments in this field, the interested reader should refer to the bibliography, in particular Refs. 12 and 13.

In computational quantum chemistry quantities such as molecular ground state energies or equilibrium geometries must be calculated with an accuracy greater than the experimental one. The energies which really matter in this type of calculations are not absolute energies, but relative differences as for instance the dissociation energy, which is a difference between ground state energies. The typical average energy to break an atomic bound ranges in the interval $10^2 - 10^3$ kcal/mol. Consequently for a theoretical calculation to be useful its accuracy should be of the order of 1 kcal/mol, or 1 millihartree. Typically, a HF calculation is able to give more than 99% of the energy, but this is far from the chemical accuracy because the errors are of the order of a few tenths of hartree. CCM calculations generally give, as compared with other *ab initio* methods, more acurate energies and properties of molecules with comparable or less computational effort.

In Table 1 are displayed several results for the molecules BH, HF, and $H_2O$ molecule, for which full CI results exist. The reader should be aware that unfortunately there is no universally accepted notation in the CCM world. Our succesive n=2, 3, 4, ... approximations correspond respectively to CCSD, CCSDT, CCSDTQ, ... approximations in quantum chemistry, where S, D, T, Q, ... indicate single, double, triple, quadruple, ... excitations. Moreover, in quantum chemistry CC2, CC3, etc, means specific approximations to CCSD, CCSDT, etc. Along these lectures we are employing SUB(n) and CCn as equivalent. All the data in Table 1 have been adapted from Ref. 16. The results refer also to distorted geometries, increasing the bond lengths 1.5 and 2 times the equilibrium value $R_e$. The differences among the FCI results and the first SUB(n) approximations are also displayed. It is worth noting that in some cases these differences are negative, and consequently CCM energies are lower than the upper variational CI bounds to the ground state energies. One should keep in mind that the SUB(n) approximation is not variational. SUB(4), or CCSDTQ, is the highest level of theory applied in the hierarchy of

Table 2. Experimental ionization potentials (IP) and excitation energies (EE) of $Sc^+$, and errors of the results obtained in a multireference CCM calculation. Energies are in eV, and errors in meV.

| | State | | Expt. | CC error |
|---|---|---|---|---|
| IP | $3d4s$ | $^3D_1$ | 12.800 | 29 |
| EE | $3d4s$ | $^3D_2$ | 0.009 | 1 |
| | | $^3D_3$ | 0.022 | 1 |
| | | $^1D_2$ | 0.315 | -161 |
| | $3d^2$ | $^3F_2$ | 0.596 | 17 |
| | | $^3F_3$ | 0.606 | 18 |
| | | $^3F_4$ | 0.618 | 22 |
| | | $^1D_2$ | 1.357 | 34 |
| | $4s^2$ | $^1S_0$ | 1.455 | -56 |
| | $3d^2$ | $^3P_0$ | 1.497 | 27 |
| | | $^3P_1$ | 1.500 | 28 |
| | | $^3P_2$ | 1.507 | 29 |
| | | $^1G_4$ | 1.768 | 14 |
| | $3d4p$ | $^1D_2$ | 3.234 | 3 |
| | | $^3F_2$ | 3.403 | 10 |
| | | $^3F_3$ | 3.422 | 12 |
| | | $^3F_4$ | 3.452 | 14 |

SUB(n) approximations, and it is worth recalling that it provides exact results for four-electron clusters.

To calculate atomic and molecular open-shell states with a good approximation, the reference state cannot be restricted to a single Slater determinant, and a multireference state must be constructed considering the most important determinants. There are many varieties of multireference CCM; a review may be found in Ref. 17. Usually, one starts from a closed-shell reference state, and by adding and/or removing electrons to it, the states of interest are reached. The CC excitation operator is written in terms of operators of the form $S^{(m,n)}$, where the superscripts $(m,n)$ refer to the states reached by removing $m$ and adding $n$ electrons to the reference. In practice, multireference states with $m+n \leq 2$ are usually considered. The equations are first solved for $(m,n) = (0,0)$, then moving to (0,1) or (1,0), and continuing as needed. As an example, in Table 2 the ionization potential and excitation energies of $Sc^+$ calculated by Kaldor [18] using a multireference state in an active space which includes $4s$, $3d$, and $4p$ atomic states. There is an excelent agreement with experiment, with an average error of the order of 20 meV.

A huge range of atomic and molecular properties have been accurately calculated within CCM, for a variety of systems with up to about 100 active electrons. For the heavier atoms, a relativistic CCM has been developed [19] in the framework of the four-component Dirac-Coulomb-Breit Hamiltonian, which includes relativistic terms trhough second order in the fine-structure constant. At the SUB(2) level, a large number of heavy-atom states have been calculated with unprecedent accuracy, and reliable predictions for superheavy elements have been made.

Let us now turn to a many-electron system. The "jellium" is one of the theoretically most well studied quantum many-body systems. It was invented to model the electrons in a metal, replacing the ionic lattice by a uniform positive charge, which acts as an inert neutralizing background. In momentum representation the interaction between two electrons is given by

$$V(q) = \frac{4\pi e^2}{\Omega q^2} (1 - \delta_{q,0}) \quad , \tag{3.12}$$

where $\Omega$ is the normalization volume, and $\delta$ is the Kronecker delta. Note that the value $q = 0$ is supressed, as it is neutralized by the positive charge. [20]

The so-called one-component Coulomb plasma is the thermodynamic limit of a system of $N$ fermions or bosons interacting via this two-body potential. The density is expressed in terms of the only independent dimensionless coupling constant $r_s = r_0/a_0$ which characterizes the system expressing the average interparticle distance in units of the Bohr radius. One writes the density as

$$\rho = \frac{3}{4\pi r_s^3 a_0^3} \quad . \tag{3.13}$$

Between the high-density or weak-coupling limit ($r_s \to 0$), and the low-density or strong-coupling limit ($r_s \to \infty$) there is the interesting metallic region ($1 \leq r_s \leq 5$). It is convenient to define the ground state energy per particle in units of Rydbergs ($e^2/2a_0$), and separate this energy in two parts: the Hartree-Fock contribution and the correlation energy. Similarly to what has been mentioned for quantum chemistry calculations, a great deal of work is required to calculate accurately the correlation energy.

In Table 3 are displayed correlation energies of the uncorrelated electron gas, obtained from various calculations. The benchmark results come from the GFMC calculations of Ceperley and Adler. [21] In increasing degree of complexity there are the RPA results, [22] the full CC2 [22] results and CC[4] results. [23] The number 4 in brackets indicates that contributions from three- and four-body correlations have been included in an approximate way. It can be seen that over the entire metallic density regime, the CCM results for the correlation energy differ less than 1% respect to the GFMC results. This accuracy has never been equalled in any other *ab initio* calculation of this system.

In conclusion, the use of CCM techniques to calculate properties of systems interacting with a Coulomb interaction leads to very satisfactory results. To increase the accuracy of the calculations one simply considers successively one-body, two-body, etc, correlations until the desired convergence is reached. Very often, including correlations up to two-body level suffices to obtain accurate results. In Sections 4 and 5 we shall consider other systems, as finite nuclei and helium droplets, in which the two-body interaction typically is characterized by a strong repulsion at

Table 3. Correlation energies (in mRy) for the unpolarized electron gas. Adapted from Ref. 22.

| | 1 | 2 | 3 | 4 | 5 | 10 | 20 |
|---|---|---|---|---|---|---|---|
| RPA | 158 | 124 | 106 | 93.6 | 84.9 | 61.3 | 42.8 |
| CC2 | 123 | 91.7 | 75.1 | 64.4 | 56.8 | | |
| CC[4] | 122 | 90.4 | 73.8 | 63.4 | 56.0 | 37.0 | 23.6 |
| GFMC | 121 | 90.2 | | | 56.3 | 37.22 | 23.0 |

short distances. The extreme case of an infinite repulsion is analyzed in the next subsection.

### 3.3. *The HCSUB(n) approximation*

The hierarchy of SUB(n) approximations has proven to work very well in atoms, molecules or the Coulomb plasma. However, if the potential contains a hard core this approximation simply does not work and leads to divergent results. Consider the SUB(1) approximation. As one can see from Eq. (3.7), the description of the wave function in terms of solely single particle wave functions is not longer possible, and one has to include at least two-body correlations. We shall see that the two-body correlations must screen the hard core in a particular way. Consider the full Eq. (3.2) to determine the amplitudes $S_1$. The second term in the left hand side

$$\langle x_1x_2|V\Psi_2|\nu_1\nu\rangle_A \tag{3.14}$$

must be finite in the hard-core region, so that one must have

$$\langle x_1x_2|\Psi_2|\nu_1\nu\rangle_A = \{\langle x_1|\Psi_1|\nu_1\rangle\langle x_2|\Psi_1|\nu\rangle\}_A + \langle x_1x_2|S_2|\nu_1\nu\rangle_A \equiv 0 \quad . \tag{3.15}$$

This means that inside the hard core $S_2$ is determined completely by the single-particle wave functions. Analogously, in order the term in the second line of Eq. (3.2)

$$\langle\alpha_1\nu\nu'|V(23)\chi_3(1;23)|\nu_1\nu\nu'\rangle \tag{3.16}$$

remains finite, the function $\chi_3$ should vanish inside the hard-core region. Consequently, $S_3$ inside the hard-core region is determined by $S_2$ and $S_1$, and we see that there is no choice: in the case of a hard core, one has to neglect all this combination altogether even if we have selected the SUB(2) approximation. Similar considerations apply to higher order amplitudes. Looking at the equation which determines the $S_2$ amplitude we see that the combination $\chi_4$ should vanish inside the hard-core region, and so on and so forth. This is the so-called Hard-Core SUB(n) or HCSUB(n) approximation: [7,10] not only we neglect amplitudes $S_m$ with $m > n$, but also the amplitudes $S_n$ contained in the combinations $\chi_n$. For instance, the complete HCSUB(2) equations are:

$$\langle\alpha_1|T\Psi_1|\nu_1\rangle + \sum_\nu\langle\alpha_1\nu|V\Psi_2|\nu_1\nu\rangle_A + \sum_\nu\langle\alpha_1\nu|T(2)S_2|\nu_1\nu\rangle_A = \sum_\nu h_{\nu\nu_1}\langle\alpha_1|\Psi_1|\nu\rangle \quad , \tag{3.17}$$

and

$$\begin{aligned}
&\langle \rho_1\rho_2|[T(1)+T(2)]S_2|\nu_1\nu_2\rangle_A \\
&+\langle \rho_1\rho_2|V\Psi_2|\nu_1\nu_2\rangle_A + \frac{1}{2}\sum_{\nu\nu'}\langle \rho_1\rho_2|S_2|\nu\nu'\rangle_A\langle \nu\nu'|V\Psi_2|\nu_1\nu_2\rangle_A \\
&= \sum_{\nu}(h_{\nu\nu_1}\langle \rho_1\rho_2|S_2|\nu\nu_2\rangle_A + h_{\nu\nu_2}\langle \rho_1\rho_2|S_2|\nu_1\nu\rangle_A \quad ,
\end{aligned} \tag{3.18}$$

with the condition that $\Psi_2$ vanish inside the hard-core region.

***EXERCISE 3.*** *Show that for a homogeneous system of bosons the HCSUB(2) equation in coordinate representation may be written as*

$$-\Delta S_2(r) + V(r)(1+S_2(r)) = 4\epsilon S_2(r) \quad ,$$

*where*

$$\epsilon = \frac{1}{2}\rho\int \mathrm{d}\mathbf{r}V(r)\,(1+S_2(r))$$

*is the energy per particle, $\rho$ is the particle density, and the units used are such that $\hbar^2/m = 1$.*
*Suppose the bosons interact through a hard sphere potential: $V(r) = \infty$ if $r < a$, and $V(r) = 0$ otherwise. In the low-density limit, the energy per particle is given by the series* [20]

$$\epsilon = 2\pi\rho a\left[1 + \frac{128}{15}\left(\frac{\rho a^3}{\pi}\right)^{1/2} + \ldots\right] \quad .$$

*Calculate the HCSUB(2) energy $\epsilon$ in the low-density limit.*

## 4. Applications to light nuclei

In this section we shall consider the nucleon-nucleon interaction. For our purposes, the nuclear interaction is more complicated than the Coulomb one, for two main reasons. On the one side, it contains a complicated operatorial structure including central, tensor, spin-orbit, and other terms depending on the interaction considered. On the other side, it contains a repulsion at short distances, which may be very strong in some cases. The operatorial structure represents a great piece of technical work, but in principle does not affect the reliability of the approximations made within CCM. However, as we shall see with some detail, a very strong short-range repulsion implies that the simplest CC2 approximation does not provide reliable results for the ground state energy.

We have to mention some previous calculations on finite nuclei. The Bochum group results [10] refer to the $^4$He, $^{16}$O and $^{40}$Ca atomic nuclei, using several realistic interactions and a series of CC2, CC3 and CC4 approximations. The calculated ground state energies and charge radius differ appreciably from the experimental results, and we could say "as it should". Indeed it is well known that two-body interactions alone do not suffice, and three-body interactions are needed to reproduce

Table 4. Binding energies ($E$) and charge radii ($r$) of $^{16}$O from two different configuration-space CC3 calculations. The Reid soft-core results are extracted from Ref. 10, and the Argonne AV14 and AV18 results from Ref. 25. The last column shows the experimental values.

| | Reid | AV14 | AV18 | Experiment |
|---|---|---|---|---|
| $E$ (MeV) | 5.36 | 6.1 | 5.9 | 8.0 |
| $r$ (fm) | 2.57 | 2.86 | 2.81 | 2.73 |

the experimental data up to the alpha particle. A schematic three-body interaction, deduced from two-pion exchange, was also introduced in Ref. 10, obtaining energies in qualitative agreement with experiment. Their results for $^{16}$O using the Reid soft-core [24] interaction are showed in Table 4, together with some more recent results by Heisenberg and Mihaila. [25] The latter calculation included full two-body and approximated three-body correlations and was performed using the Argonne AV14 [26] and AV18 [27] interactions.

Let us anticipate that the two mentioned calculations are more sophisticated than the ones we are going to describe in the following pages, because they include three-body terms and make use of complete realistic interactions. On the other hand, the calculations showed here will be focused on two different aspects: the translational invariance and the formulation in coordinate space. Thus the calculations explained in this section cannot be directly compared to the ones that produced the results showed in Table 4.

Now we shall analyze systematically successive approximations to calculate the ground state energy of a nucleus, imposing translational invariance from the very beginning. In the previous sections nothing has been said about the center-of-mass (CM) motion of the system. A proper treatment of the CM requires a translationally invariant (TI) wave function, and in our case is convenient to require it separately for both the reference state and the CC correlation part. The CM motion is not a major problem when dealing with atomic or molecular systems, because in such cases the whole mass of the system is practically concentrated into the CM; thus placing the origin of coordinates in the CM does not represent a big mistake as a first step, and the small corrections to the energies can be safely treated in approximated ways. In contrast, a proper treatment of the CM motion is required when dealing with small systems of self-bound identical particles, as it is the case of finite atomic nuclei. A common approximation in nuclear physics is to estimate the intrinsic energy by using the Hamiltonian $H - T_{CM}$, where $T_{CM}$ is the kinetic energy operator of the CM, still using a wave function which is not TI. This procedure is exact only when dealing with an uncorrelated ground state wave function built up from harmonic oscillator (HO) single-particle wave functions. In such a case, the ground-state wave function of the $N$-particle system factors out as a product of a TI part and a CM wave function, and the energy associated to the CM can be unambiguously removed. However, dressing this uncorrelated state with correlations has the danger of spoiling the factorization property, unless the correlations are also TI. In a Jastrow variational approach this is simply accomplished by using two-body correlations which depend on inter-particle distances. When dealing with CC type wave functions, the method is not so straightforward, and the purpose of this chapter is to discuss in some detail this point. As a specific application we shall consider light atomic nuclei, and discuss at some length results for the $^4$He nucleus. The TI formulation of CCM was initiated some years ago [28,29] at the SUB(2) level of approximation applied to the $^4$He nucleus considered as a system of four bosons. Indeed, restricting ourselves to Wigner-type forces, the $^4$He ground-state wave function may be described as the product of a totally symmetric spatial part and a completely antisymmetric spin-isospin part. From this viewpoint, we may consider the $^4$He as a system of four bosons, and this is the reason why we shall start by considering the case of $N$ interacting bosons. Additional complications related to Fermi statistics, spin-dependence, tensor forces, etc, will be treated later.

### 4.1. *The TICC2 approximation in configuration representation*

As it was said before, a natural choice to deal properly with the CM is to build up the reference state $|\Phi\rangle$ from HO single-particle wave functions. The single-particle states can be defined through creation operators acting on the vacuum

$$|nlm\rangle = a^+_{nlm}|0\rangle \quad . \tag{4.1}$$

The reference state of a $N$-boson system is

$$|\Phi\rangle = \frac{1}{\sqrt{N!}}(a^+_{000})^N|0\rangle \quad , \tag{4.2}$$

and its coordinate representation can be written as

$$\langle \mathbf{r}_1, \ldots, \mathbf{r}_N|\Phi\rangle = \frac{\alpha^{3N/2}}{\pi^{3N/4}} \exp\left(-\frac{1}{2N}\alpha^2 \sum_{i<j} r^2_{ij}\right) \exp\left(-\frac{N}{2}\alpha^2 R^2\right) \quad , \tag{4.3}$$

where $\alpha = (m\omega/\hbar)^{1/2}$ is the HO inverse length parameter, and $\mathbf{R}$ is the CM coordinate. This shows explicitly the above mentioned feature that such a state is the product of a $1s$ state for the CM and an intrinsic wave function. This factorization property also holds for the ground-state wave function of a system of fermions described by a Slater determinant with HO single-particle wave functions. Although the wave function is not translationally invariant, this factorization property allows one to eliminate the CM motion unambiguously.

Let us now consider the correlations up to the CC2 or SUB(2) level, *i. e.* correlations described in terms of 1p-1h and 2p-2h, or equivalently in terms of one- and two-body operators. But one- and two-body operators are not in general TI, and suitable linear combinations of them must be considered for this purpose. For example, neither the one-body operators $\mathbf{r}_i$ and $\mathbf{r}_j$ nor the two-body operator $\mathbf{r}_i \cdot \mathbf{r}_j$

are TI. However, combining them in the form $(\mathbf{r}_i - \mathbf{r}_j)^2$ one gets a TI operator, which is also rotationally invariant. This example illustrates that at the SUB(2) level the correlations should be described by a new cluster operator $S^{(1,2)}$ which combines both $S_1$ and $S_2$ to preserve the TI.

The translational invariance may be imposed in the following way. To build the CC2 correlations we have to destroy a pair of particles in occupied states and create another pair in unoccupied states. In both cases, we shall recouple the product of single-particle HO states into sums of comparable products of HO states for the relative and center-of-mass of the pair. Afterwards we shall impose that the center-of-mass state of the destroyed pair of particles in the occupied subspace is the same as that of the created pair of particles in the unoccupied subspace. In such a way the CM remains untouched, and we shall deal only with correlations described in terms of relative coordinates. As we are using HO single-particle wave functions this is accomplished by using the Brody-Moshinsky brackets. [30]

For the case of a $N$ boson system, one can write the cluster operator as [28]

$$S^{(1,2)} = \sum_{n=1}^{\infty} \mathcal{S}(n) \sum_{n_1,n_2,l} \langle n0, 00, 0|n_1 l, n_2 l, 0\rangle \left[a^+_{n_1 l} \times a^+_{n_2 l}\right]^0_0 a^2_{000} \quad , \tag{4.4}$$

where $\langle nl, NL, \lambda|n_1 l_1, n_2 l_2, \lambda\rangle$ is a Brody-Moshinsky bracket. This expression shows that two bosons in occupied states (0,0,0) are destroyed, and a pair of particles is created in states $(n_i, l_i, m_i)$. The relative state of the pair has zero angular momentum (so that $l_1 = l_2$), and its center-of-mass is in a 1$s$ HO state. Note that the term with both $(n_1, l) = (0,0)$ and $(n_2, l) = (0,0)$ is excluded, as it simply reproduces the uncorrelated state $|\Phi\rangle$. However, the terms with either $(n_1, l) = (0,0), n_2 \neq 0$ or $n_1 \neq 0, (n_2, l) = (0,0)$ must be included. These terms give precisely the 1p-1h excitations in the admixture, required to obtain a TI combination. The main conclusion of this analysis is that the $S_1$ and $S_2$ clusters are not independent, and the CCM expansion must be based on appropriate combinations of 1p-1h and 2p-2h, which have been given in Eq. (4.4), where the $c$-number amplitudes to be determined are denoted by $\mathcal{S}(n)$. It is worth stressing the enormous simplification imposed by the TI, because the argument of $\mathcal{S}(n)$ is just a single number which counts the number of oscillator quanta (in fact, divided by 2) globally excited.

The TI of $S^{(1,2)}$ does not imply the invariance of the exponential. To show this explicitly, let us use the following simplified notation for the operator $S^{(1,2)}$:

$$S^{(1,2)} = s(p) a^+_p a^+_0 a^2_0 + s(p,q) a^+_p a^+_q a^2_0 \quad , \tag{4.5}$$

where $p \equiv (n_p, l_p, m_p)$, $0 \equiv (0,0,0)$, and a sum over repeated particle indices is to be assumed. The coefficients are defined as

$$s(p) = \delta(l_p, 0)\delta(m_p, 0) 2\langle n_p 0, 00, 0|n_p 0, 00, 0\rangle \mathcal{S}(n_p) \quad , \tag{4.6}$$

and

$$\begin{aligned} s(p,q) = {} & C(l_p, l_q, 0; m_p, m_q, 0) \\ & \langle n_p + n_q + l_p 0, 00, 0|n_p l_p, n_q l_p, 0\rangle \mathcal{S}(n_p + n_q + l_p) \quad . \end{aligned} \tag{4.7}$$

Note that the combination $s(p)a_p^+a_0^+ + s(p,q)a_p^+a_q^+$ is precisely the required mixture leading to a translationally invariant correlation. Consider now the square of $S^{(1,2)}$. After some manipulations with the creation and destruction operators, this square can be written in the form

$$\begin{aligned} S^{(1,2)}S^{(1,2)} &= [s(p_1)a_{p_1}^+a_0^+ + s(p_1,q_1)a_{p_1}^+a_{q_1}^+] \\ &\quad [s(p_2)a_{p_2}^+a_0^+ + s(p_2,q_2)a_{p_2}^+a_{q_2}^+]a_0^4 \\ &+ [s(p_1)a_{p_1}^+a_0^+ + s(p_1,q_1)a_{p_1}^+a_{q_1}^+]2s(p_2)a_{p_2}^+a_0^3 \quad . \end{aligned} \tag{4.8}$$

Clearly, the second term is not translationally invariant, and should be removed. The clue is given by observing that the first term is simply the normal-ordered form of the quadratic term. This device also applies for all the powers of $S^{(1,2)}$, so that the SUB(2) ansatz for the wave function can be finally written as

$$|\Psi\rangle = \ : \mathrm{e}^{S^{(1,2)}} : |\Phi\rangle \quad , \tag{4.9}$$

with the correlation operator given in Eq. (4.4). This is the so-called TICC2 ansatz.

To determine the correlation amplitudes $\mathcal{S}(n)$ implicit in $s(p)$ and $s(p,q)$, we must project the Schrödinger equation onto 1p-1h and 2p-2h states taking care that the CM motion is the same as in the reference state. The same arguments used to construct the operator $S^{(1,2)}$ apply to the construction of the excited state. Let us characterize the excitation by $2N_x$ oscillator quanta. The excited bra state having the CM in the $1s$ HO state can be written as

$$\langle\Phi|\left[C(m)(a_0^+)^2a_0a_m + C(m,n)(a_0^+)^2a_ma_n\right] \quad , \tag{4.10}$$

where the excitation amplitudes are given by

$$\begin{aligned} C(m) &= \delta(l_m,0)\delta(m_m,0)2\langle N_x0,00,0|n_m0,00,0\rangle \\ C(m,n) &= C(l_m,l_n,0;m_m,m_n,0)\langle N_x0,00,0|n_ml_m,n_nl_m,0\rangle \quad . \end{aligned} \tag{4.11}$$

Obviously, the indices entering these expressions must fulfill the selection rules implied by the Clebsh-Gordan and Brody-Moshinsky coefficients. The equation to determine the correlation amplitudes is obtained from the projection

$$\begin{aligned} &\langle\Phi|\left[C(m)(a_0^+)^2a_0a_m + C(m,n)(a_0^+)^2a_ma_n\right]H : \mathrm{e}^{S^{(1,2)}} : |\Phi\rangle = \\ &\quad E\langle\Phi|\left[C(m)(a_0^+)^2a_0a_m + C(m,n)(a_0^+)^2a_ma_n\right] : \mathrm{e}^{S^{(1,2)}} : |\Phi\rangle \quad , \end{aligned} \tag{4.12}$$

and Eq. (2.28) giving the energy reads now $E = \langle\Phi|H : \exp(S^{(1,2)}) : |\Phi\rangle$. The structure of the SUB(2) equations for the amplitudes can be written in the schematic form

$$\sum_{i_1,i_2,i_3,i_4} F(N_x,i_1,i_2,i_3,i_4)\mathcal{S}(i_1)\mathcal{S}(i_2)\mathcal{S}(i_3)\mathcal{S}(i_4) = 0 \quad , \tag{4.13}$$

with the convention $\mathcal{S}(0) = 1$. The function $F(N_x,i_1,i_2,i_3,i_4)$ contains combinations of matrix elements of kinetic and potential energies, and the independent term in this equation corresponds to $F(N_x,0,0,0,0)$.

One can see that the TICC2 method involves the solution of a coupled set of non-linear multinomial equations in the truncated set of coefficients $\{\mathcal{S}(n); n = 1, \ldots, N_{\rm max}\}$. We can solve this set of equations progressively, starting from a low value of $N_{\rm max}$, and using the obtained solutions as an input for the following ones. At the beginning, one can guess values for the relevant amplitudes, and then solve the system by a Raphson-Newton method to obtain the zeros of the system. The procedure stops when the calculated energy reaches the desired convergence. The use of a HO reference state introduces a free parameter $\alpha$, and a supplementary minimization of the ground-state energy could also be envisaged. It is interesting to consider a simplified approach, the so-called TICI2 approximation, where only the functions $F(N_x, i_1, 0, 0, 0)$ are retained in the general system of equations. CI stands for configuration interaction, and this approach corresponds to a shell-model like description restricted to a simple set of configurations. Its practical application may be equally carried out in a variational framework or in the general CC2 scheme. Note that TICC2 is a non-linear problem, and there is no a priori way to know whether all the solutions have been determined or if the obtained ones are physically relevant. A possible way to handle these non-linear equations is to introduce a quenching factor $q$, in such a way that when $q = 0$ one gets the linear or TICI2 equations, and $q = 1$ corresponds to the full equations. In this way, one can solve first the linear equations and then slowly increase the value of $q$ to get the non-linear solutions.

We shall show now the results obtained for systems containing 4 and 16 bosons with a mass equal to the nucleon mass, [32] and interacting through the Wigner part of the Afnan-Tang S3 nucleon-nucleon potential. [31] For the moment, the precise values of the ground-state energies are of no importance. We shall discuss now questions related to the convergence when the number $N_{\rm max}$ is increased, and to the non-linearity of the equations. Figs. 2 and 3 contain the TICI2 and TICC2 results for the above-mentioned systems. On the left we can see the ground-state energy as a function of the truncation in the excitation energy considered, and the graphs on the right show the correlation functions. We can see that the convergence trend is slow in both TICI2 and TICC2. Also, the results do not differ too much for the 4-boson system, which might indicate that the TICI2 is an accurate description. The proper answer is that this could be something specific to a system with 4 particles, and indeed this is what happens. Looking at the results for the system with 16 bosons we can see that the non-linear terms give here an appreciable contribution. In this new system the third- and fourth-order terms are no longer null, but we shall see later on that these terms are negligible and the contribution to the energy comes from the quadratic terms. Thus the fact that there is little difference between TICI2 and TICC2 for the four-particle system is no longer true for heavier systems.

The results are almost (but not completely) converged for a maximum excitation energy of $30\hbar\omega$ when using the S3 interaction, but one should go to even higher energies when considering the MTV interaction, which has a Yukawian core. This might indicate (although the systems treated are different) that the recent results of Heisenberg and Mihaila are not fully converged, since they use a realistic interaction

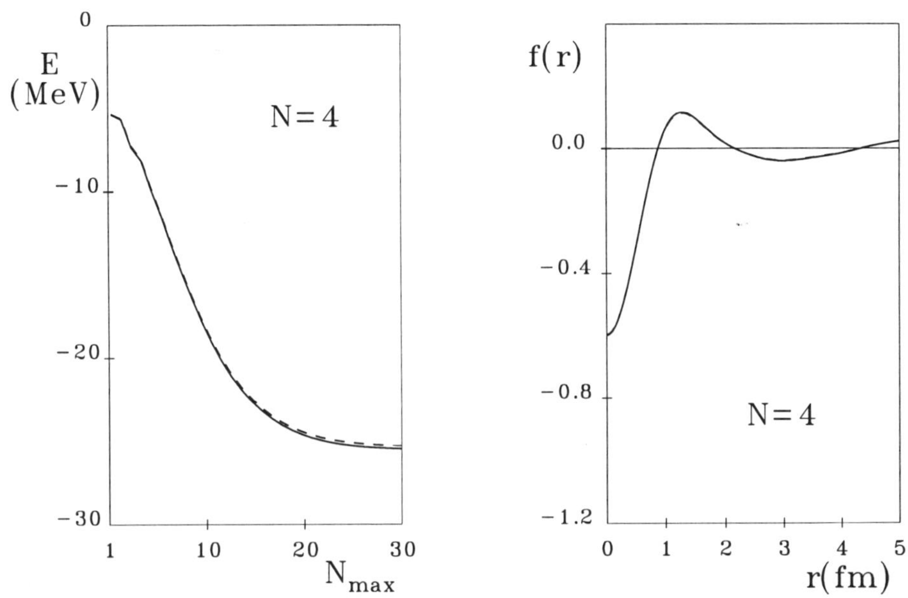


Fig. 2. TICI2 (dashed lines) and TICC2 (solid lines) ground-state energy as a function of the truncation in the excitation energy (left), and correlation functions (right) for a 4-boson system interacting trough the Wigner part of the S3 potential.

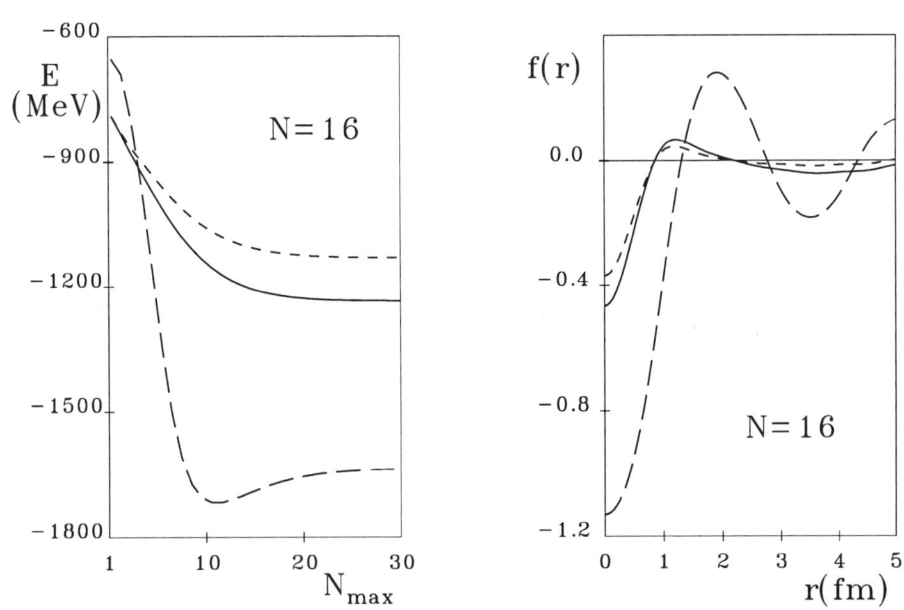


Fig. 3. Same as previous figure, but for a 16-boson system. An additional spurious solution is also shown (long-dashed lines).

containing a strong repulsion. In principle this would require very high excitations energies to reach convergence.

Fig. 3 shows, in addition to the TICI2 and TICC2 solutions, another solution found when solving the system of non-linear equations, which corresponds to the long-dashed lines. There are some comments regarding these three solutions. All of them tend to stability with increasing values of $N_{\rm max}$, but the long-dashed one reaches the minimum around $N_{\rm max} = 10$ and then the value raises again to reach a converged value which is greater. This is pretty odd since one expects to get a better value with increasing excitation energies, as it happens with the other two solutions. Besides, solving the equations in the quenching scheme, there is a connection between the solid line and the TICI2 line, and this in both directions. This means that starting from the short-dashed solution (with $q = 0$), and slowly increasing the quenching factor $q$, one arrives to the solid line when $q = 1$, and conversely. However, there is not a connection between the short-dashed and the long-dashed lines. Finally, the corresponding long-dashed correlation function has not the right behavior for a pair-correlation function. This means that the long-dashed line corresponds to a spurious solution, which is not physically relevant. As a supplementary check, we have also performed a stochastic calculation. Using the respective correlation functions, which are displayed in these three figures, the CCM wave function can be constructed and used to calculate the expectation value of the Hamiltonian. Only with the correlation function corresponding to the solid line we have obtained an energy in agreement with the value obtained by the direct solution of the non-linear problem. The value obtained with the long-dashed line strongly disagrees with the direct solution value, and has also an unacceptable variance.

The conclusion of this part is rather a warning: we should be very careful when trying to solve a non-linear problem. Regarding the problems related to the convergence trend, we shall immediately see that they can be avoided writing and solving the equations in coordinate representation.

### 4.2. *TICC2 in coordinate representation*

Actual calculations are simpler in coordinate space. Consider the coordinate representation of the two-particle operator $S^{(1,2)}$ given by Eq. (4.4), which is obtained by projecting $S^{(1,2)}|\Phi\rangle$ onto the symmetrized $N$-particle position bra vector $\langle \mathbf{r}_1, \ldots, \mathbf{r}_N|$. After some standard transformations, the coordinate representation of the correlation operator $S^{(1,2)}$ is shown to be

$$\langle \mathbf{r}_1 \ldots \mathbf{r}_N | S^{(1,2)} | \Phi \rangle = 2 \sum_{n=1}^{\infty} \mathcal{S}(n) \left[ \frac{2^n n!}{(2n+1)!!} \right]^{1/2} \sum_{i<j} L_n^{1/2} \left( \frac{1}{2} \alpha^2 r_{ij}^2 \right) \langle \mathbf{r}_1 \ldots \mathbf{r}_N | \Phi \rangle \quad . \tag{4.14}$$

Instead of dealing with the amplitudes $\mathcal{S}(n)$ we can define a function $f(r)$ using the completeness of the Laguerre polynomials

$$f(r) = 2\sum_{n=1}^{\infty} \mathcal{S}(n) \left[\frac{2^n n!}{(2n+1)!!}\right]^{1/2} L_n^{1/2}\left(\frac{1}{2}\alpha^2 r^2\right) \quad . \tag{4.15}$$

A similar procedure gives for the quadratic term

$$\langle \mathbf{r}_1, \ldots, \mathbf{r}_A | : S^{(1,2)} S^{(1,2)} : |\Phi\rangle = \sum_{i<j} f(r_{ij}) \sum_{k<l}{}' f(r_{kl}) \langle \mathbf{r}_1, \ldots, \mathbf{r}_A | \Phi\rangle \quad , \tag{4.16}$$

where the primed sum means that the indices $kl$ in the second summation are different from the indices $ij$ in the first one. In this manner one arrives at the following expression for the TICC2 ground-state wave function in coordinate representation

$$\begin{aligned} \Psi(\mathbf{r}_1, \ldots, \mathbf{r}_A) &= \langle \mathbf{r}_1, \ldots, \mathbf{r}_A | : \exp S^{(1,2)} : |\Phi\rangle \\ &= \Big(1 + \sum_{i<j} f(r_{ij}) + \frac{1}{2!}\sum_{i<j} f(r_{ij}) \sum_{k<l}{}' f(r_{kl}) \\ &\quad + \frac{1}{3!}\sum_{i<j} f(r_{ij}) \sum_{k<l}{}' f(r_{kl}) \sum_{m<n}{}'' f(r_{mn}) + \ldots \Big) \Phi(\mathbf{r}_1, \ldots, \mathbf{r}_A) \quad , \end{aligned} \tag{4.17}$$

with the convention that repeated indices in the products of primed sums are excluded. This equation shows clearly the fact that TICC2 only considers *independent* pair correlations; in general, CC$n$ deals only with independent $n$-body correlation operators. To this respect, it is interesting to compare this ansatz to a Jastrow correlation factor with a two-body correlation function. The expansion of such an exponential gives rise to a series similar to the TICC2 ansatz, but without restrictions in the sums. Therefore, the TICC2 ansatz looks like a Jastrow factor with independent two-body correlations. [50] Later on we shall see that in the case of fermions the general expression for the function $f$ contains an operatorial structure, and one should be aware that repeated indices imply non-trivial technical problems if these functions contain non-commuting operators.

The number of distinct terms in the TICC2 ansatz with $\nu = 1, 2, \ldots, A/2$ correlation functions is $A!/[2^\nu \nu!(A-2\nu)!]$, where the factor $1/\nu!$ prevents from multiple counting of the product of pair functions and can be replaced in practice by the further ordering restriction $i < k < m < \ldots$ in the sums of Eq. (4.17). An expectation value with this ansatz wave function will then consist of a large number of terms; however it will be seen that the number of different contributions may be reduced in practical calculations by classifying the terms according to their *topology*.

The intermediate normalization condition inherent to the CCM treatment

$$\langle \Phi | \Psi \rangle = 1 \tag{4.18}$$

implies that the correlation function is restricted by the relation

$$\langle \Phi | \Big( \sum_{i<j} f(r_{ij}) + \frac{1}{2!}\sum_{i<j} f(r_{ij}) \sum_{k<l}{}' f(r_{kl}) + \ldots \Big) | \Phi \rangle = 0 \quad . \tag{4.19}$$

In the case of a bosonic system this condition is immediately deduced from the vanishing of *each* of the terms in the summation, *i. e.*

$$\langle \phi_{000}(\mathbf{r}_1)\phi_{000}(\mathbf{r}_2)|f(r_{12})|\phi_{000}(\mathbf{r}_1)\phi_{000}(\mathbf{r}_2)\rangle = 0 \quad , \tag{4.20}$$

which is always satisfied. Indeed, the reference state is built up from ground-state HO single-particle wave functions

$$\phi_{000}(\mathbf{r}_i) = \left(\frac{\alpha}{\sqrt{\pi}}\right)^{3/2} \exp\left(-\tfrac{1}{2}\alpha^2 r_i^2\right) \quad , \tag{4.21}$$

and when introducing the relative $\mathbf{r}$ and centre-of-mass $\mathbf{R}$ coordinates of the pair $\{\mathbf{r}_1, \mathbf{r}_2\}$, it results the following chain of identities

$$\begin{aligned} &\langle \phi_{000}(\mathbf{r}_1)\phi_{000}(\mathbf{r}_2)|f(r_{12})|\phi_{000}(\mathbf{r}_1)\phi_{000}(\mathbf{r}_2)\rangle \\ &= \langle \phi_{000}(\mathbf{R})\phi_{000}(\mathbf{r})|f(r)|\phi_{000}(\mathbf{R})\phi_{000}(\mathbf{r})\rangle = \langle \phi_{000}(\mathbf{r})|f(r)|\phi_{000}(\mathbf{r})\rangle \\ &= 2\sum_{n=1}^{\infty} \mathcal{S}(n)\left(\frac{2^n n!}{(2n+1)!!}\right)^{1/2} \langle \phi_{000}(\mathbf{r})|L_n^{1/2}(\tfrac{1}{2}\alpha^2 r^2)|\phi_{000}(\mathbf{r})\rangle \quad , \end{aligned} \tag{4.22}$$

where the HO wave function for the relative coordinate corresponds to a parameter $\alpha/\sqrt{2}$. The last expression is identically zero due to the orthogonality properties of the associated Laguerre polynomials. The physical meaning is clear: the pair correlation function $f(r_{12})$ projects the product $\phi_{000}(\mathbf{r}_1)\phi_{000}(\mathbf{r}_2)$ of occupied orbitals onto the unoccupied subspace.

Once we have defined the correlated wave function, the typical CCM procedure allows one to find equations for the ground-state energy and the coefficients in the wave function by projecting the Schrödinger equation onto the reference state and all the other states that are present in the $S$ operator used to build the wave function. In fact, as we have already seen in previous sections, the traditional CCM way is to premultiply first the Schrödinger equation by $e^{-S}$ to take advantage of the linked cluster expansion of the $e^{-S}He^{S}$ operator. [10] This is not feasible in our case because we are working within the coordinate-space representation, and even if we were using the configuration-space representation, Eq. (4.9), the use of the normal ordering prevents us from using the linked cluster expansion. Hence the projection, both in the configuration-space and the coordinate-space calculations, is performed directly onto the Schrödinger equation, without any previous manipulation. If we project onto the reference state and onto a generic excitation with the same characteristics as $f$, and which is described by multiplying $|\Phi\rangle$ with a function $g$, we get

$$\langle \Phi|\, H\, |\Psi\rangle = E \quad , \tag{4.23}$$

$$\langle \Phi|\, gH\, |\Psi\rangle = E\,\langle \Phi|\, g\, |\Psi\rangle \quad , \tag{4.24}$$

where the non-relativistic Hamiltonian $H = T + V$ is the sum of the kinetic energy operator and a two-body nucleon-nucleon interaction

$$H = -\frac{\hbar}{2m}\sum_i \nabla_i{}^2 + \sum_{i<j} V_{ij} \quad . \tag{4.25}$$

Eqs. (4.23) and (4.24) provide us with either an integro-differential equation for arbitrary $g$ and $f$, or a set of non-linear coupled equations if we expand $g$ and $f$ using a suitable basis, Gaussians for example. Note that this is completely different from the expansion in Laguerre polynomials deduced in Eq. (4.15). This technique, largely employed in atomic and molecular problems, has proven to give an excellent representation of the correlation function $f$ if we consider negative as well as positive values for the Gaussian parameters. The Gaussian expansion is particularly adequate for the HO description of the single-particle states, since the required space integrals can be shown to be of the form of the product of an exponential of a positive-definite quadratic form with a polynomial. The equations can be most succinctly presented using a diagrammatic notation, [33,34] where a circle represents a particle, solid lines denote the correlation function $f$, dashed lines stand for the interaction, crosses indicate the action of the kinetic energy operator, and wavy lines denote the function $g$. The equation corresponding to the projection onto a generic excitation takes the form [35]

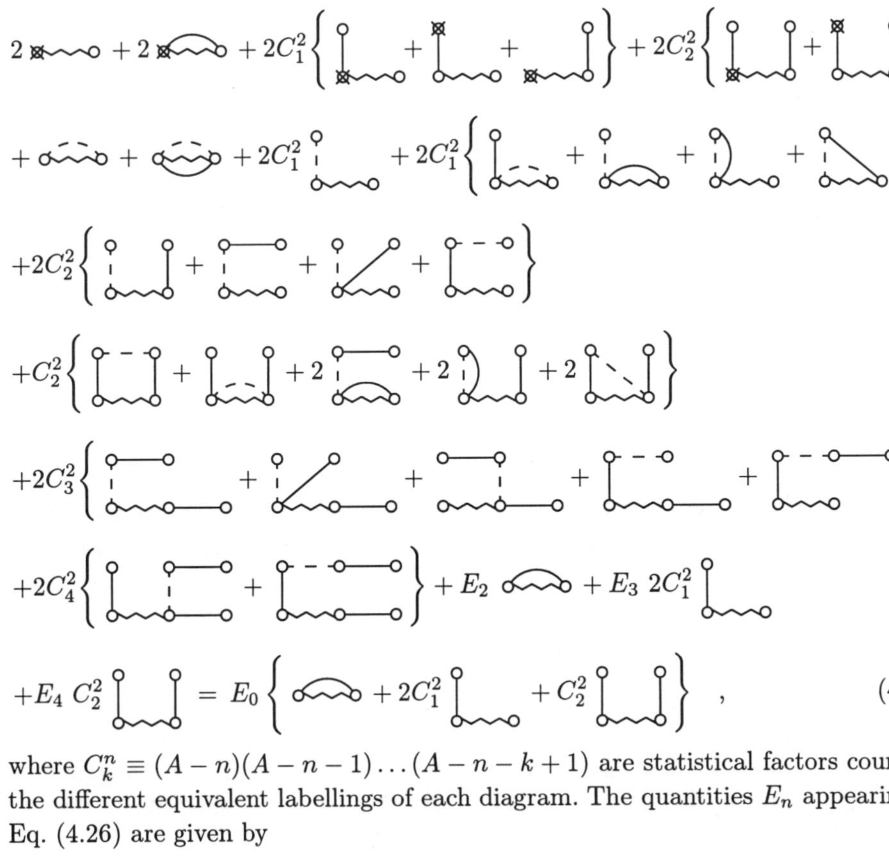

(4.26)

where $C_k^n \equiv (A-n)(A-n-1)\ldots(A-n-k+1)$ are statistical factors counting the different equivalent labellings of each diagram. The quantities $E_n$ appearing in Eq. (4.26) are given by

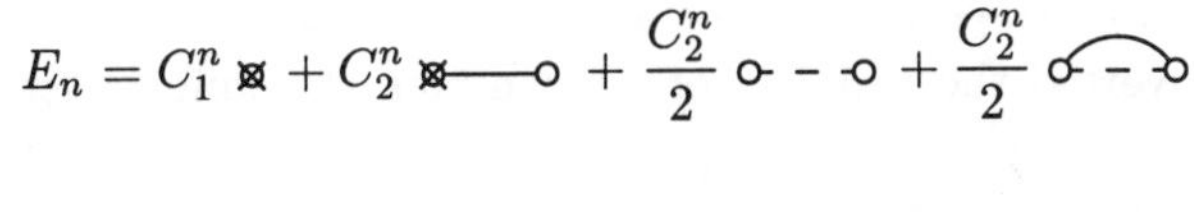

Table 5. Binding energies (in MeV) for various "bosonic nuclei" interacting trough the Wigner part of the S3 interaction as a function of the truncation of the equations in powers of $f$. Order 1 corresponds to the TICI2 case, and Order 4 is the full TICC2 result. [35] The last row shows the full TICC2 configuration-space results. [32]

| | $A = 4$ | $A = 8$ | $A = 16$ | $A = 40$ |
|---|---|---|---|---|
| Order 1 | 25.42 | 225.46 | 1131.2 | 7495.3 |
| Order 2 | 25.60 | 235.12 | 1235.7 | 8457.1 |
| Order 3 | | 235.12 | 1235.1 | 8458.5 |
| Order 4 | | 235.12 | 1235.1 | 8458.7 |
| Config. space | 25.49 | 235.03 | 1234.9 | 8456.6 |

$+C_3^n$ 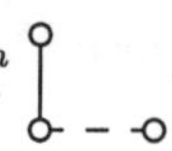 $+\frac{C_4^n}{2}$ 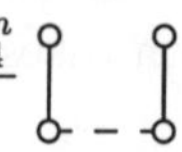 (4.27)

and they are related to unlinked diagrams. In particular, $E_0$ corresponds to the ground-state energy.

To clarify the meaning of the diagrammatic notation consider the following diagram,

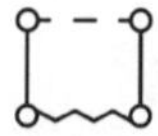 $$\equiv \int \Phi^* g^*(r_{12}) V(r_{34}) f(r_{13}) f(r_{24})\, \Phi \, d\mathbf{r}_1 \ldots d\mathbf{r}_A \quad . \tag{4.28}$$

From this expression we see that Eq. (4.26) is the integral of the function $g^*(r_{12})$ times a non-linear functional, depending on the pair correlation $f(r)$, equated to zero

$$\int g^*(r_{12}) F[f] \, d\mathbf{r}_1 \ldots d\mathbf{r}_A = 0 \quad . \tag{4.29}$$

Given that $g(r)$ is not a fully arbitrary function, because of the required orthogonality of the excited states with respect to the reference state, one cannot obtain straightforwardly an integro-differential equation for $f(r)$. We could introduce a Lagrange multiplier, but it is easier to introduce a completely arbitrary function $G(r)$, and replace $g$ by $G - \langle\Phi|G|\Phi\rangle$ in Eq. (4.26), which obviously satisfies the constraint. After this substitution, the integro-differential equation for $f$ is easily obtained.

***EXERCISE 4.*** *A generic TI (1p-1h + 2p-2h) excitation has been represented by a function $g(ij)$ with the property $\langle\Phi|g|\Phi\rangle = 0$. Let us call TICC2[g] the corresponding TICC2 equation. To deal with a completely arbitrary function, the change $g \to G - \langle\Phi|G|\Phi\rangle$ has been made. Write the new TICC2[G] equation and compare with TICC2[g].*

Summarizing, by carrying out the transformation to the coordinate representation we have converted the problem of determining the infinite set of amplitudes

$\{\mathcal{S}(n)\}$ in configuration space into the determination of a single pair-correlation function $f(r)$ which satisfies an integro-differential equation. Solving this equation avoids, in principle, convergence problems that in configuration representation are related to the finite truncation of the set of amplitudes.

In order to make the calculation manageable we expand the unknown correlation function in a set of Gaussians

$$f(r) = \sum_i c_i \exp(-\beta_i r^2) \quad , \tag{4.30}$$

with a set $c_i$ of unknown coefficients, and a set $\{\beta_i\}$ of pre-determined exponents. Note that this is completely different from the expansion in Laguerre polynomials deduced in Eq. (4.15). Since the different components in this decomposition are not orthogonal, the expectation value of the energy can be given in the form

$$E(\{c_i\}) = \frac{\sum_{ij} c*_i \mathcal{H}_{ij} c_j}{\sum_{ij} c_i^* \mathcal{N}_{ij} c_j} \quad . \tag{4.31}$$

The optimization with respect to $c_i^*$ leads us to the eigenvalue problem

$$\sum_j \mathcal{H}_{ij} c_j = \lambda \sum_j \mathcal{N}_{ij} c_j \quad , \tag{4.32}$$

and the lowest eigenvalue $\lambda$ gives an upper bound to the ground state energy. As an example, in Table 5 we list results for different systems of nucleons considered as bosons and interacting through the Wigner part of the Afnan-Tang S3 interaction. [31] There is an excellent agreement between the configuration-space calculations [32] and the coordinate-space ones. [35] It is worth noting that the coordinate-space numbers are fully converged with the use of between 10 and 14 Gaussians. By contrast, in configuration space up to 30 amplitudes were used (corresponding to HO single-particle excitation energies up to $60\hbar\omega$), not reaching full convergence in some cases. Moreover, working in coordinate-space results in a much faster computation of the required expansion coefficients.

Another interesting feature of the method is revealed when we compare the different results obtained as a function of the truncation of the equations in powers of $f$, as shown in Table 5. In this truncation, the first order corresponds to the linear approach (TICI2) and the fourth order to the full TICC2 approach. Although in the light $A = 4$ system the contribution of non-linear terms is not very relevant, for the heavier systems the improvement in the energy is indeed quite remarkable. It is interesting to note that the third- and fourth-order terms give an almost negligible contribution, so that one may safely simplify Eq. (4.26) by keeping only up to quadratic terms in $f(r)$.

### 4.3. *The nuclei $^4$He and $^{16}$O in the TICI2 approximation*

Let us consider the simplest approximation to the wave function within the TICC2 scheme, limiting the expansions in Eqs. (4.17) or (4.26) to their linear terms. This is the meaning of replacing CC in TICC2 by CI (configuration interaction). It gives rise either to a standard linear problem to determine the amplitudes $\mathcal{S}(n)$, or to an equivalent integro-differential equation for the two-body correlation function $h(r_{ij})$, using from the very beginning the equivalent ansatz

$$|\Psi\rangle = \sum_{i<j} h(r_{ij})|\Phi\rangle \quad , \tag{4.33}$$

where the missing $n = 0$ terms in the $\mathrm{e}^S$ expansion have been absorbed in $h(r_{ij})$.

It is important to realize that this ansatz is not exclusive of the bosonic nature of the particles we have assumed to obtain it. The bosonic or fermionic character of the particles is contained in the reference state, and what TICI2 says is that the simplest way to consider pair correlations translationally invariant is to determine a correlation function $h(r_{ij})$ depending on the relative coordinate of a pair. In fermionic systems $h(r_{ij})$ may be generalized to deal with discrete degrees of freedom. For example, suppose we are dealing with a finite nucleus, with a nucleon-nucleon interaction with the central V4 operatorial form

$$V_{ij} = \sum_{p=1}^{4} V^{(p)}(r_{ij})\Theta^{(p)}(ij) \quad , \tag{4.34}$$

where the operators $\Theta^{(p)}$ refer to the familiar spin and isospin exchange operators

$$\Theta^{(1)}(ij) = 1,\ \Theta^{(2)}(ij) = P^{\sigma}_{ij},\ \Theta^{(3)}(ij) = P^{\tau}_{ij},\ \Theta^{(4)}(ij) = P^{\sigma}_{ij}P^{\tau}_{ij} \quad , \tag{4.35}$$

In correspondence with this V4 form it seems natural to consider an operatorial structure for the pair correlation operator also of the form

$$h_{ij} = \sum_{p=1}^{4} h^{(p)}(r_{ij})\Theta^{(p)}(ij) \quad . \tag{4.36}$$

As it was done before, the unknown correlation functions are expanded in a set of Gaussians

$$h^{(p)}(r) = \sum_{i} c_i^{(p)} \exp(-\beta_i r^2) \quad , \tag{4.37}$$

with a set $\{c_i^{(p)}\}$ of unknown coefficients, and a set $\{\beta_i\}$ of pre-determined exponents. Since the different components in this decomposition are not orthogonal, the expectation value of the energy can be given in the form

$$E(\{c_i^{(p)}\}) = \frac{\displaystyle\sum_{(ip),(jq)} c_i^{(p)*}\mathcal{H}_{ij}^{(pq)}c_j^{(q)}}{\displaystyle\sum_{(ip),(jq)} c_i^{(p)*}\mathcal{N}_{ij}^{(pq)}c_j^{(q)}} \quad . \tag{4.38}$$

Table 6. Binding energies (in MeV) of $^4$He and $^{16}$O using several interactions. SI and SD stand for state independent and state dependent correlations respectively. [47]

| | $^4$He | | $^{16}$O | |
|---|---|---|---|---|
| Interaction | SI | SD | SI | SD |
| B1 | 37.86 | 37.86 | 145.94 | 167.30 |
| S3 | 25.41 | 28.19 | 141.64 | 164.88 |
| MS3 | 25.41 | 27.99 | 85.56 | 105.64 |
| MT I/III | 29.45 | 30.81 | 194.10 | 207.52 |
| MT V | 29.45 | 29.45 | 966.65 | 973.67 |

If we now optimize with respect to $c_i^{(p)*}$ we end up with a generalized eigenvalue problem

$$\sum_{(jq)} \mathcal{H}_{ij}^{(pq)} c_j^{(q)} = \lambda \sum_{(jq)} \mathcal{N}_{ij}^{(pq)} c_j^{(q)} \quad , \tag{4.39}$$

and the lowest eigenvalue $\lambda$ gives an upper bound to the ground state energy.

We shall present and discuss now the TICI2 results obtained for the nuclei $^4$He and $^{16}$O using several V4-type interactions. We refer the reader to Ref. 47 for further details about the calculations and also for results for other nuclei. The interactions considered are B1, [36] S3, [31] MS3, [37] MT I/III and MT V. [38] There are two reasons for this choice. On the one hand, it allows for the best comparison with existing results in the literature. On the other hand, some of these interactions are semi-realistic in the sense that they fit some of the two-nucleon properties. Two types of calculations are considered, according to the selected two-body correlation operator, which can be state-independent (SI), *i. e.* restricted to the first term in Eq. (4.36), or state-dependent (SD), in which case four correlation functions are to be determined. The value of the HO parameter, $\alpha$, has been optimized for the SI case, and this value is kept fixed for the ensuing SD calculations.

In Table 6 are displayed the results so obtained. It is worth keeping in mind that TICI2 is a variational calculation. Since the SD calculation deals with a variational space which is larger than the SI one, it should produce a lower value for the ground state energy. We can appreciate that the effect of SD correlations is in the 0 to 4 MeV range for $^4$He, and in the 10 to 20 MeV range for $^{16}$O. Note that the interactions B1 and MTV contain only Wigner and Majorana terms ($p = 1$ and $p = 4$ respectively in Eq. 4.34), and then do not couple to the spin-dependent piece of the correlation operator in $^4$He. Consequently, there is no difference in the SD and SI energies for these interactions in $^4$He. However, even for those potentials, spin-dependent correlations do play a role for $^{16}$O, so long as the uncorrelated state is not fully space symmetric.

In Table 7 the TICI2 results are compared with those obtained employing several different technologies. Here are quoted results obtained in a Green's function Monte Carlo (GFMC) method, a diffusion Monte Carlo (DMC) procedure, a series

Table 7. A comparison of the binding energies (in MeV) of $^4$He and $^{16}$O using various techniques with the interactions MTV, B1 and MS3.

| | $^4$He (MTV) | $^{16}$O (MTV) | $^{16}$O (B1) | $^{16}$O (MS3) |
|---|---|---|---|---|
| SI | 29.45 | 966.65 | 145.94 | 85.56 |
| SD | 29.45 | 973.67 | 167.30 | 105.64 |
| GFMC [44] | $31.3 \pm 0.2$ | $1194 \pm 20$ | | |
| DMC | $31.32 \pm 0.02$ [39] | $1189 \pm 1$ [40] | | |
| VMC | | $1138.5 \pm 0.2$ [40] | $150.9 \pm 0.3$ [41] | |
| FHNC/0 | | 987/1152 [45] | 167.2 [45] | 113.5 [45] |
| | | 1059/1055 [43] | | 108.1/145.5 [46] |
| FHNC-1 [45] | | | 152.4 | 105.3 |
| IDEA [42] | 30.7-31.2 | 1021-1027 | 164.4-165.2 | 102.9-103.2 |

of results based on a Jastrow-correlation form, and finally the results obtained in the hyperspherical harmonics approach called IDEA. Both GFMC and DMC calculations provide results for the ground state energy which are essentially exact within statistical errors. The interaction MTV does not saturate nuclear matter and consequently it produces high-density systems. Consequently, this is an ideal case to test the importance of three- and more-body correlations. The $^4$He results compare favorably with the Green's function and diffusion Monte Carlo results. We can also see that TICI2, with only two-body correlations, is able to reproduce over 80% of the DMC binding energy for $^{16}$O. This comparison suggests that the effects due to the non-linear terms neglected in TICI2 are comparatively more important for heavier systems.

The TICI2 method also appears to be competitive with several of the methods based on Jastrow-correlated non-interacting wave functions of a single Slater determinant form, where the Jastrow correlation operator is a product over all pairs of particles of a scalar pair function of the inter-particle distance only. The many-body energy expectation value for such trial Jastrow wave function is calculated using variational Monte Carlo (VMC) techniques. Each of the VMC results quoted in Table 7 is of this form, with the differences between them resulting from different choices of single-particle wave functions and from different forms of the Jastrow correlation function over which the variational search is made. As an alternative to the stochastically exact VMC calculations of the energy expectation value, the various Fermi Hypernetted Chain (FHNC) approximations cited in Table 7 represent massively resummed cluster expansions of the same quantity, in which some cluster terms (or chain of ladder form) are summed to all orders but various so-called elementary diagrams are neglected or approximated. The resulting FHNC approximants for the ground-state energy are no longer strict variational bounds, insofar as the higher-order terms have been neglected.

The so-called Integro-differential equation approach (IDEA) is formulated purely in terms of the hyperradius and the relative coordinate of the two particles needed to describe the pair correlations incorporated in the method. In this method, as

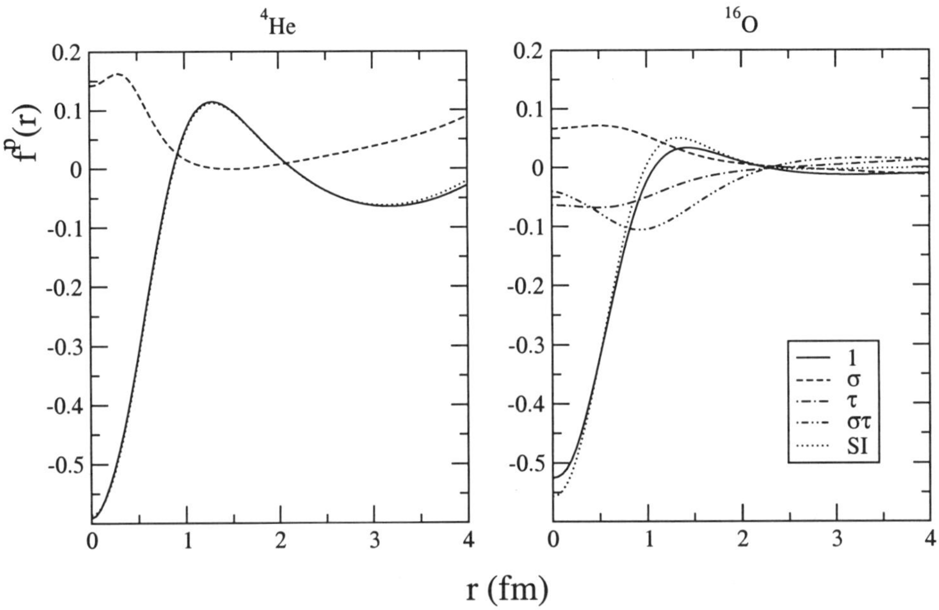


Fig. 4. The calculated TICI2 correlation functions in the case of the MS3 potential. Dotted lines correspond to state independent (SI) correlations.

currently practiced, several uncontrolled and often unjustified approximations are made. As a consequence, the results are non-variational, and usually one has a range of values depending on which of many approximations has been made, none of which is fundamentally preferred to another. A detailed comparison between IDEA and TICI2 for bosons has been made in Refs. 56, 57.

It is interesting to have a look on the obtained two-body correlation functions. They are shown in Fig. 4. The labels $1, \sigma, \tau$, and $\sigma\tau$ correspond to the terms $p = 1, 2, 3$, and 4 in the parameterization of Eq. (4.36). For the $^4$He nucleus, only the first two are required, due to the spatial symmetry of the wave function. One can see that the central scalar correlation function changes little from the SI to SD calculation. This may be due to some kind of perturbation argument: in general, the spin- and/or isospin-dependent correlation functions are much smaller than the central one. The behaviour of the spin and isospin correlation functions for $^{16}$O reflects the fact that, for the MS3 interaction, the correspondent terms in the potential are equal in magnitude but with opposite signs. The shape of the correlation functions at long distances (specially for $^4$He) should not worry us, because it is due to the inclusion of negative values in the Gaussian expansion and it is corrected by the harmonic-oscillator single-particle wave functions, which are not included in the graphs.

Table 8. TICI2 binding energies (in MeV) for $^4$He and $^{16}$O using several V6 interactions.

| | GPDT | SSC | AV14 | AV18 | Reid-V6 |
|---|---|---|---|---|---|
| $^4$He | 27.36 | 24.12 | 14.77 | 15.40 | 5.67 |
| $^{16}$O | 128.68 | 63.55 | 14.97 | 23.76 | |

Now we consider interactions of the V6 form, which means that we add two more operatorial terms to the previous interaction

$$\Theta^{(5)}(ij) = S_T(ij) \quad , \quad \Theta^{(6)}(ij) = S_T(ij)P_{ij}^{\tau} \quad , \tag{4.40}$$

where $S_T$ is the tensor two-body operator. We have used the V6 part of the interactions GPDT, [53] SSC, [54] AV14, [26] AV18 [27] and Reid, [24] although they contain more operatorial terms, like spin-orbit, quadratic angular momentum, etc. (for example, the interactions AV14 and AV18 are so called because of the number of operatorial terms they contain). In this way, only the four central and the two tensor terms are retained. There is no "renormalization" of the parameters, except for the Reid-V6 interaction, which is in fact a V6 adaptation [55] of the original Reid interaction. All these interactions are realistic, in the sense that they contain at least the one-boson exchange part of the NN interaction, plus a phenomenological short range part, and they reproduce the measured phase shifts, as well as the deuteron properties. In correspondence with this V6 form we consider an operatorial structure for the pair correlation operator also of the V6 form, *i. e.* we have to add two more terms to Eq. (4.36) with the operators to include the tensor operator $S_T$ into the correlations. The introduction of tensor correlations increases dramatically the computer time needed in the calculations, mainly because the tensor operator contains also a spatial dependence which increases greatly the number of integrals to be considered and computed.

The obtained V6 results [48] are displayed in Table 8. The interaction GPDT gives the best results, being the softer of the interactions considered. On the opposite side, the results with Reid V6 are very disappointing: the $^{16}$O nucleus is not bound. With the remaining interactions, this nucleus is unstable against $\alpha$ disintegration. The trend of the results is clear: the harder the core the worse the results. From a detailed analysis of the calculations it comes out that there is a competition between central and tensor forces. Most of the binding is obtained from the off-diagonal tensor interaction, *i. e.* the matrix element between the central- and tensor-correlated channels. But this matrix element competes with the strong repulsion in the central channel, which is the origin of the problem. The conclusion is that TICI2 does not contain enough correlations to deal with strongly repulsive interactions. Therefore, something else should be included in the wave function to control the short-range repulsion of the realistic nucleon-nucleon interactions.

### 4.4. *Beyond TICI2*

In the previous section we have seen that the TICI2 wave functions give good results for semi-realistic interactions. However, when realistic nucleon-nucleon interactions are used the correlations provided by the TICI2 wave function are not able to tackle with their short-range phenomenological part which is highly repulsive. The HCSUB(n) approximation has been derived precisely to deal with this type of interactions. However, we shall discuss here other alternative possibilities which are currently being explored.

A first way is to incorporate more correlations into the TICCM scheme, and consider the higher orders of the TICC2 wave function, of which only the linear ones were included in the TICI2 scheme. One can use the same procedure explained in Sect. 4.2 for a bosonic system, and obtain equations for a closed-shell fermionic system, providing that the reference state used has the correct symmetry. Since we have already seen that the relevant terms in the TICC2 wave function are the linear and quadratic ones, the next step after TICI2 is to use the following wave function

$$\Psi(\mathbf{r}_1,\ldots,\mathbf{r}_A) = \left(1+\sum_{i<j} f(r_{ij}) + \frac{1}{2!}\sum_{i<j} f(r_{ij}) \sum_{k<l}{}' f(r_{kl})\right)\Phi(\mathbf{r}_1,\ldots,\mathbf{r}_A) \quad . \quad (4.41)$$

We can represent the different terms in the equations by their associated diagrams, as previously explained. In this diagrammatic expansion, a significant difference between the bosonic and fermionic cases can be clearly seen. In the bosonic case the unlinked diagrams factorise because in the reference state all the particles are in the lowest HO state $(nlm) = (000)$, and thus a contribution like $\langle\Phi| V(r_{12})f(r_{34})|\Phi\rangle$ evaluates to $\langle\Phi| V(r_{12})|\Phi\rangle \times \langle\Phi| f(r_{34})|\Phi\rangle$. Since $\langle\Phi| f(r_{ij})|\Phi\rangle = 0$, all the diagrams containing a disconnected $f$ give a null contribution. This means that, even without truncating the wave function, the equations are of finite order because there is an upper limit to the number of linked $f$ correlations that can appear in a given diagram (four with two-body interactions: two linked by $G$ and two by the interaction). But this is not the case with fermions because the permutations in the reference state involve different HO states. Now we are not able to compute $\langle\Phi| V(r_{12})f(r_{34})|\Phi\rangle$ as $\langle\Phi| V(r_{12})|\Phi\rangle \times \langle\Phi| f(r_{34})|\Phi\rangle$, and therefore such contributions are not null anymore. Thus if the wave function is not truncated, the number of diagrams to be considered is, in principle, infinite. Even when the wave function is truncated (to second order, in the present case) the number of diagrams to consider is much greater than in the bosonic case due to the inclusion of contributions with disconnected correlation functions. Nevertheless, it turns out that although these disconnected diagrams are not null, their contribution is almost negligible, at least to compute the ground state energies of light nuclei. Table 9 shows the results obtained including the quadratic terms in the wave function, and considering state-dependent correlations. [52] As compared to TICI2 results, the improvement is quite remarkable, specially for the MS3 interaction which has a stronger short-range repulsion than the BB1 potential. Calculations with realistic V6 interactions have not yet been performed.

Table 9. TICC2 binding energies (in MeV) for nuclei in the $p$-shell, using the B1 and MS3 interactions. SI and SD stand for state-independent or V4 state-dependent correlations.

| | | $^4$He | $^8$Be | $^{12}$C | $^{16}$O |
|---|---|---|---|---|---|
| B1 | SI | 37.92 | 50.04 | 86.16 | 149.36 |
| | SD | 37.92 | 65.09 | 109.47 | 173.76 |
| MS3 | SI | 25.59 | 28.37 | 50.53 | 94.71 |
| | SD | 28.21 | 42.93 | 74.30 | 123.79 |

Let us discuss now another possibility to deal with a short-range repulsion. The idea is to use a correlated reference state, including for instance a Jastrow factor; this is the starting point of a mixed scheme, Jastrow-TICI2 [48,49] or J-TICI2 in short. It combines the short-range Jastrow correlations with the medium- and long-range state-dependent TICI2 ones. The wave function is therefore written as

$$\Psi(1\ldots A) = \left(\prod_{i<j} g(ij)\right)\left(\sum_{k<l}\sum_{p} h^{(p)}(kl)\Theta^{(p)}(kl)\right)\Phi(1\ldots A) \quad . \tag{4.42}$$

This is similar to the Correlated Basis Function, in which are combined Jastrow correlations and non-orthogonal perturbation theory. Obviously, to keep on the spirit of CCM one should determine the Jastrow factor from the very beginning, avoiding thus the complicated task of a simultaneous determination of the TICI2 and the Jastrow correlation functions. As an exploratory study, let us consider a simple Jastrow factor consisting of a single Gaussian function

$$g(r) = 1 + a e^{-br^2} \quad . \tag{4.43}$$

We have to solve a variational problem, minimizing the ground state energy with respect to the Jastrow parameters $a$, and $b$, with respect to the TICI2 correlation functions $h^{(p)}$ (which in the general case are state dependent), and with respect to the HO parameter $\alpha$.

Table 10 displays the results obtained for the $^4$He nucleus using several V4 and V6 interactions. Let us first consider the V4 interactions. We can see that there is an appreciable difference in the resulting ground state energies, and the more repulsive is the potential the more important is the difference between TICI2 and J-TICI2. The same conclusions applies for the V6 interactions, but here the effects are more impressive. Looking at the results with the interaction Reid-V6, which has the stronger repulsion, one can see that the HO parameter changes appreciably from TICI2 to J-TICI2. At the end, the value of the parameter is nearly the same for all interactions when the approximation J-TICI2 is used. This means that we get nearly the same size. For comparison, the last column in the table shows some results coming from DMC or GFMC calculations. Note that the calculated energies are very close to the exact ones for the V4 interactions B1 and MTV. However,

Table 10. TICI2 and J-TICI2 binding energies (in MeV) for $^4$He using several V4 and V6 interactions. The harmonic-oscillator inverse-length parameter $\alpha$ (in fm$^{-1}$) is quoted beside the energies, as well as the parameters of the Jastrow factor $a$, and $b$ (in fm$^{-2}$). The last column shows the results of some Monte Carlo calculations (DMC for B1 and MT V, and GFMC for AV14 and Reid-V6) as a comparison.

| | TICI2 | | J-TICI2 | | | | MC |
|---|---|---|---|---|---|---|---|
| | E | $\alpha$ | E | $\alpha$ | $a$ | $b$ | |
| B1 | 37.86 | 0.73 | 38.28 | 0.77 | -0.41 | 1.8 | $38.32 \pm 0.01$ |
| S3 | 28.19 | 0.72 | 30.16 | 0.78 | -0.70 | 2.1 | |
| MS3 | 27.99 | 0.71 | 29.97 | 0.74 | -0.70 | 2.1 | |
| MT I/III | 30.81 | 0.74 | 32.70 | 0.75 | -0.88 | 5.0 | |
| MT V | 29.45 | 0.74 | 31.21 | 0.75 | -0.87 | 5.6 | $31.31 \pm 0.02$ |
| GPDT | 27.36 | 0.70 | 27.58 | 0.72 | -0.29 | 1.9 | |
| SSC | 24.12 | 0.68 | 26.74 | 0.75 | -0.66 | 2.1 | |
| AV14 | 14.77 | 0.59 | 20.37 | 0.71 | -0.93 | 2.7 | $24.79 \pm 0.20$ |
| AV18 | 15.40 | 0.61 | 21.08 | 0.74 | -0.92 | 3.2 | |
| Reid-V6 | 5.67 | 0.43 | 22.70 | 0.72 | -1.05 | 2.4 | $28.30 \pm 0.12$ |

there are still important differences in the case of V6 interactions. The J-TICI2 has been also successfully applied to heavier nuclei such as $^{16}$O considering only V4 interactions. [49]

These results for $^4$He and $^{16}$O confirm our previous idea that the Jastrow factor is more effective to screen the short-range repulsion. There is a division of roles in the J-TICI2 ansatz: the Jastrow factor takes care of the short-range correlations, and the TICI2 factor is responsible for the medium and long range correlations.

## 5. Helium droplets

We have seen in the previous Section that an interaction having a strong repulsion at short-distances causes some troubles in the SUB(n) approximation of CCM. In a diagrammatic representation of TICI2 equations it is easily seen where the problem lies: the additive TICI2 correlations leave terms where the potential has free legs, resulting in unbalanced contributions to the total energy if the potential has a strong repulsion. One possibility to deal with such interactions was followed by the Bochum group many years ago, and it is called HCSUB(n) approximation, which we have considered in Sect. 3.2. In the previous Section we have shown an alternative cure to the strong repulsion problem, namely the use of a correlated reference states, and a mixture of Jastrow and CCM correlations has to be considered. The purpose of this Section is to discuss in some depth this hybrid method. As the specific physical system to deal with we have chosen drops with a finite number of helium atoms. There are several reasons for this choice. First of all, this is a relatively new field which is attracting an increasing interest. A second reason, and more specific, is that the atom-atom interaction is well known, and relatively simple as it is purely central. We shall thus avoid the complications related to the complex structure of the nucleon-nucleon interaction. And finally, stochastic techniques are available allowing one to obtain essentially exact solutions for the ground state energies of droplets formed by $^4$He atoms, as they are bosons. A comparison with these results will precise the validity of the approximation.

### 5.1. *The J-TICI3 approximation*

The basic idea is to start with a conveniently correlated reference state, putting on top of it CC correlations, which will be restricted here to TICI3 level. The correlations to be considered for the reference state must be of short-range, because it is for such a case that the usual CCM has to be modified, and we shall consider to introduce them by means of a Jastrow factor. The reference state is written as

$$\Phi_J(\mathbf{R}) = \prod_{i<j} g(r_{ij})\ \Phi(\mathbf{R}) \quad , \tag{5.1}$$

where $\mathbf{R}$ represents all single-particle coordinates, $\Phi$ is an uncorrelated reference state and the function $g(r_{ij})$ is a two-body Jastrow correlation. We can imagine a double minimization process to determine Jastrow and TICI3 correlations, but this will result in a complicated numerical task. In fact, the purpose of these Jastrow correlations is to control conveniently the short-range part of the wave function, and this can be reasonably obtained by looking at the short-range part of the interaction. We shall use the following function

$$g(r) = \exp\left[-\frac{1}{2}\left(\frac{b}{r}\right)^{\nu}\right] \quad , \tag{5.2}$$

introduced by McMillan many years ago for a Lennard-Jones interaction. Although the helium-helium interaction is not of a Lennard-Jones type, many Variational Monte Carlo calculations have used this McMillan form and have shown that it is also appropriate when dealing with Aziz type interactions, with parameters $\nu = 5.2$ and $b = 2.90$ Å and 2.81 Å, for $^4$He and $^3$He respectively. In the spirit of our use of Jastrow correlations, these values should be used independently of the size of the system under consideration, because its behaviour is only determined by the two-body interaction potential at very short distances.

> ***EXERCISE 5.*** *Consider two helium atoms, interacting through a Lenard-Jones potential $V(r) = 4\epsilon[(\sigma/r)^{12} - (\sigma/r)^6]$, with $\epsilon = 10.22K$ and $\sigma = 2.556$ Å. Obtain the very short-range behaviour of the relative wave function, thus justifying the choice of the Jastrow factor in Eq. 5.2. To be quantitative, use the values $\hbar^2/m =$ 12.1194 K Å$^2$ and 16.084 K Å$^2$, respectively for $^4$He and $^3$He.*

For the uncorrelated part of the reference state we shall employ again a HO wave function, to take advantage of the CM factorization of this type of functions. To complete the description of our wave function we have to include TICI correlations.

Up to the linear third-order correlation, the total wave function is written as [58]

$$\Psi_{JTICI3}(\mathbf{R}) = \Phi_J(\mathbf{R})\Phi_{TICI3}(\mathbf{R}) \equiv \Phi_J(\mathbf{R}) \left(1 + \sum_{i<j} f_2(r_{ij}) + \sum_{i<j<k} f_3(r_{ij}, r_{ik}, r_{jk})\right) \quad . \tag{5.3}$$

In the previous Section we have applied a similar ansatz for the wave function of the atomic nucleus of $^4$He, using a Jastrow factor more adapted to the nuclear case, and restricting the CCM part to a TICI2 description. Once fixed the reference state, the wave function is completely determined by the correlation functions $f_2$ and $f_3$ and the HO parameter $\alpha$.

In practice we have expanded again the TICI3 correlation terms on a Gaussian basis

$$\Phi_{TICI3}(\mathbf{R}) = \sum_{p<q<r}^{N_\beta} C_{\{\mu\}} G_{\{\mu\}}(\mathbf{R}) \quad , \tag{5.4}$$

with

$$G_{\{\mu\}}(\mathbf{R}) = \mathcal{S} \left( \sum_{i<j<k} \mathrm{e}^{-\beta_p r_{ij}^2} \mathrm{e}^{-\beta_q r_{ik}^2} \mathrm{e}^{-\beta_r r_{jk}^2} \right) \quad , \tag{5.5}$$

where $\{\mu\} \equiv \{p, q, r\}$. There are now three relative coordinates. The function $G$ is the symmetric combination of Gaussians (as indicated by the symmetriser operator $\mathcal{S}$), and the coefficients $C$ are the unknown quantities to be determined. Only five Gaussians have been used, with the following widths:

$$(\beta_p/\alpha^2) = (0, -0.05, 0.5, 1, 4) \quad (^4\mathrm{He}) \quad , \tag{5.6}$$

$$= (0, 0.5, 1, 2, 4) \quad (^3\mathrm{He}) \quad . \tag{5.7}$$

The interest of using one null value is that we obtain easily the limit where there are no CC correlations, when the three labels are put equal to 1, and the J-TICI2 case, putting two of the labels equal to 1.

Now we have to solve a generalized eigenvalue problem

$$\sum_{\mu_2} (\mathcal{K}_{\mu_1,\mu_2} + \mathcal{V}_{\mu_1,\mu_2}) C_{\mu_2} = E \sum_{\mu_2} \mathcal{N}_{\mu_1,\mu_2} C_{\mu_2} \quad , \tag{5.8}$$

to obtain the energy. The norm and potential matrices are given by

$$\mathcal{N}_{\mu_1,\mu_2} = \int \mathrm{d}\mathbf{R} |\Phi_J(\mathbf{R})|^2 G^*_{\mu_1}(\mathbf{R}) G_{\mu_2}(\mathbf{R}) \quad , \tag{5.9}$$

$$\mathcal{V}_{\mu_1,\mu_2} = \int \mathrm{d}\mathbf{R} |\Phi_J(\mathbf{R})|^2 G^*_{\mu_1}(\mathbf{R}) \sum_{m<n} V(r_{mn}) G_{\mu_2}(\mathbf{R}) \quad . \tag{5.10}$$

We choose to write the kinetic energy matrix as

$$\mathcal{K}_{\mu_1,\mu_2} = \int \mathrm{d}\mathbf{R} |\Phi_J(\mathbf{R})|^2 G^*_{\mu_1}(\mathbf{R}) \frac{1}{\Phi_J(\mathbf{R})} \left( -\frac{\hbar^2}{2m} \sum_n \Delta_n \right) G_{\mu_2}(\mathbf{R}) \Phi_J(\mathbf{R}) \quad , \tag{5.11}$$

because the multidimensional integrals can be calculated by means of the Monte Carlo method, using the positive definite function $|\Phi_J(\mathbf{R})|^2$ as the guide of a Metropolis random walk. Note that no substraction of the CM is needed because we are using a TI wave function.

The number of unknown amplitudes is related to the number of Gaussians in the expansion, and it is given by

$$\binom{N_\beta + 2}{3} \ .$$

With five Gaussians the dimension of the matrix to diagonalize is 35.

## 5.2. *Ground state of* $^4$*He droplets*

As $^4$He atoms are bosons the total wave function must be completely symmetrical with respect to the exchange of any pair of atoms. We shall use an HO uncorrelated wave function. Once removed the CM, the intrinsic part may be written as

$$\prod_{i<j} \exp\left(-\frac{\alpha^2}{2N} r_{ij}^2\right) \ . \tag{5.12}$$

This means that all the bosons are in the lowest $1s$ HO state, which is the boson condensate. This part of the wave function depends on a size parameter $\alpha$, and it is similar to a a Jastrow form. Thus, we may write the reference wave function for a system of $N$ atoms of $^4$He as

$$\Phi_J(\mathbf{R}) = \prod_{i<j} \exp\left[-\frac{1}{2}\left(\frac{b}{r_{ij}}\right)^\nu - \frac{\alpha^2}{2N} r_{ij}^2\right] \ . \tag{5.13}$$

This simple form has been used as trial function in many variational Monte Carlo calculations. The best variational calculations include more terms into the exponent, in particular a three-body correlation term.

In Table 11 some J-TICI3 results [58] are displayed together with the corresponding DMC results, [59,60] both for the energy per particle and the unit radius. As there are no experiments on these systems, we use the DMC results as the benchmark to compare with. The agreement is quite encouraging. As expected both results coincide, within the statistical errors, for three particles, because in this case the TICI3 ansatz includes all correlations.

The convergence in the ground state energy when going from a pure Jastrow wave function to the full J-TICI3 ansatz is shown in Fig. 5. Again, DMC results are also displayed. It is seen that three-body correlations represent an appreciable improvement with respect to a simple pair-correlation description. However, the agreement with the DMC results deteriorates when the number of atoms is increased. This should be related to the size extensivity problem that we have seen in the context of the CIM discussion. In practical terms, this means that at least quadratic terms in the pair correlation should be included.

Table 11. Ground state energies per particle and unit radii of $^4$He droplets, as obtained within a J-TICI3 approximation and the DMC calculation. In parentheses are indicated the sampling errors. The atom-atom interaction is HFD-B(HE).

| | −E/N (K) | | $r_0$(Å) | |
|---|---|---|---|---|
| N | J-TICI3 | DMC | J-TICI3 | DMC |
| 3 | 0.0430(10) | 0.0436(2) | 5.4(2) | 5.59 |
| 4 | 0.1398(15) | 0.1443(2) | 4.11(11) | 4.13 |
| 5 | 0.2616(13) | 0.2670(3) | 3.66(7) | 3.65 |
| 6 | 0.3868(11) | 0.3950(2) | 3.42(6) | |
| 7 | 0.5081(12) | 0.5206(4) | 3.31(5) | 3.22 |
| 8 | 0.6289(13) | 0.6417(4) | 3.18(4) | |
| 9 | 0.7392(12) | 0.7563(6) | 3.11(3) | |
| 10 | 0.8484(19) | 0.8654(7) | 3.01(3) | |
| 14 | 1.215(2) | 1.2478(12) | 2.91(2) | 2.83 |
| 20 | 1.6336(15) | 1.688(2) | 2.727(14) | 2.69 |
| 40 | 2.4563(14) | 2.575(3) | 2.578(8) | |

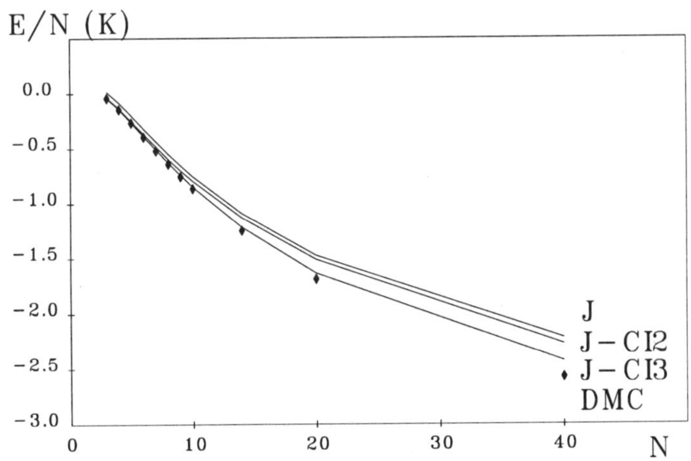


Fig. 5. Ground state energies per particle of $^4$He droplets, as obtained with the approximations J, J-TICI2 and J-TICI3. Diamonds indicate the DMC results.

Fig. 6 shows the dependence of the energy per particle with the HO parameter for an eight-atoms drop. As a reference, a VMC calculation, including two- and three-body Jastrow functions, and the DMC results are also displayed. Obviously, these calculations do not depend on the HO parameter, and this is why their results appear as two constant lines. It may be seen that neither J-TICI2 nor J-TICI3 energies are very sensitive to the value of this parameter, at least in a large interval of values. This means that the results are rather independent of the choice of the uncorrelated reference system, which is a proof of the quality of the wave function.

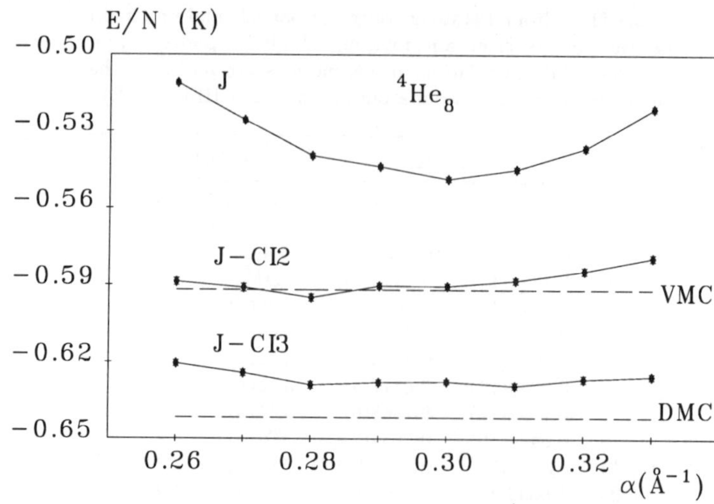


Fig. 6. The calculated J, J-TICI2 and J-TICI3 ground state energies per particle of the drop with 8 atoms of $^4$He are represented as a function of the HO parameter $\alpha$.

We can further check the ground state wave function by computing, for instance, the static structure factor

$$S(q) = 1 + \frac{2}{N}\langle \sum_{i<j} e^{i\vec{q}\cdot\vec{r}_{ij}} \rangle - |\langle \sum_i e^{i\vec{q}_i\cdot(\vec{r}_i - \vec{R})}\rangle|^2 \quad , \tag{5.14}$$

which is related to the Fourier transform of the two-body distribution function. This factor, calculated within the J-TICI3 ansatz, is displayed in Fig. 7 for $^4$He droplets with 20 and 40 atoms. In the same figure, the VMC and the DMC static structure factors of Chin and Krotscheck [62] are also represented. It can be appreciated that for 20 atoms, the J-TICI3 static structure factor lies between the VMC and DMC ones, whereas for 40 atoms the J-TICI3 and VMC structure factor are practically undistinguishable. In conclusion, the J-TICI3 ansatz is in good agreement with DMC calculations, for $^4$He droplets with up to 40 atoms, not only for the ground state energy, but also for the static structure factor, which is another test of the quality of the J-TICI3 wave function. The sizeable differences at $N = 40$ are an indication of the limits of the J-TICI3 approach. As mentioned before, quadratic terms in the pair correlation function could improve the results.

### 5.3. *Collective states of $^4$He droplets*

With the only knowledge of the ground state wave function it is possible to obtain useful information about the excitation spectrum using sum rules. Let us briefly recall some features of sum rules. [63] For an arbitrary transition operator $Q$ the sum

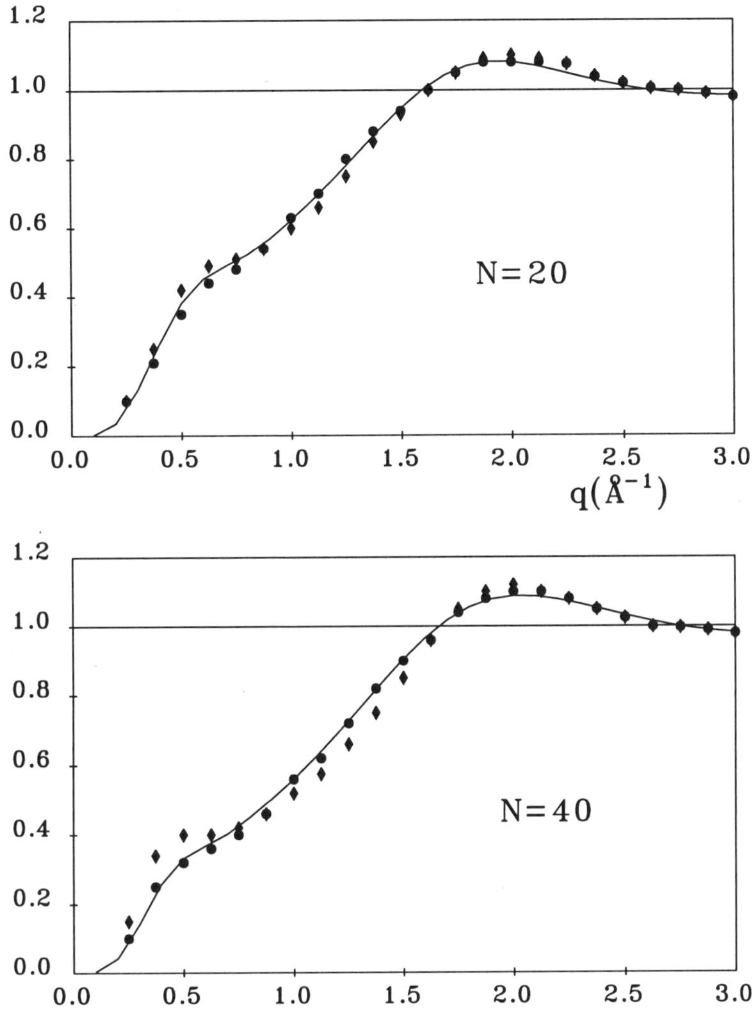


Fig. 7. Static structure factors of the drops with 20 and 40 atoms of $^4$He, calculated within J-TICI3 approximation, and compared to the VMC results (circles) and DMC results (diamonds) obtained in Ref. 62

rule of order $k$ is defined as

$$M_k(Q) = \sum_{n \neq 0} (E_n - E_0)^k |\langle n|Q|0\rangle|^2 \quad , \tag{5.15}$$

where $E_n$ is the energy of the state $|n\rangle$ and the index 0 refers to the ground state. It follows that an upper bound to the energy $E_1$ of the first excited state, kinematically accesible from the ground state through the $Q$ operator, may be written in the following way:

$$E_1 - E_0 \leq \frac{M_1(Q)}{M_0(Q)} \quad . \tag{5.16}$$

Table 12. Centroids $\hbar\omega_L$ calculated with the J-TICI3 trial function. The excitation energies $E^*_{(2)}$ and $E^*_{(4)}$ of Fig. 8 are also displayed for comparison. All energies are in K.

| $N$ | $\hbar\omega_0$ | $\hbar\omega_2$ | $E^*_{(2)}$ | $\hbar\omega_4$ | $E^*_{(4)}$ |
|---|---|---|---|---|---|
| 6 | $1.699 \pm 0.011$ | $1.434 \pm 0.008$ | $1.377 \pm 0.015$ | $2.70 \pm 0.02$ | $2.286 \pm 0.018$ |
| 8 | $2.156 \pm 0.016$ | $1.569 \pm 0.011$ | $1.487 \pm 0.014$ | $3.03 \pm 0.03$ | $2.74 \pm 0.04$ |
| 10 | $2.63 \pm 0.02$ | $1.695 \pm 0.012$ | $1.582 \pm 0.019$ | $3.40 \pm 0.03$ | $3.08 \pm 0.02$ |
| 12 | $2.78 \pm 0.02$ | $1.646 \pm 0.014$ | $1.58 \pm 0.02$ | $3.20 \pm 0.03$ | $3.12 \pm 0.02$ |
| 14 | $3.00 \pm 0.03$ | $1.640 \pm 0.016$ | $1.51 \pm 0.03$ | $3.21 \pm 0.03$ | $2.97 \pm 0.03$ |
| 16 | $3.19 \pm 0.04$ | $1.62 \pm 0.02$ | $1.47 \pm 0.03$ | $3.21 \pm 0.04$ | $3.15 \pm 0.03$ |
| 18 | $3.32 \pm 0.05$ | $1.68 \pm 0.03$ | $1.55 \pm 0.04$ | $3.33 \pm 0.05$ | $3.30 \pm 0.04$ |
| 20 | $3.57 \pm 0.07$ | $1.78 \pm 0.04$ | $1.58 \pm 0.04$ | $3.62 \pm 0.06$ | $3.33 \pm 0.04$ |
| 30 | $3.92 \pm 0.09$ | $1.85 \pm 0.05$ | $1.37 \pm 0.09$ | $3.55 \pm 0.06$ | $3.27 \pm 0.09$ |
| 40 | $3.93 \pm 0.10$ | $1.79 \pm 0.06$ | $1.5 \pm 0.2$ | $3.38 \pm 0.09$ | $3.1 \pm 0.2$ |

Assuming $Q$ real, the non-energy weighted sum rule is written as

$$M_0(Q) = \langle 0|Q^2|0\rangle - |\langle 0|Q|0\rangle|^2 \quad . \tag{5.17}$$

Moreover, if $|0\rangle$ is the *exact* ground state wave function, the energy-weighted sum rule may be written as

$$M_1(Q) = \frac{1}{2}\langle 0|[Q,[H,Q]]|0\rangle \quad . \tag{5.18}$$

An estimate of the centroid

$$\hbar\omega = \frac{M_1(Q)}{M_0(Q)} \quad , \tag{5.19}$$

can be obtained using the trial J-TICI3 wave function. For finite droplets it is convenient to consider multipolar excitation operators $Q_L$ of the form

$$Q_0 = \sum_{i<j} r_{ij}^2 \quad , \tag{5.20}$$

$$Q_L = \sum_{i<j} r_{ij}^L Y_{L0}(\hat{r}_{ij}) \quad . \tag{5.21}$$

Let us see now that it is possible to generalize the J-TICI3 trial function to calculate variational upper bounds to the energy of the low-lying excited states with angular momentum different from zero. We can build a non-zero angular momentum wave function by placing $L \neq 0$ components either in the reference state, in the TICI operator, or in both. The first possibility is much simpler, and thus the reference state (5.13) is changed to

$$\Phi_J^{(L)}(\mathbf{R}) = \prod_{i<j} \exp\left(-\frac{1}{2}\left(\frac{b}{r_{ij}}\right)^\nu - \frac{\alpha^2}{2N} r_{ij}^2\right) \sum_{i<j} r_{ij}^L Y_{L0}(\hat{r}_{ij}) \quad , \tag{5.22}$$

where $Y_{LM}$ is a spherical harmonic depending on the angles of the relative coordinate $\vec{r}_{ij}$. Although we have not placed any angular momentum in the TICI3 correlation term, it is different from the one calculated for the ground state. Indeed, the functions $f_2$ and $f_3$ are determined by minimizing the expectation value of the Hamiltonian in the new trial state with angular momentum $L$. Consequently, these functions depend implicitly on the value of the angular momentum.

In Fig. 8 are displayed the excitation energies

$$E^*_{(L)} = E(L) - E(L = 0) \tag{5.23}$$

as well as the chemical potential $\mu(N)$, calculated as the difference between the ground state energies of clusters with $N$ and $N - 1$ atoms. It is worth keeping in mind that although each absolute energy is a variational upper bound, the relative differences $E^*_{(L)}$ and $\mu(N)$ are not rigorous upper bounds. These results indicate that excited states with angular momentum $L = 2$ are bound for $N \geq 10$ atoms, whereas the threshold for bound states with $L = 4$ lies around $N = 30$ atoms. Moreover one may see that the excitation energies are roughly constant within the size interval considered here: around 1.5 K for states with $L = 2$ and around 3.1 K for states with $L = 4$.

In Table 12 the centroids $\hbar\omega_L$ for $L$ =0, 2 and 4 are shown, as well as the excitation energies $E^*_{(L)}$ displayed in Fig. 8. We can see that in all cases the centroids provide an upper bound to the corresponding excitation energies, and this may be considered again as a supplementary test of the goodness of the trial function. It turns out that there is a huge contribution of the first excited state to the sum rules $M_0$ and $M_1$ (more than 80%), which is an indication of the colectivity of these excitations. This may also be inferred qualitatively from the fact that the excitation energies are rather close to their respective centroids.

Previous theoretical descriptions of the excited states [62,65,66] are based on the Feynman treatment [64] of compressional excitations in liquid $^4$He. Once determined the ground state wave function, $\Psi_0(\mathbf{R})$, a trial function for the excited state is constructed as

$$\Psi_{\text{exc}}(\mathbf{R}) = F(\mathbf{R})\Psi_0(\mathbf{R}) \tag{5.24}$$

where $F$ is an excitation operator to be determined. If $\Psi_0$ is the *exact* ground state wave function the excitation energy is precisely given by

$$\hbar\omega = \frac{1}{2}\frac{\langle\Psi_0|[F,[H,F]]|\Psi_0\rangle}{\langle\Psi_0|F^2|\Psi_0\rangle - |\langle\Psi_0|F|\Psi_0\rangle|^2} \quad . \tag{5.25}$$

One is naturally led to determine the excitation operator $F$ by minimizing this excitation energy, even when $\Psi_0$ is not the true ground state function. This scheme has been used by Chin and Krotscheck considering the monopolar and quadrupolar components of a translationally invariant one-body function for $F(\mathbf{R})$. [66] It is worth mentioning that the J-TICI3 approach bears some similarity with Feynman's treatment for excited states. It is easy to realize that the J-TICI3 trial function for the excited state with angular momentum $L$ may be formally written as

$$\Phi^{(L)}_{JTICI3}(\mathbf{R}) = F(\mathbf{R})\Phi^{(L=0)}_{JTICI3}(\mathbf{R}) \tag{5.26}$$

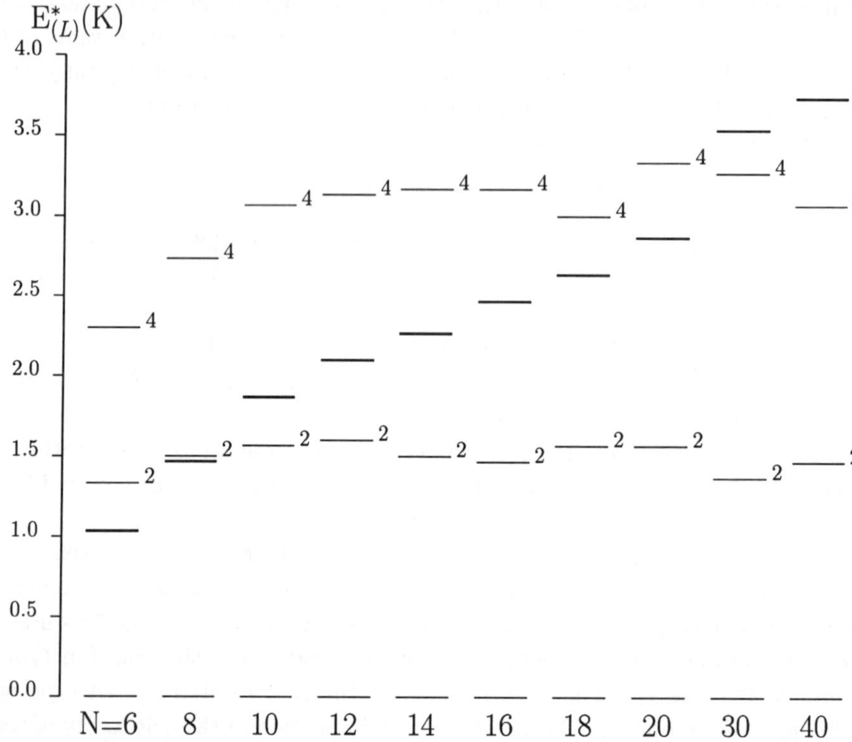


Fig. 8. Excitation energies $E^*_{(L)} = E(L) - E(L=0)$ and chemical potentials $\mu(N)$ as a function of the number $N$ of atoms in a cluster. The former are labelled by the value of $L$, the latter are the thicker lines. All energies are in K.

where the Feynman-like excitation function is

$$F(\mathbf{R}) = \frac{1 + \sum f_2^{(L)}(r_{ij}) + \sum f_3^{(L)}(r_{ij}, r_{ik}, r_{jk})}{1 + \sum f_2^{(L=0)}(r_{ij}) + \sum f_3^{(L=0)}(r_{ij}, r_{ik}, r_{jk})} \sum_{i<j} r_{ij}^L Y(\hat{r}_{ij}) \tag{5.27}$$

*i. e.*, it is a many-body function which is formally determined in two steps. First the correlation functions $f_2^{(L=0)}$ and $f_3^{(L=0)}$ of the ground state are obtained, and then those of the excited state with $L \neq 0$. In principle, this particular $F(\mathbf{R})$ is not a translationally invariant one-body function.

Table 13 shows some J-TICI3 results for the clusters with 20 and 40 atoms. The first row displays the ground state energies, the chemical potentials and the excitation energies of the states with $L = 2$, whereas the second row shows the centroid energies. The next two rows contain the results of Ref. 62. The VMC energies were obtained using a harmonic oscillator factor and a Jastrow factor with two- and three-body functions. It is similar to Eq. (5.13), apart from the Jastrow three-body function. This VMC trial function was employed in the DMC calculation as the importance sampling function.

Table 13. The ground state energies, chemical potentials, excitation energies $E^*_{(L)}$ and centroids $\hbar\omega_L$ obtained within a J-TICI3 calculation [61] are compared with the variational (VMC) and diffusion (DMC) Monte Carlo results of Ref. 62. All energies are in K.

| | $N=20$ | | | | $N=40$ | | | |
|---|---|---|---|---|---|---|---|---|
| | $-E$ | $-\mu$ | $L=0$ | $L=2$ | $-E$ | $-\mu$ | $L=0$ | $L=2$ |
| J-TICI3 | 32.73±0.02 | 2.87 | | 1.58 | 98.17±0.16 | 3.72 | | 1.5 |
| $\hbar\omega_L$ | | | 3.57 | 1.78 | | | 3.93 | 1.79 |
| VMC | 30.44±0.04 | | 2.72 | 2.26 | 93.76±0.04 | | 3.53 | 2.04 |
| DMC | 33.18±0.06 | 2.91 | 2.80 | 1.77 | 101.52±0.16 | 3.67 | 3.68 | 1.50 |

The J-TICI3 chemical potentials agree with the DMC ones, as it could be expected in view of the agreement between the ground state energies (see Table 11 and Fig. 5). Note that the J-TICI3 excitation energies are lower than the VMC value; also, for the cluster with $N = 20$ J-TICI3 produces a lower $L = 2$ energy than the DMC one. This should be attributed to the different Feynman-like excitation function (5.27), which in principle generates a larger variational space than the one corresponding to a one-body operator. Therefore, the comparison between different excitation energies is only qualitative.

### 5.4. $^3$*He droplets*

In the case of $^3$He atoms, we write the uncorrelated wave function as the product of two HO Slater determinants, one for spin up particles, the other for spin down. Moreover we also include what is often called back-flow correlations, consisting in replacing each single-particle coordinate of the Slater determinant by

$$\mathbf{r}_i \longrightarrow \mathbf{r}_i + \sum_{i\neq j} \eta(r_{ij})(\mathbf{r}_i - \mathbf{r}_j) \quad . \tag{5.28}$$

We have employed the same function $\eta(r) = \lambda/r^3$ considered in Ref. 70, with $\lambda = 5$ Å$^3$.

We expect to have important differences, as compared to $^4$He drops, for two main reasons. Firstly, the smaller mass of a $^3$He atom implies that the zero-point motion is larger in $^3$He than in $^4$He. Secondly, the fact that $^3$He atoms are fermions means that Pauli repulsion effects are present. These differences in mass and statistics produce the same effect, reducing the binding in a $^3$He drop compared with the analogous $^4$He one.

As an illustrative example, we can consider an eight-atom drop, with two possibilities for the mass (either $^4$He mass $m_4$, or $^3$H3 mass $m_3$) and for the statistics (either fermions or bosons). The J-TICI3 ground state energies, obtained using the Aziz interaction HFD-B(HE), are shown in Table 14. These results indicate that eight atoms of $^4$He form a bound system, but eight atoms of $^3$He are not bound.

Table 14. J-TICI3 ground state energies for eight atoms of Helium with different masses and statistics. The interaction is Aziz HFD-B(HE).

| | $m_4$ | $m_3$ |
|---|---|---|
| Bosons | $-4.78 \pm 0.04$ | $-0.53 \pm 0.07$ |
| Fermions | $-0.78 \pm 0.05$ | $+2.57 \pm 0.07$ |

Table 15. Comparison between ground state energies for $^3$He drops, using the interaction Aziz HFDHE-2. VMC and J-TICI3 are the results given in Refs. 70 and 69 respectively.

| $N$ | VMC | J-TICI3 |
|---|---|---|
| 20 | $4.12 \pm 0.14$ | $3.44 \pm 0.05$ |
| 40 | $-1.44 \pm 0.08$ | $-2.55 \pm 0.07$ |

Consequently, a minimum number of $^3$He atoms, greater than eight, is required to form a bound system of $^3$He atoms.

To determine the minimum size of a self-bound $^3$He drop, a non-local density functional (including finite range effects) has been employed, complemented with shell-model techniques to deal with open fermionic shells. [68] The calculation indicates that from around 30 atoms on the system is bound, and most importantly, that the ground states of these droplets have the maximum possible spin.

Let us consider the results given by a J-TICI3 calculation. [69] As already mentioned, the uncorrelated reference state is the product of two Slater determinants, one for each orientation of the spin, built up from HO single-particle wave functions. So, naturally we consider the magic numbers of the HO potential well. As far as we know there is only a previous variational calculation on $^3$He droplets, [70] with the Aziz interaction HFDHE-2. The trial wave function of Ref. 70 consists of a Jastrow factor containing pair and triplet correlation functions, plus the back-flow correlations showed at the beginning of this section. In Table 15 are compared J-TICI3 and VMC results. The J-TICI3 variational upper bounds to the ground state energies are significatively lower than the VMC ones, by 1 K, which is quite important in the threshold region we are considering. Both calculations indicate that the minimum number of atoms required to have a bound system lies between 20 and 40 atoms. This was also the conclusion of the density functional calculations. So, we have to deal with open-shell systems, and the problem may be faced in a simple way.

In a first approach, HO single-particle wave functions in the cartesian basis are used, thus breaking the spatial rotational invariance. The rotational symmetry is partially restored by restricting the occupation of levels, so as to maintain symmetry under 90 degrees rotations, in the form reported in Table 16. For instance, there

Table 16. Orbital occupation numbers for configurations of $n$ particles in the $1f2p$ open shell in Cartesian coordinates.

| $n$ | $x^3$ | $y^3$ | $z^3$ | $x^2y$ | $x^2z$ | $y^2x$ | $y^2z$ | $z^2x$ | $z^2y$ | $xyz$ |
|---|---|---|---|---|---|---|---|---|---|---|
| 10 | 1 | 1 | 1 | 1 | 1 | 1 | 1 | 1 | 1 | 1 |
| 9 | 1 | 1 | 1 | 1 | 1 | 1 | 1 | 1 | 1 | 0 |
| 7 | 0 | 0 | 0 | 1 | 1 | 1 | 1 | 1 | 1 | 1 |
| 6 | 0 | 0 | 0 | 1 | 1 | 1 | 1 | 1 | 1 | 0 |
| 4 | 1 | 1 | 1 | 0 | 0 | 0 | 0 | 0 | 0 | 1 |
| 3 | 1 | 1 | 1 | 0 | 0 | 0 | 0 | 0 | 0 | 0 |
| 1 | 0 | 0 | 0 | 0 | 0 | 0 | 0 | 0 | 0 | 1 |

is only one possibility to put four atoms with a given spin, say spin up, in the $fp$ shell; this corresponds to the occupation of the $x^3$, $y^3$, $z^3$ and $xyz$ orbitals. The computational benefit of this simple-minded approach is substantial: the reference state is the product of two Slater determinants, one for spin up and the other for spin down, instead of a linear combination of determinants. Moreover, the single-particle wave functions are real. On the other hand, the trial wave function has not good $L$ and $S$ quantum numbers in general. Regarding the spin, it will be greater than or equal to $S_z$, which is one half of the difference between the number of spins up and spins down. Nevertheless, those states with *maximum* $S_z$ do have good spin. In practice, for $N > 30$ they correspond to having a full spin up $1f2p$ shell and a partially occupied spin down $1f2p$ shell.

The calculations of Ref. 69 indicate that the threshold for stability is around 34 and 35 atoms, as the resulting ground state energies are $0.09 \pm 0.06$ and $-0.33 \pm 0.09$ K, respectively. Moreover, it turns out that for a given value of N, the lowest energy is reached when the spin is maximum, thus confirming the results obtained in a phenomenological density functional calculation. [68]

The fact that the $N = 34$ case is really on the binding threshold suggested the convenience of carrying out a further study considering full rotational symmetry, [71] although filling separately the $1f$ and $2p$ shells, so that the mixing of configurations in both shells is not allowed. Results corresponding to bound states are shown in Table 17. In two cases the Table includes two configurations, one with active $f$ shell and the other with active $p$ shell. We may summarize the the main results of this refined approach by saying that the relevant quantum number is the total spin $S$, the ground state energies are almost independent of $L$ and the balancing of particles between the $p$ and the $f$ shell seems to be of no importance. This last result should not be a surprise, because TICI2 and TICI3 are actually an unusual way of configuration mixing. Due to the self-adaptability of the TICI amplitudes and the variational procedure, calculations starting at different configurations could finally evolve to almost the same wave function.

An interesting bridge between these microscopic calculations [71] and the non-local density functional approach [68] is to make an *effective interaction* analysis of the microscopic results in the Cohen and Kurath sense. [72] The relevant quantities

Table 17. Good angular momentum states for configurations filling either the subshell $1f$ or $2p$.

| $N$ | Conf. | $L$ | $S$ | $E$ (K) |
|---|---|---|---|---|
| 34 | $p^6 f^8$ | 3 | 3 | $-0.03 \pm 0.05$ |
| | | 7 | 2 | $0.17 \pm 0.08$ |
| | | 8 | 2 | $0.35 \pm 0.12$ |
| | | 10 | 1 | $0.41 \pm 0.06$ |
| | | 11 | 1 | $0.46 \pm 0.10$ |
| | | 12 | 0 | $0.55 \pm 0.09$ |
| 35 | $p^6 f^9$ | 1 | 5/2 | $-0.52 \pm 0.09$ |
| | | 3 | 5/2 | $-0.59 \pm 0.09$ |
| | | 5 | 5/2 | $-0.39 \pm 0.06$ |
| | | 8 | 3/2 | $-0.16 \pm 0.08$ |
| | | 9 | 3/2 | $-0.23 \pm 0.08$ |
| | | 10 | 1/2 | $-0.15 \pm 0.10$ |
| | | 11 | 1/2 | $-0.03 \pm 0.06$ |
| 36 | $p^6 f^{10}$ | 0 | 2 | $-0.97 \pm 0.08$ |
| | | 2 | 2 | $-1.04 \pm 0.10$ |
| | | 3 | 2 | $-1.01 \pm 0.10$ |
| | | 4 | 2 | $-0.93 \pm 0.08$ |
| | | 6 | 2 | $-1.00 \pm 0.10$ |
| | | 8 | 1 | $-0.80 \pm 0.07$ |
| | | 9 | 1 | $-0.78 \pm 0.09$ |
| | $p^2 f^{14}$ | 1 | 1 | $-0.69 \pm 0.09$ |
| 38 | $p^6 f^{12}$ | 1 | 1 | $-2.23 \pm 0.09$ |
| | | 3 | 1 | $-2.22 \pm 0.09$ |
| | | 5 | 1 | $-2.21 \pm 0.05$ |
| | $p^4 f^{14}$ | 1 | 1 | $-2.31 \pm 0.07$ |

are the two-body matrix elements of the active shell, and the analysis shows that the matrix elements obtained from the microscopic calculations have the same trends than those used in the density functional approach, being of special relevance the preference for aligned spins (antipairing). The overall effect on the energy of these calculations is very small. Again, we see that 34 is in the threshold for a self-bound $^3$He drop, with an energy of $-0.03 \pm 0.05$ K, still larger than the density functional prediction.

## Acknowledgements

We gratefully acknowledge previous collaboration and discussions with R.F. Bishop, M. Portesi, A. Puente and N.R. Walet. One of us (JN) also would like to acknowledge interesting discussions with J. Sánchez-Marín. This work has been supported by DGI (Spain), grants PB97-1139 and BFM2001-0262.

## References

1. F. Coester, Nucl. Phys. **7**, 421 (1958).

2. F. Coester and H. Kümmel, Nucl. Phys. **17**, 477 (1960).
3. J. Čižek, J. Chem. Phys. **45**, 4256 (1966).
4. J. Čižek, Adv. Chem. Phys. **14**, 35 (1969).
5. J. Čižek and J. Paldus, Int. J. Quantum Chem. **5**, 359 (1971).
6. J. Paldus, J. Čižek, and I. Shavitt, Phys. Rev. **A5**, 50 (1972).
7. H. Kümmel, Nucl. Phys. A **176**, 205 (1971).
8. H. Kümmel and K. H. Lührmann, Nucl. Phys. A **191**, 525 (1972).
9. J. G. Zabolitzky, Nucl. Phys. A **228**, 272 (1974); *ibid* 285.
10. H. Kümmel, K. H. Lürmann and J. G. Zabolitzky, Phys. Reports **36**, 1 (1978).
11. R. F. Bishop, Theor. Chim. Acta **80**, 95 (1991).
12. R. J. Bartlett, Ann. Rev. Phys. Chem. **32**, 359 (1981); R. J. Bartlett, J. Phys. Chem. **93**, 1697 (1989); R. J. Bartlett, Theor. Chim. Acta **81**, 71 (1991).
13. J. Paldus and X. Li, Adv. Chem. Phys. **110**, 1 (1999).
14. R. F. Bishop, in *Microscopic Quantum Many-Body Theories and Their Applications,* eds. J. Navarro and A. Polls (Springer-Verlag, Berlin, 1998), p. 1.
15. R. F. Bishop and H. G. Kümmel, Phys. Today **40**, 52 (1987).
16. S. A. Kucharski and R. J. Bartlett, J. Chem. Phys. **97**, 4282 (1992).
17. D. Mukherjee and S. Pal, Adv. Quantum Chem. **20**, 292 (1989).
18. U. Kaldor, Proc. 11th Many-Body Conference, Eds. R. F. Bishop, K. Gernoth, N. R. Walet, and X. Yang, World Scientific, in press.
19. U. Kaldor, in *Microscopic Quantum Many-Body Theories and Their Applications*, eds. J. Navarro and A. Polls (Springer-Verlag, Berlin, 1998), p.71.
20. A. L. Fetter and J. D. Walecka, *Quantum Theory of Many-Particle Systems*, McGraw-Hill, New York, 1971.
21. D. M. Ceperley and B. J. Alder, Phys. Rev. Lett. **45**, 566 (1980).
22. R. F. Bishop and K. H. Lührmann, Phys. Rev. B **17**, 3757 (1978).
23. K. Emrich and J. G. Zabolitzky, Phys. Rev. B **30**, 2049 (1984).
24. R. V. Reid, Ann. Phys. (NY) **50**, 411 (1968).
25. J. H. Heisenberg and B. Mihaila, it Phys. Rev. C **59**, 1440 (1999).
26. R. B. Wiringa, R. A. Smith and T. L. Ainsworth, Phys. Rev. C **29**, 1207 (1984).
27. R. B. Wiringa, V. G. J. Stoks and R. Schiavilla, Phys. Rev. C **51**, 38 (1995).
28. R. F. Bishop, M. F. Flynn, M. C. Boscá, E. Buendía, and R. Guardiola, Phys. Rev. C **42**, 1341 (1990).
29. R. F. Bishop, E. Buendía, M. F. Flynn, and R. Guardiola, J. Phys. G **18**, 1157 (1992).
30. T. A. Brody and M. Moshinsky, *Tables of Transformation Brackets for Nuclear Shell-Model Calculations*, Gordon and Breach. New York, 1967.
31. I. R. Afnan and Y. C. Tang, Phys. Rev. **175**, 1337 (1968).
32. R. Guardiola, I. Moliner, J. Navarro, and M. Portesi, Nucl. Phys. A **628**, 187 (1998).
33. R. Guardiola, I. Moliner, J. Navarro, and M. Portesi, in *Many-Body Theory of Correlated Fermion Systems*, eds. J. M. Arias, M. I. Gallardo, and M. Lozano (World Scientific, 1998) p.25.
34. J. Navarro, R. Guardiola, M. Portesi, and I. Moliner, in *Microscopic Approaches to the Structure of Light Nuclei*, eds. R. F. Bishop and N. R. Walet, Advances in Quantum Many-Body Theory, Vol. 2 (World Scientific, Singapore, 2001), in press.
35. I. Moliner, R. F. Bishop, N. R. Walet, R. Guardiola, J. Navarro, and M. Portesi, Phys. Lett. B **480**, 61 (2000).
36. D. M. Brink and E. Boeker, Nucl. Phys. A **91**, 1 (1967).
37. R. Guardiola, A. Faessler, H. Müther and A. Polls, Nucl. Phys. A **371**, 79 (1981).
38. R. A. Malfliet and J. A. Tjon, Nucl. Phys. A **127**, 161 (1969).
39. R. F. Bishop, E. Buendía, M. F. Flynn and R. Guardiola, J. Phys. G **18**, L21 (1992).

40. S. A. Chin and E. Krotscheck, Nucl. Phys. A **560**, 151 (1993).
41. M. C. Boscá, E. Buendía and R. Guardiola, Phys. Lett. B **198**, 312 (1987); Condensed Matter Theories, Vol. 3, eds. J. Arponen, R. F. Bishop and M. Manninen (Plenum Press, New York, 1987) p. 101.
42. R. Brizzi, M. Fabre de la Ripelle and M. Lassaut, Nucl. Phys. A **596**, 199 (1996).
43. E. Krotscheck, Nucl. Phys. A **465**, 461 (1987).
44. J. G. Zabolitzky and M. H. Kalos, Nucl. Phys. A **356**, 114 (1981); U. Helmbrecht and J. G. Zabolitzky, Nucl. Phys. A **442**, 109 (1985); J. G. Zabolitzky, K. E. Schmidt and M. H. Kalos, Phys. Rev. C **25**, 1111 (1982).
45. G. Co', A. Fabrocini, S. Fantoni, and I. E. Lagaris, Nucl. Phys. A **549**, 439 (1992); G. Co', A. Fabrocini, and S. Fantoni, Nucl. Phys. A **568**, 73 (1994).
46. F. Arias de Saavedra, G. Co', A. Fabrocini, and S. Fantoni, Nucl. Phys. A **605**, 359 (1996).
47. R. Guardiola, P. I. Moliner, J. Navarro, R. F. Bishop, A. Puente, and N. R. Walet, Nucl. Phys. A **609**, 218 (1996).
48. R. F. Bishop, R. Guardiola, I. Moliner, J. Navarro, M. Portesi, A. Puente, and N. R. Walet, Nucl. Phys. A **643**, 243 (1998).
49. E. Buendía, F.J. Gálvez, J. Praena and A. Sarsa, J. Phys. G **26**, 1795 (2000).
50. J. C. Owen, Ann. Phys. (NY) **118**, 373 (1979).
51. W. C. Rheinboldt and J. Burkardt, ACM TOMS **9**, 236 (1983).
52. I. Moliner, N. R. Walet, and R. F. Bishop, in preparation.
53. D. Gogny, P. Pires and R. de Tourreil, Phys. Lett. B **32**, 591 (1970).
54. R. de Tourreil and D. W. L. Sprung, Nucl. Phys. A **201**, 193 (1973).
55. J. Carlson, Phys. Rev. C **38**, 1879 (1988).
56. R. Guardiola, P. I. Moliner and J. Navarro, *Condensed Matter Theories, Vol. 11,* eds. E. V. Ludeña, P. Vashishta and R. F. Bishop (Nova Science, New York, 1996) p. 485.
57. R. Guardiola, P. I. Moliner, and J. Navarro, Phys. Lett. B **B383**, 243 (1996).
58. R. Guardiola, M. Portesi and J. Navarro, Phys. Rev. B **60**, 6288 (1999).
59. K. B. Waley, Int. Rev. Phys. Chem. **13**, 41 (1994).
60. M. Lewerenz, J. Chem. Phys. **106**, 4596 (1997).
61. R. Guardiola, J. Navarro and M. Portesi, Phys. Rev. B **63**, 224519-1 (2001).
62. S. A. Chin and E. Krotscheck, Phys. Rev. B **45**, 852 (1992).
63. O. Bohigas, A. M. Lane, and J. Martorell, Phys. Reports **51**, 267 (1979).
64. R. P. Feynman and M. Cohen, Phys. Rev. **102**, 1189 (1956).
65. M. V. Krishna and K. B. Whaley, Phys. Rev. Lett. **64**, 1126 (1990); J. Chem. Phys. **93**, 746 (1990).
66. S. A. Chin and E. Krotscheck, Phys. Rev. Lett. **65**, 2658 (1990).
67. R. Guardiola, M. Portesi, and J. Navarro, Phys. Rev. **B63**, 224519-1 (2001).
68. M. Barranco, J. Navarro and A. Poves, Phys. Rev. Lett. **78**, 4729 (1997).
69. R. Guardiola and J. Navarro, Phys. Rev. Lett. **84**, 1144 (2000).
70. V. R. Pandharipande, S. C. Pieper and R. B. Wiringa, Phys. Rev. **B34**, 4571 (1986).
71. R. Guardiola, Phys. Rev. B **62**, 3416 (2000).
72. S. Cohen and D. Kurath, Nucl. Phys. **73**, 1 (1965).

CHAPTER 4

# EXPERIMENTS WITH A RUBIDIUM BOSE-EINSTEIN CONDENSATE

E. Arimondo, D. Ciampini, O. Morsch, J. H. Müller

*INFM, Dipartimento di Fisica, Università di Pisa, Via Buonarroti 2, I-56127 Pisa, Italy*
*E-mail: arimondo@df.unipi.it*

We report on three experiments based on a rubidium Bose-Einstein condensate. We examine the atomic micromotion of the condensate within a time dependent time orbiting potential. We have loaded Bose-Einstein condensates into one-dimensional, off-resonant optical lattices and accelerated them by chirping the frequency difference between the two lattice beams. We have studied the photoionization processes of the condensate.

## 1. Introduction

Since the first experimental realizations, many aspects of Bose-Einstein condensed atomic clouds (BECs) have been studied[1], ranging from collective excitations to superfluid properties and quantized vortices. Such a controlled realization of dense and cold atomic samples opened the way for investigations in the low and ultra-low temperature regimes not accessible with conventional techniques. Magneto optical traps and magnetic traps where atoms are laser cooled and confined in space through the applications of counterpropagating laser beams and/or magnetic fields, produce cold atomic samples with sufficient atoms and atomic density to perform accurate measurements of spectroscopy and collisional physics.

Up to now, BEC in alkali atoms within magnetic traps has been achieved in two different kinds of magnetic traps[1]: static traps of the Ioffe-Pritchard type, and dynamic ones, also called time-orbiting-potential (TOP) traps. In the latter, a quadrupole magnetic field is modified by a superimposed rotating bias field which eliminates the zero of the magnetic field at the origin and, in an adiabatic approximation, creates an effective harmonic potential when averaged over the fast rotation of the instantaneous field. This approximation accounts for most of the observed features of these traps. There are, however, residual non-adiabatic effects, leading to a micromotion of the atoms at the frequency of the rotating bias field and a dependence of their equilibrium position in the trap on the sense of rotation of the bias field. Despite the wide-spread use of TOP-traps, these effects, which can be appreciable under certain experimental conditions, have not been studied experimentally in a consistent manner so far. We have recently investigated the detailed measurements of such effects on a condensate of rubidium atoms and compared our results with analytical models and numerical simulations[2]. Those measurements will be briefly reported here.

The properties of ultra-cold atoms in periodic light-shift potentials in one, two and three dimensions have been studied extensively in the past ten years[3]. In near-resonant and far-detuned optical lattices, a variety of phenomena have been studied, such as the magnetic properties of atoms in optical lattices, revivals of wave-packet oscillations, and Bloch oscillations in accelerated lattices [4,5,6]. The properties of BECs in periodic potentials constitute a vast new field of theoretical research, for instance [7,8,9,10,11], and several experiments in the pulsed standing wave regime [12,13] as well as studies of the tunneling of BECs out of the potential wells of a shallow optical lattice in the presence of gravity [14] or by displacing the magnetic trap minimum[15] have taken the first steps in that direction. In a recent work [16] we have presented the results of experiments on BECs of $^{87}$Rb atoms in accelerated optical lattices. In particular, we have demonstrated coherent acceleration of BECs adiabatically loaded into optical lattices. For small values of the lattice depth we have observed Bloch oscillations preserving the condensate wavefunction. Those oscillations exhibited Landau-Zener breakdown when the lattice depth was further reduced and/or the acceleration increased. The properties of a Bose-Einstein condensate located in a periodic optical lattice are described through the Gross-Pitaevskii equation valid for the single-particle wavefunction. Because the nonlinear term in that equation reproduces the spatial periodicity of the wavefunction, the condensate ground state also is periodic. We have demonstrated that the role of the nonlinear interaction term may be described through an effective potential in a noninteracting gas model.

In the past few years much attention has been concentrated on the photoionization of cold alkali atoms[17]. More recently ultra-cold plasmas have been created making use of laser cooled atomic samples and laser excitation to high Rydberg states[18]. The realization of Bose-Einstein condensates (BEC) of alkali atom vapors have attracted much interest to new aspects of photon-matter interaction arising from the coherent nature of the atomic ensemble. Theoretical attention has been paid recently to the photoionization of a BEC by monochromatic laser light [19]. The products of the photoionization process (electrons and ions) obey Fermi-Dirac statistics. Due to the coherent nature of the initial atomic ensemble and the narrow spectral width of the laser ionization source, the occupation number of the electron/ion final states can become close to unity, especially for photoionization close to threshold. Under those conditions the condensate ionization rate should be slowed down by a Pauli blockade process and finally determined by a balance between the laser light acting on the atoms and the rate of escape of the charged products from the system. Stimulated by these theoretical predictions an experiment investigating the photoionization process within a rubidium condensate has been started. Few experimental observations have been performed based on pulse laser sources. The progress realized up to now with the photoionization of cold rubidium atoms and of the rubidium condensate will be schematically reported.

Section II discusses the experimental apparatus developed for the Bose-Einstein condensation of rubidium and the micromotion observations. The following two Sections report the results of the experimental investigations on the condensate properties within an optical lattice, and the condensate photoionization.

## 2. Micromotion

Our experimental apparatus and procedure for creating BECs are described in detail elsewhere [20]. Briefly, we use a double magneto-optical trap (MOT) system with two cells connected by a glass tube to capture, cool and transfer atoms into a triaxial TOP magnetic trap. The latter consists of a pair of quadrupole coils oriented along a horizontal axis and two pairs of bias-field coils, one incorporated into the quadrupole coils and the other along a horizontal axis perpendicular to that of the quadrupole coils. After transferring the atoms into the TOP-trap, we evaporatively cool them first by reducing the circle-of-death radius (which describes the location of the zero of field at which Majorana spin-flips are induced) and finally by applying an RF-field. Once Bose-Einstein condensation has occurred, we adiabatically change the quadrupole gradient and the bias field to the desired values. The instantaneous magnetic field seen by the atoms can be written as

$$\vec{B} = \hat{\mathbf{i}}\big[2b'x + B_o \sin(\Omega_{TOP}t)\big] + \hat{\mathbf{j}}\big[b'y + B_o \cos(\Omega t)\big] + \hat{\mathbf{k}}b'z \tag{2.1}$$

where $b'$ is the quadrupole gradient along the $z$-axis, and $B_o$ and $\Omega$ are the magnitude and the frequency of the bias field, respectively.

In our apparatus $5 \times 10^7$ atoms captured in the lower MOT were transferred into the TOP [2]. Subsequently, the atoms were evaporatively cooled down to the transition temperature for Bose-Einstein condensation, and after further cooling we obtained condensates of $\approx 10^4$ atoms without a discernible thermal component in a magnetic trap with frequencies around $15 - 30\,\mathrm{Hz}$.

The time dependent potential experienced by the atoms under the action of the magnetic field of Eq. 2.1 is

$$U = -\vec{\mu} \cdot \Big(\hat{\mathbf{i}}[2b'x + B_o sin(\Omega t)] + \hat{\mathbf{j}}\big[b'y + B_o cos(\Omega t)\big] + \hat{\mathbf{k}}b'z\Big). \tag{2.2}$$

In order to analyze the atomic motion within this time-dependent potential, we impose the time-dependent atomic position $\vec{r}(t)$ to be composed of a slow variable $\vec{R}(t)$ and a fast variable $\vec{\zeta}(t)$:

$$\vec{r}(t) = \vec{R}(t) + \vec{\zeta}(t). \tag{2.3}$$

In the time averaged approximation, *i.e.* limiting to the slow variable component, and applying the adiabatic approximation of the magnetic moment aligned along the local magnetic field, the potential becomes,

$$U_o = -\mu \frac{b'^2}{B_o}\left(x^2 + \frac{1}{4}y^2 + \frac{1}{2}z^2\right). \tag{2.4}$$

The atoms thus experience a triaxial harmonic potential with frequency ratios $\omega_x : \omega_y : \omega_z$ of $2 : 1 : \sqrt{2}$. In the next order approximation, keeping the fast variable, the potential contains time-dependent components at frequency $\Omega$:

$$\Delta U(t) = \mu b' \left[ -2x cos(\Omega t) + y sin(\Omega t) \right], \tag{2.5}$$

producing a condensate motion along an ellipse with time-dependent $x$ and $y$ variables

$$x(t) = -\frac{2\mu b'}{M\Omega^2} cos(\Omega t), \tag{2.6}$$

$$y(t) = \frac{\mu b'}{M\Omega^2} sin(\Omega t). \tag{2.7}$$

In order to observe this micromotion of the condensate at the TOP-frequency (usually $2\pi \times 10\, s^{-1}$ for our apparatus), the trap fields were switched off in a time that was short compared to the period of the TOP-field. Our apparatus allowed us to switch off the quadrupole in less than $50\mu s$. The instant at which the trapping fields were switched off could be determined to better than 1% of a TOP-cycle. After switching off the quadrupole field at a well-defined phase of the rotating bias field, we imaged the atoms after several ms of free fall. Because the atoms executed an elliptic micromotion at the frequency of the bias field, we could deduce from the measured position of the atoms after the free fall the initial velocity component along the $x$-direction. We have found a spatial amplitude of the micromotion of the order of 100 nm for our trap parameters and hence below our resolution limit of 5 $\mu$m, so that the position of the atoms along the $x$-axis after the time-of-flight was determined solely by their initial velocity component in that direction. In ref.[2] measurements of the $x$-component of the micromotion velocity as a function of the bias-field phase presented an oscillating behavior at frequency $\Omega$, with a typical velocity amplitude around $5\,\mathrm{mm\,s^{-1}}$.

## 3. BEC in 1D optical lattice

In the experiments on optical lattices, a horizontal 1-D optical lattice was switched on, the interaction between the condensate and the lattice taking place both inside the magnetic trap and after switching it off. The lattice beams were created by a 50 mW diode slave-laser injected by a grating-stabilized master-laser blue-detuned by $\Delta \approx 28 - 35$ GHz from the $^{87}$Rb resonance line. The laser light was split and passed through two acousto-optic modulators (AOMs) that were separately controlled by two phase-locked RF function generators operating with a frequency separation $\delta$. The beams at the AOM outputs generated the optical lattice. An acceleration of the lattice was effected by applying a linear ramp to the detuning $\delta$. For the values of the detuning and laser intensity used in our experiment, the spontaneous photon scattering rate ($\approx 10\,\mathrm{s^{-1}}$) was negligible during the interaction times of a few milliseconds. The lattice constant of the periodic optical potential is $d = \pi / \sin(\theta/2)k$, with $k$ the laser wavenumber and $\theta$ the angle between the two laser beams creating the 1-D optical lattice. In our experimental setup, we realized a counter-propagating lattice geometry with $\theta = 180\,\text{deg}$ and an angle geometry with $\theta = 29\,\text{deg}$, leading to optical lattice constants $d$ of 0.39 and 1.56 $\mu$m, respectively, with the large lattice constant meaning also that only a few potential wells were occupied by the condensate.

In a preliminary optical lattice experiment aimed to determine the depth of the periodic optical potential, and performed after switching off the magnetic field, we flashed on the counterpropagating lattice with a detuning $\delta = \hbar k^2/M = 15.08\,\text{kHz}$ between the two lasers making up the optical lattice for a time of $10 - 400\,\mu\text{s}$. The detuning chosen corresponded to the first Bragg resonance, causing the condensate to undergo Rabi oscillations between the momentum states $|p = 0\rangle$ and $|p = 2\hbar k\rangle$. From the measured Rabi frequency we could then determine the lattice depth. The results of those measurements fell short by a factor of about 2 of the theoretical prediction, which was mainly due to the $10 - 15\%$ uncertainty in our intensity measurements and imperfections in the alignment and polarization of the lattice beams.

In order to accelerate the condensate, we adiabatically loaded it into the lattice by switching one of the lattice beams on suddenly and ramping the intensity of the other beam from 0 to its final value in a time between $200\,\mu\text{s}$ and few ms. Thereafter, the linear increase of the detuning $\delta$ provided a constant acceleration $a = \frac{\lambda}{2}\frac{d\delta}{dt}$ of the optical lattice. After a few milliseconds of acceleration, the lattice beams were switched off and the condensate was imaged after another $10 - 15\,\text{ms}$ of free fall. As the lattice can only transfer momentum to the condensate in units of $2\hbar k$, the acceleration of the condensate showed up as diffraction peaks corresponding to higher momentum classes as time increased (Fig. 1). The optical lattice potential depth was $U_0 \approx 0.29\,E_\text{B}$ with $E_\text{B} = \hbar^2(2\pi)^2/Md^2$ the lattice Bloch energy. Since for our magnetic trap parameters the initial momentum spread of the condensate was much less than a recoil momentum of the optical lattice (the Bloch momentum $p_\text{B} = \hbar 2\pi/d$) and since the adiabatic switching transfers the spread into lattice quasimomentum, the different momentum classes $|p = \pm n p_\text{B} = \pm 2n\hbar k\rangle$ (where $n = 0, 1, 2, \ldots$) occupied by the condensate wavefunction could be resolved directly after the time-of-flight.

The average velocity of the condensate was derived from the occupations of the different momentum states. Figure 2(a) shows the results of the acceleration of a condensate with $U_\text{o} = 0.29\,E_\text{B}$, and $a = 9.81\,\text{m}\,\text{s}^{-2}$. In the rest-frame of the lattice (Fig. 2 (b)), one clearly sees Bloch oscillations of the condensate velocity corresponding to a Bloch-period $\tau_B = \frac{h}{Mad} = 1.2\,\text{ms}$. The shape of these oscillations agrees well with the theoretical curve calculated from the lowest energy band of the lattice. As described in [5], the acceleration process within a periodic potential can also be viewed as a succession of adiabatic rapid passages between momentum states $|p = 2n\hbar k\rangle$ and $|p = 2(n+1)\hbar k\rangle$. Up to $6\,p_\text{B}$ momentum could be transferred to the condensate preserving the phase-space density of the condensate, a result indicating that no heating or reduction of the condensed fraction occurred.

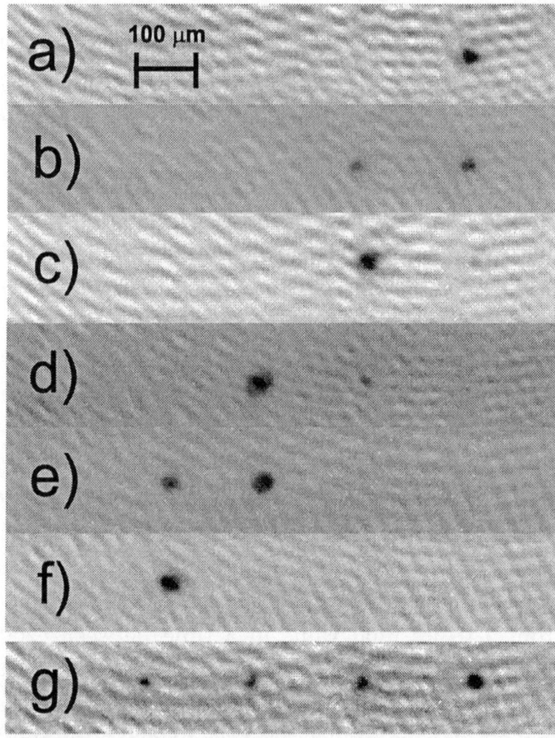


Fig. 1. Coherent acceleration of a Bose-Einstein condensate. In (a)-(f) $U_0 = 0.29\,\mathrm{E_B}$, $a = 9.81\,\mathrm{m\,s^{-2}}$, the condensate accelerated for $0.1, 0.6, 1.1, 2.1, 3.0$ and $3.9\,\mathrm{ms}$, respectively. In (g) the result of $2.5\,\mathrm{ms}$ acceleration with the same lattice depth as above, but with $a = 25\,\mathrm{m\,s^{-2}}$. In this case, a fraction of the condensate undergoes Landau-Zener tunneling out of the lowest band each time a Bragg-resonance is crossed. The separations between the different spots vary because detection occurred after different time delays.

Similar experiments were performed in the angle-tuned lattice geometry. Because of the reduced Bloch velocity $v_{\mathrm{B}} = p_{\mathrm{B}}/M$ in this geometry, the acceleration process was extremely sensitive to any initial velocity of the condensate, which in our TOP trap is intrinsically given by the micromotion at the frequency of the bias field. In order to counteract these effects, we performed experiments inside the magnetic trap, eliminating the velocity of the condensate relative to the lattice by phase-modulating one of the lattice beams at the same frequency and in phase with the rotating bias field of the TOP trap. In this way, in the rest frame of the lattice the micromotion was compensated. In Fig. 2 (c), the results of the acceleration of the condensate with a nominal lattice depth of $1.38E_{\mathrm{B}}$ are shown together with the theoretical curves for $U_0 \approx 1.38\,E_{\mathrm{B}}$ and the assumed effective potential $U_{\mathrm{eff}} \approx 0.88\,E_{\mathrm{B}}$. As expected from the band-structure calculations, the Bloch oscillations are much less pronounced for these parameter values

The properties of a Bose-Einstein condensate located in a periodic optical lattice with depth $U_0$ are described through the Gross-Pitaevskii equation valid for the

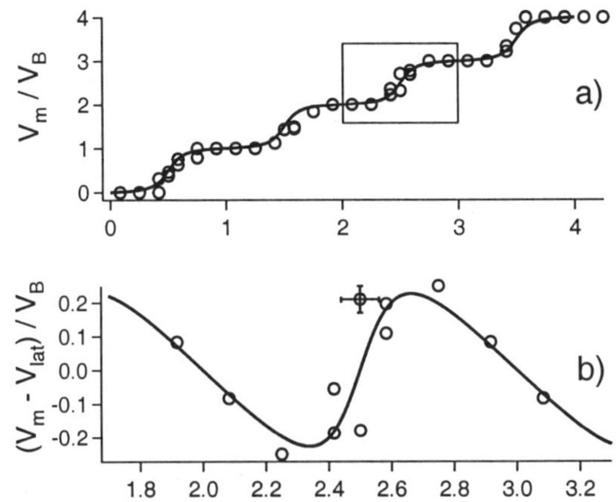


Fig. 2. Bloch oscillations of the condensate mean velocity $v_m$ in an optical lattice. (a) Acceleration in the counter-propagating lattice with $d = 0.39\,\mu$m, $U_0 \approx 0.29\,E_B$ and $a = 9.81\,\mathrm{m\,s^{-2}}$. Solid line: theory. (b) Bloch oscillations in the rest frame of the lattice, along with the theoretical prediction (solid line) derived from the shape of the lowest Bloch band.

single-particle wavefunction. The nonlinear interaction of the condensate may be described through a dimensionless parameter [9,10,11] $C = g/E_B$ corresponding to the ratio of the nonlinear interaction term $g = 4\pi n_p \hbar^2 a/M$ and the lattice Bloch energy $E_B$. The parameter $C$ contains the peak condensate density $n_p$ in the Thomas-Fermi limit, the s-wave scattering length $a$, the atomic mass $M$, and the lattice constant $d$ of the 1-D optical lattice. From this it follows that a small angle $\theta$ should result in a large lattice constant, whence a large interaction term $C$.

The role of the nonlinear interaction term of the Gross-Pitaevskii equation may be described through an effective potential in a non-interacting gas model [9,10]. In the perturbative regime of shallow optical lattices the effective potential is [9]

$$U_{\text{eff}} = \frac{U_o}{1 + 4C}, \tag{3.1}$$

with the same spatial structure as the optical lattice. We therefore expect that for large values of $C$, *i.e.* large mean-field effects, the effective optical lattice potential acting on the condensate should be significantly reduced.

In order to measure more accurately the variation of the effective potential $U_{\text{eff}}$ with the interaction parameter $C$, we studied Landau-Zener tunneling out of the lowest Bloch band for small lattice depths in both geometries. To this end, the acceleration of the lattice was increased in such a way that when the condensate crossed the edge of the Brillouin zone, an appreciable fraction of the atoms tunneled across the band gap into the first excited band, see Fig. 1 (c). According to Landau-Zener theory within the effective potential $U_{\text{eff}}$, this fraction is [5]

$$r = \exp\left(-\frac{\pi U_{\text{eff}}^2}{8\hbar p_B a}\right), \tag{3.2}$$

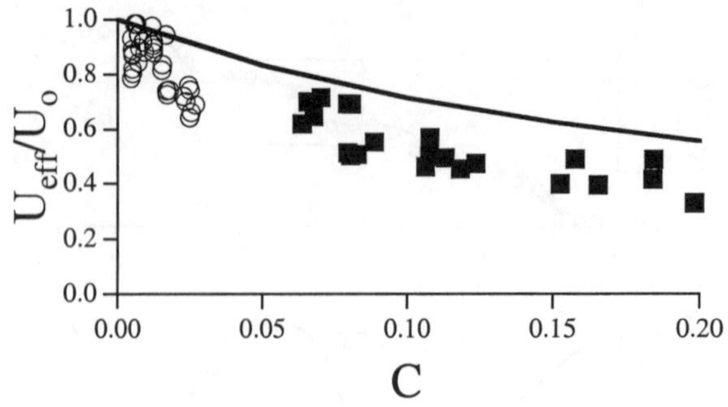


Fig. 3. Dependence of the effective potential $U_{\text{eff}}$ on $C$ parameter for the two lattice geometries. The experimental results for the counter-propagating (circles) and angle geometries (squares) are plotted together with the theoretical prediction (solid line). Parameters in these experiments were $a = 23.4\,\text{m}\,\text{s}^{-2}$ and $U_0 = 0.28\,E_B$ for the counter-propagating lattice and $a = 3.23\,\text{m}\,\text{s}^{-2}$ and $U_0 = 0.71\,E_B$ for the angle geometry.

giving a velocity $v_m = (1 - r)v_B$ of the condensate at the end of the acceleration process for a final velocity $v_B$ of the lattice. We verified that this formula correctly described the tunneling of the condensate in our experiment by varying both the potential depth and acceleration in the conditions where the mean-field interactions where negligible. Thereafter, we studied the variation of the final mean velocity $v_m$ as a function of the condensate density for the two lattice geometries. The density was varied by changing the mean frequency of the magnetic trap (from $\approx 25$ Hz to $\approx 100$ Hz). From the mean velocity the effective potential was then calculated using the Landau-Zener formula given above. Fig. 3 shows the ratio $U_{\text{eff}}/U_0$ as a function of the dimensionless parameter C for the counter-propagating geometry and the angle-geometry. As expected, the reduction of the effective potential is much larger in the angle geometry. The theoretical prediction of Eq. (3.1) is also shown in the figure, with the potential $U_0$ calculated taking into account losses at the cell windows and imperfections of the polarizations of the lattice beams. The residual combined error due to uncertainties in the absolute intensity measurements, the position of the beam axes relative to the position of the BEC and a small initial velocity due to sloshing of the BEC in the magnetic trap, as well as a systematic error due to the difference in position of about 300 $\mu$m of the condensate between the weakest and the strongest trap used, was estimated at about 20%. Because the theoretical analysis of ref. [9] supposes a uniform atomic density, while the experiment shows a spatial dependence of the atomic density given by the magnetic trapping, in dealing with the mean-field effect, we supposed the peak atomic density to represent the most appropriate value to insert into the $C$ parameter of the theoretical comparison.

## 4. Condensate photoionization

Photoionization of the cold atomic cloud was mainly effected by pulsed lasers. The energy of 4.177 eV required for the rubidium ionization from the ground state was provided either by the one-photon absorption at a wavelength of 296 nm, or by a two photon non-resonant transition at 593 nm. The absorption of one 296 nm photon or two 593 nm photons by the ground state rubidium atom produced an electron with kinetic energy around 10 meV. The two-photon 594 nm radiation was produced by an excimer-pumped dye laser, with intensities in the tens of MW $cm^{-2}$ range and a pulse duration of 16 ns. Laser pulse repetition rates up to 12 Hz were used. The 297 nm radiation was generated from the 594 nm radiation through a frequency doubling process within a BBO crystal, with intensity in the 1 MW $cm^{-2}$ range. The photoionizing laser was focused on the cloud of rubidium atoms, whose dimensions were in the 5 $\mu$m range for the condensate cloud and in the 0.2-1 mm range for the MOT cloud. The dye laser was focused down to 50 $\mu$m, while for the pulsed ultraviolet radiation, owing to the beam astigmatism, beam waists of 200 $\mu$m and 400 $\mu$m in the two directions were achieved. For a photoionization experiment with the transverse section of the photoionizing laser smaller than the cold cloud transverse area, the ionization efficiency is reduced by the ratio between the transverse laser area and the cloud area.

Because the ionization products obey Fermi-Dirac statistics, the photoionization rate depends on the final state occupation through $1 - n_e$, with $n_e$ the electron density. Thus, in presence of $n_e$ values close to unity, the ionization process is blocked by the previous electron production. The photoionization experiment is not yet at the stage of checking the predictions of this phenomenon of Pauli blockade introduced in ref. [19]. However our preliminary observations have already produced some interesting results. We measured the photoionization rates from the losses of the magnetic trapped cold atoms or trapped condensate by monitoring the atoms remaining in the magnetic trap after application of the photoionization laser. The major result up to now is the observation that the ionization process does not destroy the condensate itself and could be used for a fine control of the production of very cold positive and negative charges, without modifying the atoms remaining in the condensate. However large losses were measured for the condensate following the photoionization within the magnetic trap. Additional investigations are required in order to test the source of these large losses. In the experiment with a single ionizing pulse on the BEC sample, we measured also the density profile modifications of the condensed cloud produced by the photoionization laser. We noticed that the photoionization pulse ejected a secondary condensate from the original one. After its creation, the secondary condensate separated in space from the original one. We have realized that starting from a condensate in the $(F = 2, m = 2)$ state the secondary cloud was constituted by atoms in the $(F = 2, m = 1)$ Zeeman state. The process taking place within the condensate could be described as an atom laser emission from the original condensate following the excitation by the photoionizing laser. However we don't know which process has produced the coherent transfer of condensate atoms from the initial Zeeman level to the final one. The motion of charged particles within a gaseous condensate should be compared to the motion of positive and negative charges within superfluid helium, a research area quite important for the determination of the properties of superfluid liquid helium. Thus the study of the motion of the positive and negative particles within a gaseous condensate could probe the condensate excitations and the superfluid properties.

## 5. Conclusions and acknowledgments

In summary, we have investigated micromotion of the condensate within the time-dependent potential of the TOP trap. Furthermore we investigated the coherent acceleration of Bose-Einstein condensates adiabatically loaded into a 1-D optical lattice as well as Bloch oscillations and Landau-Zener tunneling out of the lowest Bloch band. The results obtained are in good agreement with the available theories and extend the corresponding work on ultra-cold atoms in optical lattices into the domain of Bose-Einstein condensates. We have verified that Bloch oscillations and Landau-Zener tunneling are produced in a condensate without a noticeable modification of the phase-space and condensate depletion. Finally we have initiated experiments of photoionization of the condensate by pulsed laser radiation.

The collaboration of M. Anderlini, M. Cristiani, F. Fuso and R. Mannella to parts of the research here presented is gratefully acknowledged. This work was supported by the INFM PRA PhotonMatter Project, by the MURST through a PRIN200 Initiative, and by the European Community through Cold-Quantum Gases Network, contract HPRN-CT-2000-00125. O.M. gratefully acknowledges a Marie-Curie Fellowship from the European Union.

## References

1. See reviews by W. Ketterle *et al.,* and by E. Cornell *et al.* in *Bose-Einstein condensation in atomic gases,* edited by M. Inguscio, S. Stringari and C. Wieman (IOS Press, Amsterdam) (1999).
2. J. H. Müller, *et al.*, Phys. Rev. Lett. **85**, 4454 (2000).
3. For reviews see P. S. Jessen and I. H. Deutsch, Adv. At. Mol. Opt. Phys. **37**, 95 (1996); G. Grynberg and C. Triché, in *Coherent and Collective Interactions of Particles and Radiation Beams* edited by A. Aspect; W. Barletta, and R. Bonifacio, (IOS Press, Amsterdam) (1996); D. R. Meacher, Cont. Phys. **39**, 329 (1998).
4. Qian Niu, *et al.*, Phys. Rev. Lett. **76**, 4504 (1996); S. R. Wilkinson, *et al.*, Phys. Rev. Lett. **76**, 4512 (1996); C. F. Bharucha, *et al.*, Phys. Rev. A **55**, R857 (1997).
5. M. Ben-Dahan *et al.*, Phys. Rev. Lett. **76** 4508 (1996); E. Peik *et al.*, Phys. Rev. A **55**, 2989 (1997).
6. M. Raizen, C. Salomon, and Q. Niu, Phys. Today **50**, 7, 30 (1997).
7. K. Berg-Sørensen and K. Mølmer, Phys. Rev. A **58**, 1480 (1998).
8. D. Jaksch, *et al.*, Phys. Rev. Lett. **81**, 3108 (1998); J. Javanainen, Phys. Rev. A **60**, 4902 (1999); A. Brunello, *et al.*, Phys. Rev. Lett. **85**, 4422 (2000); O. Zobay and B. M. Garraway, Phys. Rev. A **61**, 033603 (2000); S. Pötting, *et al.*, Phys. Rev. A **64**, 023604 (2001).
9. Dae-Il Choi and Qian Niu, Phys. Rev. Lett. **82**, 2022 (1999).
10. M. L. Chiofalo, M. Polini, and M. Tosi, Eur. Phys. J. D **11**, 371 (2000).
11. A. Trombettoni and A. Smerzi, Phys. Rev. Lett. **86**, 2353 (2001).

12. M. Kozuma *et al.*, Phys. Rev. Lett. **82**, 871 (1999); E.W. Hagley *et al.*, *ibidem* **83**, 3112 (1999)); J. E. Simsarian *et al.*, *ibidem* **85**, 2040 (2000); Yu. B. Ovchinnikov, *et al.*, *ibidem* **83**, 284 (1999).
13. J. Stenger *et al.*, Phys. Rev. Lett. **82**, 4569 (1999); and erratum: *ibidem* **84**, 2283 (2000).
14. B. P. Anderson and M. Kasevich, Science **282**, 1686 (1998); C. Orzel *et al.*, Science **231**, 2386 (2001).
15. S. Burger, *et al.*, Phys. Rev. Lett. **86**, 4447 (2001); F. S. Cataliotti, *et al*, Science **293**, 843 (2001).
16. O. Morsch *et al.*, Phys. Rev. Lett. **87**, 140402 (2001).
17. See for instance the Pisa work: O. Maragó, *et al.*, Phys. Rev. A **57**, R4110 (1998); F. Fuso, *et al.*, Opt. Comm. **173**, 223 (2000); E. Arimondo, *et al.* Appl. Surf. Sci. **154–155**, 527 (2000).
18. T. C. Killian, *et al.*, Phys. Rev. Lett. **83**, 4776 (1999); S. Kulin, *et al.*, Phys. Rev. Lett. **85**, 318 (2000); M. P. Robinson, *et al.*, Phys. Rev. Lett. **85**, 4466 (2000); T. C. Killian, *et al.*, *Phys. Rev. Lett.* **86**, 3759 (2001); Dutta S. K., *et al.*, Phys. Rev. Lett. **86**, 3993 (2001).
19. I. E. Mazets, Quantum Semiclass. Opt. **10**, 675 (1998).
20. J. H. Müller *et al.*, J. Phys. B: Atom. Mol. Opt. Phys. **33**, 4095 (2000).
21. E. W. Hagley, *et al.*, Science **283**, 1706 (1999).

CHAPTER 5

# THEORETICAL ASPECTS OF BOSE–EINSTEIN CONDENSATION

A. Fabrocini

*Dipartimento di Fisica "E.Fermi", Università di Pisa,*
*and Istituto Nazionale di Fisica Nucleare, sezione di Pisa,*
*I-56100 Pisa, Italy*
*E-mail: adelchi.fabrocini@df.unipi.it*

I review some aspects of Bose–Einstein condensation in infinite and confined systems and discuss the differences between the dense and dilute cases. As an example of dense and strongly interacting system I focus on the zero temperature behavior of atomic $^4$He. The condensate fraction, the momentum distribution and the natural orbits of the homogeneous liquid and of the clusters are studied. I then address a dilute gas of hard–spheres, in the standard thermodynamic limit or trapped in an external potential. The Gross–Pitaevski equation is used to describe a large but finite number of alkali atoms in harmonic traps. The limitations of this approach when the gas parameter becomes large is finally discussed.

## 1. Bosons and condensation

Bose–Einstein Condensation (BEC, for Many–Body physics addicted) has been awarded a great deal of attention in the last years, following its actual visualization in systems of very cold alkali atoms confined in a harmonic magnetic trap. [1,2] However, BEC is hardly a new guest in quantum mechanics. It was predicted to take place since the early times of the theory, [3,4] when it was shown that if a large number, $N$, of bosons are cooled below some critical temperature, $T_c$, then the occupation number, $N_0$, of the lowest energy single particle state, $\phi_0$, becomes macroscopic, $N_0 \sim N$. The bosons are said to *condense* in the $\phi_0$ state. Superfluidity in Helium was interpreted as a possible manifestation of BEC. [5]

BEC in a many–body system is a consequence of the statistics and it occurs for bosons , whose wave function is globally symmetric. Antisymmetric fermions do not undergo it, unless they couple to form bosonic structures (zero momentum Cooper pairs of electrons or $^3$He atoms, electron–hole excitons). So, a qualitative understanding of this phenomenon may be gathered by any simple non interacting (NI) model, where the interparticle interaction has been switched off, but the symmetry of the wave function is properly accounted for. It is clear that the interaction must be introduced when a quantitative description is required. However, the necessary degree of accuracy may largely vary with the diluteness of the system itself. In dilute gases, approximations as the mean field (MF) one are expected to provide reasonable answers. Life becomes harder when the interparticle distance competes with the range of the interaction. In this case the full complexity of the Schrödinger equation must be somehow approached.

BEC manifests itself in rather different ways in going from infinite to finite systems. In the first case it appears as a long range order (LRO) in the density matrix, providing a macroscopic occupation of the zero momentum state, with a condensate fraction, $n_0 = N_0/N$, as low as 7–10% in dense $^4$He. [6] Bosons condense in momentum space in translationally invariant, infinite quantum fluids. BEC for particles constrained in some confined geometry, as clusters of $^4$He atoms or magnetically trapped atomic gases, shows up also in coordinate space as a very large, distinguishable peak of condensed bosons in the density distribution.

These very distinct behaviors for different geometries appear already in the non interacting models. Let's consider, for simplicity, $N$ bosons at $T = 0$ temperature enclosed in a volume $\Omega$ in the uniform matter thermodynamic limit ($N \to \infty$, $\Omega \to \infty$, keeping the one–body density, $\rho_0 = N/\Omega$, constant). For NI bosons the normalized lowest single particle state is simply $\phi_0 = \Omega^{-1/2}$. The one–body density matrix (OBDM),

$$\rho_1(\mathbf{r}_1, \mathbf{r}_1') = N \int d\mathbf{r}_2 ... d\mathbf{r}_N \Psi_0^\dagger(\mathbf{r}_1, \mathbf{r}_2, ... \mathbf{r}_N) \Psi_0(\mathbf{r}_1', \mathbf{r}_2, ... \mathbf{r}_N) , \tag{1.1}$$

where $\Psi_0(\mathbf{r}_1, \mathbf{r}_2, ... \mathbf{r}_N)$ is the many–body ground state wave function, is then a flat constant in the NI case, identical everywhere to $\rho_0$. The NI momentum distribution (MD), given by the Fourier transform of the OBDM

$$n_{\mathbf{k}} = \int d\mathbf{r}_1 \int d\mathbf{r}_1' \; e^{i\mathbf{k}\cdot(\mathbf{r}_1 - \mathbf{r}_1')} \rho_1(\mathbf{r}_1, \mathbf{r}_1') , \tag{1.2}$$

exhibits a delta peak at $k = 0$, $n_{\mathbf{k}}^{unif} = \delta_{\mathbf{k}}$, implying that all bosons have condensed in the zero–momentum state.

As an example of confined geometry we may take $N$ particles trapped by an external spherical harmonic potential,

$$V_{ext}(r) = \frac{1}{2} m \omega_{HO}^2 r^2 , \tag{1.3}$$

characterized by its trap frequency, $\omega_{HO}$, and the harmonic oscillator length, $a_{HO}^2 = \hbar / m\omega_{HO}$. In actual experiments, $N$ may be as large as $10^{8-9}$ and $a_{HO} \sim \mu$m. If we neglect the particle–particle interaction, the many–body Hamiltonians is a sum of single–particle Hamiltonians, with eigenvalues

$$\epsilon_{n_x, n_y, n_z} = \left( n_x + n_y + n_z + \frac{3}{2} \right) \hbar \omega_{HO} , \tag{1.4}$$

$\{n_x, n_y, n_z\}$ being non–negative integers. In the condensed ground state all bosons occupy the $\epsilon_0 = \frac{3}{2} \hbar \omega_{HO}$ state, whose wave function is:

$$\phi_0(r_1) = \left( \frac{m\omega_{HO}}{\pi\hbar} \right)^{3/4} \exp\left[ -\frac{m\omega_{HO}}{2\hbar} r_1^2 \right] = \left( \frac{1}{a_{HO}^2 \pi} \right)^{3/4} \exp\left[ -\frac{r_1^2}{2a_{HO}^2} \right] . \tag{1.5}$$

The corresponding OBDM and MD are:

$$\rho_1^{HO}(\mathbf{r}_1, \mathbf{r}'_1) = N \left( \frac{1}{a_{HO}^2 \pi} \right)^{3/2} \exp \left[ -\frac{1}{2a_{HO}^2} \left( r_1^2 + r_1'^2 \right) \right] , \tag{1.6}$$

$$n_{\mathbf{k}}^{HO} = N \left( 4\pi a_{HO}^2 \right)^{3/2} \exp \left[ -k^2 a_{HO}^2 \right] , \tag{1.7}$$

giving a Gaussian one–body density (OBD) distribution,

$$\rho_0^{HO}(\mathbf{r}_1) = \rho_1^{HO}(\mathbf{r}_1, \mathbf{r}_1) = N \left( \frac{1}{a_{HO}^2 \pi} \right)^{3/2} \exp \left[ -\frac{r^2}{a_{HO}^2} \right] , \tag{1.8}$$

with an average width corresponding to $a_{HO}$. In confined systems, BEC appears as a peak in the coordinate space density as well as in the momentum distribution. This feature is solely due to the spatial confinement of the bosons. The condensate has been indeed detected experimentally as a sharp peak in the density distribution, qualitatively reproduced by the NI model.

At $T = 0$ all the NI bosons belong to the condensate and $n_0 = 1$. At $T \neq 0$, but below the transition temperature $T_c$, states other than the lowest one can be thermally occupied. The transition temperature for an uniform NI gas, $T_c^0$, is given by

$$k_B T_c^0 = 2\pi \frac{\hbar^2}{m} \left[ \frac{\rho_0}{\zeta(3/2)} \right]^{2/3} , \tag{1.9}$$

where $\zeta(n)$ is the Riemann $\zeta$ function. $T_c^0$ is independent on the number of particles and the condensate fraction follows the law: $n_0(T) = 1 - (T/T_c^0)^{3/2}$. A similar analysis for the NI harmonically trapped bosons leads to

$$k_B T_c^0 = \hbar \omega_{HO} \left[ \frac{N}{\zeta(3)} \right]^{1/3} . \tag{1.10}$$

This result provides a natural extension of the thermodynamic limit concept to a finite system. The limit is obtained by letting $N \to \infty$ and $\omega_{HO} \to \infty$, but keeping $N\omega_{HO}^3$ constant. With this definition, $T_c^0$ is well defined in the thermodynamic limit and independent on $N$. The temperature dependence of the condensate fraction is now $n_0(T) = 1 - (T/T_c^0)^3$. The experimental temperature dependence of the condensate fraction closely follows the predictions of the NI model, and the corresponding estimates of the transition temperature have been a quantitative guide to experimentalists searching for the proper temperature domain. The diluteness of the atomic gases is the reason of the success of this seemingly naive approach. A comprehensive review of BEC for trapped atoms may be found in Ref. 7.

The non interacting and mean field models patently fail if the diluteness condition is no longer satisfied. In the next Section we will briefly discuss the occurrence of BEC in dense atomic $^4$He, both in infinite and confined geometries. In the last Section we will come back to the dilute regime and present the standard mean field approach, represented by the Gross–Pitaevskii equation (GPE), as well as attempts to overcome it.

## 2. BEC in $^4$He

The atomic gases diluteness is usually expressed in terms of the *gas parameter*, $x = \rho_0 a^3$, where $a$ is the $s$–wave scattering length, coinciding with the core radius, $c$, in a purely repulsive hard core model. The largest $x$–values attained in recent experiments at JILA [8] are of the order of $10^{-2}$, and correspond to an average interatomic spacing $r_0 = (3/4\pi\rho_0)^{1/3}$ of $\sim 24\ a$. The equilibrium density of atomic $^4$He liquid at zero pressure and temperature is $\rho_0(^4\text{He}) = 0.02186$ Å$^{-3}$. The $^4$He–$^4$He interaction has a strong short–range repulsion whose radius is $c \sim 2.6$ Å, followed by a weak intermediate range attraction. As a consequence, we obtain $r_0(^4\text{He}) \sim 2.2$ Å, and comparable to the core. The scattering length is $a(^4\text{He}) \sim 90$ Å, [9] providing a huge gas parameter value.

This back of the envelope analysis points to the fact that the interatomic interaction, $v(r_{12})$, is an unavoidable ingredient of the many–body Hamiltonian

$$H = -\frac{\hbar^2}{2m}\sum_i \nabla_i^2 + \sum_{i<j} v(r_{ij}) \tag{2.1}$$

in dense atomic Helium. The Helium pair potential is known to a good accuracy [10] and depends on the distance between the particles only, not from their relative state (as for nuclear potentials). The relative simplicity of the interaction and the high density have made $^4$He systems the ideal playground and comparison field for any many–body theory that pretends to be realistic.

BEC occurs in momentum space in $^4$He homogeneous fluid. In fact, the momentum distribution takes the form

$$n_{\mathbf{k}} = \rho_0 \left( n_0 \delta_{\mathbf{k}} + \int d\mathbf{r}\ e^{\imath \mathbf{k}\cdot\mathbf{r}} \left[ \frac{\rho_1(r)}{\rho_0} - n_0 \right] \right), \tag{2.2}$$

where the OBDM depends on the relative distance and $n_0 = \rho_1(r \to \infty)/\rho_0$ provides the long range order of the density matrix.

*Ab initio* calculations of $\rho_1(r)$ have been difficult to carry on for these dense systems within field–theoretical approaches. [11] Quantitative estimates began with the development of Variational Monte Carlo (VMC) [12] and Hypernetted Chain (HNC) [13,14] methods. Both theories employed Jastrow correlated wave functions,

$$\Psi_J(\mathbf{r}_1, \mathbf{r}_2, ...\mathbf{r}_N) = \prod_{i<j} f(r_{ij}), \tag{2.3}$$

$f(r)$ being a pair correlation function depending on the relative distance and taking care of the short range repulsion, $f(r \to 0) \sim 0$, as well as of the asymptotic non interacting limit, $f(r \to \infty) \to 1$. Richer correlated wave functions are obtained by considering triplet [15] and backflow correlations, [16] these ones becoming particularly important when excited states are considered. All these approaches are variational, since the correlation function is fixed by minimizing the ground state energy, and differ between each other in the techniques employed to evaluate the expectation values of the operator of interest (stochastic average for VMC and cluster expansion with infinite terms summation for HNC). Actually, the correlated variational method is the zeroth order of the more general Correlated Basis Function perturbation theory. [17] Some contributions to this book are devoted to a more detailed presentation of the VMC, HNC and CBF theories.

The $T = 0$ momentum distribution of $^4$He has been also computed by the Green Function Monte Carlo (GFMC) [18] and Diffusion Monte Carlo (DMC) [19] methods. Both methods exactly solve, within the statistical accuracy of the Monte Carlo sampling, the many–body Schrödinger equation. Path Integral Monte Carlo (PIMC) has been used to compute the $^4$He momentum distribution at finite temperature. [20]

In Figure 1 I report the one–body density matrix and momentum distribution of liquid $^4$He at T=0 provided by the DMC calculation of Ref. 19 at the equilibrium density, $\rho_0 = 0.02186$ Å$^{-3}$, and at $\rho_0 = 0.02622$ Å$^{-3}$, using the HFDHE2 potential of Aziz *et al.*. [10] The LRO of the OBDM is clearly visible, corresponding to a condensate fraction of $n_0 = 0.072$ at $\rho_0 = 0.02186$ Å$^{-3}$, and of $n_0 = 0.027$ at $\rho_0 = 0.02622$ Å$^{-3}$. The MD shows the Gavoret and Nozieres [21] singular behavior at low $k$ values, induced by the long range correlations,

$$\lim_{k \to 0} k n_{\mathbf{k}} = \frac{n_0 m c}{2\hbar}, \tag{2.4}$$

where $c$ is the sound velocity. The HNC method gives $n_0$(HNC)=0.092 and 0.043 at $\rho_0 = 0.02186$ Å$^{-3}$ and 0.02622 Å$^{-3}$, respectively, [15] close to the GFMC results of Ref. 18. Empirical estimates of $n_0$ are obtained by fitting a model MD to the measured Compton profile in neutron scattering experiments. [22] This procedure is, however, strongly model dependent. In fact, different choices of the MD give equivalently good description of the Compton profile, while leading to largely different $n_0$ values, ranging from 0 to 0.1 [6]. The DMC momentum distribution itself gives a profile in agreement with the experiments. The extracted values of Ref. 22 are $n_0$(expt)=0.100±0.012, at $\rho_0 = 0.0224$ Å$^{-3}$ and $n_0$(expt)=0.055±0.012, at $\rho_0 = 0.0258$ Å$^{-3}$, at T=0.75 K.

In order to look for signatures of BEC in $^4$He clusters one has to dip into the details of the finite systems one–body density matrix. The diagonalization of the OBDM provides the natural orbits (NO), $\psi_\alpha$,

$$\rho_1(\mathbf{r}_1, \mathbf{r}_1') = \sum_\alpha N_\alpha \psi_\alpha^\dagger(\mathbf{r}_1) \psi_\alpha(\mathbf{r}_1'), \tag{2.5}$$

where $N_\alpha$ is the occupation number of the $\psi_\alpha$ orbital. *The occupation of the lowest orbit, $\psi_0$, divided by the total number of particles gives the finite system condensate fraction, $n_0$.* In a mean field model all bosons lie in the lowest orbit, $N_0 = N$ and $N_{\alpha \neq 0} = 0$, with $n_0 = 1$.

The one–body density and the momentum distribution are immediately written in terms of the NO as: $\rho_0(\mathbf{r}_1) = \sum_\alpha N_\alpha |\psi_\alpha(\mathbf{r}_1)|^2$ and $n_{\mathbf{k}} = \sum_\alpha N_\alpha |\tilde{\psi}_\alpha(\mathbf{k})|^2$, where $\tilde{\psi}_\alpha(\mathbf{k}) = (2\pi)^{-3/2} \int d\mathbf{r} \;\; e^{\imath \mathbf{k} \cdot \mathbf{r}} \psi_\alpha(\mathbf{r})$.

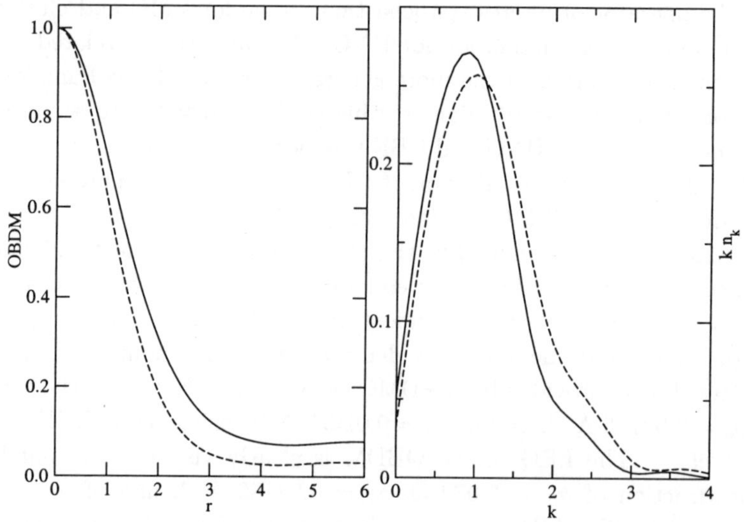


Fig. 1. The T=0 one–body density matrix (left panel), normalized to $\rho_0$, and the momentum distribution (right panel) of liquid $^4$He in DMC at $\rho_0$ =0.02186 Å$^{-3}$ (solid lines) and $\rho_0$ =0.2622 Å$^{-3}$ (dashed lines). Distances in Å and momenta in Å$^{-1}$.

Spherical $^4$He droplets, containing up to 240 atoms, have been studied by VMC in Ref. 23. The variational wave function included two– and three–body correlations and the NO were obtained by numerical diagonalization of the OBDM.

Because of the spherical symmetry, the density matrix may be expanded as:

$$\rho_1(\mathbf{r}_1, \mathbf{r}_1') = \sum_l \frac{2l+1}{4\pi} P_l(\hat{\mathbf{r}}_1 \cdot \hat{\mathbf{r}}_1') \rho_l(r_1, r_1'), \tag{2.6}$$

where $P_l(x)$ are the Legendre polynomials. Similarly, the natural orbitals can be written as:

$$\psi_{\alpha=nlm}(\mathbf{r}) = \psi_{nl}(r)\, Y_{lm}(\hat{\mathbf{r}}), \tag{2.7}$$

$Y_{lm}(\hat{\mathbf{r}})$ being the spherical harmonics. Finally, one gets

$$\rho_l(r_1, r_1') = \sum_n N_{nl}\, \psi_{nl}(r_1)\, \psi_{nl}(r_1'), \tag{2.8}$$

and the NO normalization is: $\int r^2\, dr |\psi_{nl}(r)|^2 = 1$.

Table 1 gives the occupation numbers of the $nl$–natural orbits in the $N$=70 atoms $^4$He droplet, as computed in Ref. 23. A large fraction ($n_0 \sim 36\%$) of particles are condensed in the 1$s$ NO, to be compared with the $n_0 \sim 10\%$ value in the homogeneous liquid. Figure 2 shows the total density of the droplet and the condensate density,

$$\rho_c(r) = N_{1s} |\psi_{1s}(r)|^2 . \tag{2.9}$$

Table 1. Occupation numbers of natural orbits of the $N$=70 $^4$He drop from Ref. 23.

| $nl$ | $N_{nl}$ | $nl$ | $N_{nl}$ | $nl$ | $N_{nl}$ |
|---|---|---|---|---|---|
| $1s$ | 25.33 | $1h$ | 0.24 | $1k$ | 0.104 |
| $1p$ | 0.49 | $2f$ | 0.22 | $2i$ | 0.086 |
| $1d$ | 0.44 | $3p$ | 0.22 | $3g$ | 0.078 |
| $2s$ | 0.44 | $1i$ | 0.19 | $4d$ | 0.077 |
| $1f$ | 0.37 | $2g$ | 0.17 | $5s$ | 0.100 |
| $2p$ | 0.35 | $3d$ | 0.16 | $1l$ | 0.063 |
| $1g$ | 0.30 | $4h$ | 0.19 | $2j$ | 0.060 |
| $2d$ | 0.28 | $1j$ | 0.14 | $3h$ | 0.046 |
| $3s$ | 0.30 | $2h$ | 0.12 | $4f$ | 0.049 |
| | | $3f$ | 0.11 | $5p$ | 0.046 |
| | | $4p$ | 0.11 | | |

At the center of the drop, $\rho_c(0) \sim 0.1\ \rho(0)$, close to the extended liquid value. The figure gives also the cumulative contributions to the density up to a given $l_{max}$. The condensed fractions obtained for different atom numbers are: $n_0$=0.55, 0.36, 0.29 at $N$=20, 70 and 240, respectively.

The $1s$ NO of the $N$=70 drop can be accurately approximated by:

$$\psi_{1s}(r) \sim C\left[1 - 1.68\frac{\rho(r)}{\rho_0}\right]\psi_{1s}^{\rm MF}(r)\,, \tag{2.10}$$

where $C$ is a renormalization constant, $\rho_0$ is the liquid $^4$He equilibrium density and $\psi_{1s}^{\rm MF}(r) = \sqrt{\rho(r)/N}$ is the mean field fully occupied, $n_0^{\rm MF}$=1, lowest orbital. The factor multiplying $\psi_{1s}^{\rm MF}(r)$, $(1 - 1.68\rho/\rho_0)$, can then be interpreted as the square root of the condensed fraction of liquid $^4$He at density $\rho$ and the full expression as a local density approximation to the condensate natural orbit of the drop.

## 3. BEC in dilute systems

The energy per particle of a low density homogeneous gas of bosonic hard–spheres, whose diameter coincides with the scattering length, may be expanded as: [11]

$$e_{HS}^{LD} = 2\pi\frac{\hbar^2}{ma^2}x\left[1 + \frac{128}{15}\sqrt{\frac{x}{\pi}} + 8(\frac{4}{3}\pi - \sqrt{3})x\ln(x) + O(x)\right]\,. \tag{3.1}$$

All potentials having the same scattering length give identical results at this level of the expansion. DMC calculations [24] have confirmed this universal behavior. The first term of the expansion is accurate only at very low $x$–values, whereas the addition of the first correction in $\sqrt{x}$ well reproduces the exact energies up to $x = 10^{-2}$. [25,26] In contrast, the logarithmic term spoils this agreement already for

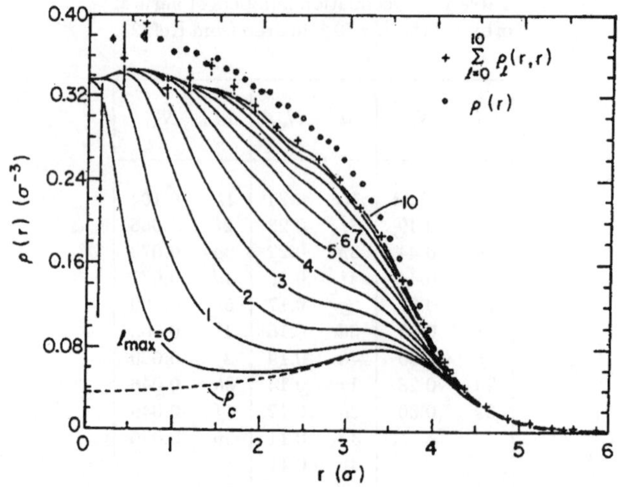

Fig. 2. Densities of the $N$=70 $^4$He droplet from Ref. 23. Circles are the total density. The curves show the cumulative contributions of the natural orbits up to a given $l_{max}$. The dashed line gives the condensate density, $\rho_c(r)$. $\sigma$=2.556 Å.

intermediate values of the gas parameter. A similar expansion for the condensed fraction gives, at the first order in $\sqrt{x}$, [11]

$$n_0^{LD} = 1 - \frac{8}{3}\sqrt{\frac{x}{\pi}} . \tag{3.2}$$

A variational approach, that employs the Jastrow correlated wave function of Eq. (2.3) and the HNC equation method, [25] is in even better agreement with the DMC results than the low density expansion. A reliable choice for the correlation function, $f(r)$, is the one minimizing the two–body cluster energy, with a healing condition at a distance $d$, such that$f(r \geq d) = 1$ and $f'(r = d) = 0$. The healing distance is then taken as the only variational parameter to minimize the full many–body energy. The corresponding correlation function for a hard sphere potential is

$$f(r \leq a) = 0 \; ; \; f(a \leq r \leq d) = \frac{d}{r}\frac{\sin[k(r-a)]}{\sin[k(d-a)]} , \tag{3.3}$$

where $k$ is solution of: $\mathrm{tg}[k(d-a)] = kd$.

For $N$ bosons trapped by an external potential, $V_{ext}$, at T=0 and described by the Hamiltonian

$$H = -\frac{\hbar^2}{2m}\sum_i \nabla_i^2 + \sum_i V_{ext}(\mathbf{r}_i) + \sum_{i<j} v(r_{ij}) , \tag{3.4}$$

the energy may be written as:

$$E = \frac{\hbar^2}{2m}\int d\mathbf{r}_1 [\vec{\nabla}_1 \rho_1^{1/2}(\mathbf{r}_1)]^2 + \int d\mathbf{r}_1 \rho_1(\mathbf{r}_1) V_{ext}(\mathbf{r}_1) + E_{corr} , \tag{3.5}$$

where $E_{corr}$ is a *correlation energy*, [25] being absent in a mean field model. $E_{corr}$ may be estimated starting from a correlated wave function and using CBF theory. However, if the one body density is a smooth function of the distance, the local density approximation (LDA) provides a viable and easy approach. In fact, in LDA $E_{corr}$ is approximated by:

$$E_{corr} \sim E_{corr}^{LDA} = \int d\mathbf{r}_1 \rho_1(\mathbf{r}_1) e[\rho_1(\mathbf{r}_1)] \,, \tag{3.6}$$

where $e[\rho_1(\mathbf{r})]$ is the energy per particle of the homogeneous system at density $\rho_1(\mathbf{r})$. A minimization of the energy with respect to $\rho_1(\mathbf{r})$ is obtained by solving a correlated Hartree–like equation. [25]

The energy expression associated with the Gross–Pitaevskii theory is obtained in LDA by using in (3.6) the low density functional (3.1) truncated at the first order. In fact, one gets:

$$E_{GP}[\psi] = \int dr \left[ \frac{\hbar^2}{2m} \mid \nabla\psi(r) \mid^2 + \frac{m}{2}\omega_{HO}^2 r^2 \mid \psi \mid^2 + \frac{2\pi\hbar^2 a}{m} \mid \psi \mid^4 \right] , \tag{3.7}$$

where we have used a spherically symmetric harmonic trapping potential and the *wavefunction* $\psi$ is related to the density by: $\rho_1(r) = \mid \psi(r) \mid^2$. A functional variation of $E_{GP}[\psi]$ gives the Euler-Lagrange equation, known as the Gross-Pitaevskii equation,

$$\left[ -\frac{\hbar^2}{2m}\nabla^2 + \frac{m}{2}\omega_{HO}^2 r^2 + \frac{4\pi\hbar^2 a}{m} \mid \psi \mid^2 \right] \psi = \mu\psi, \tag{3.8}$$

where $\mu$ is the chemical potential introduced to guarantee the correct normalization of $\psi$. Modified (and, hopefully, more accurate) GPE [25] are obtained via the inclusion of the successive terms in the expansion of $e_{HS}^{LD}$.

An alternative choice consists in using the HNC variational energy in Eq. (3.6). The minimization procedure gives a HNC correlated Hartree equation,

$$\left[ -\frac{\hbar^2}{2m}\nabla^2 + \frac{m}{2}\omega_{HO}^2 r^2 + e_{HS}^{HNC}(x_{loc}) + x_{loc}\frac{\partial e_{HS}^{HNC}(x_{loc})}{\partial x_{loc}} \right] \psi = \mu\psi, \tag{3.9}$$

where $e_{HS}^{HNC}(x_{loc})$ is the HNC energy per particle of the homogeneous boson gas at the local gas parameter value $x_{loc}(r) = \mid \psi(r) \mid^2 a^3$. This second approach has the advantage of not being limited to the hard sphere potential.

The need to develop theories beyond the GPE is becoming clearer and clearer with the increase of the average gas parameter value, $x_{av}$, attained in the latest experiments on magnetically trapped atomic gases. As long as $x_{av} \leq 10^{-3}$ the GPE is reliable and accurate. However, the gas parameter can be brought outside the GPE regime of validity either by increasing the number of atoms or (and more efficiently) by changing the scattering length. In a recent experiment performed at JILA, [8] about $10^4$ atoms of $^{85}$Rb were confined in a cylindrical trap, and the scattering length was varied from negative to very high positive values by exploiting a Feshbach resonance at a magnetic field of $B \sim 155$ Gauss. As a result, gas parameter values

Table 2. Chemical potential, energy per atom and peak gas parameter of $N = 10^4$ $^{85}$Rb atoms in the cylindrical trap of Ref. 8. Energies in units of the HO energy, $\hbar\omega_\perp$; scattering lengths in units of the Bohr radius of the Hydrogen atom, $a_0$.

| | $a/a_0$ | GP | TF | HNC |
|---|---|---|---|---|
| $\mu$ | 3000 | 13.25 | 13.15 | 14.38 |
| | 8000 | 19.55 | 19.47 | 24.37 |
| | 10000 | 21.36 | 21.29 | 28.09 |
| $E/N$ | 3000 | 9.52 | 9.39 | 10.23 |
| | 8000 | 14.00 | 13.91 | 16.98 |
| | 10000 | 15.29 | 15.21 | 19.42 |
| $x_{pk} \times 10^2$ | 3000 | 0.39 | 0.39 | 0.32 |
| | 8000 | 4.10 | 4.09 | 2.53 |
| | 10000 | 7.00 | 6.98 | 3.86 |

of the order of $10^{-2}$ and larger were obtained. To explore with confidence such regimes is of particular relevance since the analyses of the experiments are often based upon the Thomas–Fermi (TF) approximation, amounting to disregard the kinetic energy term in the GPE. The TF results are mostly analytical and are very close to the GP ones for large $N$ and/or $a$ values. So, the accuracy of these methods must be maintained under strict control in order to avoid misinterpretations of the experimental data.

The GP and HNC equations have been solved [27] for the cylindrical trap of Ref. 8 and we give in Table 2 the results for $N = 10^4$ $^{85}$Rb atoms. The radial and axial frequencies associated with the external potential are $\omega_\perp/2\pi = 17.5$Hz and $\omega_z/2\pi = 6.9$ Hz, respectively. The Table provides the chemical potential and energy per atom (in units of the HO energy, $\hbar\omega_\perp$) and the peak gas parameter, $x_{pk} = \rho_1(0)a^3$, for several values of the scattering length and for different approaches (GP, TF and HNC).

At these values of $a$ and $N$ the TF and GP estimates are close to each other, whereas the HNC ones are substantially different. In fact, the HNC corrections to the chemical potential go from $\sim 9\%$ to as much as $\sim 32\%$ at the highest $a$ value. Even larger are the corrections to $x_{pk}$, going from $\sim 18\%$ to $\sim 45\%$. Results similar to HNC are provided by the solution of the modified GPE obtained by considering also the second term in the expansion (3.1).

The structural differences between the GP and HNC results are manifest in the column densities, shown in the left panel of Figure 3 at $a/a_0 = 10000$. The

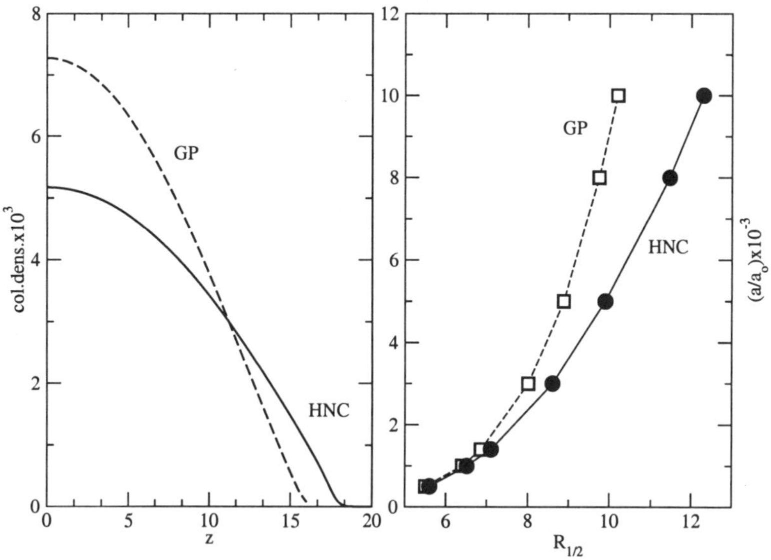


Fig. 3. Column densities (right panel) of $N = 10^4$ $^{85}$Rb atoms at $a/a_0 = 10000$ and half–density radius (left panel) in GP and HNC. Distances in units of the harmonic oscillator length, $a_{HO,\perp}$.

experimentally measured column density is defined as $\rho_{col}(z) = \int dx\ \rho_1(x, 0, z)$, where the $x$ direction coincides with that of the light beam used to image the atomic cloud. The quantity $\hat{\rho}_{col}(z) = (a_{HO,\perp}/N)\rho_{col}(z)$ is reported in the figure, where the radial harmonic oscillator length is $a^2_{HO,\perp} = \hbar/m\omega_\perp$.

In this regime the GP and TF densities practically coincide and the HNC corrections are sizeable and considerably deplete the density at the center of the trap as a consequence of the repulsive nature of the extra terms. The half–density radius, $R_{1/2}$, defined by $\rho_c(z = R_{1/2}) = \rho_c(0)/2$, may be taken as a measure of the size of the atomic cloud. The radii in function of the scattering length are shown in the right panel of Figure 3 for the GP and HNC approaches in the JILA experiment setup. It appears clear that these gas parameter regimes cannot any longer be analyzed in terms of he GP (and/or TF) equations. Effects beyond the mean field capabilities come into play and a truly microscopic description of the system is asked for in order to correctly analyze the experiments.

## 4. Conclusions

I have given a short review of some of the properties of systems that undergo Bose-Einstein condensation. Even though BEC is just a consequence of the statistics, the theoretical tools necessary for its quantitative interpretation largely vary according to the diluteness of the bosonic fluid. BEC in dilute, magnetically trapped atomic gases (that has attracted the world interest in the last years) is accurately approached by means of the simple, mean-field based Gross–Pitaevskii equation. Dense atomic liquid $^4$He, and the related systems call for modern and powerful many-body theories in order to achieve a reliable description of their behavior. The correlated basis function theory, together with the techniques needed to implement it (as infinite cluster expansions, integral equations and Monte Carlo sampling), is an example of successful story in Helium physics. The advances in the stochastic methods have even allowed to exactly solve the many–body Schrödinger equation for the ground state of strongly interacting Bose fluids. Since a new wave of experiments with trapped atoms is rapidly approaching the low diluteness regime, where the validity of the time honored mean field description is heavily questioned, the same modern quantum many-body theories, tested in Helium, are the natural and accurate tools to address this rich and fascinating field.

## 5. Acknowledgments

Part of the results presented in this contribution have been obtained in collaboration with Artur Polls. This research was partially supported by the Progetto di Ricerca di Interesse Nazionale: *Fisica Teorica del Nucleo Atomico e dei Sistemi a Molti Corpi* from MURST.

## References

1. M. H. Anderson, J. R. Ensher, M. R. Matthews, C. E. Weiman, and E. A. Cornell, *Science* **269**, 198 (1995).
2. Davis K. B. *et al.*, Phys. Rev. Lett. **75**, 3969 (1995).
3. S. N. Bose, *Z. Phys.* **26**, 178 (1924).
4. A. Einstein, *Sitzber. Kgl. Preuss. Akad. Wiss.* 261 (1924); 3 (1925).
5. F. London, *Nature* (London) **141**, 643 (1938).
6. P. Sokol, in *Bose Einstein Condensation*, edited by A. Griffin, D. W. Snokee, and S. Stringari (Cambridge University Press), 51 (1995).
7. F. Dalfovo, S. Giorgini, L. P. Pitaevskii, and S. Stringari, Rev. Mod. Phys. **71**, 463 (1999).
8. S. L. Cornish, N. R. Clausen, J. L. Roberts, E. A. Cornell, and C. E. Wieman, Phys. Rev. Lett. **85**, 1795 (2000).
9. P. Barletta, and A. Kievsky, Phys. Rev. A **64**, 042514 (2001).
10. R. A. Aziz, V. P. S. Nain, J. S. Carley, W. L. Taylor, and G. T. Conville, *J. Chem. Phys.* **70**, 4330 (1979).
11. A. L. Fetter, and J. D. Walecka, *Quantum Theory of Many–Particle Systems* (McGraw–Hill, New York, 1971).
12. D. M. Ceperley, and M. H. Kalos, in *Monte Carlo Methods in Statistical Physics*, edited by K. Binder (Springer, New York, 1979).
13. S. Fantoni, Nuovo Cimento A **44**, 191 (1978).
14. M. L. Ristig, in *From Particles to Nuclei*, Proc. Int. Schoolof Phys. E. Fermi, Course LXXIX, edited by A. Molinari (North Holland, Amsterdam, 1981).
15. E. Manousakis, V. R. Pandharipande and Q. N. Usmani, Phys. Rev. B **31**, 7022 (1985).
16. E. Manousakis, and V. R. Pandharipande, Phys. Rev. B **31**, 7029 (1985).
17. E. Feenberg, *Theory of Quantum Liquids* (Academic, New York, 1969).
18. P. A. Whitlock, and R. Panoff, *Can. J. of Phys.* **65**, 1409 (1987).
19. S. Moroni, G. Senatore, and S. Fantoni, Phys. Rev. B **25**, 1040 (1997).
20. D. M. Ceperley, and E. L. Pollock, Phys. Rev. Lett. **56**, 351 (1986).

21. J. Gavoret, and P. Noziéres, *Ann. Phys.* **28**, 349 (1964).
22. W. N. Snow, Y. Wang, and P. E. Sokol, *Europhys. Lett.* **19**, 403 (1992).
23. D. S. Lewart, V. R. Pandharipande, and S. C. Pieper, Phys. Rev. B **37**, 4950 (1988).
24. S. Giorgini, J. Boronat, and J. Casulleras, Phys. Rev. A **60**, 5129 (1999).
25. A. Fabrocini, and A. Polls, Phys. Rev. A **60**, 2319 (1999).
26. A. Polls, and A. Fabrocini, *Condensed Matter Theories*, Vol. XVI (Nova Science Pub., New York, 1999).
27. A. Fabrocini, and A. Polls, Phys. Rev. A **60**, 2319 (2001).

## CHAPTER 6

# ELEMENTARY EXCITATIONS AND DYNAMIC STRUCTURE OF QUANTUM FLUIDS

M. Saarela
*Dept. of Physical Sciences, P.O. Box 3000, FIN-90014 University of Oulu, Finland*
*E-mail: Mikko.Saarela@oulu.fi*

The equations of motion method for studying excitations and dynamic structure of quantum fluids is reviewed in this series of lectures. The method is based on the least action principle where one minimizes the action integral of the dynamic system. As a result one gets the continuity equations, which connect the density fluctuations and currents to an external driving force. The external force is assumed to infinitesimal and the response of the system to that is linear. The real poles of the linear response function determine the elementary excitation modes and the imaginary part of the self energy defines the continuum limit and gives the finite lifetime of the decaying modes. Our dynamic wave function contains time-dependent one- and two-particle correlation functions, which includes couplings between three modes. Thus one mode can split into two modes if energy and momentum are conserved. We begin with the Feenberg's $\beta$-derivative formulation of the optimized ground state and then derive general equations of motion for the dynamic system from the least action principle. We show how the simplest one-body approximation leads to the Feynman theory of excitations. By including the fluctuating two-body correlation function within the uniform limit one recovers the correlated basic function approximation. The fully consistent theory gives a good account of the elementary excitations and we show results on current patterns in the maxon-roton regions and on the precursor of the liquid-solid phase transition. Finally we apply the method to the excitations of the impurity and derive the hydrodynamic effective mass of the $^3$He impurity in $^4$He and the $^3$He dynamic structure function.

## 1. Introduction

Theoretical understanding of the microscopic structure of excitations and of the physical processes involved in the dynamics of quantum Bose fluids is relatively incomplete in comparison to the understanding of the ground state properties. Monte Carlo simulations give in principle exact results for the ground state energy and structure function when the interaction between the bosonic constituents of the many–body system is known.[1] Also energies of the lowest lying elementary excitation modes like the phonon–roton spectrum in the liquid $^4$He and the plasmon mode in charged Bose gas are reasonably well produced in simulations.[2,3] Yet, understanding of dynamic processes like phonon decay,[4] multiphonon excitations responsible for the higher frequency behavior of the dynamic structure function or topological excitations, vortices and vortex–antivortex pairs in thin superfluid films and rotating systems require microscopic descriptions, which are sensitive to these processes.

On the experimental side the cleanest, well studied quantum Bose fluid is the superfluid $^4$He. The dynamic structure function along with the elementary excitation spectrum with the temperature and pressure dependence is known to a high precision from x–ray and neutron scattering experiments. (A review of the subject with references to earlier experimental and theoretical work is given by Glyde in his book.[5]) Thus, modern many–body methods face a demanding challenge of reproducing the correct spectrum; a solid agreement would create confidence in the applied tools and indicate that the essential physical ingredients have been understood and included into the calculations. A fully microscopic approach is needed here, because these phenomena defy a simple hydrodynamic description.

Let us quickly summarize the milestones on the path to the present comprehension. The preliminary work on the subject was done by Bijl and Landau in the 1940's.[6,7,8,9] Landau proposed that there are two separate collective excitation modes in liquid $^4$He: phonons, thought of as collective density (sound) modes having linear dispersion, and rotons, assumed to be a collective rotation of the fluid having a separate dispersion curve. Later on, he joined these excitations into a single collective mode dispersion curve continuous in the wave vector.[9] Phonons and rotons were then interpreted as the low– and high–$k$ regions of the same collective excitation. Between them we have what is called the maxon region. This seemed to be in qualitative agreement with experimental data and, at the same time, consistent with the continuous dispersion curve for excitations in a dilute Bose gas derived microscopically by Bogoljubov in his seminal paper.[10]

After these early developments Feynman took the first steps towards a microscopic description by suggesting a specific trial excited–state wave function.[11] He wrote the wave function of the excited state $\Psi_{\mathbf{k}}$ of momentum $\hbar\mathbf{k}$ as a product $\Psi_{\mathbf{k}} = \rho_{\mathbf{k}}\Psi_0$ of the ground–state wave function $\Psi_0$ and of a density–fluctuation operator $\rho_{\mathbf{k}} = \Sigma_j \exp(i\mathbf{k}\cdot\mathbf{r}_j)$, offering thus a microscopic explanation for phonons and rotons as collective density excitations. The spectrum calculated variationally using the proposed wave function, leading to the dispersion relation

$$\varepsilon_F(k) = \frac{\hbar^2 k^2}{2mS(k)}\,, \tag{1.1}$$

contains much of the relevant physics and is exact in the long–wavelength limit. It also provides an upper bound for the lowest–lying excitations, but it has severe deficiencies at shorter wavelengths. For example, the computed roton energy is by a factor of two larger that the experimentally observed value. Owing to this discrepancy, the theory was subsequently supplemented by Feynman and Cohen[12] to include so–called backflow corrections which increased the flexibility of the wave function and, thus, lowered the roton energy significantly towards measured values. The term backflow is used to describe the correlated motion of neighboring particles around a given reference atom. The initial picture put forward by Feynman and Cohen assumed that particles move in a dipolar flow field, behaving in a sense like a smoke ring. After Feynman's original arguments the method of correlated basis functions (CBF) for density oscillations was developed along the same lines to further improve the agreement between theoretical predictions and experimental data, most notably by Feenberg and his collaborators. [13,14,15,16,17,18] In the CBF theory the excited–state wave function $\Psi_{\mathbf{k}}$ is written as $\Psi_{\mathbf{k}} = F_{\mathbf{k}}\Psi_0$, and the excitation operator $F_{\mathbf{k}}$ is further expressed as a polynomial in the density–fluctuation operators $\rho_{\mathbf{k}}$. Thus, in the lowest order we have the usual Feynman form for the excited states, and terms beyond the linear one introduce the backflow effects. Attempts to calculate the dynamic structure function were also made. [19]

More recently the shadow wave function (SWF) method has been extended to permit the investigations of excited states.[20,21] By the provision that the momentum–carrying factor in $\Psi_{\mathbf{k}}$ is a density fluctuation in the subsidiary (shadow) variables, one has, in principle, a parameter–free wave function of the Feynman form in which the fluctuations in the subsidiary variables allow for the presence of backflow effects in the particle variables. Again, this backflow is represented by terms of all orders in the density fluctuation $\rho_{\mathbf{k}}$ of the real variables. The CBF and SWF methods, along with the application of release–node Monte Carlo simulations[22], all give results which agree reasonably well with experimental data.

The excitation spectrum of liquid $^4$He is in two dimensions qualitatively very similar to the three–dimensional spectrum, as one might expect since the physics of liquid $^4$He is dominated by short–range correlations. What makes the 2D system especially interesting, however, is that topological excitations, i.e. the vortex–antivortex pair excitations, appear there naturally as low–lying elementary excitation modes and form the driving mechanism into the Kosterlitz–Thouless phase transition. [23,24,25,26,27] Therefore, this system forms an ideal framework within which it is possible to study the specific differences between the roton and vortex excitations.

In this series of lectures we review the equation motion method for density fluctuations for quantum Bose fluids and fluid mixtures. [28–32] The goal is to find the optimized density response of the many–body system to an infinitesimal external perturbation, which drives excitations into the system. The time dependence of the external potential creates currents, which are calculated together with the density fluctuations from the equations of motion. We limit ourselves into the linear response and thus calculate the linear response function and from that the dynamic structure function. The method has been applied to homogeneous and inhomogeneous quantum fluids and their mixtures.

Here the main emphasis is in the analytic structure of the linear response function, dynamic structure function and the self–energy. We present three approximate ways to solve the continuity equations starting from the Feynman–like approximation and finishing with the solution of the coupled set of one– and two–particle continuity equations. At the end we apply the method to the excitations of the impurities in a quantum fluid.

## 2. Ground state of a quantum Bose fluid

Before entering the discussion of the excitation we begin with a general formulation of the variational ground state of a Bose system and derive the Euler equation which optimize the many–particle distribution functions following Feenberg's prime–derivative technique. We will arrive at a practical set of numerically solvable equations by making the Jastrow–Feenberg ansatz for the wave function and summing the hypernetted chain diagrams together with the elementary diagrams. These equations can be solved iteratively and give very accurate results for the Bose fluids.

### 2.1. *Optimized ground state*

In the microscopic, variational theory the calculation of the ground state properties of the system of $N$ particles with the mass $m$ starts from the empirical Hamiltonian

$$H_0 = -\frac{\hbar^2}{2m}\sum_{i=1}^{N}\nabla_i^2 + \frac{1}{2}\sum_{i\neq j}^{N} V(|\mathbf{r}_i - \mathbf{r}_j|)\,. \tag{2.1}$$

One assumes that the two–particle interaction $V(|\mathbf{r}_i - \mathbf{r}_j|)$ is either known like the Coulomb interaction between charged particles or it is empirically determined as is the case with the interaction between helium atoms.

The variational problem is to minimize the total energy

$$E_0 = \frac{\langle\Psi_0|H_0|\Psi_0\rangle}{\langle\Psi_0|\Psi_0\rangle} \tag{2.2}$$

with respect to the $n$–particle densities of the system

$$\rho_n(\mathbf{r}_1,\ldots\mathbf{r}_n) = \frac{N!}{(N-n)!}\frac{1}{\langle\Psi_0|\Psi_0\rangle}\int d^3r_{n+1}\ldots d^3r_N|\Psi_0(\mathbf{r}_1,\ldots\mathbf{r}_N)|^2\,. \tag{2.3}$$

In the homogeneous system with a constant density $\rho_0$ it is more convenient to use the $n$–particle distribution functions, which are normalized to unity when all particles are far apart,

$$g_n(\mathbf{r}_1,\ldots\mathbf{r}_n) = \frac{\rho_n(\mathbf{r}_1,\ldots\mathbf{r}_n)}{\rho_0^n}\,. \tag{2.4}$$

Following Feenberg's book [13] we start from the quadratic form

$$X^{(n)}(\mathbf{r}_1,\ldots\mathbf{r}_n) = \frac{N!}{(N-n)!}\int d^3r_{n+1}\ldots d^3r_N\Psi_0(H-E_0)\Psi_0\,, \tag{2.5}$$

and show that the right hand side of the equation can be expressed in terms of distribution functions. If these are correctly generated from the solution of the Schrödinger equation $H_0\Psi_0 = E_0\Psi_0$ then the right hand side vanishes defining the $n$–particle Euler equation of the system.

For the Bose systems the ground state wave function is a real, positive function of coordinates and the form which enforces that is the Bijl–Dingle–Jastrow ansatz

$$\Psi_0(\mathbf{r}_1, \ldots, \mathbf{r}_N) = \exp\left[\frac{1}{2}\sum_{i<j}^{N} u_2(\mathbf{r}_i, \mathbf{r}_j)\right] . \tag{2.6}$$

For the real wave function the kinetic energy part of the Hamiltonian can be written as

$$\frac{\hbar^2}{2m}\Psi_0\nabla^2\Psi_0 = \frac{\hbar^2}{8m}\nabla^2\Psi_0^2 + \frac{\hbar^2}{8m}\Psi_0^2\nabla^2\log(\Psi_0^2) \tag{2.7}$$

and the integrand in Eq. (2.5) becomes

$$\begin{aligned}\Psi_0(H - E_0)\Psi_0 = \Bigg[&-\frac{\hbar^2}{8m}\sum_{i=1}^{N}\nabla_i^2 - E_0 \\ &+ \frac{1}{2}\sum_{i\neq j}^{N} V(|\mathbf{r}_i - \mathbf{r}_j|) - \frac{\hbar^2}{8m}\sum_{i=1}^{N}\nabla_i^2\log\Psi_0^2\Bigg]\Psi_0^2 .\end{aligned} \tag{2.8}$$

It can be further simplified by introducing the generalized normalization integral

$$I_0(\beta) = \int d^3r_1 \ldots d^3r_N\ \Psi_0^2(\mathbf{r}_1, \ldots \mathbf{r}_N)\ e^{\beta V_{JF}(\mathbf{r}_1,\ldots\mathbf{r}_N)} \tag{2.9}$$

with the Jackson–Feenberg effective potential

$$V_{JF}(\mathbf{r}_1, \ldots \mathbf{r}_N) = \frac{1}{2}\sum_{i\neq j}^{N} V(|\mathbf{r}_i - \mathbf{r}_j|) - \frac{\hbar^2}{8m}\sum_{i=1}^{N}\nabla_i^2\log\Psi_0^2 . \tag{2.10}$$

We can replace for a moment the square of the wave function with a generalized $\beta$–dependent wave function

$$\Psi_0^2(\mathbf{r}_1, \ldots \mathbf{r}_N; \beta) = \frac{1}{I_0(\beta)}\ e^{\beta V_{JF}(\mathbf{r}_1,\ldots\mathbf{r}_N)}\Psi_0^2(\mathbf{r}_1, \ldots \mathbf{r}_N) . \tag{2.11}$$

Quantities which depend on the wave function like distribution functions $g_n(\mathbf{r}_1, \ldots \mathbf{r}_n; \beta)$ and the energy expectation value become then $\beta$–dependent and the return to the original notation takes place by setting $\beta = 0$ at the end of the calculation.

In this notation the total energy of the system is then simply

$$E_0 = \frac{d}{d\beta}I_0(\beta)\bigg|_{\beta=0} \tag{2.12}$$

and the full integrand (2.8) assumes the form

$$\Psi_0(H-E_0)\Psi_0 = \frac{d}{d\beta}\left[\frac{1}{I_0(\beta)}e^{\beta V_{JF}(\mathbf{r}_1,\ldots\mathbf{r}_N)}\right]\bigg|_{\beta=0}\Psi_0^2(\mathbf{r}_1,\ldots\mathbf{r}_N) - \frac{\hbar^2}{8m}\sum_{i=1}^{N}\nabla_i^2\Psi_0^2(\mathbf{r}_1,\ldots\mathbf{r}_N)\,. \tag{2.13}$$

The quadratic equation can now be formally integrated giving

$$X^{(n)}(\mathbf{r}_1,\ldots\mathbf{r}_n) = \frac{d}{d\beta}\rho_n(\mathbf{r}_1,\ldots\mathbf{r}_n;\beta)\bigg|_{\beta=0} - \frac{\hbar^2}{8m}\sum_{i=1}^{n}\nabla_i^2\rho_n(\mathbf{r}_1,\ldots\mathbf{r}_n;0)\,. \tag{2.14}$$

In the case of $n=2$ if the Schrödinger equation is satisfied, then

$$X^{(2)}(\mathbf{r}_1,\mathbf{r}_2) = 0 \tag{2.15}$$

and we get the Euler equation for the radial distribution function $g(r) = \frac{1}{\rho_0}\rho_2(|\mathbf{r}_1 - \mathbf{r}_2|)$ as a function of radius $r = |\mathbf{r}_1 - \mathbf{r}_2|$

$$g'(r) \equiv \frac{d}{d\beta}g(r;\beta)\bigg|_{\beta=0} = \frac{\hbar^2}{4m}\nabla^2 g(r;0)\,. \tag{2.16}$$

We have introduced here Feenberg's prime notation and ignored $\beta$ from the list of arguments. This notation should not be mixed with the ordinary derivative with respect to the coordinate $r$. The Euler equation is valid for all real wave functions, but the main problem is to calculate the $\beta$–derivative term.

### 2.2. *Euler equation with the Jastrow wave function*

For the practical implementation of the Euler equation Eq. (2.16) we return to the Jastrow ansatz (2.6) for the wave function. Then the generalized normalization integral (2.9) becomes

$$I_0(\beta) = \int d^3r_1\ldots d^3r_N\,\exp\left[\sum_{i<j}^{N}u_2(|\mathbf{r}_i-\mathbf{r}_j|;\beta)\right]\,, \tag{2.17}$$

where we have defined the generalized correlation functions by adding the Jastrow correlation function into the Jackson–Feenberg potential (2.10),

$$u_2(r;\beta) \equiv u_2(r) + \beta\left[V(r) - \frac{\hbar^2}{4m}\nabla^2 u_2(r)\right]\,. \tag{2.18}$$

Its $\beta$–derivative is simple to calculate,

$$u_2'(r) \equiv \frac{\partial u_2(r;\beta)}{\partial\beta}\bigg|_{\beta=0} = V(r) - \frac{\hbar^2}{4m}\nabla^2 u_2(r)\,. \tag{2.19}$$

We rely on the diagrammatic summation methods for deriving the connection between the correlation and distribution functions. The hypernetted chain (HNC) equation gives in the diagrammatic terms the parallel connections

$$g(r) = e^{u_2(r)+N(r)+E(r)} \tag{2.20}$$

and the Ornstein–Zernike integral equation

$$N(|\mathbf{r}_1 - \mathbf{r}_2|) = \rho_0 \int d^3r_3 (g(|\mathbf{r}_1 - \mathbf{r}_3|) - 1) X(|\mathbf{r}_3 - \mathbf{r}_2|) \tag{2.21}$$

sums the nodal diagrams, $N(r)$, which are formed by iterating the integral equation up to infinite order with the direct correlation function $X(r)$,

$$X(r) = g(r) - 1 - N(r)\,. \tag{2.22}$$

The sum of elementary diagrams $E(r)$ must be calculated term by term. Typically one stops after the five–body diagrams.

From the HNC–equation (2.20) we can calculate the $\beta$–derivative of the radial distribution function

$$g'(r) = g(r)\left[u_2'(r) + N'(r) + E'(r)\right]\,. \tag{2.23}$$

Inserting here $u_2'(r)$ from Eq. (2.19) together with $u_2(r)$ solved from Eq. (2.20),

$$u_2(r) = \log g(r) - N(r) - E(r)\,, \tag{2.24}$$

we get

$$\begin{aligned} g'(r) &= g(r)\left(V(r) + N'(r) + E'(r)\right) \\ &\quad - \frac{\hbar^2}{4m}\left[g(r)\,\nabla^2 \log g(r) - g(r)\left(\nabla^2 N(r) + \nabla^2 E(r)\right)\right]\,. \end{aligned} \tag{2.25}$$

It is useful to define two new quantities the induced potential $w_{\text{ind}}(r)$ in terms of the sum of nodal diagrams $N(r)$,

$$w_{\text{ind}}(r) \equiv N'(r) + \frac{\hbar^2}{4m}\nabla^2 N(r)\,, \tag{2.26}$$

and the effective potential arising from the elementary diagrams,

$$V_{\text{ele}}(r) \equiv E'(r) + \frac{\hbar^2}{4m}\nabla^2 E(r)\,. \tag{2.27}$$

These simplify the expression of $g'(r)$ and if we furthermore use the identity

$$g(r)\,\nabla^2 \log g(r) = -\nabla^2 g(r) + 4\sqrt{g(r)}\,\nabla^2\sqrt{g(r)} \tag{2.28}$$

then the Euler equation (2.16) can be written in the form of a zero–energy Schrödinger equation

$$-\frac{\hbar^2}{m}\nabla^2\sqrt{g(r)} + \left(V(r) + w_{\text{ind}}(r) + V_{\text{ele}}(r)\right)\sqrt{g(r)} = 0 \tag{2.29}$$

where $\sqrt{g(r)}$ plays the role of the wave function and the normalization is such that $\sqrt{g(\infty)} = 1$. Yet, one should realize that this equation is highly non–linear because both $w_{\text{ind}}(r)$ and $V_{\text{ele}}(r)$ depend on the solution $g(r)$.

The above formulation puts weight on the short range behavior of the radial distribution function. Especially for singular potentials $V(r)$ we can derive the behavior of $g(r)$ when $r \to 0$. In the case of the Coulomb potential — $e^2/r$ with

$e^2 = q^2/(4\pi\varepsilon_0)$ where $q$ is the charge of the particles and $\varepsilon_0$ is the dielectric constant — $g(r)$ must satisfy the cusp condition

$$\frac{d}{dr}\log g(r) = \frac{e^2 m}{\hbar^2}\,. \tag{2.30}$$

In the case of the Lennard–Jones potential

$$V(r) = \epsilon\left[\left(\frac{\sigma}{r}\right)^{12} - \left(\frac{\sigma}{r}\right)^{6}\right] \tag{2.31}$$

$g(r)$ has an essential singularity at the origin

$$g(r) \overset{r\to 0}{\to} \exp\left[-\frac{A}{r^5}\right]; \qquad A = \frac{8}{5}\frac{m\sqrt{\epsilon}}{\hbar^2}\sigma^6\,. \tag{2.32}$$

Another way to solve the Euler equation (2.16), which put emphasis on the long wavelength behavior, is to define the particle–hole effective interaction,

$$V_{p-h}(r) \equiv X'(r) + \frac{\hbar^2}{4m}\nabla^2 X(r)\,, \tag{2.33}$$

in terms of the direct correlation function $X(r)$ and its $\beta$–derivative from Eq. (2.22)

$$X'(r) = g'(r) - N'(r)\,. \tag{2.34}$$

We can write Eq. (2.25) into a new form by subtracting $N'(r)$ from both sides, inserting the definitions (2.26) and (2.27) and using another identity for the logarithmic term

$$g(r)\nabla^2 \log g(r) = \nabla^2 g(r) - 4\left(\sqrt{g(r)}\right)^2\,. \tag{2.35}$$

This leads to the expression

$$\begin{aligned} X'(r) &= g(r)\left[V(r) + V_{\rm ele}(r)\right] + \left[g(r) - 1\right] w_{\rm ind}(r) \\ &\quad + \frac{\hbar^2}{4m}\left[4\left(\nabla\sqrt{g(r)}\right)^2 - \nabla^2 X(r)\right]\,. \end{aligned} \tag{2.36}$$

Combining this with the definition (2.33) we arrive at the expression for the particle–hole potential

$$V_{p-h}(r) = g(r)\left[V(r) + V_{\rm ele}(r)\right] + \left[g(r) - 1\right] w_{\rm ind}(r) + \frac{\hbar^2}{m}\left(\nabla\sqrt{g(r)}\right)^2\,. \tag{2.37}$$

For the homogeneous system it is often more convenient to work in momentum space and define the structure function $S(k)$ as the Fourier transform of $g(r) - 1$,

$$S(k) = 1 + \rho_0 \int d^3r\,\left(g(r) - 1\right) e^{-i\mathbf{k}\cdot\mathbf{r}}\,. \tag{2.38}$$

Similarly we get the Fourier transform of the $\beta$–derivative

$$S'(k) = \rho_0 \int d^3r\; g'(r)\, e^{-i\mathbf{k}\cdot\mathbf{r}}\,. \tag{2.39}$$

With these Fourier transforms the Euler equation (2.16) can be written in the momentum space

$$S'(k) = -\frac{\hbar^2 k^2}{4m}\left(S(k) - 1\right) . \tag{2.40}$$

The Ornstein–Zernike equation (2.21) contains the convolution integral, which is of a product form in the Fourier space and can be solved easily together with Eq. (2.22) giving

$$\tilde{X}(k) = 1 - \frac{1}{S(k)} . \tag{2.41}$$

Performing the $\beta$–derivative operation and using the Eq. (2.40) we get

$$\tilde{X}'(k) = \frac{S'(k)}{S^2(k)} = -\frac{\hbar^2 k^2}{4m}\frac{\left(S(k) - 1\right)}{S^2(k)} . \tag{2.42}$$

Inserting these into the Fourier transform of the definition (2.33) we find a simple expression for $\tilde{V}_{\text{p-h}}(k)$

$$\begin{aligned} \tilde{V}_{\text{p-h}}(k) &= \tilde{X}'(k) - \frac{\hbar^2 k^2}{4m}\tilde{X}(k) \\ &= -\frac{\hbar^2 k^2}{4m}\left(\frac{S(k) - 1}{S^2(k)} + \frac{S(k) - 1}{S(k)}\right) \\ &= -\frac{\hbar^2 k^2}{4m}\left(1 - \frac{1}{S^2(k)}\right) , \end{aligned} \tag{2.43}$$

which can be solved for the structure function and we get the Bogoljubov–like form of the Euler equation

$$S(k) = \frac{k}{\sqrt{k^2 + \frac{4m}{\hbar^2}\tilde{V}_{\text{p-h}}(k)}} . \tag{2.44}$$

We still need to calculate the induced potential $w_{ind}(r)$ defined in Eq. (2.26). From Eqs. (2.22) and (2.41) we get

$$\tilde{N}(k) = \frac{\left(S(k) - 1\right)^2}{S(k)} . \tag{2.45}$$

The $\beta$–derivative and the use of Eqs. (2.40) and (2.42) gives

$$\tilde{N}'(k) = S'(k) - \tilde{X}'(k) = -\frac{\hbar^2 k^2}{4m}\frac{\left(S(k) - 1\right)^2 \left(S(k) + 1\right)}{S^2(k)} . \tag{2.46}$$

Inserting these into the definition (2.26) we get the Fourier transform of $w_{ind}(r)$

$$\begin{aligned}\tilde{w}_{ind}(k) &= \tilde{N}'(k) - \frac{\hbar^2 k^2}{4m}\tilde{N}(k) \\ &= -\frac{\hbar^2 k^2}{4m}\left(S(k)-1\right)^2\left(\frac{S(k)+1}{S^2(k)} + \frac{1}{S(k)}\right) \\ &= -\frac{\hbar^2 k^2}{4m}(2S(k)+1)\left(1-\frac{1}{S(k)}\right)^2 . \end{aligned} \tag{2.47}$$

Equations (2.44), (2.37) and (2.47) form a self–consistent set of equations which can be solved iteratively.

In the the long wavelength limit the structure function of the $^4$He liquid is linear and inversely proportional to the speed of sound $c$

$$S(k) \overset{k\to 0}{\to} \frac{\hbar k}{2mc}, \tag{2.48}$$

which tells that $\tilde{V}_{\rm p-h}(0) = mc^2$. In the case of charged Bose gas the Coulomb interaction $4\pi e^2\rho_0/k^2$ dominates the small $k$ behavior of $\tilde{V}_{\rm p-h}(k)$ and the structure function becomes quadratic

$$S(k) \overset{k\to 0}{\to} \frac{\hbar k^2}{2m\omega_p} \tag{2.49}$$

where $\omega_p = \sqrt{4\pi\rho_0 e^2/m}$ is the plasma frequency.

## 3. Equation of motion method

In the above description of the ground state we have optimized the total energy and derived the Euler equations for the pair distribution and structure functions. In this section we follow a similar approach for the dynamic, time dependent system and optimize the action integral with respect to density oscillations. The time dependence is induced into the system by adding an external, driving potential into the Hamiltonian. The bosonic quantum system responds to that by changing its density. Since the Hamiltonian is known and the wave function is solved from the least action principle we can derive the continuity equations for the one– and two–particle currents and densities, which are the equations of motion of the system. In this approach all particles are treated on the same footing including those which are in the Bose condensed state. Special attention to the condensate fraction can be payed by calculating the asymptotic behavior of the time dependent density matrix after the equations of motion have been solved.

### 3.1. *Linear response*

Let us assume that a quantum fluid is driven out of the ground state by an infinitesimal external interaction $\tilde{U}_{\rm ext}(k,\omega)$, with a given frequency $\omega$ and wave number $k$.

It induces a change in the density, $\delta\tilde{\rho}_1(k,\omega)$, of an originally homogeneous system. The response of the system is assumed to be linear and the information on its dynamic properties, excitations and decay modes, is contained in the density–density linear–response function defined as

$$\chi(k,\omega) = \frac{\delta\tilde{\rho}_1(k,\omega)}{\rho_0 \tilde{U}_{\rm ext}(k,\omega)} . \tag{3.1}$$

This is a complex function and its imaginary part is connected to the dynamic and static structure functions by the fluctuation–dissipation theorem

$$S(k,\omega) = -\frac{1}{\pi}\Im m\left[\chi(k,\omega)\right] . \tag{3.2}$$

Hence, $S(k,\omega)$ is known, once the relation between the one–body density fluctuations and the perturbation has been established and one can compare the theoretical results with experiments.

The dynamic structure function is measured in (x–ray, neutron) scattering experiments and provides information of the strength, lifetime, and dispersion of excitations. At low temperatures it consists typically of a low–frequency sharp peak and of a high–frequency broad contribution.[33,5] It is therefore customary to write

$$S(k,\omega) = Z(k)\delta(\hbar\omega - \hbar\omega_0(k)) + S_{\rm mp}(k,\omega) . \tag{3.3}$$

This immediately suggests that the density–density response function can be written in the form familiar from the random phase approximation (RPA)) as we will show in detail

$$\chi(k,\omega) = S(k)\left[\frac{1}{\hbar\omega - \varepsilon(k) - \Sigma(k,\omega)} - \frac{1}{\hbar\omega + \varepsilon(k) + \Sigma^*(k,-\omega)}\right] . \tag{3.4}$$

Here $\varepsilon(k)$ is the energy of a single collective mode and $\Sigma(k,\omega)$ is the complex self–energy,

$$\Sigma(k,\omega) = \Delta(k,\omega) - i\Gamma(k,\omega) . \tag{3.5}$$

Recalling that for singular integrand one has

$$\frac{1}{\omega-\omega'} = \mathcal{P}\frac{1}{\omega-\omega'} - i\pi\delta(\omega-\omega') \tag{3.6}$$

where $\mathcal{P}$ denotes the Cauchy principal value, we see that the dynamic structure function (3.2) has contributions from the poles of the response function and from the imaginary part of the self–energy.

The elementary excitation modes are determined by the positive frequency poles $\omega = \omega_0(k)$ of Eq. (3.4). We will show later that the poles lie on the real axis

$$\hbar\omega_0(k) = \varepsilon(k) + \Delta(k,\omega_0(k)) \tag{3.7}$$

and $\Gamma(k,\omega_0(k)) = 0$. The strength $Z(k)$ of that pole can then be evaluated from the derivative of the self–energy,

$$Z(k) = S(k)\left[1 - \left.\frac{d\Delta(k,\omega)}{d(\hbar\omega)}\right|_{\omega=\omega_0}\right]^{-1} . \tag{3.8}$$

The second term in (3.3), $S_{\mathrm{mp}}(k,\omega)$, is the multi–phonon background, *i.e.* the contribution in which a projectile like a neutron probing the system exchanges energy with two or more excitation modes. This contribution is determined by the imaginary part of the self–energy

$$S_{\mathrm{mp}}(k,\omega) = \frac{1}{\pi} S(k) \frac{\Gamma(k,\omega)}{\left[\hbar\omega + \varepsilon(k) + \Delta(k,\omega)\right]^2 + \left[\Gamma(k,\omega)\right]^2} \,. \tag{3.9}$$

The dynamic structure function is a positive function for positive values of $\omega$ and that is why also $\Gamma(k,\omega)$ must be a non–negative function. For $\Gamma(k,\omega) > 0$ the energy $\hbar\omega$ transfered into the system has created such excitation modes, that can decay into other lower lying excitations within finite lifetime.

In addition, the relative weight, $Z(k)/S(k)$, gives the efficiency of scattering processes from a single collective excitation, as seen from the (zeroth–moment) sum rule

$$\int_0^\infty S(k,\omega) d(\hbar\omega) = S(k) \,. \tag{3.10}$$

In other words, it gives the fraction of available scattering processes going through a single collective mode at a given wave vector. If the only excitation in the system were the collective mode $\varepsilon(k)$, like in the Feynman approximation, then the ratio $Z(k)/S(k)$ would be one.

### 3.2. *Time–dependent correlation functions*

Let us proceed with the details of the time–dependent response of the system. If a weak, time dependent interaction perturbs the system then the ground–state wave function, $\Psi_0(\mathbf{r}_1,\ldots,\mathbf{r}_N)$, is modified accordingly and it becomes time dependent. For later use we separate in the notations the time dependent phase due to the unperturbed ground state energy

$$\Psi(\mathbf{r}_1,\ldots,\mathbf{r}_N;t) = e^{-iE_0 t/\hbar} \Phi(\mathbf{r}_1,\ldots,\mathbf{r}_N;t) \,, \tag{3.11}$$

the change in the normalization due to the perturbation

$$\Phi(\mathbf{r}_1,\ldots,\mathbf{r}_N;t) = \frac{1}{\sqrt{\mathcal{N}(t)}} \, \phi(\mathbf{r}_1,\ldots,\mathbf{r}_N;t) \tag{3.12}$$

and the time dependence of the correlation functions

$$\phi(\mathbf{r}_1,\ldots,\mathbf{r}_N;t) = e^{\frac{1}{2}\delta U(\mathbf{r}_1,\ldots,\mathbf{r}_N;t)} \Psi_0(\mathbf{r}_1,\ldots,\mathbf{r}_N) \,. \tag{3.13}$$

One can think $\delta U(\mathbf{r}_1,...,\mathbf{r}_N;t)$ as a complex–valued excitation operator

$$\delta U(\mathbf{r}_1,...,\mathbf{r}_N;t) = \sum_i \delta u_1(\mathbf{r}_i;t) + \sum_{i<j} \delta u_2(\mathbf{r}_i,\mathbf{r}_j;t) \tag{3.14}$$

expanded in terms of one– and two–body correlation functions. The time–dependent one–body function $\delta u_1(\mathbf{r}_i;t)$ must be included into the description because dynamics will normally break the translational invariance of the system, but restricting the time dependence to the one–body component only would lead directly to the Feynman theory of one–phonon excitations as we will show later. The time–dependent two–body component is significant in situations where the external field excites fluctuations of wavelengths comparable to the inter–particle distance, as explicitly demonstrated for liquid $^4$He [31,29,30,28,17] and for the bosonic Coulomb systems[34].

In the previous section we showed that the optimized ground state $\Psi_0(\mathbf{r}_1, \ldots, \mathbf{r}_N)$ satisfies the Schrödinger equation

$$H_0 \Psi_0 = E_0 \Psi_0 \tag{3.15}$$

where $H_0$ is the ground state Hamiltonian given in Eq. (2.1) and $E_0$ is the ground state energy appearing in the phase factor of definition Eq. (3.11). In order to simplify notations we assume that the ground state wave function is normalized to unity

$$\langle \Psi_0 | \Psi_0 \rangle = 1\,. \tag{3.16}$$

The normalization factor in Eq. (3.12) has changed from that due to the time dependence of the correlation factors

$$\mathcal{N}(t) = \int d^3r_1 \ldots d^3r_N |\Psi_0(\mathbf{r}_1, \ldots \mathbf{r}_N)|^2 e^{\Re e[\delta U(\mathbf{r}_1, \ldots \mathbf{r}_N; t)]}\,. \tag{3.17}$$

### 3.3. *Action integral*

The time evolution of the wave function functions is governed by the least–action principle [35,36]

$$\delta \mathcal{S} = \delta \int_{t_0}^{t} dt \mathcal{L}(t) \equiv \delta \int_{t_0}^{t} dt \left\langle \Psi(t) \left| H(t) - i\hbar \frac{\partial}{\partial t} \right| \Psi(t) \right\rangle = 0 \tag{3.18}$$

where the Hamiltonian

$$H(t) = H_0 + \hat{U}_{\rm ext}(t) \tag{3.19}$$

now contains, besides the ground–state Hamiltonian $H_0$, also the time dependent operator $\hat{U}_{\rm ext}(t) = \sum_{i=1}^{N} U_{\rm ext}(\mathbf{r}_i; t)$, which creates an external, infinitesimal disturbance into the system. This is exactly the term that provides us the relation between the one–body density fluctuations and the perturbation, needed to calculate the dynamic linear response.

Using the ground state Schrödinger equation (3.15) we can write the integrand in the form

$$\begin{aligned} \mathcal{L}(t) &= \left\langle \Phi(t) \left| H - E_0 - \frac{\hbar}{2}\left( i\frac{\partial}{\partial t} + h.c. \right) \right| \Phi(t) \right\rangle \\ &= \frac{1}{\mathcal{N}(t)} \left\langle \Psi_0 \left| e^{\frac{1}{2}\delta U^*(t)} [H_0, e^{\frac{1}{2}\delta U(t)}] \right| \Psi_0 \right\rangle \\ &+ \left\langle \Phi(t) \left| -\frac{\hbar}{2}\left( i\frac{\partial}{\partial t} + h.c. \right) + \hat{U}_{ext}(t) \right| \Phi(t) \right\rangle\,. \end{aligned} \tag{3.20}$$

With the notation h.c. we add the hermitian conjugation of the time–derivative operator.

The potential energy term commutes with $\delta U(t)$ and thus only the kinetic energy gives contribution to the commutator. This can be evaluated with a little bit of algebra,

$$\int d\tau\ \Psi_0\ e^{\frac{1}{2}\delta U^*(t)}\nabla^2\left(e^{\frac{1}{2}\delta U(t)}\ \Psi_0\right)$$

$$= \int d\tau \Psi_0 e^{\frac{1}{2}\delta U^*(t)}\nabla\cdot\left(\frac{1}{2}e^{\frac{1}{2}\delta U(t)}\Psi_0\nabla\delta U(t)+e^{\frac{1}{2}\delta U(t)}\nabla\Psi_0\right)$$

$$= \int d\tau\Big[-e^{\frac{1}{2}\delta U^*(t)}\left(\frac{1}{2}\Psi_0\nabla\delta U^*(t)+\nabla\Psi_0\right)\frac{1}{2}e^{\frac{1}{2}\delta U(t)}\Psi_0\nabla\delta U(t)$$

$$+\ \Psi_0 e^{\Re e[\delta U(t)]}\left(\frac{1}{2}\nabla\Psi_0\cdot\nabla\delta U(t)+\nabla^2\Psi_0\right)\Big]$$

$$= \int d\tau\left[-\frac{1}{4}|\Psi_0|^2 e^{\Re e[\delta U(t)]}\,|\nabla\delta U(t)|^2+\Psi_0 e^{\Re e[\delta U(t)]}\nabla^2\Psi_0\right]$$

giving the result

$$\frac{1}{\mathcal{N}(t)}\left\langle\Psi_0\left|e^{\frac{1}{2}\delta U^*(t)}[H_0,e^{\frac{1}{2}\delta U(t)}]\right|\Psi_0\right\rangle=\frac{\hbar^2}{8m}\left\langle\Phi(t)\left|\sum_{j=1}^{N}|\nabla_j\delta U(t)|^2\right|\Phi(t)\right\rangle\ . \tag{3.21}$$

The evaluation of the time derivative gives

$$-\frac{\hbar}{2}\left\langle\Psi_0\left|\frac{e^{\frac{1}{2}\delta U^*(t)}}{\sqrt{\mathcal{N}(t)}}\left(i\frac{\partial}{\partial t}+h.c.\right)\frac{e^{\frac{1}{2}\delta U(t)}}{\sqrt{\mathcal{N}(t)}}\right|\Psi_0\right\rangle=\frac{1}{2}\hbar\langle\Phi(t)|\Im m[\delta\dot{U}(t)]|\Phi(t)\rangle \tag{3.22}$$

where we have used the dot–notation $\dot{f}(t)=\frac{\partial f(t)}{\partial t}$. Collecting all together we have the integrand

$$\mathcal{L}(t)=\left\langle\Phi(t)\left|\frac{\hbar^2}{8m}\sum_{j=1}^{N}|\nabla_j\ \delta U(t)|^2+\frac{1}{2}\hbar\ \Im m[\delta\dot{U}(t)]+\hat{U}_{ext}(t)\right|\Phi(t)\right\rangle\ . \tag{3.23}$$

### 3.4. *Least action principle*

In the least–action principle we search for the correlation functions $\delta u_1(\mathbf{r}_1;t)$, $\delta u_2(\mathbf{r}_1,\mathbf{r}_2;t)$, etc, which minimize the action integral (3.18). The variation

$$\delta\mathcal{S}=\delta\int_{t_0}^{t}dt'\mathcal{L}(t')=0 \tag{3.24}$$

with respect to a general correlation function $\delta u_n(\mathbf{r}_1, \ldots, \mathbf{r}_n; t)$, which depends on $n$ coordinates and time gives

$$\int d^3r_{n+1} \ldots d^3r_N \Big[ -\frac{\hbar^2}{8m} \sum_{j=1}^{n} \nabla_j \cdot \left( |\Phi(t)|^2 \nabla_j \delta U(t) \right)$$
$$- \frac{i\hbar}{4} \frac{\partial}{\partial t} |\Phi(t)|^2 + \left( \frac{\hbar}{2} \Im m[\delta \dot{U}] + \hat{U}_{ext} \right) \frac{1}{\mathcal{N}(t)} \frac{\partial |\phi|^2}{\partial \delta U^*(t)} \Big]$$
$$- \left\langle \Phi(t) \left| \frac{\hbar}{2} \Im m[\delta \dot{U}] + \hat{U}_{ext} \right| \Phi(t) \right\rangle \frac{1}{\mathcal{N}(t)} \frac{\partial \mathcal{N}(t)}{\partial \delta U^*(t)} \,. \tag{3.25}$$

The derivatives can be calculated from the definitions (3.12)

$$\frac{\partial |\phi|^2}{\partial \delta U^*(t)} = \frac{1}{2} |\phi|^2$$
$$\frac{1}{\mathcal{N}} \frac{\partial \mathcal{N}}{\partial \delta U^*(t)} = \frac{1}{2} \int d^3r_{n+1} \ldots d^3r_N |\Phi(t)|^2$$
$$\frac{\partial}{\partial t} |\Phi(t)|^2 = |\Phi(t)|^2 \left[ \Re e \delta \dot{U} - \int d^3r_1 \ldots d^3r_N |\Phi(t)|^2 \Re e \delta \dot{U} \right]$$

and the least action principle can be written in the form

$$\int d^3r_{n+1} \ldots d^3r_N \Big[ -\frac{\hbar^2}{4m} \sum_{j=1}^{n} \nabla_j \cdot \left( |\Psi|^2 \nabla_j \delta U(t) \right) \tag{3.26}$$
$$+ |\Psi|^2 \left( -\frac{i\hbar}{2} \delta \dot{U} + \hat{U}_{ext} - \left\langle \Psi \left| -\frac{i\hbar}{2} \delta \dot{U} + \hat{U}_{ext} \right| \Psi \right\rangle \right) \Big] = 0 \,.$$

### 3.5. *Many–particle densities and currents*

In order to simplify the notations in Eq. (3.26) we generalize the definition of the $n$–particle density in Eq. (2.3) to the time dependent wave function,

$$\bar{\rho}_n(\mathbf{r}_1, \ldots \mathbf{r}_n; t) = \frac{N!}{(N-n)!} \int d^3r_{n+1} \ldots d^3r_N |\Psi(\mathbf{r}_1, \ldots \mathbf{r}_N; t)|^2 \,. \tag{3.27}$$

In the linear response theory one assumes that the time dependent perturbation is infinitesimal and hence it is natural to separate the time dependent and independent parts of the density,

$$\bar{\rho}_n(\mathbf{r}_1, \ldots \mathbf{r}_n; t) = \rho_n(\mathbf{r}_1, \ldots \mathbf{r}_n) + \delta\rho_n(\mathbf{r}_1, \ldots \mathbf{r}_n; t) \,. \tag{3.28}$$

Expanding to the first order in $\delta U(t)$ we get

$$\delta\rho_n(\mathbf{r}_1, \ldots \mathbf{r}_n; t) \approx \frac{N!}{(N-n)!} \int d^3r_{n+1} \ldots d^3r_N |\Psi_0|^2$$
$$\times \Re e \Big[ \delta U(t) - \langle \Psi_0 | \delta U(t) | \Psi_0 \rangle \Big] \,. \tag{3.29}$$

The physical density is a real quantity, but in Eq. (3.26) the terms with the time derivative can be identified with the density if we generalize the above definition to the complex density fluctuations by removing the notation for the real part.

Similarly we define the $n$–particle current

$$\mathbf{j}_n(\mathbf{r}_1,\dots\mathbf{r}_n;t) = \frac{\hbar}{2mi}\frac{N!}{(N-n)!}\int d^3r_{n+1}\dots d^3r_N \left[\Psi^*\sum_{j=1}^{n}\nabla_j\Psi - \Psi\sum_{j=1}^{n}\nabla_j\Psi^*\right], \tag{3.30}$$

expand it to the first order in $\delta U(t)$

$$\mathbf{j}_n(\mathbf{r}_1,\dots\mathbf{r}_n;t) \approx \frac{\hbar}{2m}\frac{N!}{(N-n)!}\int d^3r_{n+1}\dots d^3r_N \, |\Psi_0(\mathbf{r}_1,\dots\mathbf{r}_N)|^2 \Im m\left[\sum_{j=1}^{n}\nabla_j\,\delta U(t)\right] \tag{3.31}$$

and generalize the definition to complex currents as needed in Eq. 3.26).

### 3.6. *One– and two–particle continuity equations*

With the generalized definitions of Eqs. (3.29) and (3.31), the least action principle (3.26) can be separated into one– and two–particle continuity equations — the equations of motion of the system, [31,29,30]

$$\nabla_1\cdot\mathbf{j}_1(\mathbf{r}_1;t) + \delta\dot\rho_1(\mathbf{r}_1;t) = D_1(\mathbf{r}_1;t) \tag{3.32}$$

$$[\,\nabla_1\cdot\mathbf{j}_2(\mathbf{r}_1,\mathbf{r}_2;t) + \text{same for }\;(1\leftrightarrow 2)] + \delta\dot\rho_2(\mathbf{r}_1,\mathbf{r}_2;t) = D_2(\mathbf{r}_1,\mathbf{r}_2;t)\,. \tag{3.33}$$

Inserting the definition of the excitation operator (3.14) into the definition of the density (3.29) and using the complex notation we get the expression of the one–particle density fluctuations in terms of the two– and three–particle distribution functions (2.4)

$$\begin{aligned}\delta\rho_1(\mathbf{r}_1;t) &= \rho_0\,\delta u_1(\mathbf{r}_1;t) + \rho_0^2\int d^3r_2(g_2(\mathbf{r}_1,\mathbf{r}_2)-1)\delta u_2(\mathbf{r}_2;t)\\ &+ \rho_0^2\int d^3r_2\Big[g_2(\mathbf{r}_1,\mathbf{r}_2)\delta u_2(\mathbf{r}_1,\mathbf{r}_2;t)\\ &+ \frac{\rho_0}{2}\int d^3r_3\,(g_3(\mathbf{r}_1,\mathbf{r}_2,\mathbf{r}_3)-g_2(\mathbf{r}_2,\mathbf{r}_3))\,\delta u_2(\mathbf{r}_2,\mathbf{r}_3;t)\Big]\,.\end{aligned} \tag{3.34}$$

From the definition (3.29) it is easy to verify that the particle number is conserved in the fluctuations

$$\int d^3r\delta\rho_1(\mathbf{r}) = 0 \tag{3.35}$$

and that the sequential relation is satisfied,

$$\int d^3r_2\;\delta\rho_2(\mathbf{r}_1,\mathbf{r}_2;t) = (N-1)\;\delta\rho_1(\mathbf{r}_1;t)\,. \tag{3.36}$$

The one– and two–particle currents can be read out of Eq. (3.31). Using again the complex notation we get

$$\mathbf{j}_1(\mathbf{r}_1;t) = \frac{\hbar\rho_0}{2mi}\left\{\nabla_1\delta u_1(\mathbf{r}_1;t) + \rho_0\int d^3r_2 g_2(\mathbf{r}_1,\mathbf{r}_2)\nabla_1\delta u_2(\mathbf{r}_1,\mathbf{r}_2;t)\right\}. \quad (3.37)$$

$$\mathbf{j}_2(\mathbf{r}_1,\mathbf{r}_2;t) = \frac{\hbar\rho_0^2}{2mi}\left\{g_2(\mathbf{r}_1,\mathbf{r}_2)[\nabla_1\delta u_1(\mathbf{r}_1;t) + \nabla_1\delta u_2(\mathbf{r}_1,\mathbf{r}_2;t)]\right.$$
$$\left. + \rho_0\int d^3r_3 g_3(\mathbf{r}_1,\mathbf{r}_2,\mathbf{r}_3)\nabla_1\delta u_2(\mathbf{r}_1,\mathbf{r}_3;t)\right\}. \quad (3.38)$$

The currents also satisfy the sequential relation

$$\int d^3r_2\ \mathbf{j}_2(\mathbf{r}_1,\mathbf{r}_2;t) = (N-1)\ \mathbf{j}_1(\mathbf{r}_1;t)\,. \quad (3.39)$$

The terms which depend on the external potential are collected into the functions $D_1(\mathbf{r}_1;t)$ and $D_2(\mathbf{r}_1,\mathbf{r}_2;t)$ and they are the terms which drive excitations into the system,

$$D_1(\mathbf{r}_1;t) = \frac{2\rho_0}{i\hbar}\left\{U_{\rm ext}(\mathbf{r}_1;t) + \rho_0\int d^3r_2\Big[g_2(\mathbf{r}_1,\mathbf{r}_2)-1\Big]U_{\rm ext}(\mathbf{r}_2;t)\right\} \quad (3.40)$$

$$D_2(\mathbf{r}_1,\mathbf{r}_2;t) = \frac{2\rho_0^2}{i\hbar}\left\{g_2(\mathbf{r}_1,\mathbf{r}_2)\,[U_{\rm ext}(\mathbf{r}_1;t) + U_{\rm ext}(\mathbf{r}_2;t)]\right.$$
$$\left. +\rho_0\int d^3r_3\Big[g_3(\mathbf{r}_1,\mathbf{r}_2,\mathbf{r}_3) - g_2(\mathbf{r}_1,\mathbf{r}_2)\Big]U_{\rm ext}(\mathbf{r}_3;t)\right\}. \quad (3.41)$$

## 4. Solving the continuity equations

Up to now, we have formulated the problem in terms of a Hamiltonian, a trial wave function, the least action principle, and we have derived two coupled continuity equations (3.32) and (3.33). What we still need to do is to find a way to actually solve those continuity equations for the unknown quantities $\delta u_1(\mathbf{r};t)$ and $\delta u_2(\mathbf{r}_1,\mathbf{r}_2;t)$; assuming that all ground state quantities are known. In the following we introduce various approximation schemes.

In the homogeneous system fluctuations are weak and it is more convenient to work in the Fourier space. We define the one–body Fourier transform and its inverse as

$$\mathcal{F}[f(\mathbf{r};t)] = \rho_0\int d^3r\ dt\ e^{-i(\mathbf{k}\cdot\mathbf{r}-\omega t)}f(\mathbf{r};t) = \tilde{f}(\mathbf{k};\omega)$$
$$\mathcal{F}^{-1}[\tilde{f}(\mathbf{k};\omega)] = \int\frac{d^3k\ d\omega}{(2\pi)^4\rho_0}e^{i(\mathbf{k}\cdot\mathbf{r}-\omega t)}\tilde{f}(\mathbf{k};\omega) = f(\mathbf{r};t) \quad (4.1)$$

and the two–body Fourier transforms in the form

$$\mathcal{F}[f(\mathbf{r}_1,\mathbf{r}_2;t)] = \rho_0^2\int d^3r_1 d^3r_2\ dt\ e^{-i(\mathbf{k}\cdot\mathbf{R}+\mathbf{p}\cdot\mathbf{r}-\omega t)}f(\mathbf{r}_1,\mathbf{r}_2;t)$$
$$\mathcal{F}^{-1}[\tilde{f}(\mathbf{k},\mathbf{p};t)] = \int\frac{d^3k d^3p\ d\omega}{(2\pi)^7\rho_0^2}e^{i(\mathbf{k}\cdot\mathbf{R}+\mathbf{p}\cdot\mathbf{r}-\omega t)}\tilde{f}(\mathbf{k},\mathbf{p};\omega)\,. \quad (4.2)$$

Here $\mathbf{R} = (\mathbf{r}_1 + \mathbf{r}_2)/2$ is the center–of–mass vector and $\mathbf{r} = \mathbf{r}_1 - \mathbf{r}_2$ the relative position vector; $\mathbf{k}$ and $\mathbf{p}$ are the center–of–mass and relative momenta, respectively.

### 4.1. *Feynman approximation*

The simplest, physically consistent approximation is the Feynman approximation where we limit the time dependence to the one–body correlation function by setting $\delta u_2(\mathbf{r}_1, \mathbf{r}_2; t) = 0$. In this case, we need to solve only the first continuity equation (3.32). The one–body current reduces to

$$\mathbf{j}_1(\mathbf{r}_1; t) = \frac{\hbar\rho_0}{2mi}\nabla_1 \delta u_1(\mathbf{r}_1; t) \tag{4.3}$$

and the time–dependent part of the density is simply

$$\delta\rho_1(\mathbf{r}_1; t) = \rho_0 \delta u_1(\mathbf{r}_1; t) + \rho_0^2 \int d^3r_2 \Big[g_2(|\mathbf{r}_1 - \mathbf{r}_2|) - 1\Big]\delta u_1(\mathbf{r}_2; t)\,. \tag{4.4}$$

The Fourier transforms can readily be calculated, resulting in

$$\delta\tilde{\rho}_1(\mathbf{k}; \omega) = S(k)\rho_0 \delta\tilde{u}_1(\mathbf{k}; \omega)\,. \tag{4.5}$$

Inserting these results into the continuity equation (3.32), along with the Fourier transform of $D_1(\mathbf{r}; t)$, we end up with

$$-\frac{\hbar k^2}{2mS(k)}\delta\tilde{\rho}_1(\mathbf{k}; \omega) + \omega\delta\tilde{\rho}_1(\mathbf{k}; \omega) = \frac{2\rho_0}{\hbar}S(k)\tilde{U}_{\text{ext}}(\mathbf{k}, \omega)\,. \tag{4.6}$$

The external potential is part of the Hamiltonian and that is why it must be hermitian. For simplicity, we assume periodic time dependence

$$U_{\text{ext}}(\mathbf{r}_i; t) = U_e(\mathbf{r}_i)\cos(\omega' t)e^{\eta t}, \qquad \eta \to 0+ \tag{4.7}$$

and switch the time dependence on adiabatically using a small, positive parameter $\eta$, which can be set equal to zero at the end of the calculations. This form has the Fourier transform

$$\tilde{U}_{\text{ext}}(\mathbf{k}, \omega) = \frac{1}{2}\tilde{U}_e(\mathbf{k})\left[\delta(\omega - \omega' - i\eta) + \delta(\omega + \omega' - i\eta)\right]\,. \tag{4.8}$$

Inserting this into Eq. (4.6) we can solve for the density fluctuations,

$$\delta\tilde{\rho}(\mathbf{k}, \omega) = \rho_0 S(k)\tilde{U}_e(\mathbf{k})\left[\frac{\delta(\omega - \omega' - i\eta)}{\hbar\omega' - \varepsilon_F(k) + i\eta} + \frac{\delta(\omega + \omega' - i\eta)}{-\hbar\omega' - \varepsilon_F(k) + i\eta}\right] \tag{4.9}$$

where we have used the notation

$$\varepsilon_F(k) = \frac{\hbar^2 k^2}{2mS(k)}\,. \tag{4.10}$$

Our aim is to calculate the linear response function (3.1) and the dynamic structure function (3.2) and for that we need the transformation back to the time–coordinate

$$\delta\rho(\mathbf{k}; t) = \rho_0 S(k)\tilde{U}_e(\mathbf{k})e^{\eta t}\left[\frac{e^{-i\omega' t}}{\hbar\omega' - \varepsilon_F(k) + i\eta} + \frac{e^{i\omega' t}}{-\hbar\omega' - \varepsilon_F(k) + i\eta}\right], \tag{4.11}$$

because the physical density needed for the linear response function must be a real function

$$\Re e[\delta\rho(\mathbf{k};t)] = \rho_0 S(k)\tilde{U}_e(\mathbf{k})\ \frac{1}{2}e^{\eta t}\Big\{e^{-i\omega' t}\left[\frac{1}{\hbar\omega' - \varepsilon_F(k) + i\eta} - \frac{1}{\hbar\omega' + \varepsilon_F(k) + i\eta}\right] + e^{i\omega' t}\left[\frac{1}{\hbar\omega' - \varepsilon_F(k) - i\eta} - \frac{1}{\hbar\omega' + \varepsilon_F(k) - i\eta}\right]\Big\}. \tag{4.12}$$

This is a symmetric function of $\omega'$ and can be written in the form

$$\Re e[\delta\rho(\mathbf{k};t)] = \frac{1}{2}e^{\eta t}\left[\delta\tilde{\rho}_1(k,\omega')e^{-i\omega' t} + \delta\tilde{\rho}_1^*(k,\omega')e^{i\omega' t}\right]. \tag{4.13}$$

Similarly we can write the external potential (4.7) in the form

$$\tilde{U}_{\text{ext}}(k;t) = \frac{1}{2}e^{\eta t}\tilde{U}_{\text{e}}(k)\left[e^{-i\omega' t} + e^{i\omega' t}\right]. \tag{4.14}$$

We are studying the time evolution of the ground state and the system can be driven only with positive frequencies, $\omega' > 0$. Using this fact we can finally define the linear response function,

$$\chi(k,\omega') = \frac{\delta\tilde{\rho}_1(k,\omega')}{\rho_0\tilde{U}_{\text{e}}(k)} = S(k)\left[\frac{1}{\hbar\omega' - \varepsilon_F(k) + i\eta} - \frac{1}{\hbar\omega' + \varepsilon_F(k) + i\eta}\right]. \tag{4.15}$$

The poles determine the collective excitation mode of the system, known as the Feynman approximation

$$\hbar\omega' = \varepsilon_F(k) = \frac{\hbar^2 k^2}{2mS(k)} \tag{4.16}$$

and in the limit $\omega' = 0$ we obtain the static response function

$$\chi(k,0) = -\frac{4mS^2(k)}{\hbar^2 k^2}. \tag{4.17}$$

In liquid $^4$He (and, in fact, also in systems where the particles interact through the screened Coulomb, or Yukawa, interaction), the excitation mode is linear in the long–wavelength limit and proportional to the speed of sound $c$,

$$\varepsilon_F \to \hbar kc\ ,\ \text{as}\ \ k \to 0\,. \tag{4.18}$$

Then the structure function must also be linear at low momenta,

$$S(k) \to \frac{\hbar k}{2mc}\ ,\ \text{as}\ \ k \to 0\,, \tag{4.19}$$

as pointed out earlier in Eq. (2.48), and the inverse of the static response function determines the incompressibility

$$-\chi^{-1}(k,0) \to mc^2\,. \tag{4.20}$$

Due to the long range of the Coulomb interaction, the long wavelength limits in charged systems depend strongly on the dimensionality of the system. For example, there is a gap in the three–dimensional plasmon spectrum contrary to the two–dimensional case. That is to say, in a three–dimensional system (regardless of the statistics) the energy of the collective excitation does not go to zero at low momenta but to a constant known as the plasma frequency and the static structure function becomes quadratic in the long wavelength limit pointed out in Eq. (2.49).

Everything put together, we have now solved our original problem: the imaginary part of the density–density response function in the single Feynman pole approximation and using Eq. (3.6) we get the dynamic structure function (3.2)

$$S(k,\omega) = S(k)\delta(\hbar\omega - \varepsilon_F(k))\,. \tag{4.21}$$

The strength of the pole is given by $Z(k) = S(k)$. The fact that $S(k,\omega)$ consists only of a single non–decaying excitation branch is where the Feynman approximation misses much of the physics. The Feynman approximation gives results that are exact in the long wavelength limit, but in the roton region the excitation energy is too large by a factor of two. Further more, the single–mode spectrum does not have an upper limit with increasing wave number, contrary to what is observed.[37] We return to these points when we discuss our results for the dynamic structure.

The elementary excitation modes of the system can also be obtained directly by setting $U_{ext} = 0$ in the continuity equations. Working still within the Feynman approximation and using the results (4.3) and (4.4), we get the differential equation

$$\frac{\hbar\rho_0}{2mi}\nabla_1^2\delta u_1(\mathbf{r}_1;\omega) - i\omega\Big[\rho_0\delta u_1(\mathbf{r}_1;\omega) + \rho_0^2\int d^3r_2\,[g(\mathbf{r}_1,\mathbf{r}_2)-1]\,\delta u_1(\mathbf{r}_2;\omega)\Big] = 0\,. \tag{4.22}$$

This has the solution (4.16) with

$$\delta u_1(\mathbf{r}_1;\omega) = e^{i\mathbf{k}\cdot\mathbf{r}_1}\,. \tag{4.23}$$

The excitation operator now takes the Feynman form,

$$\delta U = \sum_j \delta u_1(\mathbf{r}_j;\omega) = \sum_j e^{i\mathbf{k}\cdot\mathbf{r}_j}\,. \tag{4.24}$$

The full excitation operator of Eq. (3.14) can then be viewed as a generalization of this phonon–creation operator.

## 5. CBF–approximation

The next step is to allow also the two–particle correlation function to vary with time. Then the first continuity equation (3.32) can be written in momentum space in a general form

$$\hbar\omega - \varepsilon_F(k) - \Sigma(k,\omega) = 2S(k)\rho_0\tilde{U}_{\rm ext}(\mathbf{k},\omega) \tag{5.1}$$

where we have collected all the contributions from the two–particle correlation function into the self–energy $\Sigma(k,\omega)$. Following the derivation of the linear response function leading to Eq. (4.15) we can write the linear response function in the form

$$\chi(k,\omega) = S(k)\left[\frac{1}{\hbar\omega - \varepsilon_F(k) - \Sigma(k,\omega)} - \frac{1}{\hbar\omega + \varepsilon_F(k) + \Sigma^*(k,-\omega)}\right]. \tag{5.2}$$

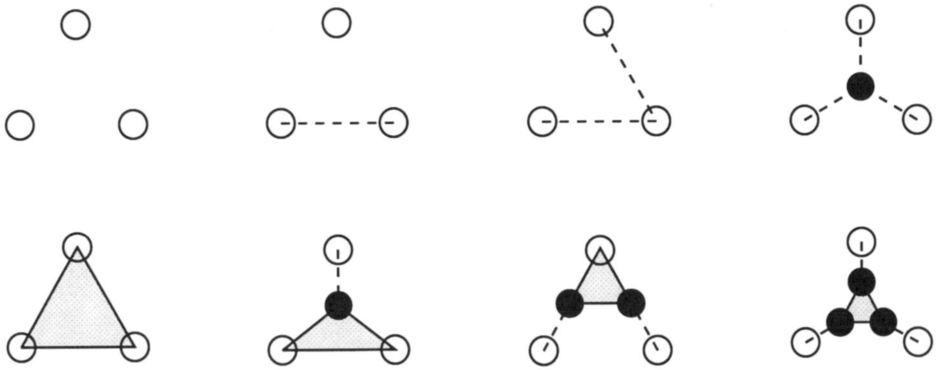

Fig. 1. Convolution approximation of $g_3(\mathbf{r}_1, \mathbf{r}_2, \mathbf{r}_3)$. Circles are particle positions, black circles are integrated and open circles not. Dashed lines are functions $h(\mathbf{r}_1, \mathbf{r}_2)$ and triangles are triplet correlation functions $u_3(\mathbf{r}_1, \mathbf{r}_2, \mathbf{r}_3)$. The second, third, sixth and seventh diagrams have three of the same kind, but with different particle coordinates.

For real values of the self–energy the response function can have poles which define the collective, elementary excitations. When the decay of the excited modes becomes possible then the self–energy acquires imaginary part and the sharp $\delta$–function in the imaginary part of the response function spreads into a broader peak.

### 5.1. *Convolution approximation*

The derivation of the self–energy starts [31,32] with a so called convolution approximation of the three–particle distribution function, but including also a special set of diagrams with the triplet correlation function $u_3(\mathbf{r}_1, \mathbf{r}_2, \mathbf{r}_3)$. The terms included are shown in Fig. 1.

In the algebraic form it becomes

$$
\begin{aligned}
g_3(\mathbf{r}_1, \mathbf{r}_2, \mathbf{r}_3) &= 1 + h(\mathbf{r}_1, \mathbf{r}_2) + h(\mathbf{r}_1, \mathbf{r}_3) + h(\mathbf{r}_2, \mathbf{r}_3) \\
&\quad + h(\mathbf{r}_1, \mathbf{r}_2)h(\mathbf{r}_1, \mathbf{r}_3) + h(\mathbf{r}_1, \mathbf{r}_2)h(\mathbf{r}_2, \mathbf{r}_3) + h(\mathbf{r}_1, \mathbf{r}_3)h(\mathbf{r}_2, \mathbf{r}_3) \\
&\quad + \int d^3r_4 h(\mathbf{r}_1, \mathbf{r}_4)h(\mathbf{r}_2, \mathbf{r}_4)h(\mathbf{r}_3, \mathbf{r}_4) \\
&\quad + \text{ terms with triplet correlations functions}\,. \qquad (5.3)
\end{aligned}
$$

Here we have also introduced the short–hand notation $h(\mathbf{r}_1, \mathbf{r}_2) = g_2(\mathbf{r}_1, \mathbf{r}_2) - 1$. After the Fourier transform this simplifies to a product form

$$
\mathcal{F}[g_3(\mathbf{r}_1, \mathbf{r}_2, \mathbf{r}_3) - 1] = S(\mathbf{k}_1)S(\mathbf{k}_2)S(\mathbf{k}_3)\,[1 + u_3(\mathbf{k}_1, \mathbf{k}_2, \mathbf{k}_3)] - 1\,. \qquad (5.4)
$$

We ignore triplet correlations for a moment and return to them at the end of this section.

### 5.2. *Two–particle equation*

Our aim is to get an approximation for $\delta u_2(\mathbf{r}_1,\mathbf{r}_2;t)$ using Eq. (3.33). The simplest term to approximate is $D_2(\mathbf{r}_1,\mathbf{r}_2;t)$ in Eq. (3.41) using the convolution approximation (5.3)

$$D_2(\mathbf{r}_1,\mathbf{r}_2;t) = \rho_0\Big\{ g_2(\mathbf{r}_1,\mathbf{r}_2)\left(D_1(\mathbf{r}_1;t) + D_1(\mathbf{r}_2;t)\right) \tag{5.5}$$
$$+ \frac{2\rho_0^2}{i\hbar}\int d^3r_3 U_{\rm ext}(\mathbf{r}_3;t)\left[h(\mathbf{r}_1,\mathbf{r}_3)h(\mathbf{r}_2,\mathbf{r}_3) + \int h(\mathbf{r}_1,\mathbf{r}_4)h(\mathbf{r}_2,\mathbf{r}_4)h(\mathbf{r}_3,\mathbf{r}_4)\right]\Big\}.$$

The last two lines can be written in the form

$$\rho_0\int d^3r_3 Y(\mathbf{r}_1,\mathbf{r}_2;\mathbf{r}_3)D_1(\mathbf{r}_3;t) \tag{5.6}$$

with $Y(\mathbf{r}_1,\mathbf{r}_2;\mathbf{r}_3) = h(\mathbf{r}_1,\mathbf{r}_3)h(\mathbf{r}_2,\mathbf{r}_3)$. The triplet correlation function will give an additional contribution to that. We can express now the two particle driving term entirely in terms of $D_1(\mathbf{r};t)$.

$$D_2(\mathbf{r}_1,\mathbf{r}_2;t) = \rho_0\Big\{ g_2(\mathbf{r}_1,\mathbf{r}_2)\left(D_1(\mathbf{r}_1;t) + D_1(\mathbf{r}_2;t)\right)$$
$$+ \rho_0\int d^3r_3 Y(\mathbf{r}_1,\mathbf{r}_2;\mathbf{r}_3)D_1(\mathbf{r}_3;t)\Big\}. \tag{5.7}$$

Similarly we can write the expression for the time dependent two–particle density using Eqs. (3.29) and (3.34)

$$\delta\rho_2(\mathbf{r}_1,\mathbf{r}_2;t) = \rho_0\Big\{ g_2(\mathbf{r}_1,\mathbf{r}_2)\left(\delta\rho_1(\mathbf{r}_1;t) + \delta\rho_1(\mathbf{r}_2;t)\right)$$
$$+ \rho_0\int d^3r_3 Y(\mathbf{r}_1,\mathbf{r}_2;\mathbf{r}_3)\delta\rho_1(\mathbf{r}_3;t)\Big\}$$
$$+ \rho_0^2 g_2(\mathbf{r}_1,\mathbf{r}_2)\delta u_2(\mathbf{r}_1,\mathbf{r}_2;t) + \mathcal{F}[\delta u_2] \tag{5.8}$$

where we have removed the dependence on $\delta u_1(\mathbf{r},t)$ in favor of $\delta\rho_1(\mathbf{r},t)$. The functional $\mathcal{F}[\delta u_2]$ contains all the rest of the terms with $\delta u_2(\mathbf{r}_1,\mathbf{r}_2;t)$. They can be written explicitly using the definition (3.29), but they are not included in the CBF–approximation.

The two particle current has a term with one–particle current, but also structure which comes from the time–dependent two–particle correlations,

$$\mathbf{j}_2(\mathbf{r}_1,\mathbf{r}_2;t) = \rho_0 g_2(\mathbf{r}_1,\mathbf{r}_2)\,\mathbf{j}_1(\mathbf{r}_1;t) + \frac{\hbar\rho_0^2}{2mi}\Big\{ g_2(\mathbf{r}_1,\mathbf{r}_2)\nabla_1\delta u_2(\mathbf{r}_1,\mathbf{r}_2;t)$$
$$+ \rho_0\int d^3r_3\left[g_3(\mathbf{r}_1,\mathbf{r}_2,\mathbf{r}_3) - g_2(\mathbf{r}_1,\mathbf{r}_2)g_2(\mathbf{r}_1,\mathbf{r}_3)\right]\nabla_1\delta u_2(\mathbf{r}_1,\mathbf{r}_3;t)\Big\}.$$

The final steps of the derivation are the approximations necessary to bring the two–body equation in a numerically tractable form. Our scheme follows the general strategy of the uniform limit approximation [13] which has been quite successful for the calculation of the optimal static three–body correlations [38,39,40]. The essence of the approximation is to consider *all products of two or more two–body functions small in coordinate space.*

In our specific case, the uniform limit approximation amounts to taking $g_2(\mathbf{r}_1,\mathbf{r}_2)\delta u_2(\mathbf{r}_1,\mathbf{r}_2;t) \approx \delta u_2(\mathbf{r}_1,\mathbf{r}_2;t)$ and a similar expression for $\nabla_1\delta u_2(\mathbf{r}_1,\mathbf{r}_2;t)$. While this approximation places more emphasis on the structure of $\delta u_2(\mathbf{r}_1,\mathbf{r}_2)$ it is physically appealing since it simply removes the *redundant* relevant short–range structure shared by $g_2(\mathbf{r}_1,\mathbf{r}_2)$ and $\delta u_2(\mathbf{r}_1,\mathbf{r}_2)$. Invoking the equivalent uniform limit for the three–body distribution function, the terms in Eq. (5.9) which depend on $\delta u_2(\mathbf{r}_1,\mathbf{r}_2;t)$ become

$$
\begin{aligned}
&\frac{\hbar\rho_0^2}{2mi}\Big[g_2(\mathbf{r}_1,\mathbf{r}_2)\nabla_1\delta u_2(\mathbf{r}_1,\mathbf{r}_2;t)\\
&+\rho_0\int d^3r_3\left[g_3(\mathbf{r}_1,\mathbf{r}_2,\mathbf{r}_3)-g_2(\mathbf{r}_1,\mathbf{r}_3)g_2(\mathbf{r}_1,\mathbf{r}_2)\right]\nabla_1\delta u_2(\mathbf{r}_1,\mathbf{r}_3;t)\Big]\\
&\approx\frac{\hbar\rho_0^3}{2mi}\int d^3r_3\left[\delta(\mathbf{r}_3-\mathbf{r}_2)+h(\mathbf{r}_3,\mathbf{r}_2)\right]\nabla_1\delta u_2(\mathbf{r}_1,\mathbf{r}_3;t)\,.
\end{aligned}
\tag{5.9}
$$

We can now put together the approximate two–particle continuity equation

$$
\begin{aligned}
\nabla_1\cdot&\left[g_2(\mathbf{r}_1,\mathbf{r}_2)\,\mathbf{j}_1(\mathbf{r}_1;t)+\frac{\hbar\rho_0^2}{2mi}\int d^3r_3\left[\delta(\mathbf{r}_3-\mathbf{r}_2)+h(\mathbf{r}_3,\mathbf{r}_2)\right]\nabla_1\delta u_2(\mathbf{r}_1,\mathbf{r}_3)\right]\\
&+\text{same with }\ (1\leftrightarrow 2)\\
&=g_2(\mathbf{r}_1,\mathbf{r}_2)\Big(D_1(\mathbf{r}_1;t)-\delta\dot\rho_1(\mathbf{r}_1;t)+D_1(\mathbf{r}_2;t)-\delta\dot\rho_1(\mathbf{r}_2;t)\Big)\\
&+\rho_0\int d^3r_3Y(\mathbf{r}_1,\mathbf{r}_2;\mathbf{r}_3)\left(D_1(\mathbf{r}_3;t)-\delta\dot\rho_1(\mathbf{r}_3;t)\right)+\rho_0\delta\dot u_2(\mathbf{r}_1,\mathbf{r}_2;t)\,.
\end{aligned}
\tag{5.10}
$$

From the terms containing the time derivative $\delta\dot u_2(\mathbf{r}_1,\mathbf{r}_2;t)$ we have kept only the leading term in accordance with the uniform limit approximation and left out the term $\mathcal{F}[\delta\dot u_2]$.

If we furthermore use the one–particle continuity equation to replace the one–particle quantities with one–particle currents we arrive at our final approximate form

$$
\begin{aligned}
&\left[\frac{\hbar\rho_0^2}{2mi}\int d^3r_3\left[\delta(\mathbf{r}_3-\mathbf{r}_2)+h(\mathbf{r}_3,\mathbf{r}_2)\right]\nabla_1^2\delta u_2(\mathbf{r}_1,\mathbf{r}_3;t)\right.\\
&\left.+\text{same with }\ (1\leftrightarrow 2)\right]-\rho_0\delta\dot u_2(\mathbf{r}_1,\mathbf{r}_2;t)\\
&=\mathbf{j}_1(\mathbf{r}_1;t)\cdot\nabla_1 g_2(\mathbf{r}_1,\mathbf{r}_2)+\mathbf{j}_1(\mathbf{r}_2;t)\cdot\nabla_2 g_2(\mathbf{r}_1,\mathbf{r}_2)\\
&+\rho_0\int d^3r_3Y(\mathbf{r}_1,\mathbf{r}_2;\mathbf{r}_3)\nabla_3\cdot\mathbf{j}_1(\mathbf{r}_3;t)\,.
\end{aligned}
\tag{5.11}
$$

The last step is to decouple the equations of motion. We can do this by approximating the one–particle current, given in Eq. (3.37) by the Feynman current of Eq. (4.3)

$$\begin{aligned}\mathbf{j}(\mathbf{r}_1;t) &= \frac{\hbar\rho_0}{2mi}\nabla_1\delta u_1(\mathbf{r}_1;t)\\ &= \frac{\hbar\rho_0}{2mi}\nabla_1\left[\delta\rho_1(\mathbf{r}_1;t) - \rho_0\int d^3r_2 X(\mathbf{r}_1,\mathbf{r}_2)\delta\rho_1(\mathbf{r}_2;t)\right]. \qquad (5.12)\end{aligned}$$

On the second line we have solved $\delta u_1(\mathbf{r}_1;t)$ in terms of $\delta\rho_1(\mathbf{r}_1;t)$ from Eq. (4.4) using the direct correlation function $X(\mathbf{r}_1,\mathbf{r}_2)$.

The fluctuating two–particle correlation function can now be expressed, in closed form, as a functional of one–body quantities alone. The integrals in Eq. (5.11) are of convolution form and can be integrated in momentum space leading to the result

$$\begin{aligned}\left[\hbar\omega - \varepsilon_{\rm F}\left(\left|\tfrac{\mathbf{k}}{2}+\mathbf{p}\right|\right) - \varepsilon_{\rm F}\left(\left|\tfrac{\mathbf{k}}{2}-\mathbf{p}\right|\right)\right]\; S\left(\left|\tfrac{\mathbf{k}}{2}+\mathbf{p}\right|\right) S\left(\left|\tfrac{\mathbf{k}}{2}-\mathbf{p}\right|\right)\delta u_2(\mathbf{k},\mathbf{p};\omega)&\\ + \varepsilon_{\rm F}(k)\;\sigma_{\mathbf{k}}(\mathbf{p})\;\delta\rho_1(\mathbf{k};\omega) = 0 \qquad (5.13)\end{aligned}$$

where we have introduced the notation $\sigma_{\mathbf{k}}(\mathbf{p})$,

$$\begin{aligned}\sigma_{\mathbf{k}}(\mathbf{p}) = -\frac{1}{k^2}&\left[\mathbf{k}\cdot\left(\tfrac{\mathbf{k}}{2}+\mathbf{p}\right) S\left(\left|\tfrac{\mathbf{k}}{2}-\mathbf{p}\right|\right) + (\mathbf{p}\leftrightarrow -\mathbf{p})\right]\\ &+ S\left(\left|\tfrac{\mathbf{k}}{2}+\mathbf{p}\right|\right) S\left(\left|\tfrac{\mathbf{k}}{2}-\mathbf{p}\right|\right)\left[1+\tilde{u}_3\left(\tfrac{\mathbf{k}}{2}+\mathbf{p},\tfrac{\mathbf{k}}{2}-\mathbf{p},-\mathbf{k}\right)\right]. \qquad (5.14)\end{aligned}$$

We have also written down the triplet correlation function $\tilde{u}_3$ which we did not carry along in the derivations. The reader can easily verify that adding them brings only length into the equations, and that everything finally results in the form given above. It is now evident that we can solve from Eq. (5.13) the fluctuating two–body correlation function directly in terms of the one–body density fluctuation,

$$\frac{\delta u_2(\mathbf{k},\mathbf{p};\omega)}{\delta\rho_1(\mathbf{k};\omega)} = -\frac{\varepsilon_{\rm F}(k)\sigma_{\mathbf{k}}(\mathbf{p})\left[S\left(\left|\tfrac{\mathbf{k}}{2}+\mathbf{p}\right|\right) S\left(\left|\tfrac{\mathbf{k}}{2}-\mathbf{p}\right|\right)\right]^{-1}}{\hbar\omega - \varepsilon_{\rm F}\left(\left|\tfrac{\mathbf{k}}{2}+\mathbf{p}\right|\right) - \varepsilon_{\rm F}\left(\left|\tfrac{\mathbf{k}}{2}-\mathbf{p}\right|\right)}, \qquad (5.15)$$

needed for the self–energy.

### 5.3. *One–particle equation*

Let us now return to the first continuity equation (3.32) and to the one–particle current (3.37). We want again to remove $\delta u_1(\mathbf{r}_1;t)$ in favor of $\delta\rho_1(\mathbf{r}_1;t)$ within the convolution approximation. In that approximation the one–particle density (3.34) can be written in the form

$$\delta\rho_1(\mathbf{r}_1;t) = \rho_0\;\delta v_1(\mathbf{r}_1;t) + \rho_0^2\int d^3r_2 h(\mathbf{r}_1,\mathbf{r}_2)\delta v_1(\mathbf{r}_2;t) \qquad (5.16)$$

with

$$\begin{aligned}\delta v_1(\mathbf{r}_1;t) = \delta u_1(\mathbf{r}_1;t) + \rho_0&\int d^3r_2 g_2(\mathbf{r}_1,\mathbf{r}_2)\delta u_2(\mathbf{r}_1,\mathbf{r}_2;t)\\ &+\frac{1}{2}\rho_0^2\int d^3r_2 d^3r_3 Y(\mathbf{r}_2,\mathbf{r}_3;\mathbf{r}_1)\delta u_2(\mathbf{r}_2,\mathbf{r}_3;t)\Big]. \qquad (5.17)\end{aligned}$$

Eq. (5.16) can then be readily solved for $\delta v_1(\mathbf{r}_1;t)$

$$\rho_0\delta v_1(\mathbf{r}_1;t) = \delta\rho_1(\mathbf{r}_1;t) - \rho_0\int d^3r_2 X(\mathbf{r}_1,\mathbf{r}_2)\delta\rho_1(\mathbf{r}_2;t)\,. \tag{5.18}$$

From that we can solve the one–particle correlation function,

$$\begin{aligned}\rho_0\delta u_1(\mathbf{r}_1;t) = {} & \delta\rho_1(\mathbf{r}_1;t) - \rho_0\int d^3r_2 X(\mathbf{r}_1,\mathbf{r}_2)\delta\rho_1(\mathbf{r}_2;t)\\ & -\rho_0^2\int d^3r_2 g_2(\mathbf{r}_1,\mathbf{r}_2)\delta u_2(\mathbf{r}_1,\mathbf{r}_2;t)\\ & -\frac{1}{2}\rho_0^3\int d^3r_2 d^3r_3 Y(\mathbf{r}_2,\mathbf{r}_3;\mathbf{r}_1)\delta u_2(\mathbf{r}_2,\mathbf{r}_3;t)\,.\end{aligned} \tag{5.19}$$

Taking its gradient and inserting into equation (3.37) we get the one–particle current

$$\begin{aligned}\mathbf{j}_1(\mathbf{r}_1;t) = \frac{\hbar}{2mi}\Bigg\{ & \nabla_1\left[\delta\rho_1(\mathbf{r}_1;t) - \rho_0\int d^3r_2 X(\mathbf{r}_1,\mathbf{r}_2)\delta\rho_1(\mathbf{r}_2;t)\right]\\ & -\rho_0^2\int d^3r_2\delta u_2(\mathbf{r}_1,\mathbf{r}_2;t)\nabla_1 g_2(\mathbf{r}_1,\mathbf{r}_2)\\ & -\frac{1}{2}\rho_0^3\int d^3r_2 d^3r_3\nabla_1 Y(\mathbf{r}_1,\mathbf{r}_2;\mathbf{r}_3)\delta u_2(\mathbf{r}_2,\mathbf{r}_3;t)\Bigg\}\,.\end{aligned} \tag{5.20}$$

We can now collect the expressions for the current and density from Eqs. (5.20) and (5.19) and insert them into the continuity equation (3.32). In momentum space it has the form

$$[\hbar\omega-\varepsilon_{\rm F}(k)]\delta\rho_1(\mathbf{k};\omega)+\frac{\hbar^2k^2}{4m}\int\frac{d^3p}{(2\pi)^3\rho_0}\sigma_{\mathbf{k}}(\mathbf{p})\delta u_2(\mathbf{k},\mathbf{p};\omega)=2\rho_0 S(k)U_{ext}(k,\omega) \tag{5.21}$$

where $\sigma_{\mathbf{k}}(\mathbf{p})$ is the same as in the two–particle equation (5.14).

## 5.4. *The self–energy and the linear response function*

As the final step we define the self–energy by dividing the left hand side of Eq. (5.21) with $\delta\rho_1(\mathbf{k};\omega)$. That gives the result

$$[\hbar\omega-\varepsilon_{\rm F}(k)-\Sigma(\mathbf{k},\omega)]\,\delta\rho_1(\mathbf{k};\omega) = 2\rho_0 S(k)U_{ext}(k,\omega) \tag{5.22}$$

from which the self–energy emerges

$$\Sigma(\mathbf{k},\omega) = -\frac{\hbar^2k^2}{4m}\int\frac{d^3p}{(2\pi)^3\rho_0}\sigma_{\mathbf{k}}(\mathbf{p})\frac{\delta u_2(\mathbf{k},\mathbf{p};\omega)}{\delta\rho_1(\mathbf{k};\omega)}\,. \tag{5.23}$$

But the ratio $\delta u_2(\mathbf{k},\mathbf{p};\omega)/\delta\tilde\rho_1(\mathbf{k};\omega)$ was exactly what we got out of the two–body equation (5.15) in closed form. The expression of the self–energy turns into an more familiar form [17] if we change the variables $\frac{\mathbf{k}}{2}+\mathbf{p}\to-\mathbf{p}$ and $\frac{\mathbf{k}}{2}-\mathbf{p}\to-\mathbf{q}$ and then introduce the Dirac $\delta$–function to insure the momentum conservation, $\mathbf{p}+\mathbf{q}+\mathbf{k}=0$,

$$\Sigma(k,\omega) = \frac{1}{2}\int\frac{d^3p\,d^3q}{(2\pi)^3\rho_0}\delta(\mathbf{k}+\mathbf{p}+\mathbf{q})\frac{|V_3(\mathbf{k};\mathbf{p},\mathbf{q})|^2}{\hbar\omega-\varepsilon_{\rm F}(p)-\varepsilon_{\rm F}(q)} \tag{5.24}$$

where the three–plasmon/phonon coupling matrix element

$$\begin{aligned} V_3(\mathbf{k};\mathbf{p},\mathbf{q}) &= -\frac{\hbar^2}{2m}\frac{1}{\sqrt{S(p)S(q)S(k)}} \\ &\times\left[\mathbf{k}\cdot\mathbf{p}S(p)+\mathbf{k}\cdot\mathbf{q}S(q)+k^2S(p)S(q)\left(1+\tilde{u}_3(\mathbf{k},\mathbf{p},\mathbf{q})\right)\right] \\ &= \frac{\hbar^2}{2m}\sqrt{\frac{S(p)S(q)}{S(k)}}\left[\mathbf{k}\cdot\mathbf{p}\tilde{X}(p)+\mathbf{k}\cdot\mathbf{q}\tilde{X}(q)-k^2\tilde{u}_3(\mathbf{k},\mathbf{p},\mathbf{q})\right] \end{aligned} \tag{5.25}$$

is given in terms of the ground–state structure function $S(k)$, the direct correlation function $\tilde{X}(k) = 1 - S(k)^{-1}$, and the three–body correlation function $\tilde{u}_3(\mathbf{k},\mathbf{p},\mathbf{q})$.

Using the explicit time dependence of Eq. (4.7) for the external potential and following the derivation of the linear response function presented there we can write $\chi(k,\omega)$ in the form of Eq. (3.4).

$$\chi(k,\omega) = S(k)\left[\frac{1}{\hbar\omega-\varepsilon_F(k)-\Sigma(k,\omega)} - \frac{1}{\hbar\omega+\varepsilon_F(k)+\Sigma^*(k,-\omega)}\right]. \tag{5.26}$$

### 5.4.1. *Numerical evaluation of the self–energy*

The integrand in the expression of the self–energy in Eq. (5.24) can have poles, which makes the self–energy a complex function. Let us look next in detail how it can be calculated numerically. After integrating the $\delta$–function and the $\phi$–coordinate we are left with the double integral

$$\Sigma(k,\omega) = \frac{1}{2}\frac{1}{(2\pi)^2\rho_0}\int_0^\infty q^2dq\int_{-1}^1 dx\frac{|V_3(k;q,x)|^2}{\hbar\omega-\varepsilon_{\rm F}(|\mathbf{k}+\mathbf{q}|)-\varepsilon_{\rm F}(q)} \tag{5.27}$$

where we have chosen the following variables

$$\begin{aligned} \mathbf{p} &= -(\mathbf{k}+\mathbf{q}) \\ p^2 &= k^2+2\mathbf{k}\cdot\mathbf{q}+q^2 = k^2+q^2+2kqx \\ x &= \frac{p^2-k^2-q^2}{2kq}. \end{aligned} \tag{5.28}$$

Replacing $x$ with $p$ we can write the integral in the form

$$\Sigma(k,\omega) = \frac{1}{8\pi^2\rho_0 k}\int_0^\infty qdq\int_{|k-q|}^{k+q} pdp\frac{|V_3(k;q,p)|^2}{\hbar\omega-\varepsilon_{\rm F}(p)-\varepsilon_{\rm F}(q)}. \tag{5.29}$$

This integral has a pole when

$$\hbar\omega = \varepsilon_{\rm F}(p)+\varepsilon_{\rm F}(q). \tag{5.30}$$

In other words when the energy of the excitation is equal to the energy of two elementary Feynman modes. In such a case the self–energy becomes a complex function. Assuming that this pole is the only pole in the integrand and that the integrand converges fast enough at infinity we can separate the real and imaginary parts

$$\Sigma(k,\omega) = \Delta(k,\omega) - i\Gamma(k,\omega) \tag{5.31}$$

by remembering that

$$\frac{1}{\omega' - \omega} = \mathcal{P}\frac{1}{\omega' - \omega} - i\pi\delta(\omega' - \omega)\,. \tag{5.32}$$

The imaginary part can then be calculated with one integration only

$$\begin{aligned}\Gamma(k,\omega) = {} & \frac{1}{8\pi\rho_0 k}\int_0^\infty q dq \int_{|k-q|}^{k+q} p dp \\ & \times |V_3(k;q,p)|^2 \delta(\hbar\omega - \varepsilon_{\rm F}(p) - \varepsilon_{\rm F}(q))\,. \end{aligned}\tag{5.33}$$

The real part could be calculated from the above principle value integral, but it is much more convenient for numerics to calculate it from the imaginary part using Kramers–Kronig relations which connect the real and imaginary parts. If $f(\omega)$ is an analytic complex function

$$f(\omega) = a(\omega) + ib(\omega) \tag{5.34}$$

then

$$\begin{aligned} a(\omega) &= \frac{1}{\pi}\mathcal{P}\int_{-\infty}^\infty d\omega' \frac{b(\omega')}{\omega' - \omega} \\ b(\omega) &= -\frac{1}{\pi}\mathcal{P}\int_{-\infty}^\infty d\omega' \frac{a(\omega')}{\omega' - \omega}\,. \end{aligned}\tag{5.35}$$

Provided that $a(\omega)$ and $b(\omega)$ converge fast enough at large $\omega$.

Using the first relation we can write the real part of the self–energy in the form

$$\Delta(k,\omega) = -\frac{1}{\pi}\mathcal{P}\int_0^\infty d\omega' \frac{\Gamma(k,\omega')}{\omega' - \omega}\,. \tag{5.36}$$

The imaginary part is non–zero only when $\omega' > 0$. In the numerical integration of the principle value one distributes the integration mesh symmetrically around $\omega$ and leaves out the point $\omega' = \omega$.

## 5.5. *Analytic structure of the self–energy*

In this section we study the analytic structure of the self–energy (5.24) in the regions where the poles of the response functions (5.2) lie very close to the continuum. We want to show that in some cases the structure of the self–energy guarantees that the pole solution exists. That happens when there is a near by minimum like the roton minimum in the spectrum. In other cases the existence of the pole solution can be determined only numerically as is the case with the anomalous dispersion relation in the low momentum and low frequency region in three–dimensional $^4$He. The other issue we want to stress is that the existence of the solution does not tell much about the strength of the mode and we will show that the closer the mode is to the continuum limit the less strength it has.

Let us begin with the proof that all poles of the response function must lie on the real $\omega$–axis and thus they determine the collective, elementary excitation modes of the system. The pole solution of Eq.(3.7) must satisfy the implicit equation

$$\hbar\omega_0(k) = \varepsilon_F(k) + \Sigma(k, \omega_0(k)) \tag{5.37}$$

We start the proof by assuming that $\omega_0$ is complex,

$$\omega_0 = x_0 + iy_0\,, \tag{5.38}$$

and show that this leads to a contradiction.[41]

The general requirement that the dynamic structure function must be positive implies that the imaginary part of the self–energy $\Gamma(k,\omega) \geq 0$ when $\omega \geq 0$ and zero otherwise. Using the Kramer–Kronig relation of Eq. (5.36) we can write the self–energy in the spectral form

$$\Sigma(k,\omega) = \frac{1}{\pi}\int_0^\infty d\omega' \frac{\Gamma(k,\omega')}{\omega - \omega'}\,. \tag{5.39}$$

If we make the assumption of Eq. (5.38) then the poles are given by the solutions of the equation

$$x_0 + iy_0 - \varepsilon_F(k) = \frac{1}{\pi}\int_0^\infty d\omega' \frac{\Gamma(k,\omega')}{x_0 + iy_0 - \omega'}\,. \tag{5.40}$$

Setting imaginary parts equal on both sides of the equation we get

$$1 = -\int_0^\infty \frac{d\omega'}{\pi} \frac{\Gamma(k,\omega')}{(x_0 - \omega')^2 + y_0^2}\,. \tag{5.41}$$

The right–hand side is always negative and the equation can never be satisfied. We can conclude that the poles must lie on the real $\omega$–axis.

From our definition (5.24) of the self–energy we can deduce the inequality

$$\Sigma(k,\omega) \leq \Sigma(k,0) \leq 0 \tag{5.42}$$

from which we immediately obtain that the lowest collective mode satisfies the exact inequality [42]

$$\hbar\omega_0(k) \leq -\frac{2S(k)}{\chi(k,0)}\,. \tag{5.43}$$

In writing down Eq. (5.37) we have to assume that $\Sigma(k,\omega_0(k))$ is real. This is the case when the energy denominator in Eq. (5.24) does not change sign, which is true when the collective energy is *below* the critical value

$$\hbar\omega_0 < \hbar\omega_{\rm crit}(k) \equiv \min_{\mathbf{p}}\left[\varepsilon_F(|\tfrac{1}{2}\mathbf{k} - \mathbf{p}|) + \varepsilon_F(|\tfrac{1}{2}\mathbf{k} + \mathbf{p}|)\right] \tag{5.44}$$

determining the continuum boundary. We have returned here to the original notation $\mathbf{p} \to -\frac{\mathbf{k}}{2} - \mathbf{p}$ and $\mathbf{q} \to -\frac{\mathbf{k}}{2} + \mathbf{p}$. Above the critical energy the self–energy is complex. Moreover, for $\hbar\omega < \hbar\omega_{\rm crit}(k)$, it follows from Eq. (5.24) that

$$\frac{d\Sigma(k,\omega)}{d\omega} < 0\,. \tag{5.45}$$

In order to determine if Eq. (5.37) has a solution, we must find out whether $\Sigma(k,\omega)$ becomes singular at the branch–cut $\omega = \omega_{\rm crit}(k)$ or not. This depends, of course, on the details of the reference spectrum $\varepsilon_F(k)$ in the energy denominator of Eq. (5.37). We shall study here two relevant cases.

### 5.5.1. *Anomalous dispersion in liquid $^4$He*

The first case is that the reference spectrum $\varepsilon_F(k)$ is convex.[31,34] This refers typically to the regime of low momentum transfer in $^4$He where the sound mode has an anomalous dispersion or to high momentum transfer where the spectrum approaches the single particle kinetic energy. When $\hbar\omega_{\rm crit}(k) = 2\varepsilon_F(k/2) < \varepsilon_F(k)$, this critical energy is *below* the reference energy. In order to determine whether Eq. (5.37) has a solution, we must therefore study the analytic behavior of $\Sigma(k,\omega)$ as a function of $\omega$ near the branch–point $\omega = \omega_{\rm crit}(k)$. We shall treat only the simplest cases here, assuming a monotonically growing, convex spectrum $\varepsilon'_F(k/2) > 0$ and $\varepsilon''_F(k/2) > 0$ and we are interested in the singular behavior only.

We want to evaluate the integral (5.24)

$$\Sigma(k,\omega) = \frac{1}{2}\int \frac{d^3p}{(2\pi)^3\rho_0}\;\frac{\left|V_3(\mathbf{k}; -\mathbf{p}-\frac{\mathbf{k}}{2}, \mathbf{p}-\frac{\mathbf{k}}{2})\right|^2}{\hbar\omega - \varepsilon_F(|\frac{\mathbf{k}}{2}+\mathbf{p}|) - \varepsilon_F(|\frac{\mathbf{k}}{2}-\mathbf{p}|) + i\eta} \tag{5.46}$$

for $\omega \to \omega_{\rm crit}$ from below, which keeps the self–energy real. The minimum critical value $\omega_{\rm crit}(k) = 2\varepsilon_F(k/2)$ is reached at $\mathbf{p} = 0$ and we expand the energy denominator around that point up to second order.

$$\hbar\omega - \varepsilon_F(|\tfrac{\mathbf{k}}{2}+\mathbf{p}|) - \varepsilon_F(\tfrac{\mathbf{k}}{2}-\mathbf{p}|) = a(k,\omega) + b(k,x)\,p^2 \tag{5.47}$$

where $x = \mathbf{k}\cdot\mathbf{p}/(kp) = \cos(\theta)$ and we have introduced the notations

$$a(k,\omega) = \hbar\omega - 2\varepsilon_F(k/2) < 0 \tag{5.48}$$

$$b(k,x) = A(k) + B(k)x^2 \tag{5.49}$$

$$A(k) = -\frac{2}{k}\,\varepsilon'_F(k/2) \quad \text{and} \quad B(k) = \frac{2}{k}\,\varepsilon'_F(k/2) - \varepsilon''_F(k/2)\,. \tag{5.50}$$

We assume that the numerator is a slowly varying function around the singular point of the denominator and evaluate it at $\mathbf{p} = 0$. The integration is then carried out over the energy denominator only and we perform first the momentum integration and then the angle integration

$$\int_{-1}^{1} dx \int_0^R \frac{p^2dp}{a(p,\omega) + b(p,x)\,p^2} = \frac{\pi}{2}\frac{k}{\varepsilon'_F(k/2)}\left[\sqrt{\frac{2\varepsilon_F(k/2) - \hbar\omega}{\varepsilon''_F(k/2)}} + O(k,R)\right]. \tag{5.51}$$

The first term vanishes at the critical frequency and the second term is a negative term independent of $\omega$, but it depends on the asymptotic behavior of the numerator far from $p = 0$. Inserting that result into Eq. (5.46) we get

$$\Sigma(k,\omega) \overset{\hbar\omega\to 2\varepsilon_F(k/2)}{\to} \frac{|V_3(\mathbf{k};-\mathbf{k}/2,-\mathbf{k}/2)|^2}{8\pi\rho_0} \frac{k\sqrt{2\varepsilon_F(k/2)-\hbar\omega}}{2\varepsilon'_F(k/2)\sqrt{\varepsilon''_F(k/2)}} - C(k,R)\,. \quad (5.52)$$

This result shows that there is no guarantee that a real pole exists. It depends on the size of the term $C(k, R)$, which must be evaluated numerically.

We can also calculate the strength of the possible pole near $\omega_{\rm crit}$. For that we need the derivative

$$\frac{d\Sigma(k,\omega)}{d(\hbar\omega)} \approx -\frac{|V_3(\mathbf{k};-\mathbf{k}/2,-\mathbf{k}/2)|^2}{8\pi\rho_0} \frac{k}{4\sqrt{2\varepsilon_F(k/2)-\hbar\omega}\;\varepsilon'_F(k/2)\;\sqrt{\varepsilon''_F(k/2)}}\,. \quad (5.53)$$

At $\hbar\omega = 2\varepsilon_F(k/2)$ it diverges and thus if there is a pole near by $2\varepsilon_F(k/2)$ its strength

$$Z(k) = \frac{S(k)}{1 - \left.\dfrac{d\Sigma(k,\omega)}{d(\hbar\omega)}\right|_{\hbar\omega_0}} \overset{\omega_0\to\omega_{\rm crit}}{\to} 0 \quad (5.54)$$

becomes vanishingly small.

A similar calculation in two dimensions leads to a logarithmically diverging integral at the critical frequency, which enforces the pole solution to exist

$$\int_0^{2\pi} d\phi \int_0^R \frac{p\,dp}{a(p,\omega)+b(p,\cos\phi)p^2} = \pi\frac{k\log(2\varepsilon_F(k/2)) - \hbar\omega)}{\sqrt{2\varepsilon'_F(k/2)\varepsilon''_F(k/2)}} + O(k\log(R))\,. \quad (5.55)$$

Inserting that to the self–energy we get

$$\Sigma(k,\omega) \overset{\hbar\omega\to 2\varepsilon_F(k/2)}{\to} \frac{|V_3(\mathbf{k};-\mathbf{k}/2,-\mathbf{k}/2)|^2}{4\pi\rho_0} \frac{k\log(2\varepsilon_F(k/2)-\hbar\omega)}{\sqrt{2\varepsilon'_F(k/2)\varepsilon''_F(k/2)}}\,. \quad (5.56)$$

In Fig. 2 we show the sum of the Feynman energy and the self–energy in the CBF approximation $\varepsilon_F(q)+\Sigma(q,\omega)$, for a low density two–dimensional case, as a function of energy $\hbar\omega$ for different momenta. The logarithmic singularity at $\omega = \omega_{\rm crit}(q)$ is clearly seen, also that this singularity is, for low momenta, very sharp and permits under certain circumstances three solutions. The solutions are the crossings where $\hbar\omega = \varepsilon_F(q) + \Sigma(q,\omega)$ and they are shown in Fig. 3. There, the dash–dotted line is the boundary $\hbar\omega_{\rm crit}(k)$ above which all solutions are complex, and the dashed loop is the solution of the *real part* of Eq. (5.37). It is also clearly seen that the *imaginary* part of the complex solution peaks just before the "mode" disappears.

### 5.5.2. *Absolute minimum in the spectrum*

The second case is when the reference spectrum has an absolute, roton–like minimum. These results are relevant for the phonon–roton spectrum above the roton minimum and for the low density plasmon spectrum in three dimensions. In this

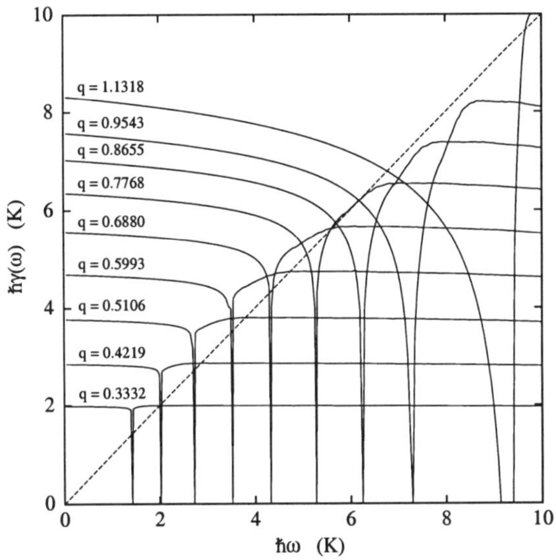


Fig. 2. The CBF energy $\hbar\gamma(\omega) = \varepsilon_F(q)+\Sigma(q,\omega)$ is shown, for two-dimensional $^4$He, as a function of $\omega$ for a sequence of momentum values at the density = 0.035 Å$^{-2}$. Solutions of Eq. (5.37) correspond to the points $\gamma_q(\omega) = \hbar\omega$ along the dotted line. Clearly, for $q < 0.77$ Å$^{-1}$ one has three solutions. It is also clearly seen that the logarithmic singularity is extremely narrow at small momenta. (The original figure in Ref. 31.)

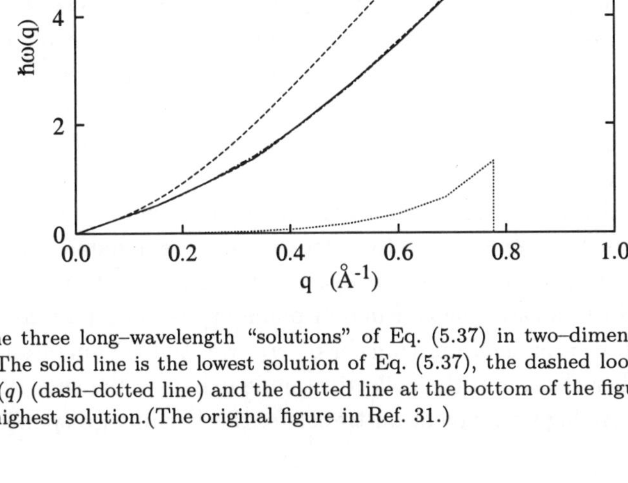


Fig. 3. The three long-wavelength "solutions" of Eq. (5.37) in two-dimensional $^4$He for $n$ = 0.035 Å$^{-2}$. The solid line is the lowest solution of Eq. (5.37), the dashed loop the two solutions above $\hbar\omega_{\rm crit}(q)$ (dash-dotted line) and the dotted line at the bottom of the figure is the imaginary part of the highest solution.(The original figure in Ref. 31.)

case, we have $\omega_{\rm crit}(k) = 2\omega_r$, where $\hbar\omega_r$ is the "roton energy" located at the wave number $k_r$. Expanding the energy denominator about this point yields the result

$$\lim_{\hbar\omega\to 2\hbar\omega_r} \Sigma(k,\omega) = \frac{|V_3(\mathbf{k};-\mathbf{k}-\mathbf{k}_r,\mathbf{k}_r)|^2}{8\pi\rho_0}\,\frac{k_r^2 \log(2\hbar\omega_r - \hbar\omega)}{k\;\varepsilon_F''(k_r)} \tag{5.57}$$

where $\mathbf{k}_r$ is a vector of length $k_r$ oriented such that the three vectors $\mathbf{k}$, $-\mathbf{k}-\mathbf{k}_r$, and $\mathbf{k}_r$ form an isosceles triangle. In two dimensions, one finds similarly a logarithmic singularity.

The strength $Z(k)$ of this pole depends again strongly on "how close" the solution of the implicit equation (5.37) is to the critical energy $2\hbar\omega_{\rm pl}$. Exact quantitative statements are difficult to make. From our numerical results we find that the strength is weak, $Z(k) << S(k)$, and diminishes with increasing momentum.[31,32] In $^4$He that mode has been recently identified in measurements by Fåk and Bossy. [37]

To summarize the analysis of this section, we presented analytic arguments that the pole solution can exist in the case of a nearby "roton" minimum in the spectrum and in two dimensions for the anomalous dispersion, whereas in three dimension the analytic arguments give no definite answer for the anomalous dispersion, but numerically no pole solution was found in $^4$He. An upper bound for these excitations is the reference spectrum (Feynman spectrum here) or the continuum boundary $\hbar\omega_{\rm crit}(k)$, whichever is lower.

### 5.6. *Dynamic structure function in the CBF–approximation*

The results for the dynamic structure function we show here have been calculated for liquid $^4$He [43,31] and for the charged Bose gas. [34] In liquid $^4$He, shown in Fig. 4, the theory gives collective modes, which are somewhat too high in energy as compared with the experimental data, yet the overall qualitative features of the spectrum are well reproduced. At low momenta one sees the anomalous dispersion of the phonon spectrum and above the roton minimum the spectrum flattens below the two roton limit and the strength of the pole slowly vanishes before the mode merges into the continuum. The most apparent shortcoming is that excitations decay into two Feynman excitations, not into two excitations with self–energy–corrected energies. As a result, the two–roton limit appears at an energy higher than the limit taken from the phonon–roton spectrum given by the theory itself. The results satisfy the lowest energy weighted sum rules exactly. In fact, this feature of the CBF theory was first proven by Jackson [41].

For the charged Bose gas Fig. 5 shows the dynamic structure with the plasmon mode at $r_s = 10$ in three dimensions. In order to make the dynamic structure function more visible at small momentum transfers, the figure shows a grey–scale–plot of $S(k,\omega)/S(k)$. In this case, the continuum boundary is almost twice the plasma frequency, and up to a momentum transfer of $kr_0 \approx 4$, the Feynman and the CBF spectrum are rather close, and the collective mode exhausts the sum rules ([41]) almost completely. At higher momentum transfers, the Feynman dispersion relation

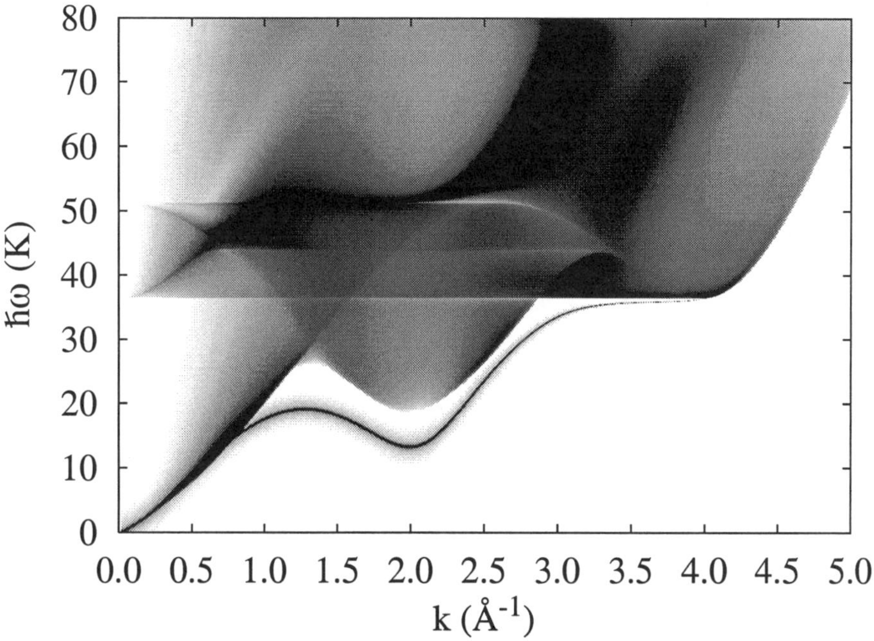


Fig. 4. The dynamic structure function $S(k,\omega)$ for three–dimensional liquid $^4$He in the CBF–approximation at the saturation density 0.022 Å$^{-3}$.

turns into a broad ridge along the single–particle kinetic energy. The rigorously "collective" mode loses its strength over a rather small momentum regime, and disappears for $kr_0 > 6$. Note that this corresponds to a regime where no real solutions of Eq. (5.37) exist due to the anomalous dispersion. Figure 5 also gives some justification for the use of the Feynman spectrum since the actual strength is indeed concentrated around the Feynman spectrum, and not around the solution of Eq. (5.37) or some extrapolation of it to higher momenta.

It was pointed out by Gold [44] that a roton–like structure — as it is familiar from the helium liquids — should also appear in the plasmon dispersion curve of the charged Bose gas at low densities. The existence of such a minimum follows immediately from the negative plasmon dispersion coefficient; a structure weakly reminiscent of a roton minimum is indeed seen even in the Feynman approximation at $r_s = 10$ see Fig. 5. A much more pronounced roton minimum is seen at $r_s = 50$, in Fig 6. In this case, the multi–pair regime of the dynamic structure function begins slightly above the plasma frequency. At high momentum–transfers, $kr_0 > 6.5$, the collective excitation again rapidly loses weight and ultimately disappears. However, we see a that the main strength of $S(k,\omega)$ appears to be concentrated along a straight extrapolation of the discrete mode towards higher energies and momenta.

Significant changes happen when the roton minimum drops below half the plasma frequency. Now, the plasmon is, in the long–wavelength limit, *in principle*

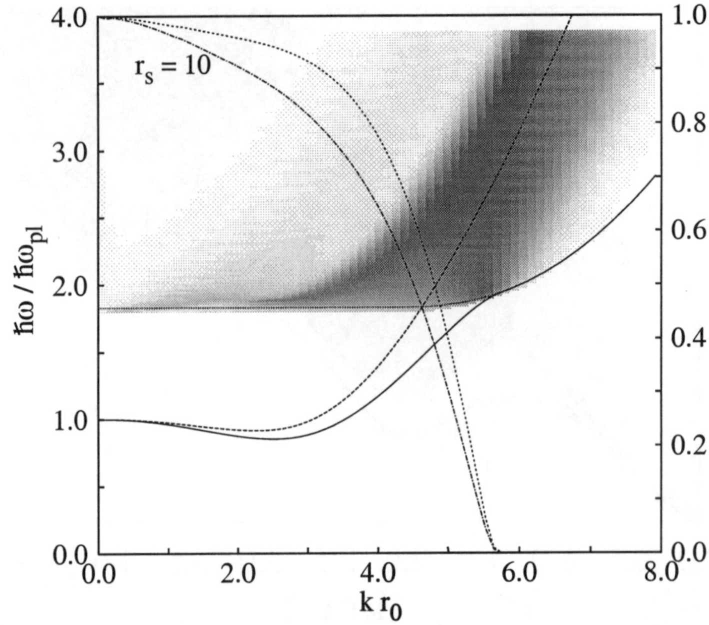


Fig. 5. The normalized dynamic structure function $S(k,\omega)/S(k)$ is shown, for the 3D charged Bose gas, at $r_s = 10$. The strength of $S(k,\omega)/S(k)$ is indicated by the grey–scale. Also shown are the Feynman dispersion relations $\varepsilon_F(k)$, (long–dashed line), the self–energy corrected dispersion relation (solid line), and the continuum boundaries $\hbar\omega_{\rm crit}(k)$ (dotted line). The scale on the right refers to the *relative weight* $Z(k)/S(k)$ (dotted line) and $2m\omega_0(k)Z(k)/(\hbar k^2)$ (dash–dotted line) of the collective mode to the $\omega^0$ and $\omega^1$ sum rules. (The original figure in Ref. 34.)

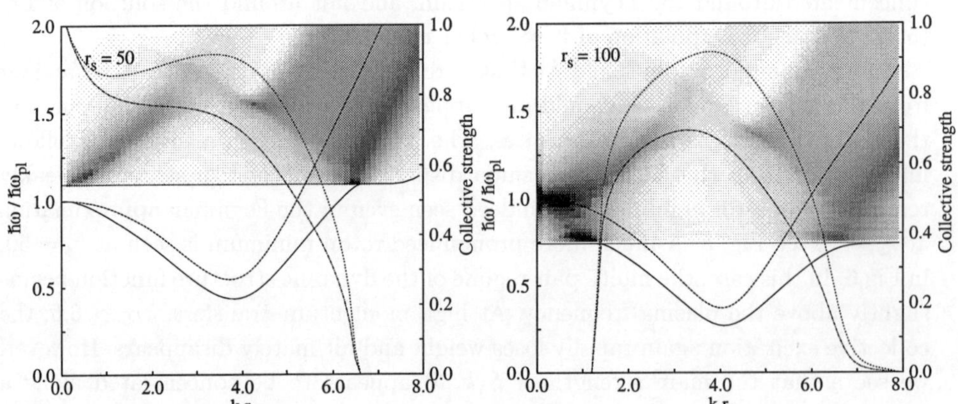


Fig. 6. Same as Fig. 5 for the 3D charged Bose gas with $r_s = 50$ and $r_s = 100$. (The original figures in Ref. 34.)

*no longer the lowest mode*. A second pole of the response function appears at $2\hbar\omega_r$, where $\hbar\omega_r$ is the roton energy. However, the residue of this pole is, while not exactly zero, practically negligible. The actual strength of $S(k,\omega)$ is located around the plasma frequency at small $k$ The collective mode has a minimum around $k = 4r_0$ and its strength disappears again at high momentum transfers. The main strength continues there in a broad ridge around the extrapolation of the collective excitation.

### 5.7. *Summary*

To summarize this part, we went beyond the Feynman approximation by allowing for fluctuating two–body correlations. From the physical point of view, this means adding three–phonon scattering processes into the description: an excitation can decay into two, provided that energy and momentum are conserved. Thus, we were apple to study the structure of the anomalous dispersion of phonons in liquid $^4$He in two and three dimensions and the plasmon modes in charged Bose gas [34]. We introduced first the convolution approximation to get rid of the three–body distribution function, then the uniform–limit approximation to be able to solve the two–body equation. This gave us the fluctuating two–body correlation function in terms of the fluctuating one–body density. We were then able to solve the first continuity equation and extract the self–energy and the dynamic response function. The added flexibility lowers the excitation energies considerably towards measured values and gives a qualitatively correct behavior of the collective spectrum for the full momentum range.

## 6. The full solution

In the spirit of the ground–state calculation, we next look for ways to change variables from $\delta u_2$ to the fluctuating pair–distribution function $\delta g_2$. Namely, simple analysis shows that had we chosen to optimize both of the correlation functions and not replaced $\delta u_1$ with the one–body density fluctuation, we would have ended up with the free–particle spectrum as our reference spectrum. This suggests that we can do quantitatively better and bring the results into an even closer agreement with the experimental data by applying the least–action principle to observable quantities.

### 6.1. *Continuity equations*

Let us return to the definition of the density (3.29) and calculate the general expressions for the gradients of the one and two-particle densities,

$$\begin{aligned}\nabla_1\delta\rho_1(\mathbf{r}_1;t) &= \rho_0\nabla_1\delta u_1(\mathbf{r}_1;t)\\ &\quad+\rho_0^2\int d^3r_2 g_2(\mathbf{r}_1,\mathbf{r}_2)\nabla_1\delta u_2(\mathbf{r}_1,\mathbf{r}_2;t)\\ &\quad+\int d^3r_2\delta\rho_2(\mathbf{r}_1,\mathbf{r}_2;t)\nabla_1 u_2(\mathbf{r}_1,\mathbf{r}_2)\end{aligned} \tag{6.1}$$

and

$$\begin{aligned}\nabla_1 \delta\rho_2(\mathbf{r}_1, \mathbf{r}_2; t) &= \delta\rho_2(\mathbf{r}_1, \mathbf{r}_2; t)\nabla_1 u_2(\mathbf{r}_1, \mathbf{r}_2) \\ &+ \rho_0^2 g_2(\mathbf{r}_1, \mathbf{r}_2)\nabla_1 \left[\delta u_1(\mathbf{r}_1; t) + \delta u_2(\mathbf{r}_1, \mathbf{r}_2; t)\right] \\ &+ \rho_0^3 \int d^3r_3 g_3(\mathbf{r}_1, \mathbf{r}_2, \mathbf{r}_3)\nabla_1 \delta u_2(\mathbf{r}_1, \mathbf{r}_3; t) \\ &+ \int d^3r_3 \delta\rho_3(\mathbf{r}_1, \mathbf{r}_2, \mathbf{r}_3; t)\nabla_1 u_2(\mathbf{r}_1, \mathbf{r}_3) . \end{aligned} \tag{6.2}$$

Here, we have assumed that also the ground state contains only the pair correlation function.

Those terms which depend on $\delta u_1(\mathbf{r}; t)$ and $\delta u_2(\mathbf{r}_1, \mathbf{r}_2; t)$ in Eqs. (6.1) and (6.2) appear also in the expressions for the one– and two–body currents in Eqs. (3.37) and (3.38). Thus we can eliminate them and write the currents solely in terms of fluctuating densities.

$$\begin{aligned}\mathbf{j}_1(\mathbf{r}_1; t) &= \frac{\hbar}{2mi}[\nabla_1 \delta\rho_1(\mathbf{r}_1; t) \\ &- \int d^3r_2 \delta\rho_2(\mathbf{r}_1, \mathbf{r}_2; t)\nabla u_2(\mathbf{r}_1, \mathbf{r}_2)] , \end{aligned} \tag{6.3}$$

$$\begin{aligned}\mathbf{j}_2(\mathbf{r}_1, \mathbf{r}_2; t) &= \frac{\hbar}{2mi}[\nabla_1 \delta\rho_2(\mathbf{r}_1, \mathbf{r}_2; t) \\ &- \delta\rho_2(\mathbf{r}_1, \mathbf{r}_2; t)\nabla_1 u_2(\mathbf{r}_1, \mathbf{r}_2) \\ &- \int d^3r_3 \delta\rho_3(\mathbf{r}_1, \mathbf{r}_2, \mathbf{r}_3; t)\nabla_1 u_2(\mathbf{r}_1, \mathbf{r}_3)] . \end{aligned} \tag{6.4}$$

Eq. (6.4) contains the three–body density variation, $\delta\rho_3(\mathbf{r}_1, \mathbf{r}_2, \mathbf{r}_3; t)$ which must be formulated in terms of one– and two–particle density fluctuations.

Without loss of generality we can write the density fluctuations in terms of relative changes

$$\begin{aligned}\delta\rho_1(\mathbf{r}_1; t) &\equiv \rho_0 \xi(\mathbf{r}_1; t) , \\ \delta\rho_2(\mathbf{r}_1, \mathbf{r}_2; t) &\equiv \rho_0^2 g_2(\mathbf{r}_1, \mathbf{r}_2)\Big[\xi(\mathbf{r}_1; t) + \xi(\mathbf{r}_2; t) + \Delta(\mathbf{r}_1, \mathbf{r}_2; t)\Big] , \\ \delta\rho_3(\mathbf{r}_1, \mathbf{r}_2, \mathbf{r}_3; t) &\equiv \rho_0^3 g_3(\mathbf{r}_1, \mathbf{r}_2, \mathbf{r}_3)\Big[\xi(\mathbf{r}_1; t) + \xi(\mathbf{r}_2; t) + \xi(\mathbf{r}_3; t) \\ &+ \Delta(\mathbf{r}_1, \mathbf{r}_2; t) + \Delta(\mathbf{r}_1, \mathbf{r}_3; t) + \Delta(\mathbf{r}_2, \mathbf{r}_3; t) \\ &+ \Delta_3(\mathbf{r}_1, \mathbf{r}_2, \mathbf{r}_3; t)\Big] . \end{aligned} \tag{6.5}$$

which define the function $\xi(\mathbf{r}_1; t)$, $\Delta(\mathbf{r}_1, \mathbf{r}_2; t)$ and $\Delta_3(\mathbf{r}_1, \mathbf{r}_2, \mathbf{r}_3; t)$ in terms of one- two- and three-body density fluctuations.

From the definition of the $n$–particle density (2.3) we can derive the Born–Green–Yvon (BGY) equations by applying the gradient operator

$$\nabla_1 g_2(\mathbf{r}_1,\mathbf{r}_2) = g_2(\mathbf{r}_1,\mathbf{r}_2)\nabla_1 u_2(\mathbf{r}_1,\mathbf{r}_2) + \rho_0 \int d^3r_3 g_3(\mathbf{r}_1,\mathbf{r}_2,\mathbf{r}_3)\nabla_1 u_2(\mathbf{r}_1,\mathbf{r}_3) \tag{6.6}$$

and since our ground state is homogeneous, gradient with respect to one particle density vanishes provided that

$$\rho_0 \int d^3r_2 g_2(\mathbf{r}_1,\mathbf{r}_2)\nabla_1 u_2(\mathbf{r}_1,\mathbf{r}_2) = 0\,. \tag{6.7}$$

With these substitutions we get

$$\mathbf{j}_1(\mathbf{r}_1;t) = \frac{\hbar\rho_0}{2mi}\Big\{\nabla_1\xi(\mathbf{r}_1;t) - \rho_0 \int d^3r_2 g_2(\mathbf{r}_1,\mathbf{r}_2)\left[\xi(\mathbf{r}_2;t)+\Delta(\mathbf{r}_1,\mathbf{r}_2;t)\right]\nabla_1 u_2(\mathbf{r}_1,\mathbf{r}_2)\Big\} \tag{6.8}$$

$$\begin{aligned}\mathbf{j}_2(\mathbf{r}_1,\mathbf{r}_2;t) = \frac{\hbar\rho_0^2}{2mi}\Big\{ & g_2(\mathbf{r}_1,\mathbf{r}_2)\nabla_1\left[\xi(\mathbf{r}_1;t)+\Delta(\mathbf{r}_1,\mathbf{r}_2;t)\right] \\ & - \rho_0 \int d^3r_3\, g_3(\mathbf{r}_1,\mathbf{r}_2,\mathbf{r}_3)\Big[\xi(\mathbf{r}_3;t)+\Delta(\mathbf{r}_1,\mathbf{r}_3;t) \\ & + \Delta(\mathbf{r}_2,\mathbf{r}_3;t)+\Delta_3(\mathbf{r}_1,\mathbf{r}_2,\mathbf{r}_3;t)\Big]\nabla_1 u_2(\mathbf{r}_1,\mathbf{r}_3)\Big\}\,.\end{aligned} \tag{6.9}$$

These results are exact within the assumption that the wave function contains only one– and two–particle correlations, but the currents contain now three unknown quantities. A natural truncation is to set $\Delta_3(\mathbf{r}_1,\mathbf{r}_2,\mathbf{r}_3;t) = 0$ since we have so far ignored also the fluctuating three–particle correlation functions. That is consistent with making the superposition approximation for the triplet distribution function

$$g_3(\mathbf{r}_1,\mathbf{r}_2,\mathbf{r}_3) = g_2(\mathbf{r}_1,\mathbf{r}_2)g_2(\mathbf{r}_1,\mathbf{r}_3)g_2(\mathbf{r}_2,\mathbf{r}_3)\,. \tag{6.10}$$

Inserting these approximation into the expressions for the currents and by introducing a simplifying notation $\mathbf{V}(\mathbf{r}_1,\mathbf{r}_2) \equiv g_2(\mathbf{r}_1,\mathbf{r}_2)\nabla_1 u_2(\mathbf{r}_1,\mathbf{r}_2)$, which is a ground state quantity and can be calculated using the HNC–equations, we are ready to collect the expressions for the continuity equations (3.32) and (3.33). They become now equations for fluctuations in the one–body density $\xi(\mathbf{r};t)$ and in the pair distribution function $\Delta(\mathbf{r}_1,\mathbf{r}_2;t)$

$$\frac{\hbar}{2mi}\nabla_1\Big\{\nabla_1\xi(\mathbf{r}_1;t) - \rho_0\int d^3r_2 \mathbf{V}(\mathbf{r}_1,\mathbf{r}_2)\left[\xi(\mathbf{r}_2;t)+\Delta(\mathbf{r}_1,\mathbf{r}_2;t)\right]\Big\} + \dot{\xi}(\mathbf{r}_1;t) = \frac{1}{\rho_0}D_1(\mathbf{r}_1;t) \tag{6.11}$$

and

$$\frac{\hbar}{2mi}\nabla_1\Big\{h(\mathbf{r}_1,\mathbf{r}_2)\nabla_1\Big[\xi(\mathbf{r}_1;t)+\Delta(\mathbf{r}_1,\mathbf{r}_2;t)\Big]+\nabla_1\Delta(\mathbf{r}_1,\mathbf{r}_2;t)$$
$$-\rho_0\int d^3r_3\mathbf{V}(\mathbf{r}_1,\mathbf{r}_3)\Big\{\Delta(\mathbf{r}_2,\mathbf{r}_3;t)+[h(\mathbf{r}_1,\mathbf{r}_2)h(\mathbf{r}_2,\mathbf{r}_3)+h(\mathbf{r}_1,\mathbf{r}_2)+h(\mathbf{r}_2,\mathbf{r}_3)]$$
$$\Big[\xi(\mathbf{r}_3;t)+\Delta(\mathbf{r}_1,\mathbf{r}_3;t)+\Delta(\mathbf{r}_2,\mathbf{r}_3;t)\Big]\Big\}\Big\}+\text{same with}(1\leftrightarrow 2)$$
$$+h(\mathbf{r}_1,\mathbf{r}_2)[\dot{\xi}(\mathbf{r}_1;t)+\dot{\xi}(\mathbf{r}_2;t)+\dot{\Delta}(\mathbf{r}_1,\mathbf{r}_2;t)]+\dot{\Delta}(\mathbf{r}_1,\mathbf{r}_2;t)$$
$$=\frac{1}{\rho_0^2}\left[D_2(\mathbf{r}_1,\mathbf{r}_2;t)-D_1(\mathbf{r}_1;t)-D_2(\mathbf{r}_1;t)\right]. \tag{6.12}$$

## 6.2. *Continuity equations in momentum space*

The first of the continuity equation (6.11) can be written in momentum space in the familiar form

$$\hbar\omega-\varepsilon(k)-\Sigma(k,\omega)=\frac{2S(k)\tilde{U}_{\rm ext}(k,\omega)}{\tilde{\xi}(k,\omega)}. \tag{6.13}$$

Following again the derivation of the linear response function presented after Eq. (4.7) we can write $\chi(k,\omega)$ in the form,

$$\chi(k,\omega)=S(k)\left[\frac{1}{\hbar\omega-\varepsilon(k)-\Sigma(k,\omega)}-\frac{1}{\hbar\omega+\varepsilon(k)+\Sigma^*(k,-\omega)}\right], \tag{6.14}$$

but now the collective reference mode $\varepsilon(k)$ is not a bare Feynman mode. It contains a correction term which becomes important around the roton region and higher momenta.

To see that we need to calculate the Fourier transform

$$k^2Q(k)=\mathcal{F}\left[\nabla_1^2\xi(\mathbf{r}_1;t)-\rho_0\int d^3r_2\nabla_1\cdot\mathbf{V}(\mathbf{r}_1,\mathbf{r}_2)\xi(\mathbf{r}_2;t)\right]. \tag{6.15}$$

From the HNC–equation

$$g_2(\mathbf{r}_1,\mathbf{r}_2)=\exp[u_2(\mathbf{r}_1,\mathbf{r}_2)+N(\mathbf{r}_1,\mathbf{r}_2)+E(\mathbf{r}_1,\mathbf{r}_2)] \tag{6.16}$$

we get

$$\begin{aligned}\mathbf{V}(\mathbf{r}_1,\mathbf{r}_2)&=g_2(\mathbf{r}_1,\mathbf{r}_2)\nabla_1u_2(\mathbf{r}_1,\mathbf{r}_2)\\&=\nabla_1g_2(\mathbf{r}_1,\mathbf{r}_2)+g_2(\mathbf{r}_1,\mathbf{r}_2)(\nabla_1N(\mathbf{r}_1,\mathbf{r}_2)+\nabla_1E(\mathbf{r}_1,\mathbf{r}_2))\\&=\nabla_1X(\mathbf{r}_1,\mathbf{r}_2)-\nabla_1E(\mathbf{r}_1,\mathbf{r}_2)-h(\mathbf{r}_1,\mathbf{r}_2)(\nabla_1N(\mathbf{r}_1,\mathbf{r}_2)+\nabla_1E(\mathbf{r}_1,\mathbf{r}_2))\end{aligned} \tag{6.17}$$

where again $N(\mathbf{r}_1,\mathbf{r}_2)$ is the sum of nodal diagrams, $X(\mathbf{r}_1,\mathbf{r}_2)$ the direct correlation function and $E(\mathbf{r}_1,\mathbf{r}_2)$ the sum of elementary diagrams.

The Fourier transform of $Q(k)$ is then

$$Q(k)=1-\frac{1}{S(k)}+I(k), \tag{6.18}$$

where $I(k)$ stands for the integral

$$I(k) = -E(k) - \int \frac{d^3q}{(2\pi)^3\rho} \frac{\mathbf{k}\cdot\mathbf{q}}{k^2} \left[S(|\mathbf{k}-\mathbf{q}|) - 1\right]\left[N(q) + E(q)\right]. \tag{6.19}$$

This leads to the expression

$$\varepsilon(k) = \frac{\hbar^2 k^2}{2mS(k)} \left[1 - S(k)I(k)\right], \tag{6.20}$$

for the reference collective mode. The correction term $I(k)$ to the Feynman energy is positive and therefore it lowers the reference energy.

Note that in case the three–body correlations are included one should add the term

$$\frac{\rho}{2} \int d^3r g_3(\mathbf{r}_1, \mathbf{r}_2, \mathbf{r}_3) \nabla_1 u_3(\mathbf{r}_1, \mathbf{r}_2, \mathbf{r}_3) \tag{6.21}$$

into the elementary diagrams.

The self–energy $\Sigma(k,\omega)$ is given by the integral

$$\Sigma(k,\omega) = -\frac{\hbar^2}{2m} \int \frac{d^3p}{(2\pi)^3\rho}\, \mathbf{k}\cdot\left(\tfrac{\mathbf{k}}{2} + \mathbf{p}\right) Q\left(\left|\tfrac{\mathbf{k}}{2} + \mathbf{p}\right|\right) \frac{\Delta(\mathbf{k},\mathbf{p};\omega)}{\xi(\mathbf{k};\omega)}. \tag{6.22}$$

In the CBF approximation we were able to solve the second continuity equation for $\delta u_2/\delta\tilde{\rho}_1$ explicitly, and to insert that into the self–energy integral. Now the second continuity equation is an integral equation for $\Delta(\mathbf{k},\mathbf{p};\omega)/\xi(\mathbf{k};\omega)$ and it cannot be solved in closed form.

The singularity structure of the self–energy as well as the second continuity equation are made more explicit by introducing the following notation,

$$\beta_{\mathbf{k},\omega}(\mathbf{p}) \equiv \left[\hbar\omega - \varepsilon\left(\left|\tfrac{\mathbf{k}}{2} + \mathbf{p}\right|\right) - \varepsilon\left(\left|\tfrac{\mathbf{k}}{2} - \mathbf{p}\right|\right)\right] \frac{\Delta(\mathbf{k},\mathbf{p};\omega)}{\xi(\mathbf{k};\omega)}$$

and we get the self–energy in the form

$$\Sigma(k,\omega) = -\frac{\hbar^2}{2m} \int \frac{d^3p}{(2\pi)^3\rho_0} \mathbf{k}\cdot\left(\tfrac{\mathbf{k}}{2} + \mathbf{p}\right)\; Q\left(\left|\tfrac{\mathbf{k}}{2} + \mathbf{p}\right|\right)$$

$$\times \frac{\beta_{\mathbf{k},\omega}(\mathbf{p})}{\hbar\omega - \varepsilon\left(\left|\tfrac{\mathbf{k}}{2} + \mathbf{p}\right|\right) - \varepsilon\left(\left|\tfrac{\mathbf{k}}{2} - \mathbf{p}\right|\right)}. \tag{6.23}$$

The function $\beta_{\mathbf{k},\omega}(\mathbf{p})$ is solved self–consistently from the second continuity equation,

$$\beta_{\mathbf{k},\omega}(\mathbf{p}) = \left[\hbar\omega - \varepsilon(k)\right] M_{\mathbf{k}}(\mathbf{p}) + N_{\mathbf{k}}(\mathbf{p}) \tag{6.24}$$

$$+ \int \frac{d^3q}{(2\pi)^3\rho} \frac{\beta_{\mathbf{k},\omega}(\mathbf{q})\left[K_{\mathbf{k}}(\mathbf{p},\mathbf{q}) - \hbar\omega\; s(|\mathbf{p}-\mathbf{q}|)\right]}{\hbar\omega - \varepsilon\left(\left|\tfrac{\mathbf{k}}{2} + \mathbf{q}\right|\right) - \varepsilon\left(\left|\tfrac{\mathbf{k}}{2} - \mathbf{q}\right|\right)}.$$

With the aid of notations

$$s(\mathbf{k}) = S(\mathbf{k}) - 1 \tag{6.25}$$

$$s^{\mathrm{D}}(\mathbf{k}) = \delta(\mathbf{k}) + s(\mathbf{k}) \tag{6.26}$$

we can write the terms appearing in Eq. (6.24) in the form

$$M_{\mathbf{k}}(\mathbf{p}) = \frac{1}{S(k)} \int \frac{d^3q}{(2\pi)^3\rho}\, s(|\tfrac{\mathbf{k}}{2} + \mathbf{q}|)\, s(|\tfrac{\mathbf{k}}{2} - \mathbf{q}|)\, s^{\mathrm{D}}(|\mathbf{p} - \mathbf{q}|) \tag{6.27}$$

$$\begin{aligned} N_{\mathbf{k}}(\mathbf{p}) = & -\frac{\varepsilon(k)}{k^2}\mathbf{k} \cdot (\tfrac{\mathbf{k}}{2} + \mathbf{p})\, s\,(|\tfrac{\mathbf{k}}{2} + \mathbf{p}|) \\ & - \frac{\hbar^2}{2m} \int \frac{d^3q}{(2\pi)^3\rho} (\tfrac{\mathbf{k}}{2} + \mathbf{p}) \cdot (\tfrac{\mathbf{k}}{2} + \mathbf{q})\; Q(|\tfrac{\mathbf{k}}{2} + \mathbf{q}|)\, s(|\tfrac{\mathbf{k}}{2} - \mathbf{q}|)\, s^D(|\mathbf{p} - \mathbf{q}|) \\ & + (\mathbf{p} \leftrightarrow -\mathbf{p}) \end{aligned} \tag{6.28}$$

and the kernel is

$$\begin{aligned} K_{\mathbf{k}}(\mathbf{p}, \mathbf{q}) = \frac{\hbar^2}{2m} \Bigg[ & (\tfrac{\mathbf{k}}{2} + \mathbf{p}) \cdot (\tfrac{\mathbf{k}}{2} + \mathbf{q})\, \left[1 - Q\,(|\tfrac{\mathbf{k}}{2} + \mathbf{q}|)\right] s(|\mathbf{p} - \mathbf{q}|) \\ & - \mathbf{k} \cdot (\tfrac{\mathbf{k}}{2} + \mathbf{q})\, Q\,(|\tfrac{\mathbf{k}}{2} + \mathbf{q}|)\, N_{\mathbf{k}}(\mathbf{p}) \\ & - \int \frac{d^3q'}{(2\pi)^3\rho} \Big[ (\tfrac{\mathbf{k}}{2} + \mathbf{p}) \cdot (\mathbf{q} - \mathbf{q}')\, Q(|\mathbf{q} - \mathbf{q}'|)\, s\,(|\tfrac{\mathbf{k}}{2} - \mathbf{q}'|) \\ & - (\tfrac{\mathbf{k}}{2} + \mathbf{p}) \cdot (\tfrac{\mathbf{k}}{2} + \mathbf{q}')\; Q\,(|\tfrac{\mathbf{k}}{2} + \mathbf{q}'|)\; s(|\mathbf{q} - \mathbf{q}'|) \Big] \Bigg] + (\mathbf{p} \leftrightarrow -\mathbf{p})\,. \end{aligned}$$

The collective reference spectrum entering the energy denominator in Eq.(6.24) is not the Feynman spectrum (1.1), but the spectrum of Eq. (6.20), which generally lies closer to the experimental spectrum. Therefore, this approach accounts better for the energetics of the excitations, especially at large momenta.

### 6.3. *Phonon–roton spectrum*

With the full theory we have calculated the phonon–roton spectrum in liquid $^4$He. [49] The input to the theory is the static structure function $S(k)$ of the ground state. That is available either from our variational theory [50] or directly from experiments where we have used the measurements of Ref. 46. The results are shown in Fig. 7 The resulting spectra differ somewhat in the maxon–roton region. When the experimental $S(k)$ is used the roton minimum is at $k$=1.93 Å$^{-1}$ with the energy 8.74 K and the maxon maximum at $k$=1.19 Å$^{-1}$ with the energy 13.80K. That compares very well with different experiments [51]. Woods *et al.* [52] find the roton minimum at $k$=1.926±0.005 Å$^{-1}$ with the energy 8.618±0.009K and Svensson *et al.* [53] find the maxon peak at $k$=1.13 Å$^{-1}$ with the energy 13.82K. The agreement suggests that right physical ingredients are included into our approximations. Since the peak of the variational structure function is slightly lower and at slightly higher $k$ than the experimental one the use of variational $S(k)$ moves the roton minimum to the energy 9.16K at 1.98 Å$^{-1}$ and raises the maxon peak to 14.85K at 1.21 Å$^{-1}$. At wave numbers $k > 2.5$ Å$^{-1}$ we find the flattening of the dispersion relation due to opening of the decay channel into two rotons. We have also calculated the strength of the pole, $Z(k)$, of the elementary excitation, in the dynamic structure function $S(k, \omega)$.

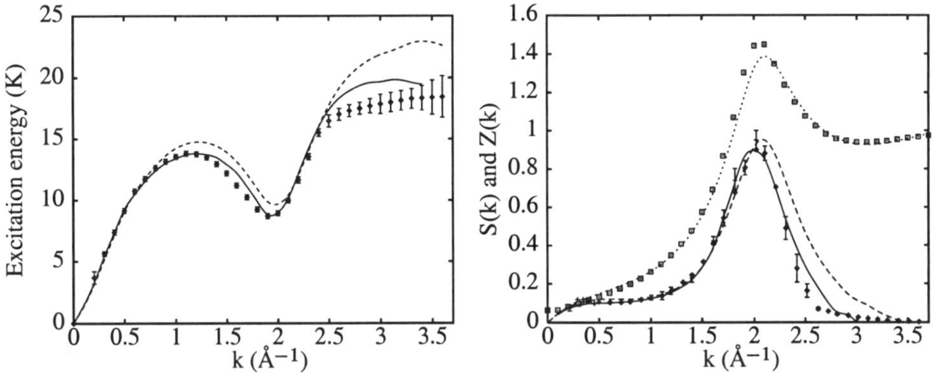


Fig. 7. Left figure compares the phonon–roton spectrum of our full theory obtained using the experimental (solid line) and the calculated $S(k)$ (dashed line) with measurements of Ref. 45. In the right figure the experimental[46] (squares) and calculated (dotted line) structure functions are shown together with the pole strength $Z(k)$ as obtained from the experimental (solid line) and calculated (dashed line) $S(k)$ and from the measurements of Ref. 45. (The original figures in Ref. 49.)

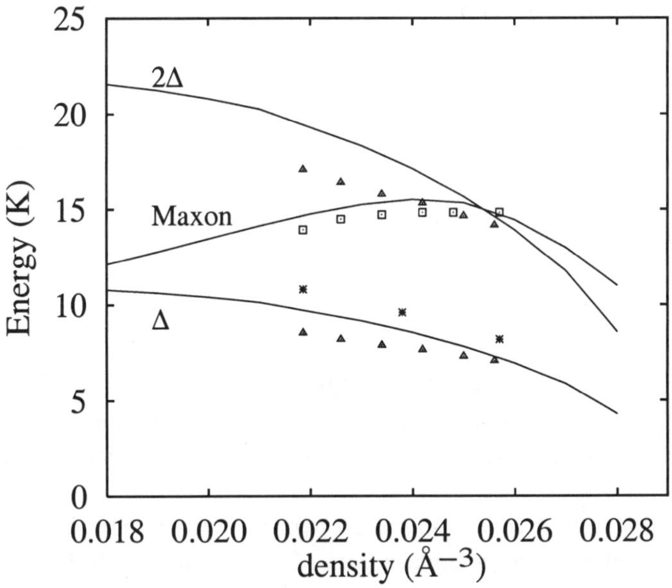


Fig. 8. The density dependence of the roton and maxon excitation energies. The squares are the measured maxon energies of Ref. 47, the triangles represent the one– and two–roton energies of Ref. 48 and the stars the Brillouin–Wigner perturbation theory results of Chang and Campbell [17]. The solid lines correspond to the results of our fully optimized theory. The theoretical maxon energy crosses the two–roton energy between 0.024 and 0.025 Å$^{-3}$, and at higher density the theoretical maxon curve drops with the two–roton curve.

That gives a more stringent test than the excitation spectrum to the solution of the continuity equations. The results agree very well with experiments for the full range of wave numbers as shown also in Fig. 7. The strength of the elementary excitation mode disappears at $k \geq 3.5$ Å$^{-1}$ due to the decay into two rotons as discussed in the previous sections. The pole strength vanishes also in the phonon region because of the anomalous dispersion, but here we have included the strength of the lowest lying complex solution which is also included in the measured strength.

In Fig. 8 we show the behavior of the maxon maximum and the roton minimum as a function of density. With increasing density the roton minimum shifts towards higher wave vectors and its energy decreases, but remains positive. The maxon maximum first increases with the increasing density until it saturates near the density 0.024 Å$^{-3}$. The two phonon decay limit pushes it down and at high densities its strength disappears and the decaying mode takes over.

### 6.4. *Sum rules*

Sum rules have frequently been applied to estimates of various physical properties of quantum liquids, such as the ground–state structure and characteristics of the response function, or used as a consistency check for the calculations. We want to see if the proposed theory satisfies known exact sum rules.

We would like to examine specifically the excited states. We therefore consider the $n$th energy–moment sum rule, defined as

$$m_n(k) = \int_0^\infty d(\hbar\omega)(\hbar\omega)^n \, S(k,\omega)\,. \tag{6.29}$$

The lowest moments are

$$S(k) = \int_0^\infty d(\hbar\omega) S(k,\omega) \tag{6.30}$$

$$\frac{\hbar^2 k^2}{2m} = \int_0^\infty d(\hbar\omega)\; \hbar\omega \; S(k,\omega) \tag{6.31}$$

$$\frac{1}{2mc^2} = \lim_{k\to 0} \int_0^\infty d\omega \; \frac{S(k,\omega)}{\omega}\,. \tag{6.32}$$

These sum rules are generally referred to as the zeroth–moment sum rule, the $f$–sum rule or the Thomas–Reiche–Kuhn sum rule, and the compressibility sum rule, respectively.

Let us first collect some general properties of the dynamic structure function. It is proportional to the imaginary part of the linear response function $S(\mathbf{k},\omega) = -\frac{1}{\pi}\Im m\chi(\mathbf{k},\omega)$ and gets contributions from the low energy pole and at higher energy from the non–zero imaginary part of the self–energy $\Sigma(k,\omega) = \Delta(\mathbf{k},\omega) - i\Gamma(\mathbf{k},\omega)$.

$$\begin{aligned} S(\mathbf{k},\omega) = Z(k)\delta(\hbar\omega - \hbar\omega_0) + \frac{S(k)}{\pi}\Bigg\{ & \frac{\Gamma(\mathbf{k},\omega)}{[\hbar\omega - \varepsilon(k) - \Delta(\mathbf{k},\omega)]^2 + \Gamma^2(\mathbf{k},\omega)} \\ & - \frac{\Gamma(\mathbf{k},-\omega)}{[\hbar\omega + \varepsilon(k) + \Delta(\mathbf{k},-\omega)]^2 + \Gamma^2(\mathbf{k},-\omega)} \Bigg\}. \end{aligned} \tag{6.33}$$

Since the dynamic structure function is a positive quantity then also $\Gamma(\mathbf{k},\omega) \geq 0$ when $\omega > \omega_0$. We have also assumed that no poles exist and no decay is possible at negative energies and thus $\Gamma(\mathbf{k},\omega) = 0$ when $\omega < 0$. We argued following Jackson [41] that all poles must lie on the real $\omega$–axis, but an infinitesimal imaginary part $\omega \to \omega + i\eta$ moves the singularities into the lower half of the complex plane and leaves $\chi(\mathbf{k},\omega)$ analytic in the upper half. With these general arguments we can drop the last term in Eq. (6.33) and define the Green's function

$$G(\mathbf{k},\omega) = \frac{1}{\hbar\omega - \varepsilon(k) - \Sigma(\mathbf{k},\omega)}\,, \tag{6.34}$$

which then fixes the dynamic structure function

$$S(\mathbf{k},\omega) = -\frac{1}{\pi} S(k)\, \Im m G(\mathbf{k},\omega)\,. \tag{6.35}$$

In the evaluations of the sum rules it is important to know the asymptotic behavior of the self–energy. Our full solution of Eq. (6.23) behaves like

$$\lim_{\omega\to\infty} \Sigma(\mathbf{k},\omega) = A(k) \tag{6.36}$$

where $A(k)$ is a function of $k$, but independent of $\omega$. In the CBF–approximation $A(k) = 0$ and the sum rules are proven to be satisfied [41]. In our full theory $A(k) \neq 0$ and the sum rules are satisfied only if that asymptotic behavior is compensated with the change in the reference spectrum $\varepsilon(k)$ as we will show shortly.

The $m_0$ sum rule can be easily calculated by closing the integration path in the upper half of the complex plane. The lower limit of the integration can be extended to $-\infty$ because the imaginary part is zero there,

$$\begin{aligned} m_0 &= -\frac{S(k)}{\pi}\Im m \int_{-\infty}^{\infty} \frac{d(\hbar\omega)}{\hbar\omega - \varepsilon(k) - \Sigma(\mathbf{k},\omega) + i\eta} \\ &= \frac{S(k)}{\pi}\Im m \lim_{R\to\infty} \int_0^{\pi} \frac{iRe^{i\theta}d\theta}{Re^{i\theta} - \varepsilon(k) - \Sigma(\mathbf{k},Re^{i\theta})}\,. \end{aligned}$$

Giving

$$m_0 = \frac{S(k)}{\pi}\Im m\; i\int_0^{\pi} \theta = S(k)\,. \tag{6.37}$$

In the calculation of the $m_{-1}$ sum rule

$$m_{-1} = \frac{1}{\pi}\lim_{k\to 0} S(k)\Im m \int_{-\infty}^{\infty} \frac{d(\hbar\omega)}{\hbar\omega} G(\mathbf{k},\omega) \tag{6.38}$$

we need to know the limits

$$\begin{aligned} \lim_{R\to\infty} G(\mathbf{k},Re^{i\theta}) &= 0 \\ \lim_{r\to 0} G(\mathbf{k},re^{i\theta}) &= -\frac{1}{\varepsilon(k) + \Sigma(\mathbf{k},0)} \\ \Sigma(\mathbf{k},0) &\stackrel{k\to 0}{\to} k^2\,. \end{aligned} \tag{6.39}$$

Closing again the integration path in the upper half of the complex plane we get

$$m_{-1} = \frac{1}{\pi} \lim_{k\to 0} S(k) \Im m \int_0^\pi \frac{i\theta}{\varepsilon(k) + \Sigma(\mathbf{k},0)} = \lim_{k\to 0} \frac{S(k)}{\varepsilon(k) + \Sigma(\mathbf{k},0)} . \tag{6.40}$$

The long wavelength limits are given in Eqs. (6.20) and (4.19) and they fulfil the compressibility sum rule

$$m_{-1} = \lim_{k\to 0} \frac{S(k)}{\varepsilon(k)} = \lim_{k\to 0} \frac{2mS^2(k)}{\hbar^2 k^2} = \frac{1}{2mc^2} . \tag{6.41}$$

The $f$–sum rule gives some problems if the self–energy does not vanish when $\omega \to \infty$. We rewrite the imaginary part by subtracting the complex conjugate

$$\begin{aligned} m_1 &= -\frac{S(k)}{\pi} \Im m \int_0^\infty d(\hbar\omega)\ \hbar\omega\ G(\mathbf{k},\omega) \\ &= \frac{S(k)}{2\pi i} \int_0^\infty d(\hbar\omega)\hbar\omega \left[G(\mathbf{k},\omega) - G^*(\mathbf{k},\omega)\right] \end{aligned} \tag{6.42}$$

and manipulate the difference a little

$$\begin{aligned} \hbar\omega[G(\mathbf{k},\omega) - G^*(\mathbf{k},\omega)] &= \frac{\hbar\omega}{\hbar\omega - \varepsilon(k) - \Sigma(\mathbf{k},\omega) + i\eta} - \frac{\hbar\omega}{\hbar\omega - \varepsilon(k) - \Sigma^*(\mathbf{k},\omega) - i\eta} \\ &= \frac{\varepsilon(k) + \Sigma(\mathbf{k},\omega)}{\hbar\omega - \varepsilon(k) - \Sigma(\mathbf{k},\omega) + i\eta} - \frac{\varepsilon(k) + \Sigma^*(\mathbf{k},\omega)}{\hbar\omega - \varepsilon(k) - \Sigma^*(\mathbf{k},\omega) - i\eta} . \end{aligned} \tag{6.43}$$

The contour integration can now be performed as before, but such that in the first term the integration closes in the upper half plane and in the second term in the lower half plane. This gives the result

$$\begin{aligned} m_1 &= \frac{S(k)}{2} \left[\varepsilon(k) + A(k) + \varepsilon(k) + A^*(k)\right] \\ &= S(k) \left[\varepsilon(k) + \Re e A(k)\right] . \end{aligned} \tag{6.44}$$

That does not satisfy the $f$–sum rule unless

$$\varepsilon(k) + \Re e A(k) = \varepsilon_F(k) = \frac{\hbar^2 k^2}{2mS(k)} . \tag{6.45}$$

In order to show that this is the case we solve the continuity equations (3.32) and (3.33) in the limit $\omega \to \infty$. This is easier to work out in the coordinate space, but with the Fourier transform of the time into the frequency space. The time derivative terms give the leading order in the limit $\omega \to \infty$ in comparison to the currents and we can drop them. This simplifies the continuity equations

$$\begin{aligned} &\hbar\omega\ \xi_1(\mathbf{r}_1;\omega) = 2U_{\rm ext}(\mathbf{r}_1;\omega) + 2\rho_0 \int d^3r_2 h(\mathbf{r}_1,\mathbf{r}_2) U_{\rm ext}(\mathbf{r}_2;\omega) \\ &\hbar\omega\ \delta\rho_2(\mathbf{r}_1,\mathbf{r}_2;\omega) = 2\ \rho_0^2\ g_2(\mathbf{r}_1,\mathbf{r}_2) \left[U_{\rm ext}(\mathbf{r}_1;\omega) + U_{\rm ext}(\mathbf{r}_2;\omega)\right] \\ &+2\rho_0^3 \int d^3r_3 [g_3(\mathbf{r}_1,\mathbf{r}_2,\mathbf{r}_3) - g_2(\mathbf{r}_1,\mathbf{r}_2)] U_{\rm ext}(\mathbf{r}_3;\omega) . \end{aligned} \tag{6.46}$$

From the first equation we get $U_{\text{ext}}(\mathbf{r}_1;\omega)$

$$\frac{2U_{\text{ext}}(\mathbf{r}_1;\omega)}{\hbar\omega} = \xi_1(\mathbf{r}_1;\omega) - \rho_0 \int d^3r_2 X(\mathbf{r}_1,\mathbf{r}_2)\xi(\mathbf{r}_2;\omega) \tag{6.47}$$

and insert that into the second equation

$$\begin{aligned}\delta\rho_2(\mathbf{r}_1,\mathbf{r}_2;\omega) &= \rho^2 g_2(\mathbf{r}_1,\mathbf{r}_2)[\xi(\mathbf{r}_1;\omega)+\xi(\mathbf{r}_2;\omega)]\\ &- \rho_0 \int d^3r_3 \xi(\mathbf{r}_3;\omega)[X(\mathbf{r}_1,\mathbf{r}_3)+X(\mathbf{r}_2,\mathbf{r}_3)]\\ &+ 2\rho_0^3 \int d^3r_3 [g_3(\mathbf{r}_1,\mathbf{r}_2,\mathbf{r}_3 - g_2(\mathbf{r}_1,\mathbf{r}_2)]\\ &\times \left[\xi(\mathbf{r}_3;\omega) - \rho \int d^3r_4 X(\mathbf{r}_3,\mathbf{r}_4)\xi(\mathbf{r}_4;\omega)\right].\end{aligned} \tag{6.48}$$

We have solved now $\delta\rho_2(\mathbf{r}_1,\mathbf{r}_2;\omega)$ entirely in terms of the one–particle density and can insert that back into the one–particle current and show that its divergence gives the Feynman spectrum

$$\mathbf{j}_1(\mathbf{r}_1;\omega) = \frac{\hbar}{2mi}[\nabla_1\delta\rho_1(\mathbf{r}_1;\omega) - \int d^3r_2 \delta\rho_2(\mathbf{r}_1,\mathbf{r}_2;\omega)\nabla u_2(\mathbf{r}_1,\mathbf{r}_2)] \tag{6.49}$$

$$\begin{aligned}&= \frac{\hbar}{2mi}[\nabla_1\delta\rho_1(\mathbf{r}_1;\omega) - \rho_0^2 \int d^3r_2 g_2(\mathbf{r}_1,\mathbf{r}_2)\nabla u_2(\mathbf{r}_1,\mathbf{r}_2)\Big[\xi(\mathbf{r}_1;\omega)+\xi(\mathbf{r}_2;\omega)\\ &- \rho_0 \int d^3r_3 \xi(\mathbf{r}_3;\omega)[X(\mathbf{r}_1,\mathbf{r}_3)+X(\mathbf{r}_2,\mathbf{r}_3)]\Big]\\ &- \rho_0^3 \int d^3r_2\, d^3r_3 [g_3(\mathbf{r}_1,\mathbf{r}_2,\mathbf{r}_3) - g_2(\mathbf{r}_1,\mathbf{r}_2)]\nabla_1 u_2(\mathbf{r}_1,\mathbf{r}_2)\\ &\left[\xi(\mathbf{r}_3;\omega) - \rho_0 \int d^3r_4 X(\mathbf{r}_3,\mathbf{r}_4)\xi(\mathbf{r}_4;\omega)\right].\end{aligned} \tag{6.50}$$

Using the BGY–equations Eqs. (6.6) and (6.7) we end up with

$$\begin{aligned}\mathbf{j}_1(\mathbf{r}_1;\omega) &= \frac{\hbar}{2mi}\Big\{\nabla_1\delta\rho_1(\mathbf{r}_1;\omega) - \rho_0^2 \int d^3r_2 \nabla_1 g_2(\mathbf{r}_1,\mathbf{r}_2)\\ &\times \left[\xi(\mathbf{r}_2;\omega) - \rho_0 \int d^3r_3 X(\mathbf{r}_2,\mathbf{r}_3)\xi(\mathbf{r}_3;\omega)\right]\Big\}.\end{aligned} \tag{6.51}$$

Finally using the Ornstein–Zernike relation (2.21) we get

$$\mathbf{j}_1(\mathbf{r}_1;\omega) = \frac{\hbar\rho_0}{2mi}\nabla_1 \left[\xi(\mathbf{r}_1;\omega) - \rho_0 \int d^3r_2 \xi(\mathbf{r}_2;\omega)X(\mathbf{r}_1,\mathbf{r}_2)\right]. \tag{6.52}$$

In the momentum space this gives exactly the Feynman spectrum

$$\mathcal{F}\left[i\hbar\nabla_1\cdot\mathbf{j}_1(\mathbf{r}_1;\omega)\right] = \frac{\hbar^2\rho_0}{2m}\frac{k^2}{S(k)} = \rho_0\varepsilon_F(k). \tag{6.53}$$

Hence we have shown that the corrections term in collective mode (6.20) cancels exactly the asymptotic behavior $A(k)$ in the self–energy.

### 6.5. *Results on two–particle currents*

The two–particle current (6.9) can be written in the form

$$\mathbf{j}_2(\mathbf{r}_1,\mathbf{r}_2;t) = \rho_0 g_2(|\mathbf{r}_1-\mathbf{r}_2|)\Big\{\mathbf{j}_1(\mathbf{r}_1;t) + \frac{\hbar\rho_0}{2mi}\tilde{\mathbf{T}}(\mathbf{r}_1,\mathbf{r}_2;t)\Big\}\,, \tag{6.54}$$

where $\tilde{\mathbf{T}}(\mathbf{r}_1,\mathbf{r}_2;t)$ depends on the fluctuations in the radial distribution function only,

$$\begin{aligned}\mathbf{T}(\mathbf{r_1},\mathbf{r_2};t) &= \nabla_1\Delta(\mathbf{r_1},\mathbf{r_2};t)\\ &-\rho_0\int d^3r_3\Delta(\mathbf{r_2},\mathbf{r_3};t)\mathbf{V}(\mathbf{r_1},\mathbf{r_3})\\ &-\rho_0\int d^3r_3 h(\mathbf{r}_2,\mathbf{r}_3)\mathbf{V}(\mathbf{r}_1,\mathbf{r}_3)\\ &\times\Big(\xi(\mathbf{r}_3;t)+\Delta(\mathbf{r_1},\mathbf{r_3};t)+\Delta(\mathbf{r_2},\mathbf{r_3};t)\Big)\,.\end{aligned} \tag{6.55}$$

We have chosen to present our results for the real part of the two–particle current of Eq. (6.54) in mixed representation as a function of center of mass momentum and relative coordinate. It can then be written in the form

$$\begin{aligned}\mathbf{j}_2(\mathbf{k},\mathbf{r},\omega) = \rho_0 g_2(r)\Big[&\mathbf{j}_1(\mathbf{k},\omega)\cos(\frac{1}{2}\mathbf{k}\cdot\mathbf{r})\\ &+\frac{\hbar\rho_0}{2mi}\tilde{\mathbf{T}}(\mathbf{k},\mathbf{r},\omega)\Big]\,.\end{aligned} \tag{6.56}$$

The radial distribution function of the ground state $g_2(r)$ gives the probability of finding another particle at the distance $r$ away from a given particle. We locate particle one into the origin and because of the repulsive core of the interaction other particles are repelled outside the radius of about 2 Å . This "correlation hole" is clearly seen in Fig. 9. We also separate out the oscillating behavior of the sound–like wave where particles move towards each other and away from each other with the wavelength determined by the center–of–mass motion in the cosine term. The more complicated flow patterns are collected into $\tilde{\mathbf{T}}(\mathbf{k},\mathbf{r},\omega)$. The coordinate system is such that the center of mass momentum points to the $z$–direction.

A detailed structure of the relative current flow is shown in Fig. 9 where we have subtracted the center of mass oscillations. The current flows to the direction of arrows and since it has cylindrical symmetry we show only the $x-z$–plane with $x \geq 0$. In the maxon and roton regions (Figs. 9.a and 9.b) the dominant feature is the oscillation of the radial distribution function, though some interesting topological structures could be identified. The pattern, however, changes completely at 2.5 Å$^{-1}$. A clear back flow loop is formed around each atom in Fig. 9.c. The radius of the circulation is of the order of atomic radius (compare with the white area in the figures). When $k$ >2.5 Å$^{-1}$ the loop gets elongated with increasing wave number and forms a tube–like structure with a diameter of atomic size. A typical case at 3.0 Å$^{-1}$ is shown in Fig. 9.d.

In our wave function we have not put in quantized vortices [27] and the structures seen in Figs. 9.c and 9.d come out of the full optimization of the action integral

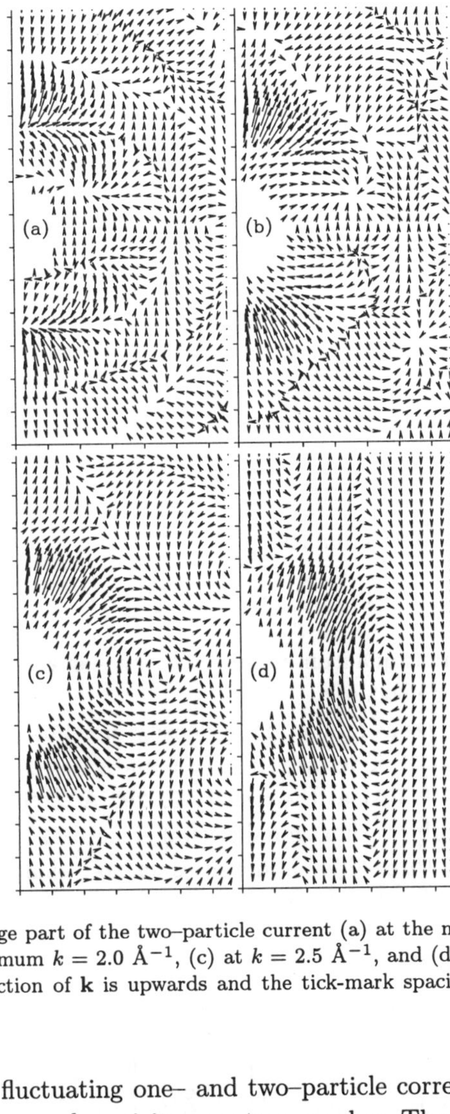


Fig. 9. The short–range part of the two–particle current (a) at the maxon region $k = 1.0$ Å$^{-1}$, (b) near the roton minimum $k = 2.0$ Å$^{-1}$, (c) at $k = 2.5$ Å$^{-1}$, and (d) in the asymptotic region $k = 3.0$ Å$^{-1}$. The direction of $\mathbf{k}$ is upwards and the tick-mark spacing is 1.0 Å. (The original figures in Ref. 49.)

with respect to the fluctuating one– and two–particle correlation functions. They do not carry any conserved vorticity quantum number. The relation between these excitations and the vortex excitations should be investigated further.

## 6.6. *Precursor of the liquid–solid transition*

The liquid–solid phase–transition is a first order phase transition. Latent heat is required for the liquid to solidify. At the same time there is an abrupt change in the density when the particles in the liquid arrange themselves into the crystal order. Some signatures of the emerging phase transition can be seen in the liquid phase by studying two–particle structures, which break the translational invariance of the liquid phase. [54,55,56]

In the ground state of the liquid phase the single particle density is constant and the system has no center of mass motion. Then the two–particle continuity equation (6.12) can be written as

$$\begin{aligned}\frac{\hbar^2}{2m}\nabla_1\Big\{g_2(\mathbf{r}_1,\mathbf{r}_2)\Big[\nabla_1\Delta(\mathbf{r}_1,\mathbf{r}_2)-\\ \rho_0\int d^3r_3 g_2(\mathbf{r}_2,\mathbf{r}_3)\mathbf{V}(\mathbf{r}_1,\mathbf{r}_3)\Big[\Delta(\mathbf{r}_1,\mathbf{r}_3)+\Delta(\mathbf{r}_2,\mathbf{r}_3)\Big]\Big]\Big\}\\ +(1\leftrightarrow 2)=\hbar\omega\; g_2(\mathbf{r}_1,\mathbf{r}_2)\Delta(\mathbf{r}_1,\mathbf{r}_2)\,.\end{aligned} \tag{6.57}$$

Again this equation is simpler to solve in momentum space and since the center of mass momentum $k = 0$ then $\tilde{\Delta}(\mathbf{p})$ depends only on the relative momentum $\mathbf{p}$, but that is a vector quantity. A convenient way to solve Eq. (6.57) is to expand $\Delta(\mathbf{p})$ in terms of Legendre polynomials in three dimension

$$\tilde{\Delta}(\mathbf{p}) = \sum_L \tilde{\Delta}_L(p)\mathrm{P}_L(\cos\theta_{\mathbf{p}}) \tag{6.58}$$

and using exponentials in two dimensions

$$\tilde{\Delta}(\mathbf{p}) = \sum_m \tilde{\Delta}_m(p)e^{im\phi}\,. \tag{6.59}$$

The equations for different quantum numbers $L$ or $m$ separate and those equations can be solved for the eigenvalues $\omega$.

The excitation energies as a function of density are shown in Fig. 10 for the 3D liquid $^4$He. The two lowest modes are the $L = 6$ and $L = 12$ modes from which the $L = 6$ mode becomes soft first at the density 0.0295 Å$^{-3}$. The $L = 6$ point group symmetry indicates a cubic or hexagonal crystal structure. In Fig. 10 we show also the behavior of $\tilde{\Delta}_6(p)$ at three different densities. It has a sharp peak which gives the

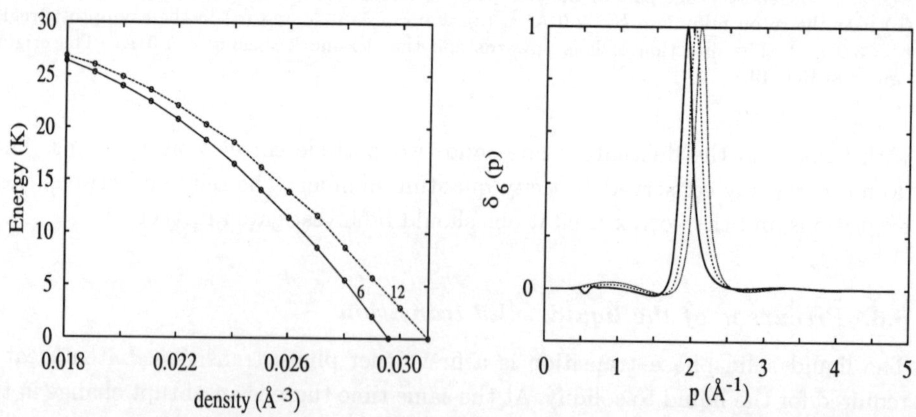


Fig. 10. In the left figure the energies of the $L = 6$ and $L = 12$ modes are shown in 3D $^4$He as a function of density. In the right figure the $L = 6$ eigenstates are shown in 3D $^4$He at densities 0.022, 0.024 and 0.029 Å$^{-3}$. The peaks of the curves move to the right with increasing density. The vertical scale of the eigenstates is arbitrary.

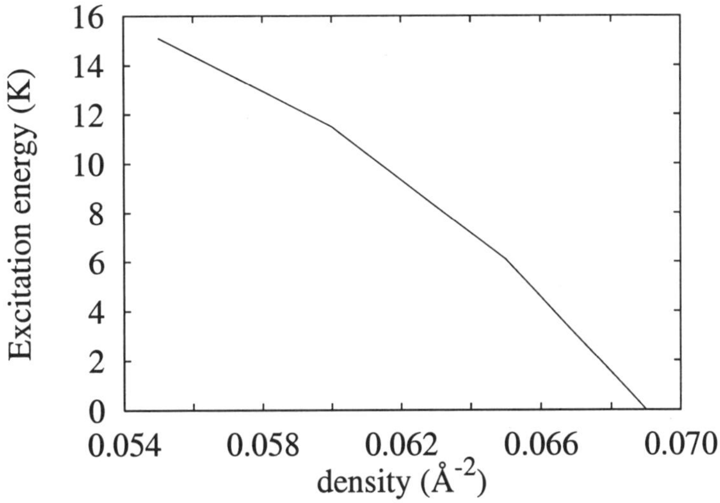


Fig. 11. The energy of the mode with the hexagonal symmetry (m=6) as a function of density.

value 2.1 Å$^{-1}$ for the reciprocal lattice vector of the emerging crystal structure. The width of the peak is related to quantum fluctuations. The excitation energy of the two dimensional $m = 6$ mode in $^4$He is shown in Fig. 11 and the energy goes through zero at 0.069 Å$^{-2}$. That point group symmetry is consistent with the hexagonal crystal structure. Our results are in reasonable agreement with experiments[57,58] in three dimension and with Monte Carlo simulations in two dimension. [59,60,61]

Finally in Fig. 12 we show the change in the pair distribution $g(r)[1 + \Delta_6(r)\cos(6\phi)]$ in 2D $^4$He. The reference particle is at the center of the figure and the gray shading shows the distribution of particles around it; the lighter areas give the higher probability for the location of atoms. The circles in the figure give the positions of the atoms in the hexagonal crystal structure at the density 0.069 Å$^{-2}$.

## 7. Dynamics of a single impurity

In this section we review the application of the equation of motion method[62,63] to the dynamics of an impurity in a quantum fluid. The specific example we use is the $^3$He impurity atom in liquid $^4$He. The basic ideas are the same as for the pure liquid $^4$He and therefore, we only summarize the necessary generalizations to the case of an impurity following the notations in Ref. 64.

### 7.1. *Continuity equations*

We start by assuming that the wave function for the ground state of a single impurity atom is optimized. As for the one–component liquid, the dynamic behavior of the impurity is driven by a weak, external time–dependent perturbation $U_{\rm ext}(\mathbf{r}_0; t)$ and

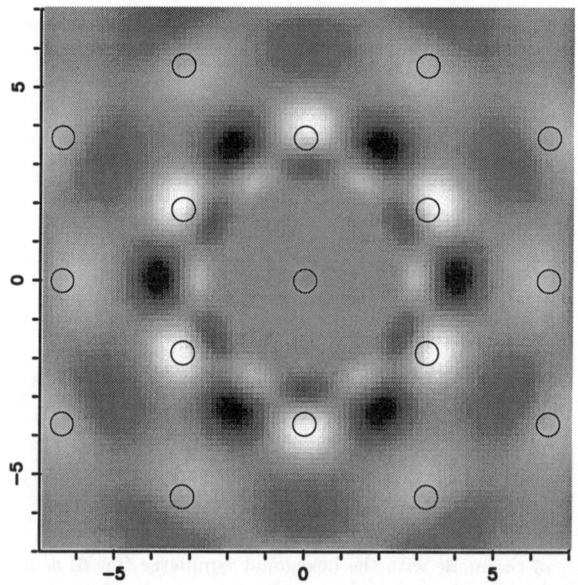


Fig. 12. The change in the relative two–body structure corresponding to the first mode (m=6) to become soft at 0.069 Å$^{-2}$ in two–dimensional $^4$He. Lighter shade indicates increased probability to find a particle at given coordinates. The gray area in the middle is the correlation hole. Superimposed rings are the lattice sites in the hexagonal crystal at the same density. (The original figure in Ref. 56.)

the response of the system shows up in the time dependence of the wave function. The kinematic and dynamic correlations are separated by writing the wave function in the form

$$\Psi^{(3)}(t) = \frac{1}{\sqrt{\mathcal{N}^{(3)}(t)}} e^{-iE_{N+1}t/\hbar}\Phi^{(3)}(\mathbf{r}_0, \mathbf{r}_1, ..., \mathbf{r}_N; t)\,. \tag{7.1}$$

Here $E_{N+1}$ is the variational ground–state energy of the system containing $N+1$ particles. The coordinates of the background particles are labeled by $\mathbf{r}_1, \ldots, \mathbf{r}_N$, the impurity coordinate is $\mathbf{r}_0$ and the superscripts 3 and 4 refer to $^3$He and $^4$He, respectively. The time dependence of the wave function $\Phi^{(3)}(\mathbf{r}_0, \mathbf{r}_1, ...\mathbf{r}_N; t)$ is in the one– and two–particle correlation functions

$$\begin{aligned}\Phi^{(3)}(\mathbf{r}_0, \ldots \mathbf{r}_N; t) &= e^{\frac{1}{2}\delta U(\mathbf{r}_0,\ldots \mathbf{r}_N;t)}\phi^{(3)}(\mathbf{r}_0, \mathbf{r}_1, ..., \mathbf{r}_N) \\ \delta U(\mathbf{r}_0 \ldots \mathbf{r}_N; t) &= \delta u^{(3)}(\mathbf{r}_0; t) + \sum_{i=1}^{N} \delta u^{(34)}(\mathbf{r}_0, \mathbf{r}_i; t)\end{aligned} \tag{7.2}$$

and $\phi^{(3)}(\mathbf{r}_0, \mathbf{r}_1, ..., \mathbf{r}_N)$ is the optimized ground state wave function normalized to unity,

$$\left\langle \phi^{(3)} | \phi^{(3)} \right\rangle = 1\,. \tag{7.3}$$

The normalization factor in Eq. (7.1) is time dependent due to the correlation factors

$$\mathcal{N}^{(3)}(t) = \int d^3r_0 \dots d^3r_N \; |\phi^{(3)}(\mathbf{r}_0,\dots,\mathbf{r}_N)|^2 e^{\Re e[\delta U(\mathbf{r}_0,\dots,\mathbf{r}_N;t)]} \,. \tag{7.4}$$

The action principle is then used to determine the correlation functions

$$\delta \mathcal{S} = \delta \int_{t_0}^{t} dt' \; \left\langle \Psi^{(3)}(t') \middle| H^{(3)} + U_{\text{ext}}(\mathbf{r}_0;t) - \frac{\hbar}{2}\left( i\frac{\partial}{\partial t'} + h.c. \right) \middle| \Psi^{(3)}(t') \right\rangle = 0\,. \tag{7.5}$$

The least action integral contains the external infinitesimal potential which operates on the impurity coordinate only and the Hamiltonian of the impurity–background system,

$$H^{(3)} = H_0 - \frac{\hbar^2}{2m_3}\nabla_0^2 + \sum_{j=1}^{N} V^{(34)}(|\mathbf{r}_0 - \mathbf{r}_1|) \tag{7.6}$$

where $H_0$ is the Hamiltonian of the background from Eq. (2.1), $m_3$ the impurity mass and $V^{(34)}(|\mathbf{r}_0 - \mathbf{r}_j|)$ is the $^3$He–$^4$He interaction.

In the previous section we showed in detail how the variation of the action integral with respect to fluctuating correlations is done. Here we vary with respect to $\delta u^{(3)}(\mathbf{r}_0;t)$ and $\delta u^{(34)}(\mathbf{r}_0,\mathbf{r}_i;t)$ and that leads to the continuity equations

$$\nabla_0 \cdot \mathbf{j}^{(3)}(\mathbf{r}_0;t) + \delta\dot{\rho}^{(3)}(\mathbf{r}_0;t) = \frac{2\rho^{(3)}(\mathbf{r}_0)}{i\hbar} U_{\text{ext}}(\mathbf{r}_0;t) \tag{7.7}$$

and

$$\begin{aligned} \nabla_0 \cdot \mathbf{j}^{(34)}(\mathbf{r}_0,\mathbf{r}_1;t) + \nabla_1 \cdot \mathbf{J}^{(34)}(\mathbf{r}_0,\mathbf{r}_1;t) &+ \delta\dot{\rho}^{(34)}(\mathbf{r}_0,\mathbf{r}_1;t) \\ &= \frac{2}{i\hbar} U_{\text{ext}}(\mathbf{r}_0;t)\rho^{(34)}(\mathbf{r}_0,\mathbf{r}_1)\,. \end{aligned} \tag{7.8}$$

The exact definition of the $n$–particle density with impurity is

$$\bar{\rho}^{(34\dots)}(\mathbf{r}_0,\dots,\mathbf{r}_{n-1};t) = \frac{N!}{(N-n+1)!} \int d^3r_n \dots d^3r_N |\Psi^{(3)}(\mathbf{r}_0,\dots \mathbf{r}_N;t)|^2 \tag{7.9}$$

and the time dependence can be separated

$$\bar{\rho}^{(34\dots)}(\mathbf{r}_0,\dots,\mathbf{r}_n;t) = \rho^{(34\dots)}(\mathbf{r}_0,\dots,\mathbf{r}_n) + \delta\rho^{(34\dots)}(\mathbf{r}_0,\dots,\mathbf{r}_n;t) \tag{7.10}$$

by expanding to the first order in $\delta U(t)$

$$\begin{aligned} \delta\rho^{(34\dots)}(\mathbf{r}_0,\dots,\mathbf{r}_{n-1};t) \approx \frac{N!}{(N-n+1)!} &\int d^3r_n \dots d^3r_N \; |\phi^{(3)}(\mathbf{r}_0,\dots,\mathbf{r}_N)|^2 \\ &\times \Re e\Big[\delta U(t) - \left\langle \phi^{(3)} \left|\delta U(t)\right| \phi^{(3)} \right\rangle\Big]\,. \end{aligned} \tag{7.11}$$

Here we have used superscripts to label the number and particle identity of the coordinates. With these definitions it is easy to verify that

$$\int d^3r_0\ \rho^{(3)}(\mathbf{r}_0) = 1$$
$$\int d^3r_0\ \delta\rho^{(3)}(\mathbf{r}_0) = 0 \tag{7.12}$$

and we can set $\rho^{(3)}(\mathbf{r}_0) = 1/\Omega$ where $\Omega$ is the integration volume.

The transition currents are defined in terms of the fluctuating one–particle density and pair correlation function,

$$\mathbf{j}^{(3)}(\mathbf{r}_0;t) = \frac{\hbar}{2m_3 i}\left[\nabla_0\delta\rho^{(3)}(\mathbf{r}_0;t) - \int d^3r_1\ \delta u^{(34)}(\mathbf{r}_0,\mathbf{r}_1;t)\nabla_0\rho^{(34)}(\mathbf{r}_0,\mathbf{r}_1)\right],$$

$$\mathbf{j}^{(34)}(\mathbf{r}_0,\mathbf{r}_1;t) = \Omega\ \mathbf{j}^{(3)}(\mathbf{r}_0;t)\rho^{(34)}(\mathbf{r}_0,\mathbf{r}_1) + \frac{\hbar}{2m_3 i}\Bigg[\rho^{(34)}(\mathbf{r}_0,\mathbf{r}_1)\nabla_0\delta u^{(34)}(\mathbf{r}_0,\mathbf{r}_1;t)$$

$$+ \int d^3r_2\left[\rho^{(344)}(\mathbf{r}_0,\mathbf{r}_1,\mathbf{r}_2) - \Omega\ \rho^{(34)}(\mathbf{r}_0,\mathbf{r}_1)\rho^{(34)}(\mathbf{r}_0,\mathbf{r}_2)\right]\nabla_2\delta u^{(34)}(\mathbf{r}_0,\mathbf{r}_2;t)\Bigg]\ , \tag{7.13}$$

and

$$\mathbf{J}^{(34)}(\mathbf{r}_0,\mathbf{r}_1;t) = \frac{\hbar}{2m_4 i}\rho^{(34)}(\mathbf{r}_0,\mathbf{r}_1)\nabla_1\delta u^{(34)}(\mathbf{r}_0,\mathbf{r}_1;t)\ . \tag{7.14}$$

The ground–state quantities needed here are the impurity–background pair and triplet densities $\rho^{(34)}(\mathbf{r}_0,\mathbf{r}_1)$ and $\rho^{(344)}(\mathbf{r}_0,\mathbf{r}_1,\mathbf{r}_2)$, respectively, and the corresponding distribution functions are defined in the usual way,

$$g^{(34)}(\mathbf{r}_0,\mathbf{r}_1) = \frac{\Omega}{\rho^{(4)}}\ \rho^{(34)}(\mathbf{r}_0,\mathbf{r}_1)\ , \tag{7.15}$$

$$g^{(344)}(\mathbf{r}_0,\mathbf{r}_1,\mathbf{r}_2) = \frac{\Omega}{\left(\rho^{(4)}\right)^2}\ \rho^{(344)}(\mathbf{r}_0,\mathbf{r}_1,\mathbf{r}_2)\ . \tag{7.16}$$

It is also convenient to define the notation

$$S^{(344)}(\mathbf{r}_0,\mathbf{r}_1,\mathbf{r}_2) = g^{(344)}(\mathbf{r}_0,\mathbf{r}_1,\mathbf{r}_2)$$
$$- g^{(34)}(\mathbf{r}_0,\mathbf{r}_1) - g^{(34)}(\mathbf{r}_0,\mathbf{r}_2) - g^{(44)}(\mathbf{r}_1,\mathbf{r}_2) + 2\,, \tag{7.17}$$

approximate the triplet distribution function with the convolution approximation and include the triplet correlation function. [50,62] Our background $^4$He liquid is homogeneous and the result is easier to write out in momentum space,

$$\tilde{S}^{(344)}(\mathbf{k}_1,\mathbf{k}_2,\mathbf{k}_3) = \delta(\mathbf{k}_1+\mathbf{k}_2+\mathbf{k}_3)\Big\{S^{(34)}(\mathbf{k}_2)\ S^{(34)}(\mathbf{k}_3)$$
$$+ S^{(34)}(\mathbf{k}_1)\left[S^{(44)}(\mathbf{k}_2)\ S^{(44)}(\mathbf{k}_3) - 1\right]$$
$$+ \tilde{u}_3^{(344)}(\mathbf{k}_1,\mathbf{k}_2,\mathbf{k}_3)\left[S^{(34)}(\mathbf{k}_1) + 1\right]\ S^{(44)}(\mathbf{k}_2)\ S^{(44)}(\mathbf{k}_3)\Big\} \tag{7.18}$$

where the impurity structure function $S^{(34)}(k)$ is the Fourier transform of the radial distribution function $g^{(34)}(r)-1$. Analogous definitions are used for the background $^4$He distribution and structure functions.

We must also define the Fourier transforms of the impurity density fluctuations $\delta\tilde{\rho}^{(3)}(k,\omega)$ and the time dependent part of the two–particle correlation function $\delta\tilde{u}^{(34)}_{\mathbf{k},\omega}(\mathbf{p})$,

$$\delta\tilde{\rho}^{(3)}(k,\omega) = \rho^{(3)} \int d^3r_0\, dt\, e^{-i(\mathbf{k}\cdot\mathbf{r}_0-\omega t)} \delta\rho^{(3)}(\mathbf{r}_0;t)\ , \tag{7.19}$$

$$\delta\tilde{u}^{(34)}_{\mathbf{k},\omega}(\mathbf{p}) = \rho^{(3)}\rho^{(4)} \int d^3r_0\, d^3r_1\, dt\, e^{-i(\mathbf{k}\cdot\mathbf{r}_0+\mathbf{p}\cdot(\mathbf{r}_0-\mathbf{r}_1)-\omega t)}\ \delta u^{(34)}(\mathbf{r}_0,\mathbf{r}_1;t)\ . \tag{7.20}$$

### 7.2. *Linear response and self–energy*

The first continuity equation (7.7) can be written using the above notations in the form

$$\hbar\omega - t^{(3)}(k) - \Sigma^{(3)}(k,\omega) = \frac{\rho^{(3)}\ \tilde{U}_{\rm ext}(k,\omega)}{\delta\tilde{\rho}^{(3)}(k,\omega)} \tag{7.21}$$

where $t^{(3)}(k) = \frac{\hbar^2k^2}{2m_3}$ is the single particle kinetic energy. The linear response function has the same general structure as in the case of the one–component system

$$\begin{aligned}\chi^{(3)}(k,\omega) &= \frac{\delta\tilde{\rho}^{(3)}(k,\omega)}{\rho^{(3)}\tilde{U}_{\rm ext}(k,\omega)} \\ &= \frac{1}{\hbar\omega - t^{(3)}(k) - \Sigma^{(3)}(k,\omega)} - \frac{1}{\hbar\omega + t^{(3)}(k) + \Sigma^{*(3)}(k,-\omega)}\ .\end{aligned} \tag{7.22}$$

Its imaginary part defines the dynamic response function $S^{(3)}(k,\omega) = -\frac{1}{\pi}\Im m\chi^{(3)}(k,\omega)$. The sign of the possible pole is chosen by adding a small imaginary part into the frequency, $\omega \to \omega + i\eta$ with $\eta > 0$ and as usual $\eta \to 0$ at the end of the calculations.

The first continuity equation defines the self–energy

$$\Sigma^{(3)}(k,\omega) = \frac{\hbar^2}{2m_3}\int \frac{d^3p}{(2\pi)^3\rho^{(4)}}\ \frac{\mathbf{k}\cdot\mathbf{p}\ S^{(34)}(p)\beta^{(34)}_{\mathbf{k},\omega}(\mathbf{p})}{\left[\hbar\omega - t^{(3)}(\mathbf{k}+\mathbf{p}) - \varepsilon^{(4)}(p)\right]} \tag{7.23}$$

where $\varepsilon^{(4)}(p)$ is the background phonon–roton spectrum. The function $\beta^{(34)}_{\mathbf{k},\omega}(\mathbf{p})$ is solved from the second continuity equation (7.8) giving the result

$$\begin{aligned}\beta^{(34)}_{\mathbf{k},\omega}(\mathbf{p}) &= \hbar\omega\ \frac{\mathbf{k}\cdot\mathbf{p}}{k^2}\ \frac{S^{(34)}(p)}{S^{(44)}(p)} \\ &\quad - \int \frac{d^3q}{(2\pi)^3\rho^{(4)}}\ \frac{\beta^{(34)}_{\mathbf{k},\omega}(\mathbf{q})K_{\mathbf{k},\omega}(\mathbf{p},\mathbf{q})}{\left[\hbar\omega - t^{(3)}(\mathbf{k}+\mathbf{q}) - \varepsilon^{(4)}(q)\right]}\end{aligned} \tag{7.24}$$

where the kernel is

$$K_{\mathbf{k},\omega}(\mathbf{p},\mathbf{q}) = S^{(44)}(q)$$
$$\Bigg\{\bigg[\Big(S^{(34)}(|\mathbf{p}-\mathbf{q}|)+1\Big)\,\tilde{u}^{(344)}(\mathbf{p}-\mathbf{q},-\mathbf{p},\mathbf{q})$$
$$+S^{(34)}(|\mathbf{p}-\mathbf{q}|)\bigg]\left[\hbar\omega - \frac{(\mathbf{k}+\mathbf{p})\cdot(\mathbf{k}+\mathbf{q})}{p^2}t^{(3)}(p)\right]\Bigg\}$$
$$-S^{(34)}(|\mathbf{p}-\mathbf{q}|)\,\frac{\mathbf{p}\cdot\mathbf{q}}{p^2}\,\varepsilon^{(4)}(p)\,. \tag{7.25}$$

The triplet correlation function $\tilde{u}^{(344)}(\mathbf{p}-\mathbf{q},-\mathbf{p},\mathbf{q})$ of the ground state is also included in the formalism.

Again, note that the self–energy can become complex in case the denominator in Eq. (7.23) is positive for some value of $\mathbf{p}$,

$$\max_{\mathbf{p}}\left[\hbar\omega - t^{(3)}(\mathbf{k}+\mathbf{p}) - \varepsilon^{(4)}(p)\right] \;>\; 0\,. \tag{7.26}$$

If this condition is satisfied, then it is kinematically possible that the $^3$He impurity loses energy by emitting a phonon–roton mode $\varepsilon^{(4)}(p)$ while making a transition into a low–energy impurity mode $t^{(3)}(\mathbf{k}+\mathbf{p})$. The singularity structure of Eq. (7.24) is the same as that of the self–energy. The imaginary part of $\beta^{(34)}_{\mathbf{k},\omega}(\mathbf{p})$ is zero if the energy denominator is negative for all values of $\mathbf{q}$.

The elementary excitation modes of the impurity are the poles of the linear response function in Eq. (7.22)

$$\hbar\omega_0 - \frac{\hbar^2k^2}{2m_3} - \Sigma^{(3)}(k,\omega_0) = 0\,. \tag{7.27}$$

One of the formal results of our analysis is that the reference phonon–roton spectrum in the calculation of the self–energy becomes the Feynman–type spectrum, $\epsilon^{(4)}_F(k) = \hbar^2k^2/[2m_4S^{(4)}(k)]$. To improve from that we could follow the same path as in the one component system and replace the fluctuating pair correlation function with the pair distribution function using BGY–equations. That lowers the reference spectrum closer to the true spectrum. Since that has not yet been done a reasonable shortcut is to modify the integral equation (7.24) and the self–energy (7.23) by using the experimental phonon–roton spectrum in $\epsilon^{(4)}(k)$ and replacing the impurity mass by its effective mass in the impurity kinetic energy term, $t^{(3)}(k)$. Our result for the elementary excitation spectrum [64] is shown in Fig. 13 and compared with measurements. Our spectrum is slightly stiffer than the experiments, but the agreement is satisfactory for the whole momentum range.

The strength of the pole $Z(k)$ can be evaluated from the derivative of the self–energy,

$$Z(k) = \left[1 - \left.\frac{d\Sigma^{(3)}(k,\omega)}{d(\hbar\omega)}\right|_{\omega=\omega_0}\right]^{-1}\,. \tag{7.28}$$

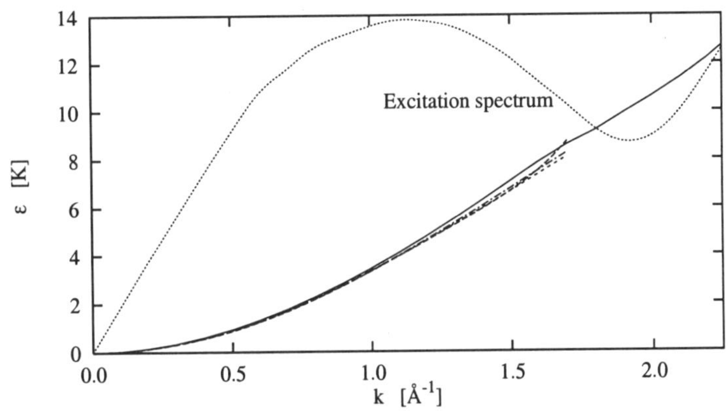


Fig. 13. The excitation spectrum of the $^3$He impurity. The solid curve is the result of the present theory. It is compared with the measurements by Greywall (Ref. 67) (long dashed line), Fåk *et al.* (Ref. 66) ( short dashed line) and Owers-Bradley *et al.* (Ref. 68) (dash-dotted line). The dotted line shows, for reference, the experimental phonon-roton spectrum.[45] (The original figure in Ref. 64.)

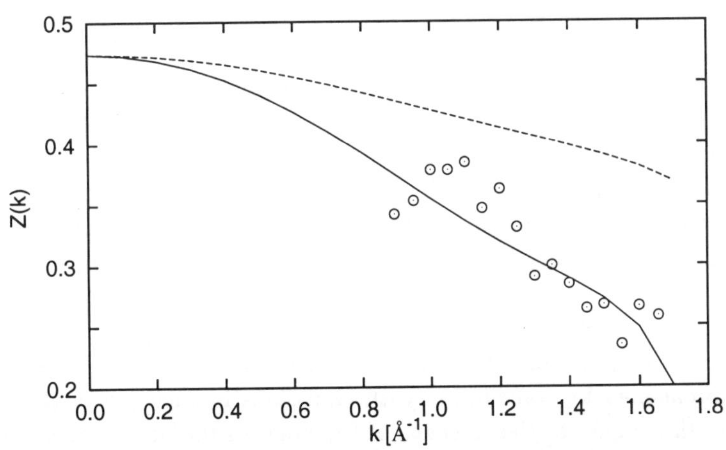


Fig. 14. The pole strength of the elementary impurity excitation mode is plotted as a function of momentum (solid line). The measured strength of the particle-hole excitation at 1% concentration and saturation vapor pressure from Fig. 11 of Ref. 66 is shown with circles. For comparison we also show the inverse of the effective mass as a function of momentum. (The original figure in Ref. 64.)

The results are plotted in Fig. 14. The long wavelength limit is inversely proportional to the effective mass as will be shown below and at about $k = 1.7$ Å$^{-1}$ the strength drops to zero because the mode crosses the boundary value where the decay becomes possible. The monotonically decreasing behavior with decreasing wavelength is partly due to the increase of the effective mass, also shown in the figure, and partly

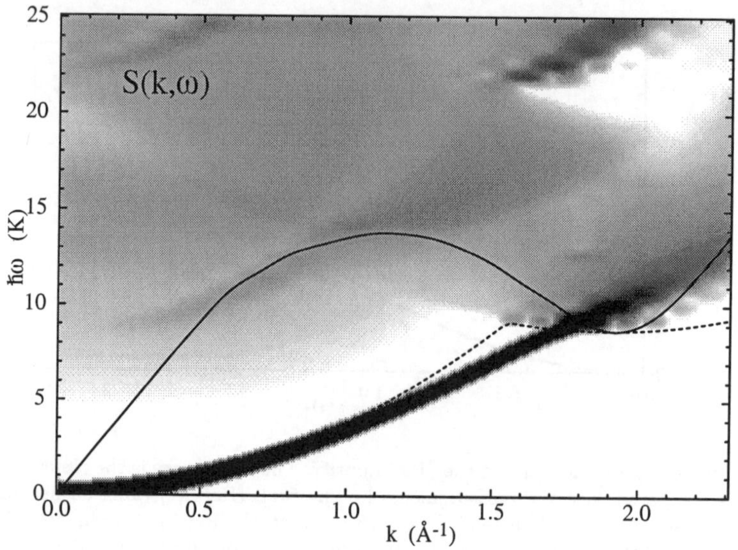


Fig. 15. The impurity dynamic structure function $S^{(3)}(k,\omega)$ plotted in the $k,\omega$ plane. Also shown are the phonon–roton spectrum of the background $^4$He (heavy solid line, the data are from 45), and the decay threshold of the impurity excitation mode (dashed line). (The original figure in Ref. 64.)

due to the increasing contribution of the phonon–impurity channels at higher frequencies. Fåk *et al.* [66] have measured the area under the particle–hole peak of the 1% concentration mixture for the momentum range $0.9\text{Å}^{-1} < k < 1.65\text{Å}^{-1}$. These results shown in Fig. 14 are in very good agreement with our calculations. The measurements with 5% concentration suggest only a weak concentration dependence and thus the comparison with the zero concentration limit calculations is justified.

In Fig. 15 we show the grey scale plot of the low energy part of the $S(k,\omega)$ for $\omega < 25\ K$. The pole contribution of the elementary excitation mode has been artificially broadened by convolution with Gaussians in such a way that the area under the peak is equal to the strength of the pole. In the figure we also show the experimental phonon–roton spectrum and the continuum boundary. The elementary excitation mode crosses that boundary at $1.72\,\text{Å}^{-1}$ and the decay into a roton becomes possible. The dominant peak in $S(k,\omega)$ crosses the phonon–roton spectrum and looses its strength to higher lying modes. The mode slightly above 20 K is clearly visible in the figure.

### 7.3. *Hydrodynamic effective mass*

In the long–wavelength limit of the elementary excitation mode both the single particle kinetic energy and the self–energy have quadratic behavior,

$$\Sigma^{(3)}(k,\omega_0) \overset{k\to 0}{\to} Ck^2 . \tag{7.29}$$

The coefficient $C$ in front of that defines the hydrodynamic effective mass $m_H^*$,

$$\hbar\omega_0 = \frac{\hbar^2 k^2}{2m_H^*}\,, \quad \text{when} \quad k \to 0\,. \tag{7.30}$$

Inserting this into Eq. (7.27) we get

$$\frac{m_H^*}{m_3} = \frac{1}{1-I} \tag{7.31}$$

with

$$I = \lim_{k\to 0+} \frac{1}{k^2} \int \frac{d^3p}{(2\pi)^3 \rho^{(4)}} \frac{\mathbf{k}\cdot\mathbf{p}\, S^{(34)}(p)\, \beta^{(34)}_{\mathbf{k},\omega_0}(\mathbf{p})}{t^{(3)}(p) + \varepsilon^{(4)}(p)}\,, \tag{7.32}$$

where $\omega_0 = \hbar k^2/2m_H^*$. Using Eqs. (7.28) and (7.30), we find that in the long–wavelength limit the pole strength is inversely proportional to the effective mass,

$$\lim_{k\to 0} Z(k) = \frac{m_3}{m_H^*}\,. \tag{7.33}$$

Our results for the effective mass are shown in Table 1 together with extrapolations to the single impurity limit of the measurements by Yorozu *et al.* [69] and Simons and Mueller [70].

Table 1. Pressure dependence of the hydrodynamic effective mass from various calculations and experiments. The second column contains the result of our microscopic calculation, the next two column contains the hydrodynamic effective mass as obtained from the fit to the experiments of Refs. 69 and 70.

| $P$ (atm) | $m_H^*/m$ | | |
|---|---|---|---|
| | This work | Ref. 69 | Ref. 70 |
| 0 | 2.09 | 2.18 | 2.15 |
| 5 | 2.22 | 2.31 | |
| 10 | 2.34 | 2.44 | 2.39 |
| 15 | 2.45 | 2.54 | |
| 20 | 2.55 | 2.64 | 2.62 |

A simple approximation for the second Euler equation is obtained by the *uniform limit approximation*[13] where one neglects all *coordinate–space* products of two functions, *i.e.* we approximate, for example,

$$\rho^{(34)}(\mathbf{r}_0,\mathbf{r}_1)\delta u^{(34)}(\mathbf{r}_0,\mathbf{r}_1) \approx \rho^{(3)}\rho^{(4)}\delta u^{(34)}(\mathbf{r}_0,\mathbf{r}_1)\,, \tag{7.34}$$

but *convolution products* are retained. Then, $\beta^{(34)}_{\mathbf{k},\omega_0}(\mathbf{p})$ has a simple form[62]

$$\beta^{(34)}_{\mathbf{k},\omega_0}(\mathbf{p}) = \frac{\hbar^2}{2m_3}\,\mathbf{k}\cdot\mathbf{p}\,\frac{S^{(34)}(p)}{S^{(44)}(p)}\,. \tag{7.35}$$

This, together with the equations (7.31) and (7.32), gives the "un–renormalized effective mass" derived by Owen.[65]

## 8. Summary

The variational Jastrow–Feenberg theory provides a powerful theoretical tool in obtaining microscopic understanding of the many–particle structure and processes present in strongly–correlated quantum fluids. A detailed derivation of the optimized ground state and the equations of motions for the dynamic system have been described. By extending the time dependence to two–particle correlations and, hence, including three–phonon processes we went beyond the conventional Feynman theory of excitations. First we showed that the CBF approximation gives qualitatively reasonable results for the dynamic structure, and then in the fully optimized approach we brought the results for the phonon–roton spectrum in liquid $^4$He into quantitative agreement with experimental data. We studied the analytic properties of the dynamic structure function and made a clear distinction between the real elementary–excitation mode and the broader complex mode, which can decay into elementary excitations. Liquid $^4$He displays anomalous dispersion at low momenta where the complex mode dominates; this is why phonons below 8 Kelvins decay rapidly into low–energy phonons. We also showed how the formalism allows us to obtain detailed information of the particle currents. In this way, we were able to shed light on the microscopic structure of the roton excitation. The softening of non–homogeneous excitations of the two–particle structure in the homogeneous liquid made it possible to study precursor of the liquid–solid phase transition and hexatic phases. Finally we applied the method to the one–impurity problem and calculated the excitation spectrum and dynamic structure of the $^3$He impurity in liquid $^4$He and derived its effective mass.

## Acknowledgments

I thank V. Apaja, J. Halinen, E. Krotscheck, J. Paaso and T. Taipaleenmäki for helping me in preparing this series of lectures. The work was supported, in part, by the Academy of Finland under project 163358.

## References

1. D. M. Ceperley, Rev. Mod. Phys. **67**, 279 (1995).
2. J. Boronat, in *Lecture notes in Physics: Microscopic Quantum Many-Body Theories and Their Applications*, edited by J. Navarro and A. Polls (Springer-Verlag, Berlin Heidelberg, 1998), Vol. 510, pp. 358-379.
3. S. Moroni, S. Conti and M. P. Tosi, Phys. Rev. B **53**, 9688 (1996).
4. A. F. G. Wyatt, M. A. H. Tucker, I. N. Adamenko, K. E. Nemchenko and A. V. Zhukov, Phys. Rev. B **62**, 9402 (2000).
5. H. Glyde, *Excitations in liquid and solid helium* (Oxford University Press, Oxford, 1994).
6. A. Bijl, Physica **7**, 869 (1940).
7. L. Landau, J. Phys. U.S.S.R. **5**, 71 (1941).
8. L. Landau, Phys. Rev. **60**, 356 (1941).
9. L. Landau, J. Phys. U.S.S.R. **11**, 91 (1947).

10. N. N. Bogoliubov, J. Phys. U.S.S.R **9**, 23 (1947).
11. R. P. Feynman, Phys. Rev. **94**, 262 (1954).
12. R. P. Feynman and M. Cohen, Phys. Rev. **102**, 1189 (1956).
13. E. Feenberg, *Theory of Quantum Fluids* (Academic, New York, 1969).
14. J. W. Clark, in *Progress in Particle and Nuclear Physics*, edited by D. H. Wilkinson (Pergamon Press Ltd., Oxford, 1979), Vol. 2, pp. 89–199.
15. H. W. Jackson and E. Feenberg, Ann. Phys. (NY) **15**, 266 (1961).
16. H. W. Jackson and E. Feenberg, Rev. Mod. Phys. **34**, 686 (1962).
17. C. C. Chang and C. E. Campbell, Phys. Rev. B **13**, 3779 (1976).
18. C. E. Campbell, in *Progress in Liquid Physics*, edited by C. A. Croxton (Wiley, London, 1978), Chap. 6, pp. 213–308.
19. S. Manousakis and V. R. Pandharipande, Phys. Rev. B **30**, 5062 (1984).
20. L. Reatto, S. A. Vitiello, and G. L. Masserini, J. Low Temp. Phys. **93**, 879 (1993).
21. S. Moroni, L. Reatto, and S. Fantoni, Czech. J. Phys. Suppl. S1 **46**, 281 (1996).
22. J. Boronat and J. Casulleras, Europhys. Lett. **38**, 291 (1997).
23. V. L. Berezinskiǐ, Sov. Phys. JETP **32**, 493 (1971).
24. J. M. Kosterlitz and D. J. Thouless, J. Phys. C **6**, 1181 (1973).
25. R. J. Donnelly, *Quantized Vortices in Helium II* (Cambridge University Press, Cambridge, 1991).
26. M. Saarela, B. E. Clements, E. Krotscheck, and F. V. Kusmartsev, J. Low Temp. Phys. **93**, 971 (1993).
27. Saarela and F. V. Kusmartsev, Phys. Lett. A **202**, 317 (1995).
28. M. Saarela, Phys. Rev. B **33**, 4596 (1986).
29. M. Saarela and J. Suominen, in *Condensed Matter Theories*, edited by J. S. Arponen, R. F. Bishop, and M. Manninen (Plenum, New York, 1988), Vol. 3, pp. 157–165.
30. J. Suominen and M. Saarela, in *Condensed Matter Theories*, edited by J. Keller (Plenum, New York, 1989), Vol. 4, p. 377.
31. B. E. Clements *et al.*, Phys. Rev. B **50**, 6958 (1994).
32. B. E. Clements, E. Krotscheck, and C. J. Tymczak, Phys. Rev. B **53**, 12253 (1996).
33. A. D. B. Woods and R. A. Cowley, Rep. Prog. Phys. **36**, 1135 (1973).
34. V. Apaja *et al.*, Phys. Rev. B **55**, 12925 (1997).
35. A. K. Kerman and S. E. Koonin, Ann. Phys. (NY) **100**, 332 (1976).
36. P. Kramer and M. Saraceno, *Geometry of the time-dependent variational principle in quantum mechanics*, Vol. 140 of *Lecture Notes in Physics* (Springer, Berlin, Heidelberg, and New York, 1981).
37. B. Fåk and J. Bossy, J. Low Temp. Phys. **112**, 1 (1998).
38. C. E. Campbell, Phys. Lett. A **44**, 471 (1973).
39. C. C. Chang and C. E. Campbell, Phys. Rev. B **15**, 4238 (1977).
40. E. Krotscheck, Phys. Rev. B **33**, 3158 (1986).
41. H. W. Jackson, Phys. Rev. A **9**, 964 (1974).
42. F. Dalfovo and S. Stringari, Phys. Rev. B **46**, 13991 (1992).
43. J. Halinen, V. Apaja, and M. Saarela, J. Low Temp. Phys. (2002), in press.
44. A. Gold, Z. Phys. B **89**, 1 (1992).
45. R. A. Cowley and A. D. B. Woods, Can. J. Phys. **49**, 177 (1971).
46. H. N. Robkoff and R. B. Hallock, Phys. Rev. B **24**, 159 (1981).
47. E. H. Graf, V. J. Minkiewicz, H. B. Möller, and L. Passell, Phys. Rev. A **10**, 1748 (1974).
48. O. W. Dietrich, E. H. Graf, C. H. Huang, and L. Passell, Phys. Rev. A **5**, 1377 (1972).
49. V. Apaja and M. Saarela, Phys. Rev. B **57**, 5358 (1998).
50. E. Krotscheck and M. Saarela, Physics Reports **232**, 1 (1993).

51. R. J. Donnelly, J. A. Donnelly, and R. N. Hills, J. Low Temp. Phys. **44**, 471 (1981).
52. A. D. B.Woods, P. A. Hilton, R. Scherm, and W. G. Stirling, J. Phys. C **10**, L45 (1977).
53. E. C. Svensson, R. Scherm, and A. D. B.Woods, J. Phys. C **39**, 6 (1978).
54. V. Apaja, J. Halinen, and M. Saarela, J. Low Temp. Phys. **113**, 909 (1998).
55. V. Apaja, J. Halinen, and M. Saarela, Physica B **284–288**, 29 (2000).
56. J. Halinen, V. Apaja, K. A. Gernoth, and M. Saarela, J. Low Temp. Phys. **121**, 531 (2000).
57. P. R. Roach, J. B. Ketterson, and C. W. Woo, Phys. Rev. A **2**, 543 (1970).
58. D. O. Edwards and R. C. Pandorf, Phys. Rev. A **140**, 816 (1965).
59. S. Giorgini, J. Boronat, and J. Casulleras, Phys. Rev. B **54**, 6099 (1996).
60. P. A. Whitlock, G. V. Chester, and M. H. Kalos, Phys. Rev. B **38**, 2418 (1988).
61. M. C. Gordillo and D. M. Ceperley, Phys. Rev. B **58**, 6447 (1998).
62. M. Saarela and E. Krotscheck, J. Low Temp. Phys. **90**, 415 (1993).
63. M. Saarela, in *Recent Progress in Many Body Theories*, edited by Y. Avishai (Plenum, New York, 1990), Vol. 2, pp. 337–346.
64. E. Krotscheck *et al.*, Phys. Rev. B **58**, 12282 (1998).
65. J. C. Owen, Phys. Rev. B **23**, 5815 (1981).
66. B. Fåk *et al.*, Phys. Rev. B **41**, 8732 (1990).
67. D. S. Greywall, Phys. Rev. B **20**, 2643 (1979).
68. J. R. Owers-Bradley *et al.*, J. Low Temp. Phys. **72**, 201 (1988).
69. S. Yorozu, H. Fukuyama, and H. Ishimoto, Phys. Rev. B **48**, 9660 (1993).
70. R. Simons and R. M. Mueller, Czekoslowak Journal of Physics Suppl. **46**, 201 (1996).

## CHAPTER 7

# THEORY OF CORRELATED BASIS FUNCTIONS

E. Krotscheck

*Institute for Theoretical Physics, Johannes Kepler University*
*A-4040 Linz, Austria*
*E-mail: Eckhard.Krotscheck@jku.at*

We describe the development of perturbation theory in a basis of correlated wave functions. The formulation of the basic theory is generic in the sense that it does not depend on a specific choice of correlation operator. We will, however, elaborate on the intricate relationships between cluster expansions for Jastrow–Feenberg wave functions on the one side, and perturbative expansions within the basis of Jastrow–Feenberg generated states.

The first part of this lecture series sets the general scene of perturbation theroy in a non–orthogonal basis, as laid out by Feshbach, Feenberg, Clark, and others. We then turn to the calculation of effective interactions, which will require the same technical tools that were developed for the Fermi-Hypernetted-Chain theory of the ground state. We then look at the interpretation of these effective interactions from various points of view and draw the connections to phenomenological theories like Landau's Fermi liquid theory, or the BCS theory of superconductivity. We also demonstrate that Fermi–sea averages of the CBF effective interaction vanish for optimized correlation functions.

Finally, we turn to cases where CBF theory to infinite order is necessary. Employing ideas of the coupled cluster theory, we develop a method for generating CBF perturbative corrections to infinite order by integral equations. Specializing to ring diagrams, we demonstrate that the "chain" diagrams of the FHNC theory are approximations for the ring diagrams of (CBF-)perturbation theory. The generalization of the time–dependent Hartree–Fock theory to correlated states will allow the formulation of a linear response theory for strongly interacting systems or, alternatively, provide the tools for mapping strong, bare interactions onto weak, effective interactions that can be used in phenomenological theories.

## 1. Introduction

This series of lectures describes the basics of correlated basis functions (CBF) theory, [1] develops its techniques, and discusses its relationship to variational ground state methods as well as to semi–phenomenological theories like Landau's Fermi–Liquid theory, [2,3] the BCS theory of superconductivity, [4,5] and the time–dependent Hartree–Fock approximation. [6]

Arturo Polls' lectures [7] were mostly concerned with the evaluation of physical properties of many–particle systems for a variational wave function chosen in a wide sense optimally within a broad class of trial functions. These discussions were based on an *ansatz* for the ground state wave function of a correlated $N$–body Fermi system:

$$\Psi_o^{(N)}(1,\ldots,N) = F_N(1,\ldots,N)\Phi_o(1,\ldots,N)\,. \tag{1.1}$$

Typically, one has for the *correlation operator* $F_N(1,\ldots,N)$ the Jastrow–Feenberg *ansatz*

$$F_N(\mathbf{r}_1,\ldots,\mathbf{r}_N) = \exp\frac{1}{2}\left[\sum_i u_1(\mathbf{r}_i) + \sum_{i<j} u_2(\mathbf{r}_i,\mathbf{r}_j) + \ldots\right], \tag{1.2}$$

in mind, and for the *model wave function* $\Phi_o(1,\ldots,N)$ either a constant for bosons, or a Slater determinant of plane waves for fermions. The Jastrow–Feenberg *ansatz* for the ground–state wave function is in principle exact for Bose systems. It is true that a *truncated* wave function employed in the Bose problem is not an exact one, but treatment at the level of three–body correlations has often been found to be sufficiently accurate.

The situation is different in the case of Fermi systems, where the local Jastrow–Feenberg *ansatz* (1.2) for the ground state has deficiencies. A most evident one is the "nodal surface" problem: The nodes of the wave function (1.1), (1.2) are the same as the nodes of the model wave function $\Phi_o(1,\ldots,N)$; but there is no reason that the exact ground state should have the same feature. The problem has many more subtle facets: The energy of a Fermi system may be considered to arise from a sum over all single–particle momentum states. Hence the two–body factor $f(r_{ij})$ appearing in the Jastrow product, being independent of momentum, cannot describe the ideal correlations of a pair of particles in given momentum states, but instead makes compromises to yield the best upper bound for the ground–state energy. The most important information missing from the Jastrow–Feenberg *ansatz* for the ground state is the explicit reference to where a given correlated particle is located within the Fermi sea. Such a treatment is perhaps acceptable for integrated quantities like the ground state energy of a Fermi liquid. However, it is not clear at all whether it is also appropriate in situations where the active particles are close to the Fermi surface and their specific correlations are essential for the determination of physical properties like the specific heat, or the magnetic susceptibility. To examine this question, it becomes necessary to go beyond a description based on a single Jastrow–correlated wave function.

Our qualitative understanding of the excitations of a quantum liquid is largely guided by results from weakly–interacting models. We therefore aim to design a theory that combines the strengths of variational theory and perturbation theories and provides a straightforward and transparent approach to full microscopic understanding of a quantum liquid.

The method to be described here is collectively known as the theory of *correlated basis functions* (CBF–theory). The successful execution of the theory requires a number of steps which will be discussed systematically. After the discussion of the basic ideas, we need to develop methods for calculating the basic ingredients of the theory. Since we will be working with Jastrow–Feenberg wave functions, we must rely heavily on the techniques that have been developed for the ground state theory, but must extend these methods to a variety of different quantities. An important aspect will be the *interpretation* of the ingredients in terms of physical observables and effects as well as the technical building blocks of other theories. These interpretations will also lead us to a better understanding of the approximations implicit to the Jastrow–Feenberg method.

We finally proceed to extend correlated basis functions theory to infinite order. A few words are in order why we would like to do that because we will first argue that "the strength of CBF theory is the weakness of its effective" interactions. Therefore, one might be led to believe that finite–order methods should be sufficient.

There are both formal and practical reasons for why one must go to infinite order CBF theory. The formal reason is that one would like to interpret the diagrammatic content of the (F)HNC–EL theory in terms of Goldstone or Feynman diagrams. Intuitively, one expects that specific (sets of) (F)HNC diagrams are approximations for specific (sets of) diagrams generated by ordinary perturbation theory, but such an intuitive interpretation must be justified.

Evidently, only an optimized theory is useful for the analysis we have in mind since the optimization defines a unique relationship between the correlation function(s) and the underlying microscopic interaction. One should roughly expect the following: If we accept the notion that the (sets of) diagrams of the (F)HNC–EL theory are approximations for (sets of) Goldstone or Feynman diagrams, then CBF perturbation theory should systematically eliminate these approximations, and replace the (F)HNC diagrams by corresponding diagrams of the perturbation series. A consequence is that approximate (F)HNC and the corresponding exact Goldstone diagrams should have some common features, like the same momentum flux, by which corresponding classes can be identified.

The most complete analysis of the resemblance between HNC and Goldstone diagrams has been carried out in the work on parquet–diagrams [8,9,10] for boson liquids. There, it was shown that the HNC–EL theory sums all "ring"– and all "ladder"–diagrams exactly. Mixed diagrams are calculated in a local approximation, whereby all diagrams are calculated exactly to third order in the bare interaction. When optimized triplet correlations are included, all diagrams to fourth order in the potential are dealt with exactly [11].

The situation is significantly more complicated for fermions, which are the subject of these lectures. In essence, *no* Goldstone diagram beyond the Hartree–Fock approximation is treated exactly by the Jastrow–Feenberg wave function. One should both understand the formal content of the approximation as well as its quantitative importance.

There is also a practical reason for why we want to examine infinite order CBF theory. There is a certain limit to which the variational method has advantages: It relies on the possibility of carrying out summations of large classes of diagrams. This is due to the simplicity of the Jastrow–Feenberg correlation functions. If this simplicity is lost, and the actual summations of high–order diagrams become too complicated, then there is no point in writing down a variational wave function. Rather, one should right away sum the correct diagrams of perturbation theory.

## 2. Basics of CBF theory

### 2.1. *Motivations, basic concepts, and definitions*

The objective of this section is to introduce and analyze the basic concepts, ingredients, and machinery involved in applying CBF theory at *finite* perturbative order. The very simple idea from which the apparatus of correlated basis functions (CBF) theory originates is that one may return to the definition (1.1) of the correlated wave function and use the correlation operator $F(1,\ldots,N)$ to generate not only a model for the ground state wave function, but a *complete* set of basis functions

$$\left|m\right\rangle \equiv \left|\Psi_m\right\rangle = I_{mm}^{-\frac{1}{2}} F(1,\ldots,N)\left|\Phi_m\right\rangle \tag{2.1}$$

for the Hilbert space.[a] The basis states are normalized by

$$I_{mm} \equiv \left\langle\Phi_m\right|F^\dagger(1,\ldots,N)F(1,\ldots,N)\left|\Phi_m\right\rangle . \tag{2.2}$$

We denote by $\left|o\right\rangle$ the correlated basis state that is created by acting with the correlation operator on the filled Fermi sea:

$$\left|o\right\rangle \equiv \left|\Psi_o\right\rangle = I_{oo}^{-\frac{1}{2}} F_N\left|\Phi_o\right\rangle . \tag{2.3}$$

The procedures for doing actual calculations in the correlated basis are collectively known as the method of correlated basis functions (CBF). The essential aim of CBF theory is to treat those effects, that are poorly described by the Jastrow–Feenberg wave function, in ways that do not sacrifice the advantages of the diagrammatic richness and the calculational simplicity that characterize the Jastrow–Feenberg variational theory.

A conspicuous and somewhat awkward feature of this basis is that the states, while normally forming a complete set, are not orthogonal. The physical interpretation of the basis formed by the correlated states (2.1) hinges largely on the choice of the correlation operator $F$. Ideally, $F$ should map the uncorrelated states $\left|\Phi_m\right\rangle$ onto the true excited states of the system. Such a mapping is certainly beyond practical reach, but we can still use the flexibility in the correlation operator to build physical intuition into the basis used to describe the physical system. This perspective stands in contrast to traditional many–body theory, where only the bulk geometry (*e.g.* translational invariance) and particle statistics enter into the choice of basis.

[a] As long as no confusion can arise, we shall omit reference to the particle number $N$.

In summary, the pragmatic strategy guiding the correlated basis functions approach is the following: One adopts a basis for the Hilbert space that embodies essential aspects of the physics of the given, nontrivial real–world many–body system that would not be present in the basis functions conventionally chosen for a noninteracting or weakly interacting system. Most prominently, we have here the core–exclusion in mind: For a strongly interacting many–body system, *all* matrix elements of the bare Hamiltonian with plane wave orbitals would be infinite, whereas all matrix elements with the correlated states (2.1) are finite. To make use of this strategy, one must, of course, have practical ways for calculating the essential ingredients of the theory with sufficient accuracy such that one does not sacrifice the accuracy gained in the ground state calculations. In practice, we will see that this means one should *optimize* the ground state correlations which essentially singles out the Jastrow–Feenberg correlation operator as the only practical choice.

For the purpose of notation, we introduce a "second–quantized" formulation of the correlated states. Creation operators ($\alpha_p^\dagger$) and annihilation operators ($\alpha_h$) of correlated states can be defined by their action on the basis states $|m\rangle$:

$$|\alpha_k^\dagger\, m\rangle = F_{N+1}a_k^\dagger|\Phi_m\rangle/\langle\Phi_m|a_k F_{N+1}^\dagger F_{N+1}a_k^\dagger|\Phi_m\rangle^{\frac{1}{2}} \tag{2.4}$$

$$|\alpha_k\, m\rangle = F_{N-1}a_k|\Phi_m\rangle/\langle\Phi_m|a_k^\dagger F_{N-1}^\dagger F_{N-1}a_k|\Phi_m\rangle^{\frac{1}{2}}\,. \tag{2.5}$$

According to these definitions, the correlated operators obey the same (anti–) commutation rules as their uncorrelated cousins, *but they are not Hermitian conjugates.* If $|m\rangle$ is an $N$–particle state, then the state in Eq. (2.4) must carry an $(N+1)$-particle correlation operator, while that in Eq. (2.5) must be formed with an $(N-1)$–particle correlation operator.

It will sometimes be convenient to label correlated states by the orbitals in which the corresponding model state, $|\Phi_m\rangle$, differs from the model ground state $|\Phi_o\rangle$. For example, we may write

$$\begin{aligned} |p\rangle &= |\alpha_p^\dagger\rangle = I_{p,p}^{-\frac{1}{2}} F_{N+1}a_p^\dagger|\Phi_o\rangle, \\ I_{p,p} &= \langle\Phi_o|a_p F_{N+1}^\dagger F_{N+1}a_p^\dagger|\Phi_o\rangle\,, \end{aligned} \tag{2.6}$$

where $a_p^\dagger$ is the familiar creation operator for an uncorrelated particle described by the quantum numbers $p$.

A number of prerequisites are necessary for the application of CBF theory, and these notes will build these prerequisites step–by–step:

- First, we need to set up the general scheme of doing perturbation theory in a non–orthogonal basis.
- Second, we must develop the tools for the calculation of the necessary ingredients, *i.e.* the diagonal and off–diagonal matrix elements of the Hamiltonian as well as overlap integrals between different correlated states.
- Third, we should seek *interpretations* of the basic ingredients of CBF theory in terms of physical observables.

- Fourth, we should examine cases where it might be necessary to carry CBF theory to infinite order. A prominent example is seen in the the coupling between single–particle and collective excitations in a Fermi liquid.
- Finally, we should take the "view from the top" and try to interpret Jastrow–Feenberg correlations in terms of ordinary perturbation theory. Consideration of these aspects will lead us to much better understanding of the physical significance, the strengths, and the weaknesses of Jastrow–Feenberg correlations, while furnishing the means to correct for these weaknesses.

As an introductory exercise, let us examine the change in energy of a Fermi liquid upon adding/removing a particle of momentum $p/h$. Consider the energy difference between the variational ground–state expectation value

$$H_{oo} = \langle o|H|o\rangle \tag{2.7}$$

and the energy obtained by inserting/removing a particle of momentum $p/h$ into/from the system, leaving the correlations unaffected: For any correlated $N+1$ (or $N-1$)–particle state

$$|p\rangle = |\alpha_p^\dagger\, o\rangle \tag{2.8}$$

$$|h\rangle = |\alpha_h\, o\rangle \tag{2.9}$$

define the correlated single particle spectrum as the energy difference

$$e_p = H_{p,p} - H_{o,o} = \langle p|\hat{H}|p\rangle - \langle o|H|o\rangle \tag{2.10}$$

$$e_h = H_{h,h} - H_{o,o} = \langle h|\hat{H}|h\rangle - \langle o|H|o\rangle. \tag{2.11}$$

We also can introduce particle–hole excitation energies

$$e_{ph} = \langle o\,\alpha_h^\dagger\alpha_p|\hat{H}|\alpha_p^\dagger\alpha_h\, o\rangle - H_{oo} = e_p - e_h + \mathcal{O}(1/N) \tag{2.12}$$

as the energy difference between a (correlated) $1p-1h$ excited state and the ground state. The term of the order $\mathcal{O}(1/N)$ will later be identified with the quasiparticle interaction.

The above examples involving correlated single–particle energies are illustrative of the role played by diagonal CBF matrix elements of the Hamiltonian

$$H_{mm} \equiv \langle m|\hat{H}|m\rangle\,. \tag{2.13}$$

These, together with off–diagonal matrix elements of the Hamiltonian and of the unit operator, $H_{mn}$ and $N_{mn}$

$$\begin{aligned} H_{mn} &\equiv \langle m|\hat{H}|n\rangle\,, \\ N_{mn} &\equiv \langle m \mid n\rangle\,, \end{aligned} \tag{2.14}$$

constitute the other essential building blocks of CBF theory. However, off–diagonal elements of $\hat{H}$ in the correlated basis, needed in perturbation theory and in the description of transitions, do not occur alone, but rather in combination with the corresponding matrix elements of the unit operator, assembled in the form

$$H'_{mn} \equiv \langle m|\hat{H} - H_{oo}|n\rangle \qquad (m \neq n)\,. \tag{2.15}$$

The diagonal matrix elements of the unit operator are just 1; it is therefore also useful to define the generically off–diagonal quantity

$$J_{mn} = \langle m \mid n\rangle \qquad (m \neq n)\,. \tag{2.16}$$

## 2.2. *Finite–order correlated basis functions theory*

Standard time–independent perturbation schemes for an orthonormal basis include the familiar Rayleigh–Schrödinger (RS) perturbation theory encountered in most textbooks, along with Brillouin-Wigner (BW) perturbation theory and the less well known Feenberg–Feshbach (FF) perturbation theory. [12,13] The RS theory consists of pure expansions of the wave function and the energy in a small parameter associated with the interaction, while in BW theory the energy being sought appears in all the energy denominators, introducing an element of self–consistency. The FF perturbation formulas, involving a continued fraction structure, are even more "highly summed" or "renormalized" than their BW counterparts. Such renormalizations, useful as they may be in some problems, come at a sacrifice. Among the three approaches under consideration, only the RS theory, from which it is possible to squeeze out a linked–cluster property, shows acceptable behavior of its energy approximants in the thermodynamic limit. Accordingly, it is natural to seek an analog of Rayleigh–Schrödinger theory for a non–orthogonal basis. Such a scheme may be derived in a number of ways. Although not fully explicated and rigorous at the level of Goldstone diagrammatic perturbation theory expansion, all known derivations [1,13–18] lead to fully consistent results.

Consider the matrix version of the Schödinger equation for an eigenstate $|\Psi\rangle$ with energy eigenvalue $E$. After projecting this equation on an arbitrary state $|m\rangle$ of the non–orthogonal basis,

$$\langle m|H - E|\Psi\rangle = 0\,, \tag{2.17}$$

we substitute the expansion

$$|\Psi\rangle = \sum_n c_n|n\rangle \tag{2.18}$$

to obtain

$$\sum_n (H_{mn} - EN_{mn})c_n = 0\,. \tag{2.19}$$

The set (2.19) of homogeneous, linear, algebraic equations for the expansion coefficients $c_n$ has a nontrivial solution if and only if the energy eigenvalue $E$ satisfies

$$\det(H_{mn} - EN_{mn}) = 0\,, \tag{2.20}$$

which is just the secular equation.

Assuming a finite number $K$ of basis states, straightforward manipulations yield $K$ solutions of Eqs. (2.20) for the coefficients $c_n$ which determine in turn the eigenvalues $E$. For $E$, this is equivalent to expanding out the secular determinant. Using induction to extend the results to $K = \infty$, infinite series are generated for the $c_n$ and corresponding $E$, which, however, are not useful for large $N$. Next, all terms are expanded formally in a "small" parameter $\lambda$ associated with the off–diagonal quantities $H'_{mn}$ of Eq. (2.15) and the resulting contributions are rearranged according to their order in $\lambda$. At appropriate stages of this process, we arrive at analogs of Feenberg–Feshbach and Brillouin–Wigner perturbation theories, and, finally, the Rayleigh–Schrödinger expansions. With some effort, one can show that the latter formulas are useful in the thermodynamic limit. [20]

Explicitly to fourth order in the perturbative strength parameter $\lambda$, the ground–state energy eigenvalue is given by the expansion

$$
\begin{aligned}
E = H_{oo} &- {\sum_m}' \frac{H'_{om}H'_{mo}}{H_{mm}-H_{oo}} + {\sum_{mn}}' \frac{H'_{om}H'_{mn}H'_{no}}{(H_{mm}-H_{oo})(H_{nn}-H_{oo})} \\
&+ {\sum_{mnp}}' \frac{H'_{om}H'_{mn}H'_{np}H'_{po}}{(H_{mm}-H_{oo})(H_{nn}-H_{oo})(H_{pp}-H_{oo})} \\
&+ {\sum_{mn}}' \left[ \frac{H'_{om}H'_{mo}H'_{on}H'_{no}}{(H_{mm}-H_{oo})^2(H_{nn}-H_{oo})} \right. \\
&\left. - \frac{J_{om}H'_{mo}H'_{on}H'_{no}}{(H_{mm}-H_{oo})(H_{nn}-H_{oo})} + \frac{H'_{om}J_{mo}H'_{on}H'_{no}}{(H_{mm}-H_{oo})(H_{nn}-H_{oo})} \right] . \qquad (2.21)
\end{aligned}
$$

The prime on the summations in this expression imply that no summation index can equal the ground–state index $o$. Accordingly, no energy denominator appearing in the series can vanish, provided that $H_{mm} \neq H_{oo}$, all $m \neq 0$.

The first and second addends on the right of Eq. (2.21) are the corrections of second and third order to the variational estimate $H_{oo}$ of the ground–state energy corresponding to the "ground" basis state $|o\rangle$. The terms within the square brackets make up the fourth–order correction. No first–order correction appears, since the unperturbed energies are taken to be the diagonal matrix elements $H_{qq} = \langle q|H|q\rangle$ of the *full* Hamiltonian. In this sense we do not have the strict analog of the conventional RS expansion, but instead have an RS theory with trivially renormalized energy denominators.

## 3. Techniques for matrix elements

### 3.1. *Definitions and notations*

The application of the theory of correlated basis functions relies heavily on the availability of techniques for the calculation of the required combinations of the matrix elements $H_{mn}$ and $N_{mn}$ introduced in Sect. 2. We now turn to the development of techniques that will provide the raw material for the application of the correlated–basis functions formalism, that is the derivation of cluster expansions and integral equation techniques for the off–diagonal matrix elements introduced in Sects. 2.1 and 2.2. For this purpose, we must specify our choice of the the correlation operator $F$.

A simple counting argument can be made for why advanced diagrammatic methods are more important for off–diagonal than for diagonal matrix–elements: From each $n$–body diagram contributing to the energy, one can construct $n(n-1)/2$ non–nodal diagrams with two external points. The construction of the off–diagonal matrix elements involves all these non–nodal diagrams, and additional nodal diagrams. There is therefore every reason to expect that the convergence of any cluster expansion for an off–diagonal quantity is much slower than that for a diagonal quantity, and that advanced integral equation techniques are important for off–diagonal matrix elements, whenever they are important for the variational energy expectation value. Hypernetted–chain and optimization methods are available for the Jastrow–Feenberg correlation operator 1.2 up to triplet correlations, but for the sake of discussion we restrict ourselves here to the Jastrow choice operator

$$F(\mathbf{r}_1,\ldots,\mathbf{r}_N) = \prod_{1\le i<j\le N} f(\mathbf{r}_i,\mathbf{r}_j) \equiv \exp\frac{1}{2}\left[\sum_{1\le i<j\le N} u_2(\mathbf{r}_i,\mathbf{r}_j)\right]. \tag{3.1}$$

Moreover, in *homogeneous* and *isotropic* systems under consideration here we have $u_1(\mathbf{r}) = 0$ and $u_2(\mathbf{r}_i,\mathbf{r}_j) = u_2(|\mathbf{r}_i - \mathbf{r}_j|)$.

A quantity that will be used repeatedly is the *generating functional*

$$G_{oo} = \ln I_{oo}. \tag{3.2}$$

The generating functional is represented diagrammatically as the sum of *all* irreducible diagrams that can be constructed from "correlation lines" $f^2(r_{ij}) - 1$ and "exchange lines" ("Slater functions") $\ell(r_{ij}k_F)$ defined in Polls' lectures. [7] We show, for further reference, a few leading diagrams of $G_{oo}$ in Fig. 1. The expansion shown in Fig. 1 is, in fact, valid for the cluster expansion of *any* generating $G_{mm} = \ln I_{mm}$ built upon a model state $|\Phi_m\rangle$, if we redefine the exchange function to

$$\ell_m(\mathbf{r}) = \frac{1}{N}\sum_{i\in m} e^{i\mathbf{k}_i\cdot\mathbf{r}}. \tag{3.3}$$

The only property that is needed for proving the irreducibility of the expansion is the *convolution property* of the exchange function,

$$\ell_m(|\mathbf{r}_1 - \mathbf{r}_2|) = \frac{\rho}{\nu}\int d^3r_3 \ell_m(|\mathbf{r}_1 - \mathbf{r}_3|)\ell_m(|\mathbf{r}_3 - \mathbf{r}_2|k_F). \tag{3.4}$$

Matrix elements between different correlated basis states $|m\rangle$ and $|n\rangle$ are conveniently classified according to the number $d$ of orbitals in which $|m\rangle$ and $|n\rangle$ differ. We will perform the explicit construction of the matrix elements in this section for the cases $d = 0$, and in the next for $d = 2$.

The program set forth in the two preceeding sections requires the calculation of the following quantities:

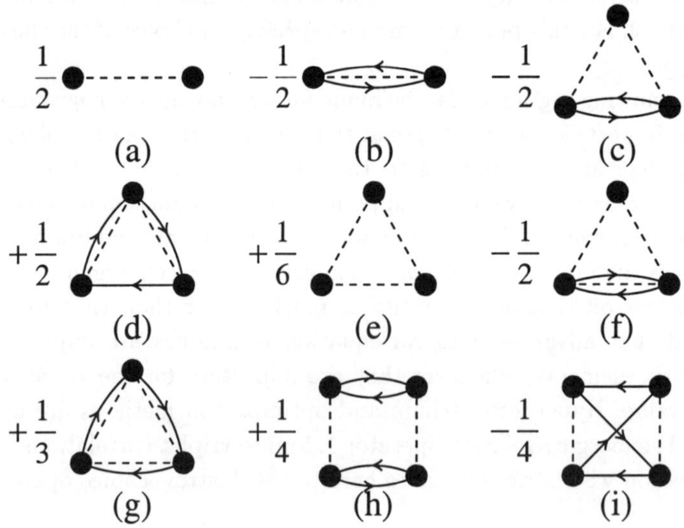


Fig. 1. Diagrammatic representation of all second and third order, as well as some fourth order diagrams of the generating function $G_{oo}$.

(1) Differences of the generating functionals

$$G_{mm} - G_{oo} = \ln\left[I_{mm}/I_{oo}\right] , \tag{3.5}$$

(2) Differences of diagonal matrix elements of the Hamiltonian

$$H_{mm} - H_{oo} , \tag{3.6}$$

(3) The off–diagonal matrix elements (2.16) of the unit operator

$$J_{mn} = \langle m \mid n \rangle \qquad (m \neq n) , \tag{3.7}$$

(4) Off–diagonal matrix elements of the Hamiltonian. These appear in the combinations (2.15)

$$H'_{mn} = \langle m | \hat{H} - H_{oo} | n \rangle \qquad (m \neq n) , \tag{3.8}$$

and

$$W_{mn} = H'_{mn} - \frac{1}{2}\left[H_{mm} + H_{nn} - 2H_{oo}\right] J_{mn} . \tag{3.9}$$

We can utilize the "prime–equation" technique introduced in and Saarela's lectures [19] to formally calculate the matrix elements of the Hamiltonian from the matrix elements of the unit operator. The "Jackson–Feenberg–identity"

$$F\nabla^2 F = \frac{1}{2}(\nabla^2 F^2 + F^2\nabla^2) + \frac{1}{2}F^2\left[\nabla, [\nabla, \ln F]\right] - \frac{1}{4}\left[\nabla, [\nabla, F^2]\right] \tag{3.10}$$

allows us to rewrite any (off–)diagonal matrix element of the kinetic energy as

$$\langle \Phi_m | F\hat{T}F | \Phi_n \rangle = -\frac{\hbar^2}{2m} \langle \Phi_m | F \sum_i \nabla_i^2 F | \Phi_n \rangle$$

$$= \frac{1}{2}\left(T_m + T_n\right)\langle \Phi_m | F^2 | \Phi_n \rangle - \frac{\hbar^2}{4m}\langle \Phi_m | F^2 \sum_{i<j} \nabla^2 u_2(\mathbf{r}_i - \mathbf{r}_j) | \Phi_n \rangle$$

$$+ \frac{\hbar^2}{8m}\langle \Phi_m | \sum_i \left[\nabla_i, \left[\nabla_i, F^2\right]\right] | \Phi_n \rangle , \tag{3.11}$$

where $T_m$ and $T_n$ are the kinetic energies of the model states $|\Phi_m\rangle$ and $|\Phi_n\rangle$, respectively. The second term on the right hand side has the same structure as the matrix element of the potential energy. We define a generalized pair correlation function

$$u_2(r_{ij};\beta) \equiv u_2(r_{ij}) + \beta \left[ v(r_{ij}) - \frac{\hbar^2}{4m}\nabla^2 u_2(r_{ij}) \right] \tag{3.12}$$

and consider all quantities as functionals of $u_2(r_{ij};\beta)$. Technically, the derivations of cluster expansions for the generating functions and overlap integrals for $u_2(r_{ij};\beta)$ are identical to those for $u_2(r_{ij})$. The matrix elements of the Hamiltonian can, at the end, be obtained by

$$H_{mm} - H_{oo} = T_F^{(m)} - T_F^{(o)} + \frac{\partial}{\partial\beta}\left[G_{mm}(\beta) - G_{oo}(\beta)\right]\Big|_{\beta=0} \tag{3.13}$$

$$W_{mn} = \frac{\partial}{\partial\beta} J_{mn}(\beta). \tag{3.14}$$

where the $T_F^{(m)}$ and $T_F^{(o)}$ are the kinetic energies of non–interacting fermions characterized by model states $\Phi_m$ or $\Phi_o$, respectively. That way, all the required quantities can be derived from cluster expansions of the generalized quantities $G_{mm}(\beta)$ and $J_{mn}(\beta)$. For the time being, we can even the suppress the $\beta$ argument of all variables because it can be introduced trivially at any time. The last term in Eq. (3.11) is a small complication, basically, corrections originating from this term are added by having gradient operators act on either the exchange function, or the plane wave orbitals.

### 3.2. *Techniques for matrix elements. I. Diagonal quantities*

This section concentrates on the development of techniques for diagonal quantities other than ground state expectation values; the next section contains our analysis to matrix elements between states which differ by exactly two orbitals. After that, the general structure of the theory will be sufficiently clear that those higher–order matrix elements that will be needed for high–order CBF perturbation corrections can be constructed by inspection.

Consider a model wave function $|\Phi_m\rangle$ which differs from the model ground state $|\Phi_o\rangle$ by exactly two orbitals, *i.e.*

$$|\Phi_m\rangle = a_p^\dagger a_h \, |\Phi_o\rangle , \tag{3.15}$$

The Pauli principle demands that $a_p^\dagger$ creates a "particle" state, and $a_h$ destroys a "hole" state. For any such model state, we define the generating functional $G_{mm} = \ln I_{mm}$.

As long as the number of states in which $\Phi_m$ and $\Phi_o$ differ is small, the difference between the exchange function $\ell_m(r)$ and the ground state exchange function $\ell(rk_F) \equiv \ell_o(r)$ is of the order of $1/N$. Consequently, the generating functional $G_{mm}$ differs from the generating functional $G_{oo}$ by terms of order $N^0$ (recall that $G_{oo}$ is of the order $N$). The difference may be expanded to first order in the exchange function

$$\begin{aligned} G_{mm} - G_{oo} &= \int d^3r \frac{\delta G_{oo}}{\delta \ell(rk_F)} \left[\ell_m(\mathbf{r}) - \ell(rk_F)\right] \\ &= \frac{1}{N} \int d^3r \frac{\delta G_{oo}}{\delta \ell(rk_F)} \left[e^{i\mathbf{p}\cdot\mathbf{r}} - e^{i\mathbf{h}\cdot\mathbf{r}}\right] \\ &= \delta G(\mathbf{p}) - \delta G(\mathbf{h}) \,, \end{aligned} \tag{3.16}$$

where $\mathbf{p}$ and $\mathbf{h}$ are the momenta associated with the particle and the hole state. In other words, the function

$$\delta G(\mathbf{k}) = \int d^3r \frac{\delta G_{oo}}{\delta \ell(rk_F)} e^{i\mathbf{k}\cdot\mathbf{r}} \tag{3.17}$$

for any wave vector $\mathbf{k}$ can be calculated by taking the variational derivative of a cluster expansion of $G_{oo}$ with respect to the exchange function $\ell(rk_F)$. Diagrammatically, this variational derivative is taken by removing, in turn, each exchange line and opening the two points it was attached to. A few lowest–order diagrams contributing to this expansion are shown in Fig. 2. Clearly, all diagrams must be so–called *cc*–diagrams since we have, in each term, removed an exchange line from a closed loop. Inspecting the series of diagrams generated by this algorithm, we find the following rules:

(1) The chaining of correlation lines begins the dressing of the bare correlation line $h(r_{ij}) \equiv f^2(r_{ij}) - 1$ to the function $\Gamma_{dd}(r)$.
(2) The parallel connection of 1–2 subdiagrams occurs with the same weight factors as the parallel connection of such diagrams in the graphical construction of $X_{cc}(r)$, and
(3) The chaining of *cc*–subdiagrams involves the weight factors of a *logarithmic series*.

The appearance of a logarithmic series can be made plausible by considering the simple ring diagrams shown in Fig. 3: The *nodal cc* diagrams arise from the variation with respect to the exchange line which is *not* superimposed to a correlation line, whereas the variation with respect to all other exchange lines generates non–nodal *cc*–diagrams. On the basis of our observations, we assert that

(a) The non–nodal subdiagrams of the series $\delta G(r)$ are just the diagrams of the set $X_{cc}(r)$ introduced in Polls' lectures. [7]

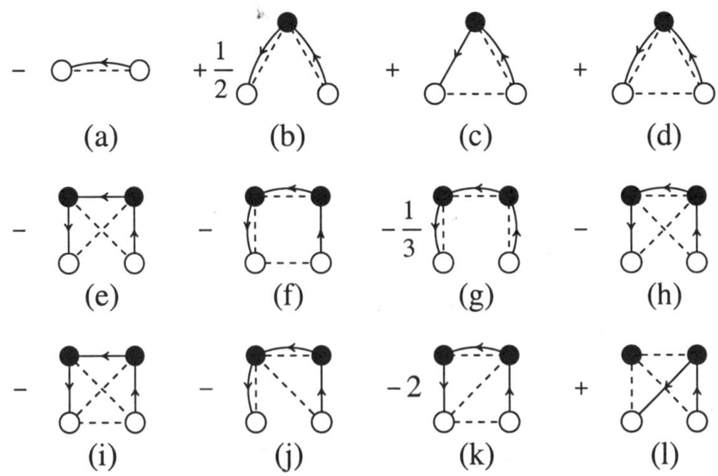

Fig. 2. A number of leading diagrams contributing to the variation $\delta G_{oo}/\delta\ell(rk_F)$ is shown. These diagrams are generated by the variation of the diagrammatic expansion of the generating functional, *cf* Fig. 1, with respect to the exchange line $\ell(rk_F)$.

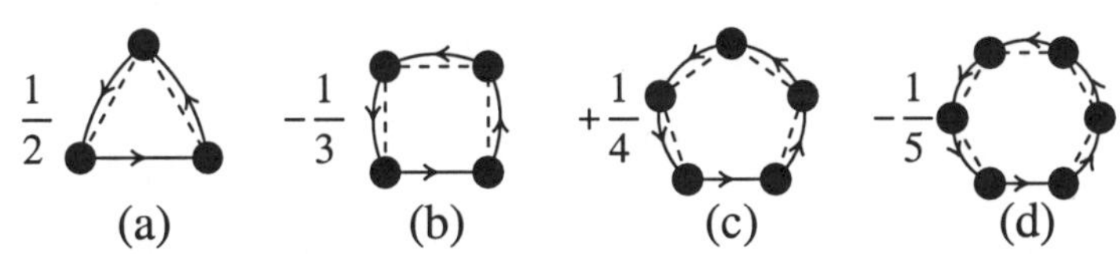

Fig. 3. Some leading cyclic–chain ring–diagrams contributing to $G_{oo}$. The variation with respect to the exchange line that is not superimposed to a correlation line generates diagrams 2b and 2g.

(b) The full function $\delta G(r)$ may be built from $X_{cc}(r)$ by adding all chains with the weight factors of a logarithmic series.

Accordingly, $\delta G(\mathbf{k})$ is expressed quite compactly as

$$\delta G(\mathbf{k}) = -\ln\left[1 - \tilde{X}_{cc}(\mathbf{k})\right] \tag{3.18}$$

The last step in our development of the diagonal quantities is to construct the particle–hole spectrum $\epsilon_{\mathbf{k}}$. The above algorithm gives us only the energy–difference (3.14), and a little extra consideration is necessary to obtain the energy differente to the ground state. According to the algorithm (3.6), we simply have

$$\begin{aligned} e_k \equiv \int d^3r \frac{\delta H_{oo}}{\delta\ell(rk_F)} e^{i\mathbf{k}\cdot\mathbf{r}} &= \frac{\hbar^2 k^2}{2m} + \frac{\tilde{X}'_{cc}(k)}{1 - \tilde{X}_{cc}(k)} + u_0 \\ &\equiv \frac{\hbar^2 k^2}{2m} + u(k) + u_0 \, , \end{aligned} \tag{3.19}$$

where $\tilde{X}'_{cc}(k)$ is to be understood in the sense of the "diagrammatic differentiation" algorithm used for deriving the Euler equations in Polls' lectures. [7] In particular, it also contains derivatives of the exchange function $\ell(rk_F)$ as well as derivatives of the plane wave orbital occurring in Eq. (3.17). The additive constant $u_0$ is determined by requiring that $e_{k_F}$ equals the chemical potential $\mu$:

$$u_0 = \mu - \frac{\hbar^2 k_F^2}{2m} - u(k_F)\,. \tag{3.20}$$

Eq. (3.19) does not spell out the JF kinetic energies explicitly, they must be inserted at the level of implementation that one would like to use "by hand".

### 3.3. *Techniques for matrix elements. II. Off–diagonal quantities*

We start our analysis with the overlap matrix elements $J_{mn}$ defined in Eq. (2.16) between two correlated states $|m\rangle$ and $|n\rangle$, which are assumed to differ in exactly $d$ orbitals. For convenience, we shall assume that the orbitals in which $m$ and $n$ differ have been shuffled to the beginning of the sequence *i.e.* , $m_i \neq n_j$ for $(i,j) \leq d$, $m_i = n_i$ for $i > d$. For our further considerations, we introduce $d$–body operators $\mathcal{N}^{(d)}[\mathbf{m}]$ and $\mathcal{W}^{(d)}[\mathbf{m}]$ which act on the $d$ orbitals in which $\mathbf{m}$ and $\mathbf{n}$ differ, *i.e.* ,

$$\begin{aligned} J_{mn} &= \Big\langle n_1 \ldots n_d \,|\, \mathcal{N}^{(d)}(1,\ldots,d;m) \,|\, m_1 \ldots m_d \Big\rangle_a \\ W_{mn} &= \Big\langle n_1 \ldots n_d \,|\, \mathcal{W}^{(d)}(1,\ldots,d;m) \,|\, m_1 \ldots m_f \Big\rangle_a \end{aligned} \tag{3.21}$$

for two states $\big|m\big\rangle$ and $\big|n\big\rangle$ which differ in the first $d$ orbitals. If the states $\mathbf{m}$ and $\mathbf{n}$ differ only by few orbitals from the ground state $\big|o\big\rangle$, we will see that

$$\begin{aligned} \mathcal{N}^{(d)}(1,\ldots d;m) &= \mathcal{N}^{(d)}(1,\ldots,d) + \mathcal{O}(N^{-1}), \\ \mathcal{W}^{(d)}(1,\ldots d;m) &= \mathcal{W}^{(d)}(1,\ldots,d) + \mathcal{O}(N^{-1})\,, \end{aligned} \tag{3.22}$$

in other words, the dependence of the operators $\mathcal{N}^{(d)}(1,\ldots d;m)$ and $\mathcal{W}^{(d)}(1,\ldots d;m)$ on the diagonal labels can normally be dropped.

We specialize in this section on states $|m\rangle$ and $|n\rangle$ which are assumed to differ in exactly two orbitals, *i.e.*

$$|m\rangle = \big|a^\dagger_{m_1} a^\dagger_{m_2} a_{n_1} a_{n_2}\, n\big\rangle\;. \tag{3.23}$$

It is sufficient to look at the off–diagoonal matrix elements of the unit operator; the corresponding matrix of the Hamiltonian can then be derived explicitly by the "graphical differentiation" method that was introduced in Saarela's lectures and that was already used in the previous section. For $d = 2$, the overlap matrix elements can be written formally as an antisymmetrized matrix element of a two–body operator (we drop the superscript (2) in this case for brevity):

$$\big\langle m\big|n\big\rangle \equiv \big\langle n_1 n_2 \big|\mathcal{N}(1,2;n)\big|m_1 m_2\big\rangle_a \equiv \mathcal{N}_{n_1 n_2, m_1 m_2} \tag{3.24}$$

(the subscript $a$ indicates antisymmetrization), and the task of constructing the overlap matrix elements (3.24) between the non–orthogonal wavefunctions is identical to the construction of a configuration–space representation of the (generally non–local) two–body operator $\mathcal{N}(1,2;n)$. The structure of the generalization of (3.24) to the case of $d(>2)$ different orbitals in $|m\rangle$ and $|n\rangle$ will then be obvious and does not need further explicit derivations.

The standard way of calculating the operator $\mathcal{N}(1,2;n)$ is to develop, analogous to the cluster expansions derived for the ground state theory

$$\mathcal{N}(1,2;n) = \sum_{s,t} (\Delta\mathcal{N})_s^{(t)}(1,2;n) . \tag{3.25}$$

Here, $(\Delta\mathcal{N})_s^{(t)}(1,2;m)$ is the contribution of all $s$–body diagrams with $t$ correlation lines. Any *individual* contribution to $\mathcal{N}(1,2;n)$ will take the analytic form of a $p$–body $(2 \le p < N)$ matrix element

$$\sum_{h_1 \ldots h_{p-2}} \big\langle n_1 n_2 h_1 \ldots h_{p-2} \big| D(\mathbf{r}_1, \ldots \mathbf{r}_p) \big| P(m_1 m_2 h_1 \ldots h_{p-2}) \big\rangle \tag{3.26}$$

$(2 \le p \le N)$, where the summation over the orbitals $h_i$ is carried out over the Fermi sea. The function $D(\mathbf{r}_1, \ldots \mathbf{r}_p)$ is the product of dynamical bond functions $h(r_{ij})$, and $P(m_1 m_2 h_1 \ldots h_{p-2})$ is some permutation of the set of orbitals $(n_1, n_2, h_1, \ldots h_{p-2})$. The place in which the two off–diagonal orbitals $n_1$, $n_2$ appear in the ket gives rise to three structurally different types of matrix elements depending on whether the off–diagonal orbitals $(n_1, n_2)$ on the left and $(m_1, m_2)$ on the right appear on the same, differ in one, or in two, particle–coordinates:

$$d_{dd}^{(p)} = \sum_{h_1 \ldots h_{p-2}} \big\langle n_1 \ldots h_{p-2} \big| D(\mathbf{r}_1 \ldots \mathbf{r}_p) \big| (m_1 m_2) P(h_1 \ldots h_{p-2}) \big\rangle , \tag{3.27}$$

$$d_{dc}^{(p)} = \sum_{h_1 \ldots h_{p-2}} \big\langle n_1 \ldots h_{p-2} \big| D(\mathbf{r}_1 \ldots \mathbf{r}_p) \big| j(m_2 m_1) P(h_1 \ldots (h_j) \ldots h_{p-2}) \big\rangle , \tag{3.28}$$

$$d_{cc}^{(p)} = \sum_{h_1 \ldots h_{p-2}} \big\langle n_1 \ldots h_{p-2} \big| D(\mathbf{r}_1 \ldots \mathbf{r}_p) \big| h_j h_k (m_1 m_2) P(h_1 \ldots (h_j\, h_k) \ldots h_{p-2}) \big\rangle . \tag{3.29}$$

The set $\{h_1 \ldots (h_k) \ldots h_{p-2}\}$ consists of the set $\{h_1 \ldots h_{p-2}\}$ with the orbital $h_k$ omitted.

With the expressions (3.27)–(3.29) in mind we conclude that the two–body operator $\mathcal{N}(1,2;m)$ has the structure of a four–point function

$$\mathcal{N}(1,2;m) = \mathcal{N}_{dd}(12) + \mathcal{N}_{dc}(12) + \mathcal{N}_{dc}(21) + \mathcal{N}_{cc}(12) \tag{3.30}$$

with

$$\mathcal{N}_{dd}(12) = \mathcal{N}_{dd}(r_{12}) \delta(\mathbf{r}_1 - \mathbf{r}_1') \delta(\mathbf{r}_2 - \mathbf{r}_2') \tag{3.31}$$

$$\mathcal{N}_{dc}(12) = \mathcal{N}_{d,cc}(\mathbf{r}_1, \mathbf{r}_2; \mathbf{r}_1', \mathbf{r}_2') \delta(\mathbf{r}_1 - \mathbf{r}_1') \tag{3.32}$$

$$\mathcal{N}_{cc}(12) = \mathcal{N}_{cc,cc}(\mathbf{r}_1, \mathbf{r}_2; \mathbf{r}_1', \mathbf{r}_2') \tag{3.33}$$

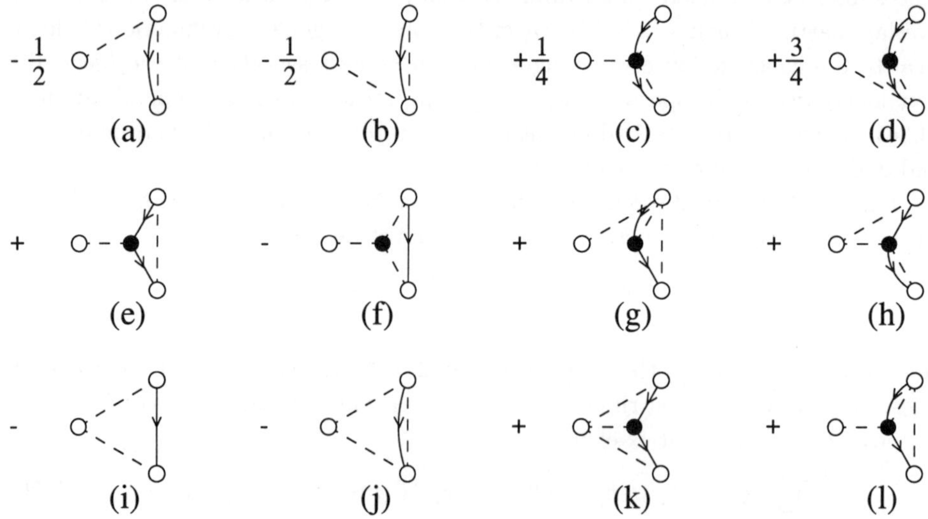


Fig. 4. Some leading diagrams contributing to $\mathcal{N}_{d,cc}(\mathbf{r}_1, \mathbf{r}_2; \mathbf{r}_1, \mathbf{r}_2')$. The dashed line is to be understood as $\Gamma_{dd}(r_{ij})$. The diagrams in the first line are factorizable.

The graphical construction of the four–point function $\mathcal{N}(12)$ is again effectively performed with the cluster–expansion and –resummation techniques of the Fermi–HNC theory. The explicit manipulations for the graphical construction of $\mathcal{N}(12)$ have been carried out in Ref. 21; we proceed here to discuss only some of the essential steps and the structure of the final result. The diagrammatic elements needed to describe any single contribution to $\mathcal{N}(12)$ are the same as the ones that are used in the representation of the radial distribution function, namely "correlation" and "exchange" lines and their "dressed" analogs $\Gamma_{ij}(r)$ (called "super–bonds" in Poll's lectures). As long as we deal with situations $d \ll N$, it is immaterial whether the Fermi–sea summations are performed over the full Fermi–sea, whether a few orbitals are taken out, or whether some orbitals lie outside the Fermi sea. This leads to the aforementioned statement that the dependence of $\mathcal{N}(12, n)$ on the underlying state $|n\rangle$ is of the order $\mathcal{O}(1/N)$. The nontrivial step after the generation of a cluster expansion is the identification of the diagrammatic substructures of $\mathcal{N}(12)$ with the compound–diagrammatical functions introduced along with the ground–state calculation.

The simplest part of the operator $\mathcal{N}_{dd}(12)$, the portion $\mathcal{N}_{dd}(r_{12})$, is readily identified with the dressed correlation line $\Gamma_{dd}(r_{12})$. The other terms are more complicated; we demonstrate the further operations with $\mathcal{N}_{d,cc}(\mathbf{r}_1, \mathbf{r}_2; \mathbf{r}_1, \mathbf{r}_2')$. A few typical diagrams contributing this portion are shown in Fig. 4. We observe that the diagrams 4a–4d and 4g–4h are factorizable; for example the matrix elements corresponding to the first two diagrams as well as their analogs obtained upon exchanging

$(\mathbf{r}_1, \mathbf{r}_1') \leftrightarrow (\mathbf{r}_2, \mathbf{r}_2')$ may be written as

$$-\frac{1}{2}\langle m_1 m_2 | \Gamma_{dd}(1,2) | n_1 n_2 \rangle_a \times$$
$$\times \sum_j \Big[ \langle m_1 j | \Gamma_{dd}(1,2) | j m_1 \rangle \;+\; \text{same for} \quad m_1 \to n_1, m_2, n_2 \Big] .$$

Diagrams 4c, 4d, 4g, and 4h are factorizable in a similar fashion; all of the factorizable diagrams are recognizable by the fact that one of the reference points is connected to the body of the diagrams through a *cc*–chain. Upon collecting all such factorizable diagrams, we can write normalization matrix elements employed in (3.31)–(3.33) as

$$\langle ij | \mathcal{N}(1,2) | kl \rangle = z(i) z(j) z(k) z(l) \langle ij | \mathcal{N}^B(1,2) | kl \rangle \tag{3.34}$$

with

$$z(i) = \frac{1}{\sqrt{1 - \tilde{X}_{cc}(p_i)}} \tag{3.35}$$

($p_i$ is the momentum associated to the orbital $i$, *etc.*). The operator $\mathcal{N}^B(1,2)$ has the same structural decomposition (3.31)–(3.33) as $\mathcal{N}(1,2)$; but it contains only "basic" diagrams in terms of the super–bonds $\Gamma_{ij}(r_{kl})$. The local portion of $\mathcal{N}^B(1,2)$ is identified with

$$\mathcal{N}^B_{dd}(1,2) = \Gamma_{dd}(r_{12}) . \tag{3.36}$$

In order to give succinct diagrammatic characterization of the reduced objects $\mathcal{N}^B_{d,cc}(\mathbf{r}_1, \mathbf{r}_2; \mathbf{r}_1, \mathbf{r}_2')$ and $\mathcal{N}^B_{cc,cc}(\mathbf{r}_1, \mathbf{r}_2; \mathbf{r}_1', \mathbf{r}_2')$, we extend the notion of "reducibility" as follows:

> A contribution to $J_{mn}$ is called reducible if it can be factorized non–trivially in such a manner that at least one of the factors depends on *only one* of the distinguished labels $m_1, m_2, n_1, n_2$.

Further:

> A diagram is called *basic* if it is irreducible and contains no $a - b$ subdiagram without exchange lines attached to $a$ or $b$ or both, except the single dashed line.

We now contend that

(1) The function $\mathcal{N}^B_{d,cc}(\mathbf{r}_1, \mathbf{r}_2; \mathbf{r}_1, \mathbf{r}_2')$ is represented by the sum of all basic diagrams having three external points 1,2,2', with an exchange path going from point 2 to 2' and no exchange lines attached to point 1.

(2) The function $\mathcal{N}^B_{cc,cc}(\mathbf{r}_1, \mathbf{r}_2; \mathbf{r}_1', \mathbf{r}_2')$ is represented by the sum of all basic diagrams having four external points 1,1',2,2', with exchange paths going from point 1 to 1' and from 2 to 2'.

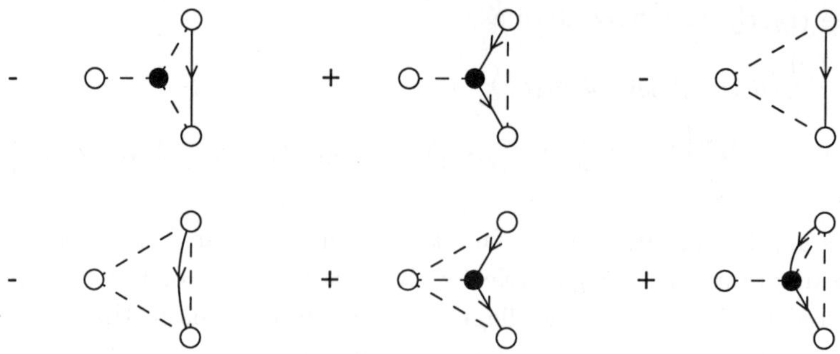

Fig. 5. Some leading diagrams contributing to $\mathcal{N}^B_{d,cc}(\mathbf{r}_1,\mathbf{r}_2;\mathbf{r}_1,\mathbf{r}_2')$. The dashed line is to be understood as $\Gamma_{dd}(r_{ij})$.

Fig. 6. Some leading diagrams contributing to $\mathcal{N}^B_{cc,cc}(\mathbf{r}_1,\mathbf{r}_2;\mathbf{r}_1',\mathbf{r}_2')$. The dashed line is to be understood as $\Gamma_{dd}(r_{ij})$.

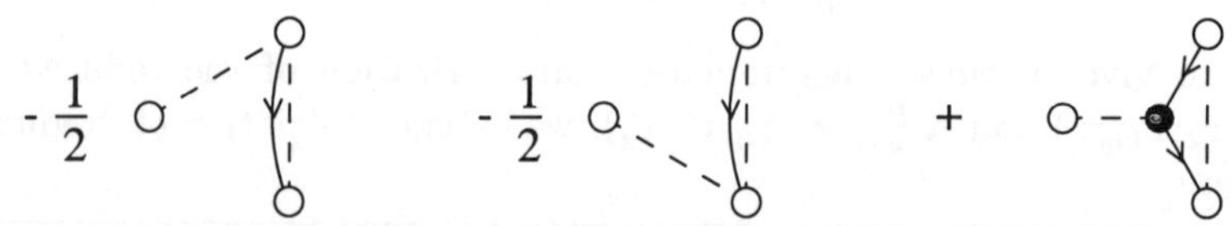

Fig. 7. Three diagrams contributing to $\mathcal{N}_{d,cc}(\mathbf{r}_1,\mathbf{r}_2;\mathbf{r}_1',\mathbf{r}_2')$. The first two diagrams are factorizable, the third is not. These diagrams *cancel* in the diagonal limit $m_1 = m_2 = n_1 = n_2$.

The graphical representation of some typical diagrams contributing to $\mathcal{N}^B_{d,cc}(\mathbf{r}_1,\mathbf{r}_2;\mathbf{r}_1,\mathbf{r}_2')$ and $\mathcal{N}^B_{cc,cc}(\mathbf{r}_1,\mathbf{r}_2;\mathbf{r}_1',\mathbf{r}_2')$ is shown in Figs. 5 and 6. With the factorization of the operator $\mathcal{N}(1,2)$ into a "basic" core and "cyclic chain" factors we have reached a highly compact representation in terms of quantities that we know already from the FHNC theory of the ground state. The factorization looks plausible and appealing, a word of caution applies: Let us look at the diagrams shown in Fig. 7

These three diagrams are easily seen to cancel each other exactly in the the diagonal limit $m_1 = m_2 = n_1 = n_2$. This can indeed occur within the context of the perturbation formula given in section 2.2. Such cancellations assume paramount importance if optimal correlation functions are used that have the asymptotic form $\Gamma_{dd}(r) \propto r^{-2}$ as $r \to \infty$.

Finally, we need to comment on how to interpret the "$\beta$–derivative" operation spelled out in Eq. (3.14) for the calculation of the matrix elements $W_{mn}$. Structurally, one must get a representation analogous to the "basic" factorization (3.34), with proper two–body term $\mathcal{W}(1,2)$ but we must pay special attention to the case where the gradient operator in the Jackson–Feenberg identity (3.10) acts on the single particle orbitals, *cf.* the last term in Eq. (3.11). When the gradient operator acts on any *internal* point, it is treated exactly as in the ground state theory, but we must now also consider the action of the gradient operator on the off–diagonal external orbitals, for example

$$\begin{aligned}&\frac{\hbar^2}{8m}\frac{1}{I_{oo}}\langle n_1 n_2 h_3 \dots h_N| \sum_{i=1,2} \left[\nabla_i, \left[\nabla_i, F^2\right]\right] |(m_1 m_2)_a (h_3 \dots h_N)_a\rangle \\ &= \frac{\hbar^2}{8m}\langle n_1 n_2| \left[\nabla_1^2 \Gamma_{dd}(r_{12}) + \nabla_2^2 \Gamma_{dd}(r_{12})\right] \left|m_1 m_2\right\rangle_a \\ &= \left\langle n_1 n_2\right| \frac{\hbar^2}{4m}\nabla^2 \Gamma_{dd}(r_{12}) \left|m_1 m_2\right\rangle_a . \end{aligned} \tag{3.37}$$

Hence

$$\mathcal{W}_{\rm loc}(r_{12}) = \Gamma'_{dd}(r_{12}) + \frac{\hbar^2}{4m}\nabla^2 \Gamma_{dd}(r_{12}) \,, \tag{3.38}$$

where, of course, the above remarks concerning the cancellation between the $z(i)$–factors (3.35) and non–factorizable diagrams apply as well as for $\left\langle ij\right|\mathcal{N}(1,2)\left|kl\right\rangle$. More complicated kinetic energy terms can be calculated when needed, but only very few numerical calculations have been carried out that went that far in the application of CBF theory.

Once the matrix elements $\left\langle ij\right|\mathcal{W}(1,2)\left|kl\right\rangle$ are available, we can also go back to the original task of calculating the off–diagonal matrix elements of the structure (3.8). In the present case where the states $|m\rangle$ and $|n\rangle$ differ in two orbitals, we can generally define a two–body operator $\mathcal{H}$ such that

$$H'_{mn} = \left\langle n_1 n_2\right|\mathcal{H}(1,2)\left|m_1 m_2\right\rangle_a \equiv \mathcal{H}_{n_1 n_2, m_1 m_2} \,, \tag{3.39}$$

and can express its matrix elements in terms of the operators $\mathcal{W}(1,2)$ and $\mathcal{N}(1,2)$ defined above, *cf.* Eq. (3.9). If $|m\rangle$ is, for example, a correlated 2–particle 2–hole state, and $|n\rangle$ is the ground state, then we obtain the explicit representation

$$\begin{aligned}&\left\langle p_1 p_2\right|\mathcal{H}(1,2)\left|h_1 h_2\right\rangle_a = \left\langle p_1 p_2\right|\mathcal{W}(1,2)\left|h_1 h_2\right\rangle_a \\ &+\frac{1}{2}\left(e_{p_1} + e_{p_2} - e_{h_1} - e_{h_2}\right)\left\langle p_1 p_2\right|\mathcal{N}(1,2)\left|h_1 h_2\right\rangle_a . \end{aligned} \tag{3.40}$$

## 4. Interpretation of effective interactions

### 4.1. *Quasiparticle interaction*

The most immediate consideration after the computation of the ground–state energy is the study of the response of the quantum system against perturbations by static external fields such as a pressure, a magnetic field, an external heat source, or a velocity field. This response can be thought of the system populating the low–lying excited states. It was one of Landau's ingenious realizations [2,3] that, in a "normal" Fermi liquid, these low–lying excited states can be classified by the quantum numbers of a free Fermi gas; in fact this feature defines a normal Fermi fluid. Since these excitations have the same quantum numbers as particles, they are called *quasiparticles.* Quasiparticles are well defined quantities in the vicinity of the Fermi surface as long as their lifetime is long. The response of the Fermi liquid may then be thought of as a manifestation of the interaction between quasiparticles.

It is not the purpose of this article to give an introduction or a survey of Landau's Fermi–Liquid theory; the material is extensively covered in the original literature [2,3], in textbooks and in review articles. [22,23,24] We shall review here only the basic arguments and the simplest results.

An excited state is characterized by the deviation of the *quasiparticles occupation numbers* $n_{\mathbf{k},\sigma}$ from the ground state configuration $n^{(0)}_{\mathbf{k},\sigma}$:

$$E = E[n_{\mathbf{k},\sigma}]\,; \qquad n_{\mathbf{k},\sigma} = n^{(0)}_{\mathbf{k},\sigma} + \delta n_{\mathbf{k},\sigma}\,. \tag{4.1}$$

The *quasiparticle spectrum* is the first variation of the energy functional with respect to the quasiparticle occupation number, taking at the ground state:

$$e^{(0)}_{\mathbf{k},\sigma} = \left.\frac{\delta E\left[n_{\mathbf{k},\sigma}\right]}{\delta n_{\mathbf{k},\sigma}}\right|_{n^{(0)}_{\mathbf{k},\sigma}}, \tag{4.2}$$

and the *quasiparticle interaction* is the second variation:

$$f_{\mathbf{k}\sigma,\mathbf{k}'\sigma'} = \left.\frac{\delta^2 E\left[n_{\mathbf{k},\sigma}\right]}{\delta n_{\mathbf{k},\sigma}\delta n_{\mathbf{k}',\sigma'}}\right|_{n^{(0)}_{\mathbf{k},\sigma}}. \tag{4.3}$$

The density of quasiparticle–states at the Fermi surface defines the effective mass

$$\frac{\hbar k_F}{m^*} \equiv \left.\frac{de^{(0)}_{\mathbf{k},\sigma}}{dk}\right|_{k=k_F}. \tag{4.4}$$

Since we are looking at small deformations of the Fermi surface, the momenta in Eq. (4.3) are on the Fermi surface, *i.e.* the quasiparticle interaction is a function of the spins of both particles and of the angle between $\mathbf{k}$ and $\mathbf{k}'$. Conventionally, one expands the quasiparticle interaction in Legendre polynomials $P_\ell(\cos\theta)$ in the angle $\theta$ between the vectors $\mathbf{k}$ and $\mathbf{k}'$ and separates the spin–dependence as follows:

$$\begin{aligned} f_{\mathbf{k}\uparrow,\mathbf{k}'\uparrow} &= f^s_{\mathbf{k},\mathbf{k}'} + f^a_{\mathbf{k},\mathbf{k}'} = \sum_{\ell=0}^{\infty}\left(f^s_\ell + f^a_\ell\right)P_\ell(\cos\theta) \equiv \frac{\pi^2\hbar^2}{\Omega m^* k_F}\sum_{\ell=0}^{\infty}\left(F^s_\ell + F^a_\ell\right)P_\ell(\cos\theta) \\ f_{\mathbf{k}\uparrow,\mathbf{k}'\downarrow} &= f^s_{\mathbf{k},\mathbf{k}'} - f^a_{\mathbf{k},\mathbf{k}'} = \sum_{\ell=0}^{\infty}\left(f^s_\ell - f^a_\ell\right)P_\ell(\cos\theta) \equiv \frac{\pi^2\hbar^2}{\Omega m^* k_F}\sum_{\ell=0}^{\infty}\left(F^s_\ell - F^a_\ell\right)P_\ell(\cos\theta)\,. \end{aligned} \tag{4.5}$$

Physical observables can then be expressed in terms of the dimensionless Landau (or "Fermi–liquid") parameters $F_\ell^s$ and $F_\ell^a$. For example, The effective mass is

$$\frac{m^*}{m} = 1 + \frac{1}{3}F_1^s\,; \tag{4.6}$$

it is related to the specific heat through

$$c_V = \frac{1}{3}m^* k_F T\,. \tag{4.7}$$

The hydrodynamic speed of sound is

$$mc^2 = \frac{\hbar^2 k_F^2}{3m^*}\left(1 + F_0^s\right)\,, \tag{4.8}$$

and the magnetic susceptibility, in units of magnetic susceptibility of the free Fermi gas $\chi_M^0$, is

$$\frac{\chi_M^0}{\chi_M} = 1 + F_0^a\,. \tag{4.9}$$

The task of many–body theory is the microscopic understanding of the nature and the properties of the quasiparticle interaction. Much work has been done in this direction, *cf.* Refs. 24, 25, and useful connections have been derived. Nevertheless, it is fair to say that a complete microscopic understanding of the Fermi–liquid interaction in $^3$He at a level that is comparable to those of the ground–state calculations of Polls' lectures is still lacking. One part of the problem is the technical aspect of the poor convergence of diagrammatic expansions. But even if all diagrams could be calculated exactly, we are facing the second problem of a more conceptional nature: The variational theory calculates the effective interactions at an average energy which is not necessarily optimal for the calculation of effects that happen at the Fermi surface. Having understood this aspect, it is plausible that the manifestly microscopic calculation of the quasiparticle interaction in $^3$He–$^4$He mixtures has been much more successful than comparable calculations for pure $^3$He: In mixtures, much of the quasiparticle interaction between $^3$He atoms is mediated by the exchange of phonons propagating in $^4$He who do not know about Fermi statistics. [26]

It is, of course, to a limited extent, possible to perform Jastrow–Feenberg variational calculations for specific facets of the quasiparticle interaction. For example, the compressibility (or the speed of the first sound) is readily obtained by calculating the energy per particle at a sequence of densities and differentiating the resulting equation of state

$$mc^2 = \frac{d}{d\rho}\left\{n^2\frac{d(E/N)}{d\rho}\right\}\,. \tag{4.10}$$

The magnetic susceptibility (and, in nuclear matter, the asymmetry energy) may be calculated using for the model state $|o\rangle$ a partially polarized Slater function; the effective mass (and, hence, the specific heat ) can be determined directly from the CBF single–particle spectrum (3.19) by means of Eq. (4.4). Especially in view of the fact that the effective interactions are anyway known after an optimized variational calculation, it is much more efficient to start from a general theory of the quasiparticle interaction and extract, following Landau [2,3] the physical observables of interest.

The calculation of the quasiparticle interaction at the level of the Jastrow–Feenberg theory is the necessary first step in a microscopic theory and needs to be calculated accurately before any further improvements are worthwhile. In particular, we shall examine how the effective interactions that have been derived in section 3.3 are related to the quasiparticle interaction.

The plausible way to implement Landau's mapping of the states of the free Fermi gas to the quasiparticles states of the interacting system is to use the correlation operator $F$ to map the non–interacting states $\{|\Phi_{\mathbf{m}}\rangle\}$ to the quasiparticle states. Instead of using the states $n^{(0)}_{\mathbf{k},\sigma}$ to form the model ground state $|o\rangle$, we have to take the plane wave orbitals according to the perturbed distribution

$$n_{\mathbf{k},\sigma} = n^{(0)}_{\mathbf{k},\sigma} + \delta n_{\mathbf{k},\sigma}\,. \tag{4.11}$$

Thus, we are led to consider the energetics of a system corresponding to a fermion occupation number $n_{\mathbf{k},\sigma}$,

$$H_{\mathbf{m},\mathbf{m}} = H_{oo}\left[F(n_{\mathbf{k},\sigma}); n_{\mathbf{k},\sigma}\right] \tag{4.12}$$

The approximation for the quasiparticle interaction is then the second variational derivative

$$f^{\rm var}_{\mathbf{k}\sigma,\mathbf{k}'\sigma'} = \frac{\delta^2 H_{oo}\left[n_{\mathbf{k},\sigma}\right]}{\delta n_{\mathbf{k},\sigma}\delta n_{\mathbf{k}',\sigma'}}\,. \tag{4.13}$$

We indicate the fact that the we deal with the contribution to the quasiparticle interaction generated by a variational energy of $H_{oo}$ by a superscript "var". Our task is now to evaluate the second–order term in $\delta n_{\mathbf{k},\sigma}$. One might in principle allow a dependence of the correlation factors $f(r_{ij})$ on the occupation number; the simplest case, a density dependence, is in fact implied by the determination of the two–body correlation through the variational principle. The best choice for the correlation operator would be to determine it by optimization

$$\left\{\frac{\delta H_{oo}\left[n_{\mathbf{k},\sigma}\right]}{\delta F}\right\} = 0 \tag{4.14}$$

for all quasiparticle occupation numbers $n_{\mathbf{k},\sigma}$. This leads to the representation of the quasiparticle interaction

$$\begin{aligned} f^{\rm var}_{\mathbf{k}\sigma,\mathbf{k}'\sigma'} &= \left\{\frac{\delta^2 H_{oo}\left[n_{\mathbf{k},\sigma}\right]}{\delta n_{\mathbf{k},\sigma}\delta n_{\mathbf{k}',\sigma'}}\right\} \\ &= \left\{\frac{\partial^2 H_{oo}}{\partial n_{\mathbf{k},\sigma}\partial n_{\mathbf{k}',\sigma'}}\right\} - \left\{\frac{\delta^2 H_{oo}\left[n_{\mathbf{k},\sigma}\right]}{\delta F^2}\frac{\delta F}{\delta n_{\mathbf{k},\sigma}}\frac{\delta F}{\delta n_{\mathbf{k}',\sigma'}}\right\}, \end{aligned} \tag{4.15}$$

where we have used Eq. (4.14); the partial derivative in Eq. (4.15) indicates that one differentiates *only* with respect to the dependence of the uncorrelated model state $|m\rangle$ on the occupation number $n_{\mathbf{k},\sigma}$. If we keep only the *explicit* dependence of the energy on the quasiparticle distribution function, we obtain an upper bound to the quasiparticle interaction compared with the one obtained with the best possible correlation operator *within the same operator space.* The omission of the last term in Eq. (4.15) does not imply an intrinsic inaccuracy of our further considerations, it reflects simply the choice of the basis states (2.1).

We are now ready to construct explicitly the variational estimate for the quasiparticle interaction. The procedure is generalization of the construction of the single–particle spectrum in Sect. 3.2.

$$f^{\mathrm{var}}_{\mathbf{k}\sigma;\mathbf{k}'\sigma'} = \frac{1}{N^2}\int d^3r_i d^3r_j d^3r_k d^3r_l \left(\frac{\delta^2 H_{oo}}{\delta\ell(r_{ij}k_F)\delta\ell(r_{kl}k_F)}\right) e^{i(\mathbf{k}\cdot\mathbf{r}_{ij}+\mathbf{k}'\cdot\mathbf{r}_{kl})} \tag{4.16}$$

Eq. (4.16) indicates the rule for the graphical construction of the variational estimate for the quasiparticle interaction: $f^{\mathrm{var}}_{\mathbf{k}\sigma,\mathbf{k}'\sigma'}$ is obtained from a given energy expression by replacing, in turn, each pair of exchange lines $l(r_{ij}k_F)$, $l(r_{kl}k_F)$ by a pair of plane wave orbitals $\exp(i\mathbf{k}\cdot\mathbf{r}_{ij})$, $\exp(i\mathbf{k}'\cdot\mathbf{r}_{kl})$. Note that we must allow for $i = j$. A configuration space representation of the quasiparticle interaction must consist of the same basic diagrammatic elements that we have encountered in the calculation of the energy, or of the two–body operators $\mathcal{W}(1,2)$ and $\mathcal{N}(1,2)$; *cf.* Sect. 3.3. However, since the quasiparticle interaction is calculated by the variation of an irreducible quantity, we can have only irreducible diagrams.

We must distinguish three generically different types of diagrams: Those where both exchange line that are subject to the variation (4.16) degenerate into density factors, those where only one degenerates that way, and those where both exchange lines subject to variation are non–trivial. The three types of variations lead to *dd*, *d, cc* and *cc, cc*–type diagrams discussed in section 3.3. We show some typical examples for the three types of diagrams in Figs. 8. In addition to the graphical elements of the FHNC theory, the plane–wave orbital $\exp(i\mathbf{k}\cdot\mathbf{r}_{ij})$ is depicted by a heavy oriented line. It is also understood that any one of the dashed lines shown in Figs. 8 may be an interaction line, or that pair of statistical links $\ell(r_{ij}k_F)$ and $\exp(i\mathbf{k}\cdot\mathbf{r}_{ik})$ with at least one common point may be a kinetic energy contribution.

The diagrams 8a and 8b are reminiscent of the Hartree–Fock contribution to the quasiparticle interaction; in fact, they are the only ones surviving in the limit $F \to 1$. Diagram 8d is an exchange term of a simple chain diagram. The corresponding direct term, diagram 8c would originate from a point–reducible contribution to the energy. It does not contribute to the quasiparticle interaction and is therefore crossed out in Fig. 8. The next two diagrams, 8e and 8f, are the direct and the corresponding exchange term of a *d, cc* structure. Diagrams 8g is crossed out because it would come from a reducible energy diagram, but the corresponding exchange diagram 8h is legitimate. Diagrams 8i–8l show finally the same situation for the *cc, cc*–type diagrams. Again, diagram 8k is crossed out because it would result from a reducible energy contribution.

Let us examine the relationship between the full effective interaction $\mathcal{W}(1,2)$ and the quasiparticle interaction defined above. For that purpose, keep, for the

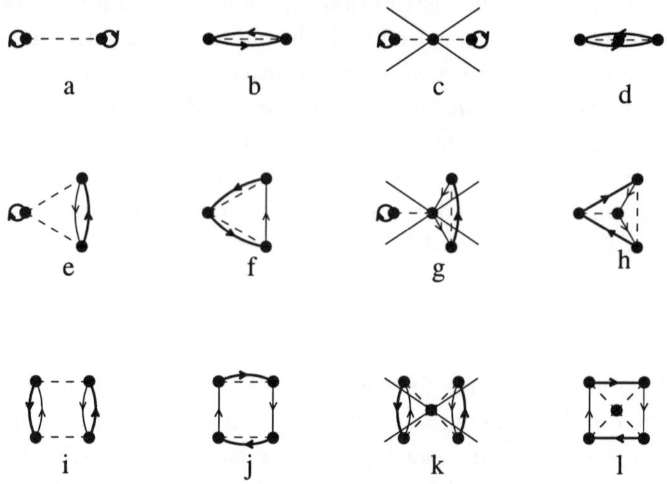


Fig. 8. A number of leading diagrams contributing to the quasiparticle interaction are shown. It is understood that the (pair of) point(s) 1,(1') are on the left, and the (pair of) point(s) 2,(2') are on the right of each individual diagram. The heavy oriented line represents a plane–wave orbital $\exp(i\mathbf{k}\cdot\mathbf{r}_{ik})$. The crossed diagrams (c), (g), and (k) do *not* contribute to the quasiparticle interaction and are drawn only for reference.

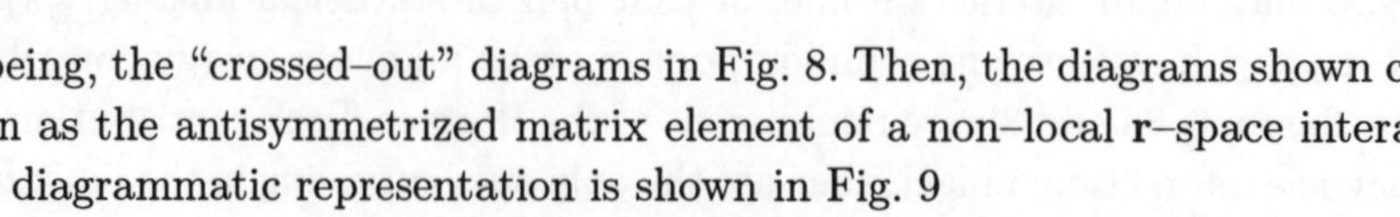

Fig. 9. The figure shown the coordinate–space representation of the effective interaction whose antisymmetrized diagonal matrix element leads to Fig. 8 *including* the crossed–out diagrams.

time being, the "crossed–out" diagrams in Fig. 8. Then, the diagrams shown can be written as the antisymmetrized matrix element of a non–local **r**–space interaction whose diagrammatic representation is shown in Fig. 9

In Fig. 9 we recover the same topological structures that were found in the derivation of $\mathcal{W}(1,2)$. Taking into account that the crossed–out diagrams and their higher–order analogs do not appear, we conclude that the variational quasiparticle interaction can be written in the form

$$f^{\rm var}_{\mathbf{k}\sigma;\mathbf{k}'\sigma'} = \left\langle \mathbf{k}\sigma\mathbf{k}'\sigma' \middle| \mathcal{V}(12) \middle| \mathbf{k}\sigma\mathbf{k}'\sigma' \right\rangle - \left\langle \mathbf{k}\sigma\mathbf{k}'\sigma' \middle| \mathcal{W}(12) \middle| \mathbf{k}'\sigma'\mathbf{k}\sigma \right\rangle \tag{4.17}$$

where $\mathcal{V}(12)$ is the subset of diagrams contributing to $\mathcal{W}(12)$ that have no nodal path between points 1 and 2.

Eq. (4.17) provides a workable prescription for the calculation of the variational estimate for the quasiparticle interaction. Unfortunately, but not unexpectedly, the Fermi liquid parameters calculated by this procedure are at best rough estimates; [27]

the predictions of Eq. (4.17) are far from quantitative and occasionally even unphysical. This shows the limitations of a purely variational treatment; while the evaluation of the above matrix elements is a necessary first step, much more refined techniques are needed for a quantitative understanding of the quasiparticle interaction of a dense Fermi liquid.

### 4.2. *Variational BCS theory*

As a further application of the techniques derived above we show in this section how the variational theory is adapted to the description of the superfluid state found in the millikelvin regime in liquid $^3$He [28] and also predicted in neutron matter. [29,30] We do not attempt here to review the rich literature on (semi)–phenomenological theories for the superfluid phase of liquid $^3$He, much less the still–proliferating theories of superconductivity. Rather, we will here, for the simplest case of $S$–wave pairing, show how the effective interactions, which enter phenomenological theories as parameters, may be calculated from first principles. The formal development will mainly follow Ref. 31 and subsequent efforts [32] to improve upon this variational approach by CBF perturbation theory.

The reader is reminded that the BCS theory of fermion superfluidity generalizes the Hartree–Fock model $\{|\Phi_m\rangle\}$ by introducing a superposition of independent particle wave functions corresponding to different particle number, in the second–quantized form

$$|\mathrm{BCS}\rangle = \prod_{\mathbf{k}}(u_{\mathbf{k}} + v_{\mathbf{k}} a^{\dagger}_{\mathbf{k},\uparrow} a^{\dagger}_{-\mathbf{k},\downarrow})|0\rangle\,. \tag{4.18}$$

The coefficient functions $u_{\mathbf{k}}$ and $v_{\mathbf{k}}$, known as Bogoljubov amplitudes, describe the distortion of the Fermi surface due to the pairing phenomenon. In particular, the square of the amplitude $v_{\mathbf{k}}$ may be interpreted as the occupation probability of fermion pairs with quantum numbers $(\mathbf{k}\uparrow, -\mathbf{k}\downarrow)$; its deviation from the Hartree–Fock distribution

$$v_{0,\mathbf{k}} = n(\mathbf{k}) = \Theta(k_F - k) \tag{4.19}$$

provides one measure of the strength of the pairing effect.

Let us first recapitulate the elements of BCS theory, considering a uniform, infinitely extended system of spin–1/2 fermions described by a nonrelativistic second–quantized Hamiltonian containing a free–space ("bare") two–body interaction $v(ij)$. The BCS trial state, as parameterized by $v_{\mathbf{k}}$ or $\eta_{\mathbf{k}}$, is designed to capture the effects of pairing correlations in momentum space, localized near the Fermi surface. Considering the BCS *ansatz* (4.18) as a trial ground state, the amplitudes $u_{\mathbf{k}}$ and $v_{\mathbf{k}}$ may be taken real without loss of generality. The condition

$$u_{\mathbf{k}}^2 + v_{\mathbf{k}}^2 = 1 \tag{4.20}$$

then determines $u_{\mathbf{k}}^2$ if $v_{\mathbf{k}}^2$ is known and serves to normalize the BCS state (4.18) to unity. It is sometimes convenient to express express $u_{\mathbf{k}}$ and $v_{\mathbf{k}}$ in terms of the real phase angle $\eta_{\mathbf{k}}$:

$$u_{\mathbf{k}} = \cos \eta_{\mathbf{k}}, \qquad v_{\mathbf{k}} = \sin \eta_{\mathbf{k}} . \tag{4.21}$$

The amplitude $v_{\mathbf{k}}$ is to be determined variationally from the extremum condition

$$\frac{\delta}{\delta v_{\mathbf{k}}} \langle \mathrm{BCS} | \hat{H} - \mu \hat{N} | \mathrm{BCS} \rangle = 0 , \tag{4.22}$$

subject to the constraint

$$\frac{\langle \hat{N} \rangle}{\Omega} = \frac{2}{(2\pi)^3} \int d^3k \, v_{\mathbf{k}}^2 = \rho \tag{4.23}$$

of a fixed particle density, where $\hat{N} = \sum_{\mathbf{k}\sigma} a_{\mathbf{k}\sigma}^\dagger a_{\mathbf{k}\sigma}$ is the particle number operator and $\mu$ is the chemical potential.

Rephrasing and explicating the variational condition (4.22) in terms of the phase angle $\eta_{\mathbf{k}}$, we find

$$\sin 2\eta_{\mathbf{k}} \left[ \varepsilon_{\mathbf{k}} - \mu \right] = -\frac{1}{2} (\cos 2\eta_{\mathbf{k}}) \sum_{\mathbf{k}}{}' \mathcal{P}_{\mathbf{k},\mathbf{k}'} (\sin 2\eta_{\mathbf{k}}') . \tag{4.24}$$

This expression contains new quantities $\varepsilon_{\mathbf{k}}$ and $\mathcal{P}_{\mathbf{k},\mathbf{k}'}$ interpreted respectively as a single particle energy and a pairing interaction:

$$\varepsilon_{\mathbf{k}} = \frac{\hbar^2 k^2}{2m} + \sum_{\mathbf{k}',\sigma'} v_{\mathbf{k}'}^2 \langle \mathbf{k}\sigma, \mathbf{k}'\sigma' | v(12) | \mathbf{k}\sigma, \mathbf{k}'\sigma' \rangle_a , \tag{4.25}$$

$$\mathcal{P}_{\mathbf{k},\mathbf{k}'} = \langle \mathbf{k} \uparrow, -\mathbf{k} \downarrow | v(12) | \mathbf{k}' \uparrow, -\mathbf{k}' \downarrow \rangle_a . \tag{4.26}$$

It is conventional to define a "gap function" connected to the phase angle and pairing interaction through

$$\Delta_{\mathbf{k}} = \frac{1}{2} \sum_{\mathbf{k}}{}' \sin 2\eta_{\mathbf{k}}' \mathcal{P}_{\mathbf{k},\mathbf{k}'} . \tag{4.27}$$

In terms of this function, the optimal $\eta_{\mathbf{k}}$, which must satisfy (4.24), is given (modulo $2\pi$ times an integer) by

$$\eta_{\mathbf{k}} = \frac{1}{2} \tan^{-1} \left[ \frac{-\Delta_{\mathbf{k}}}{\varepsilon_{\mathbf{k}} - \mu} \right] . \tag{4.28}$$

Inserting (4.27) into (4.24), determination of the optimal amplitudes

$$v_{\mathbf{k}}^2 = \frac{1}{2} \left[ 1 - \frac{\varepsilon_{\mathbf{k}} - \mu}{\sqrt{(\varepsilon_{\mathbf{k}} - \mu)^2 + \Delta_{\mathbf{k}}^2}} \right] \tag{4.29}$$

and $u_{\mathbf{k}} = (1 - v_{\mathbf{k}}^2)^{1/2}$ requires solution of the nonlinear integral equation

$$\Delta_{\mathbf{k}} = \frac{1}{2} \sum_{\mathbf{k}}{}' \mathcal{P}_{\mathbf{k},\mathbf{k}'} \frac{\Delta_{\mathbf{k}'}}{\sqrt{(\varepsilon_{\mathbf{k}'} - \mu)^2 + \Delta_{\mathbf{k}'}^2}} , \tag{4.30}$$

the so-called gap equation, self-consistently with the particle number constraint (4.23) and the relations (4.25) and (4.26).

The quantity

$$E_{\mathbf{k}} = \sqrt{(\varepsilon_{\mathbf{k}} - \mu)^2 + \Delta_{\mathbf{k}}^2} \tag{4.31}$$

is the energy needed to create a quasiparticle of momentum $\mathbf{k}$ in the superfluid state. The finite gap at the Fermi surface, $\Delta_{|\mathbf{k}|=k_F} \equiv \Delta_F$ is responsible for the novel behavior of the pair-condensed system.

Subtracting the expectation value $\langle \mathrm{BCS}|\hat{\mathrm{H}}|\mathrm{BCS}\rangle$ from its normal-state limit $v_{0,\mathbf{k}} = n(\mathbf{k})$, we obtain the *condensation energy*

$$E_c = -2\sum_{\mathbf{k}} \varepsilon_{\mathbf{k}}^o (v_{\mathbf{k}}^2 - v_{0,\mathbf{k}}^2) - \sum_{\mathbf{kk}'} u_{\mathbf{k}} v_{\mathbf{k}} \mathcal{P}_{\mathbf{kk}'} - \sum_{\mathbf{kk}'} (v_{\mathbf{k}}^2 - v_{0,\mathbf{k}}^2)(v_{\mathbf{k}'}^2 - v_{0,\mathbf{k}'}^2) V_{\mathbf{kk}'} \tag{4.32}$$

in terms of the spin-averaged Hartree-Fock matrix elements

$$V_{\mathbf{kk}'} = \frac{1}{2}\sum_{\sigma\sigma'} \langle \mathbf{k}\sigma\mathbf{k}'\sigma' | v(12) | \mathbf{k}\sigma\mathbf{k}'\sigma' \rangle_a , \tag{4.33}$$

and the Bogoljubov amplitudes $u_{\mathbf{k}}$ and $v_{\mathbf{k}}$.

The gap equation (4.30) is coupled back to itself through the presence of the function $v_{\mathbf{k}}^2$ in equation (4.25) for the single-particle energy $\varepsilon_{\mathbf{k}}$ and in the constraint equation (4.23) that determines the chemical potential $\mu$ for given density. This nonlinearity is removed by replacing $v_{\mathbf{k}}$, in Eq. (4.25), by $v_{\mathbf{k}}^{(0)} = n(\mathbf{k})$. This replacement converts $\varepsilon_{\mathbf{k}}$ to a standard Hartree-Fock single-particle energy $\varepsilon_{\mathbf{k}}^{(0)} = \epsilon_{\mathbf{k}}$ and implies $\mu = \epsilon_{\mathbf{k}_F} \equiv \epsilon_F$. Equations (4.23) and (4.25) are then decoupled from the gap equation. Errors induced by the decoupling approximation are nominally of order $(\Delta_F/\epsilon_F)^2$, a quantity that remains very small for any superfluid Fermi system known in nature. In the condensation energy (4.32), the decoupling approximation amounts to ignoring the last term.

Let us now return to the problem of making the BCS philosophy work for strongly interacting systems. The pure BCS state (4.18) is just as unsuitable for a description of strongly interacting systems like $^3$He and neutron matter as is the pure Hartree-Fock state. Once again, adequate provision must be made for the singular or near-singular nature of the two-body interaction $v(12)$, corresponding (in most models of the bare interaction) to the presence of a strong repulsion at short interparticle distances.

If we wish to pursue the variational strategy begun by BCS, it would appear that a more flexible trial state would be energetically advantageous and therefore preferable. Such a trial state should introduce *separate* variational descriptors for (i) the pairing correlations near the Fermi surface in momentum space and (ii) short-range geometrical correlations in configuration space. In addition, it should be expected that (iii) long-range collective correlations in the many-body medium, associated for example with virtual particle-hole excitations, will act to modify both the in-medium pairing interaction and the single-particle energies. Ideally, then, one should also include the latter effect in some fashion.

To build the required geometrical correlations into the microscopic description of the system, we need to devise a "correlated" BCS state. This line of argument suggests that one should consider the possibility of a hybrid variational *ansatz* incorporating both Jastrow–Feenberg and BCS correlations. Schematically, this trial state should look something like

$$\left|\Psi_o^s\right\rangle = F_{JF} \times \left|\mathrm{BCS}\right\rangle . \tag{4.34}$$

However, we are faced with a formal mismatch, which prevents us from simply applying the correlation factor $F_{JF}$ of (4.34) to $\left|\mathrm{BCS}\right\rangle$, since the former is defined in the $N$–particle Hilbert space and the latter is a vector in Fock space with indefinite particle number. At this point we should consider a more general correlation operator $F_N$ acting in the $N$–particle Hilbert space.

The most natural number non–conserving choice, which is compatible with the BCS philosophy, is formed by first projecting the bare BCS state on an arbitrary member of a complete set of independent–particle states with fixed particle numbers, then applying the correlation operator to that state, normalizing the result, and finally summing over all particle numbers. During this procedure, we must distinguish between correlation operators and normalization integrals corresponding to different particle numbers $N$, this will be done by a subscript or superscript. Our correlated BCS state is then written as

$$\begin{aligned}\left|\mathrm{CBCS}\right\rangle &= \prod_{\mathbf{k}} (u_{\mathbf{k}} + v_{\mathbf{k}} a^\dagger_{\mathbf{k},\uparrow} a^\dagger_{-\mathbf{k},\downarrow}) |0\rangle \\ &= \sum_{m,N} \left|m^{(N)}\right\rangle \langle \Phi_m^{(N)} \left|\mathrm{BCS}\right\rangle ,\end{aligned} \tag{4.35}$$

where the $\left|m^{(N)}\right\rangle$ form a complete set of correlated $N$–body wave functions built from the corresponding Slater determinants $\left|\Phi_m^{(N)}\right\rangle$. The trial state (4.35) superposes these correlated basis states with the same amplitudes with which the model states $\left|\Phi_m^{(N)}\right\rangle$ enter the corresponding expansion of the *original* BCS vector. In that sense this choice of trial state is a very natural one.

Within the class of trial ground states possessing the defining characteristics of a pair–condensed, superfluid state — off–diagonal long–range order, the continuity of the momentum distribution at the Fermi surface, existence of a gap in the single–particle spectrum — the choice yielding the largest condensation energy is considered to give the best approximation to the true ground–state wave function. Minimization of the energy expectation value determines the BCS amplitudes $u_{\mathbf{k}}$, $v_{\mathbf{k}}$. In the correlated case, we have, in principle, the additional freedom of choosing the two–body correlations separately for the projection of each independent–particle model state $\left|\Phi_m^{(N)}\right\rangle$ onto the bare BCS state, for each particle number $N$. Our choice (4.35) is plausible, but it is also dictated by practicality: it leads to a theory that can be mapped one–to–one upon the bare BCS theory and it exhibits no problems when optimized, long–ranged correlation functions are used.

To formulate the BCS theory with correlated wavefunctions, we consider the expectation value of an arbitrary (second–quantized) operator $\hat{O}$ with respect to the superfluid state:

$$\left\langle \hat{O} \right\rangle_s = \frac{\langle \mathrm{CBCS} | \hat{\mathrm{O}} | \mathrm{CBCS} \rangle}{\langle \mathrm{CBCS} | \mathrm{CBCS} \rangle} . \tag{4.36}$$

Specifically, we need, of course, the expectation value of the Hamiltonian and the number operator, $\langle \hat{H} \rangle_s$ and $\langle \hat{N} \rangle_s$.

One may pursue cluster–expansion and resummation methods of expectation values (4.36) for the superfluid trial state (4.35), this has been done successfully for the one– and two–body density matrices corresponding of a slightly different choice of the correlated BCS state. [33] We do not follow this route, but instead start by making an approximation by considering the interaction of only one Cooper pair at a time while treating the background as normal. The error introduced by this is of order $\xi = (\Delta_F/\epsilon_F)^2$ encountered in our discussion of the decoupling approximation. As indicated there, this error is negligible in all systems of physical interest, including high–$T_c$ superconductors. Thus, the stratagem of expansion in the small parameter $\xi$ is introduced at little cost. On the other hand, we will see that very much is gained, since the resulting simplifications provide us with a theory that is independent of the precise choice of the correlation operator $F_N$, it has exactly the same structure as the BCS theory for a weakly interacting system, and utilizes the effective interactions of CBF theory as input.

To implement the decoupling approximation, it is sufficient to retain the terms of *first order in the deviation* $v_{\mathbf{k}}^2 - v_{0,\mathbf{k}}^2$ and those of *second order* in $u_{\mathbf{k}}v_{\mathbf{k}}$. The calculation of $\langle \hat{H} - \mu \hat{N} \rangle$ for correlated states [31,32] is somewhat tedious, we only give the final result. It is convenient to introduce the creators and annihilators of correlated Cooper pairs,

$$\begin{aligned} \beta_{\mathbf{k}}^{\dagger} &= \alpha_{\mathbf{k}\uparrow}^{\dagger}\alpha_{-\mathbf{k}\downarrow}^{\dagger} \\ \beta_{\mathbf{k}} &= \alpha_{-\mathbf{k}\downarrow}\alpha_{\mathbf{k}\uparrow} . \end{aligned} \tag{4.37}$$

In terms of these quantities, the expectation value of an operator $\hat{O}$ is, to leading order in the small quantity $\xi$:

$$\begin{aligned} \langle \hat{O} \rangle_s = & \left\langle o \middle| O^{(N)} \middle| o \right\rangle \\ & + \sum_{k>k_F} v_{\mathbf{k}}^2 \left\langle o\,\beta_{\mathbf{k}} \middle| \left[ \hat{O}^{(N+2)} - O_{oo}^{(N)} \right] \middle| \beta_{\mathbf{k}}^{\dagger}\, o \right\rangle \\ & + \sum_{k<k_F} u_{\mathbf{k}}^2 \left\langle o\,\beta_k^{\dagger} \middle| \left[ \hat{O}^{(N-2)} - O_{oo}^{(N)} \right] \middle| \beta_{\mathbf{k}}\, o \right\rangle \\ & + \sum_{k>k_F, k'<k_F} u_{\mathbf{k}}v_{\mathbf{k}}u_{\mathbf{k}'}v_{\mathbf{k}'} \left\langle o \middle| \left[ \hat{O}^{(N)} - O_{oo}^{(N)} \right] \middle| \beta_{\mathbf{k}}^{\dagger}\beta_{\mathbf{k}'}\, o \right\rangle \\ & + \sum_{k>k_F, k'>k_F} u_{\mathbf{k}}v_{\mathbf{k}}u_{\mathbf{k}'}v_{\mathbf{k}'} \left\langle o\,\beta_{\mathbf{k}} \middle| \left[ \hat{O}^{(N+2)} - O_{oo}^{(N)} \right] \middle| \beta_{\mathbf{k}'}^{\dagger}\, o \right\rangle \end{aligned}$$

$$+ \sum_{k<k_F,k'<k_F} u_{\mathbf{k}} v_{\mathbf{k}} u_{\mathbf{k}'} v_{\mathbf{k}'} \left\langle o\, \beta^\dagger_{\mathbf{k}} \right| \left[ \hat{O}^{(N-2)} - O^{(N)}_{oo} \right] \left| \beta_{\mathbf{k}'}\, o \right\rangle$$

$$+ \sum_{k<k_F,k'>k_F} u_{\mathbf{k}} v_{\mathbf{k}} u_{\mathbf{k}'} v_{\mathbf{k}'} \left\langle o\, \beta^\dagger_{\mathbf{k}} \beta_{\mathbf{k}'} \right| \left[ \hat{O}^{(N)} - O^{(N)}_{oo} \right] \left| o \right\rangle . \tag{4.38}$$

In Eq. (4.38), the operators $\hat{O}^{(N)}$, $\hat{O}^{(N-2)}$, and $\hat{O}^{(N+2)}$ are the $N$, $N-2$, and $N+2$ –particle realizations of the operator $\hat{O}$, and $O^{(N)}_{oo}$ the expectation value of the operator $\hat{O}$ in the $N$–particle correlated ground state. Special cases for the operator $\hat{O}$ are, of course, the Hamiltonian $\hat{H}$ ( thus defining the total energy), the particle–number operator $\hat{N}$, and in particular the combination $\hat{H} - \mu\hat{N}$, where $\mu$ is a Lagrange multiplier (the chemical potential) introduced to adjust the average particle number $\langle \hat{N} \rangle_s = N$. Inserting $\hat{H} - \mu\hat{N}$ into the expansion (4.38), we recover the effective interaction and single–particle energies of section 3, *e.g.*

$$\langle o\, \beta_{\mathbf{k}} | \left[ \hat{H}^{(N+2)} - \mu(N+2) - H^{(N)}_{oo} + \mu N \right] | \beta^\dagger_{\mathbf{k}}\, o \rangle = 2[e_k - \mu] , \tag{4.39}$$

and

$$\begin{aligned} &\langle o | \left[ \hat{H}^{(N)} - \mu\hat{N} - (H^{(N)}_{oo} - \mu N) \right] | \beta^\dagger_{\mathbf{k}} \beta_{\mathbf{k}'}\, o \rangle \\ &\quad = \langle \mathbf{k}\uparrow, -\mathbf{k}\downarrow | W(1,2) | \mathbf{k}'\uparrow, -\mathbf{k}'\downarrow \rangle_a \\ &\quad + [e(k) - e(l)] \langle \mathbf{k}\uparrow, -\mathbf{k}\downarrow | N(1,2) | \mathbf{k}'\uparrow, -\mathbf{k}'\downarrow \rangle_a , \end{aligned} \tag{4.40}$$

$$\begin{aligned} &\langle o\, \beta_{\mathbf{k}} | \left[ \hat{H}^{N+2} - \mu(N+2) - (H^{(N)}_{oo} - \mu N) \right] | \beta^\dagger_{\mathbf{k}'}\, o \rangle \\ &\quad = \langle \mathbf{k}\uparrow, -\mathbf{k}\downarrow | W(1,2) | \mathbf{k}'\uparrow, -\mathbf{k}'\downarrow \rangle_a \\ &\quad + (e(k) + e(l) - 2\mu) \langle \mathbf{k}\uparrow, -\mathbf{k}\downarrow | N(1,2) | \mathbf{k}'\uparrow, -\mathbf{k}'\downarrow \rangle_a . \end{aligned} \tag{4.41}$$

Accordingly, we may write the energy of the superfluid state in the form

$$\begin{aligned} \langle \hat{H} - \mu\hat{N} \rangle_s &= H^{(N)}_{oo} - \mu N \\ &+ 2 \sum_{k>k_F} v_k^2 (e(k) - \mu) + 2 \sum_{k<k_F} u_k^2 (e_k - \mu) + \sum_{\mathbf{k},\mathbf{k}'} u_{\mathbf{k}} v_{\mathbf{k}} u'_{\mathbf{k}} v'_{\mathbf{k}} \mathcal{P}_{\mathbf{k}\mathbf{k}'} \end{aligned} \tag{4.42}$$

with the "pairing interaction"

$$\begin{aligned} \mathcal{P}_{\mathbf{k}\mathbf{k}'} &= \langle \mathbf{k}\uparrow, -\mathbf{k}\downarrow | \mathcal{W}(1,2) | \mathbf{k}'\uparrow, -\mathbf{k}'\downarrow \rangle_a \\ &+ (|e_k - \mu| + |e_{k'} - \mu|) \langle \mathbf{k}\uparrow, -\mathbf{k}\downarrow | \mathcal{N}(1,2) | \mathbf{k}'\uparrow, -\mathbf{k}'\downarrow \rangle_a . \end{aligned} \tag{4.43}$$

With the results (4.42) and (4.43), we have arrived at a formulation of the theory which is formally identical to the BCS theory for weakly interacting systems. The correlation operator serves here to tame the short–range dynamical correlations. The special correlations between Cooper–pairs are entirely decoupled. We may proceed now in the conventional way to determine the Bogoljubov– amplitudes $u_{\mathbf{k}}$, $v_{\mathbf{k}}$, by variation of the condensation energy (4.42) with respect to the phase angle $\eta_{\mathbf{k}}$; to compute the superfluid condensation energy; or to investigate the local stability of the normal ground state by second variation. Most notably, we have discovered the same effective interactions which surfaced in the calculation of CBF perturbation corrections to the energy.

### 4.3. CBF, the optimization problem, and momentum–dependent correlations

We have in the past two sections formulated two of the standard approaches of many–particle theory in the language of correlation operators and correlated basis functions: Landau's quasiparticle picture, and the BCS theory of the superfluid phase transition. In each of these cases, we have been able to identify effective interactions which can, in principle, be used to replace phenomenological models of effective interactions for the same purposes.

We turn in this section to the relationship between the optimization problem for the ground state, its consequences for the convergence of CBF perturbative expansions, and further interpretations of effective interactions.

One way of phrasing the optimization problem is that for all variations $\delta u_2(\mathbf{r},\mathbf{r}')$ around the optimal $u_2(\mathbf{r},\mathbf{r}')$ we have

$$\begin{aligned}
0 = \delta_u H_{oo} &= \frac{1}{2}\left[\langle o|(\hat{H}-H_{oo})\sum_{i<j}\delta u_2(\mathbf{r}_i,\mathbf{r}_j)|o\rangle + \text{c.c.}\right]\\
&= \frac{1}{4}\int d^3r d^3r' \delta u_2(\mathbf{r},\mathbf{r}') \times\\
&\quad\times\left[\langle o|(\hat{H}-H_{oo})(\hat{\rho}(\mathbf{r})\hat{\rho}(\mathbf{r}') - \delta(\mathbf{r}-\mathbf{r}')\hat{\rho}(\mathbf{r}))|o\rangle + \text{c.c.}\right]\\
&= \frac{1}{4I_{oo}}\int \frac{d^3q d^3q'}{(2\pi)^6}\delta u_2(\mathbf{q},\mathbf{q}') \times\\
&\quad\times\sum_{h\,h'}\left[\langle\Phi_0|F^\dagger(\hat{H}-H_{oo})F a^\dagger_{\mathbf{q}+\mathbf{h}}a^\dagger_{\mathbf{q}'+\mathbf{h}'}a_{\mathbf{h}}a_{\mathbf{h}'}|\Phi_0\rangle + \text{c.c.}\right]\,. \qquad (4.44)
\end{aligned}$$

Since $u_2(\mathbf{r},\mathbf{r}')$ is arbitrary, the sum of matrix elements in the last line must be zero; moreover, due to momentum conservation in the translationally system, we must have $\mathbf{q}' = -\mathbf{q}$. The states $\mathbf{h}$ and $\mathbf{h}'$ must both lie within the model Fermi sea $|\Phi_0\rangle$ — they must be "hole states"—, whereas the states $\mathbf{h}+\mathbf{q}$ and $\mathbf{h}'-\mathbf{q}$ must lie outside — they must be "particle states". Hence, the optimization condition is equivalent to the statement that

$$\begin{aligned}
&\frac{1}{I_{oo}}\sum_{h\,h'}\langle\Phi_0|F^\dagger(H-H_{oo})F a^\dagger_{\mathbf{q}+\mathbf{h}}a^\dagger_{\mathbf{h}'-\mathbf{q}}a_{\mathbf{h}}a_{\mathbf{h}'}\,|\Phi_0\rangle\\
&= \sum_{h\,h'}\bar{n}(\mathbf{h}+\mathbf{q})\bar{n}(\mathbf{h}'-\mathbf{q})\left(\frac{I_{mm}}{I_{oo}}\right)^{1/2}\langle\mathbf{h},\mathbf{h}'|\mathcal{H}(1,2)|\mathbf{h}+\mathbf{q},\mathbf{h}'-\mathbf{q}\rangle_a = 0\,, \qquad (4.45)
\end{aligned}$$

where $I_{mm}$ is the norm of the correlated state $F a^\dagger_{\mathbf{q}+\mathbf{h}}a^\dagger_{\mathbf{q}+\mathbf{h}'}a_{\mathbf{h}}a_{\mathbf{h}'}|\Phi_0\rangle$. In other words, the optimization condition is equivalent to the statement that the *Fermi sea average* (4.45) of the effective interaction is zero for optimized correlation functions. In the Bose liquid, all "hole" states have zero momentum, and the statement trivializes to the simple fact that, if the pair correlations are optimized, the effective two–body interaction is zero.

The statement that, the better the correlations the smaller any perturbative corrections, is *per se* quite plausible. But we can draw a number of interesting conclusions on the geometric interpretation of CBF theory and on situations where the CBF corrections should have their maximum impact. From a more general point of view, we are interested to see which correlation operator would produce CBF perturbative corrections.

Since, as seen above, the optimization of the correlations is an important aspect for the interpretation of CBF corrections, we apply a simplification of the FHNC–EL equation that shows, while not necessarily being of a sufficient accuracy for high–precision numerical applications, the essential features. The simplest version of the FHNC equations that is consistent with optimization is the approximation $\tilde{X}_{de}(q) = 0$, $\tilde{X}_{ee}(q) = S_F(q) - 1$. In this approximation, the FHNC equations reduce to two equations,

$$\Gamma_{dd}(r) = \exp[u_2(r) + N_{dd}(r)] - 1, \qquad X_{dd}(r) = \Gamma_{dd}(r) - N_{dd}(r), \tag{4.46}$$

$$\tilde{N}_{dd}(q) = \frac{\tilde{X}^2_{dd}(q)^2 S_F(q)}{1 - \tilde{X}_{dd}(q) S_F(q)} \tag{4.47}$$

$$S(q) = \frac{S_F(q)}{1 - \tilde{X}_{dd}(q) S_F(q)} = S_F(q)(1 + \tilde{\Gamma}_{dd}(q) S_F(q))\,. \tag{4.48}$$

The Euler–equation is

$$\frac{\hbar^2 q^2}{4m}\left(S(q) - 1\right) + S'(q) = 0\,, \tag{4.49}$$

where $S'(q)$ is obtained by the aforementioned "priming operation", using

$$S'_F(q) = -\frac{\hbar^2 q^2}{4m}\left(S_F(q) - 1\right)$$

$$\Gamma'_{dd}(r) = (1 + \Gamma_{dd}(r))\left(v(r) - \frac{\hbar^2}{2m}\nabla^2 u_2(r)\right) + \Gamma_{dd}(r) N'_{dd}(r)\,. \tag{4.50}$$

The pair correlation function $u_2(r)$ is eliminated in favor of $\Gamma_{dd}(r)$ using Eq. (4.46) and, after a few algebraic manipulations, the Euler equation (4.49) can be rewritten as

$$S(k) = \frac{S_{\rm F}(k)}{\sqrt{1 + \frac{4mS^2_{\rm F}(k)}{\hbar^2 k^2}\tilde{V}_{\rm p-h}(k)}}$$

$$V_{\rm p-h}(r) = [1 + \Gamma_{\rm dd}(r)]\, v(r) + \Gamma_{\rm dd}(r) w_I(r) + \frac{\hbar^2}{m}\left|\nabla\sqrt{1 + \Gamma_{\rm dd}(r)}\right|^2 \tag{4.51}$$

$$\tilde{w}_I(q) = -\frac{\hbar^2 q^2}{4m}\left[\frac{1}{S_F(q)} - \frac{1}{S(q)}\right]^2\left[2\frac{S(q)}{S_F(q)} + 1\right]. \tag{4.52}$$

As a consequence of the Euler equations, one also derives readily the relationships

$$\tilde{\Gamma}'_{\rm dd}(q) = \frac{\hbar^2 q^2}{4m}\tilde{\Gamma}_{\rm dd}(q)\left[1 - \frac{2}{S_{\rm F}(q)}\right], \tag{4.53}$$

$$\tilde{\mathcal{W}}(q) = -\frac{\hbar^2 q^2}{2mS_F(q)}\tilde{\Gamma}_{\rm dd}(q)\,. \tag{4.54}$$

Returning to CBF theory, we will see in the next chapter that subsets of CBF diagrams can be derived from a "correlation operator". There, we will show that CBF perturbative corrections can be generated from an *ansatz*

$$\Psi_o(1,\ldots,N) = \left|e^S\, o\right\rangle , \tag{4.55}$$

where the operator $S$ is written in the second quantized form (5.14). The correlations generated by that operator are evidently quite complicated. In the simplest approximation, it will be seen that

$$S_{p,p';h,h'} = -\frac{\langle p,p'|\,\mathcal{H}\,|h,h'\rangle}{e_p + e_{p'} - e_h - e_{h'}} . \tag{4.56}$$

To interpret some of the features of these correlations, we go back to the structure (3.8) of the off–diagonal matrix elements and use the local approximations (3.36). Consistent with these local approximations, we disregard the normalization ratio $I_{mm}/I_{oo}$ and use free single–particle energies. Then,

$$\begin{aligned}
&\left\langle \mathbf{q}+\mathbf{h}, -\mathbf{q}+\mathbf{h}'\middle|\mathcal{H}\middle|\mathbf{h},\mathbf{h}'\right\rangle \approx \left\langle \mathbf{q}+\mathbf{h}, -\mathbf{q}+\mathbf{h}'\middle|\mathcal{W}_{\rm loc}\middle|\mathbf{h},\mathbf{h}'\right\rangle\\
&+\frac{\hbar^2}{4m}\left((\mathbf{q}+\mathbf{h})^2 + (-\mathbf{q}+\mathbf{h}')^2 - h^2 - h'^2\right)\left\langle \mathbf{q}+\mathbf{h}, -\mathbf{q}+\mathbf{h}'\middle|\mathcal{N}_{\rm loc}\middle|\mathbf{h},\mathbf{h}'\right\rangle\\
&= \frac{1}{N}\left[\tilde{\mathcal{W}}_{\rm loc}(q) + \frac{\hbar^2}{4m}\left((\mathbf{q}+\mathbf{h})^2 + (-\mathbf{q}+\mathbf{h}')^2 - h^2 - h'^2\right)\tilde{\Gamma}_{dd}(q)\right]\\
&= \frac{1}{N}\frac{\hbar^2}{2m}\tilde{\Gamma}_{dd}(q)\left[-\frac{q^2}{S_{\rm F}(q)} + \left(q^2 + \mathbf{q}\cdot(\mathbf{h}-\mathbf{h}')\right)\right] ,
\end{aligned} \tag{4.57}$$

where we have used in the last line the result (3.38) for $\tilde{\mathcal{W}}_{\rm loc}(q)$ as well as the relationship (4.53) following from optimization.

Three observations can be made:

- The numerator of the $S$ operator consists of a local and a non–local part. However, the *Fermi–sea average* of the square bracket in Eq. (4.57) over the hole states are as defined in Eq. (4.45) vanishes:

$$\sum_{h,h'} \bar{n}(|\mathbf{h}+\mathbf{q}|)\bar{n}(|\mathbf{h}'-\mathbf{q}|)\left[-\frac{q^2}{S_{\rm F}(q)} + \left(q^2 + \mathbf{q}\cdot(\mathbf{h}-\mathbf{h}')\right)\right] = 0 . \tag{4.58}$$

  Recall that, in order to derive the above form of the effective interaction, the optimization condition has been used. Moreover, this feature comes from an *exact cancellation* between the two terms.
- The term in round brackets in Eq. (4.57) vanishes as one approaches the Fermi surface $q \to 0$ due to the restriction that $\mathbf{h}+\mathbf{q}$ and $\mathbf{h}'-\mathbf{q}$ must lie outside the Fermi sea. Therefore, one would expect that the maximum impact of CBF theory will be found when one considers effects that happen close to the Fermi surface. The above statements quantify our more intuitive view expressed earlier that the Jastrow–Feenberg variational theory should do well for average quantities like the energy, but not so well especially when one looks at effects like the Fermi–liquid interaction close the the Fermi surface.

- If we *further simplify* the $S$–operator by setting the energy denominator in Eq. (4.56) to a constant, we arrive at a correlation operator of the so–called "backflow"–form

$$\nabla\Gamma_{\rm dd}(r)\cdot\nabla\,. \tag{4.59}$$

  Such correlation operators that act like gradients on the Slater determinant have proven to be a very useful method in Monte Carlo calculations to move the nodes of the wave function. The phrase "backflow" should, however, be taken with much reservation: "Backflow" was originally introduced by Feynman and Cohen [35] to describe *excitations,* especially at the roton minimum in liquid $^4$He. It is not clear whether this has much to do with the simple observation *on the ground state* that the nodes of the Slater wave function are not expected to be identical to those of the ground state wave function.

## 5. Infinite order CBF theory

### 5.1. *Introduction*

Let us now turn to the extension of CBF theory to high orders in the off–diagonal matrix elements $H'_{mn}$. To demonstrate the technical problem, consider the perturbation formula (2.21). When applying this formula, we must specify the set of states $|\Phi_m\rangle$ over which the sums of (2.21) are to run. A natural classification is to count the number $d$ of orbitals in which the states $|\Phi_m\rangle$, $|\Phi_n\rangle$, and $|\Phi_p\rangle$ differ from each other and from the ground state $|\Phi_o\rangle$. For a state–independent Jastrow operator $F$, and $d = 2$, the operators $\mathcal{W}(1,2)$ and $\mathcal{N}(1,2)$ have been constructed in Sect. 3.3. The extension of these derivations to higher values of $d$ is thereafter reasonably obvious and does not require a repetition of the reasonably tedious derivations presented there; the properties of these operators may be verified by straightforward cluster expansion techniques. [21]

For $d = 2$ and $d = 3$, $\mathcal{W}^{(d)}(1,\ldots,d)$ and $\mathcal{N}^{(d)}(1,\ldots,d)$ are irreducible $d$–body operators in the sense that they vanish if one of the particles is removed far from the others. This property is no longer maintained for $d \geq 4$. We have, for example

$$\begin{aligned}\mathcal{N}^{(4)}(1,2,3,4) &= \mathcal{N}_c^{(4)}(1,2,3,4)\\ &+\mathcal{N}^{(2)}(1,2)\mathcal{N}^{(2)}(3,4)+\mathcal{N}^{(2)}(1,3)\mathcal{N}^{(2)}(2,4)+\mathcal{N}^{(2)}(1,4)\mathcal{N}^{(2)}(2,3)\,,\end{aligned} \tag{5.1}$$

where $\mathcal{N}_c^{(4)}(1,2,3,4)$ stands for the "connected" part of the operator $\mathcal{N}^{(4)}(1,2,3,4)$; this is the part that vanishes if any of the four particles is moved far from the others. The corresponding decompositions of $\mathcal{W}^{(4)}(1,2,3,4)$ and the $\mathcal{W}^{(d)}$ and $\mathcal{N}^{(d)}$ operators for $d > 4$ are obvious. If we include, *e.g.*, $d = 4$ combinations in the second–order CBF correction (that is the second term on the right hand side of Eq. (2.21), we obtain unlinked contributions with an unphysical dependence on the particle number $N$. These contributions are canceled by $d = 2$ contributions to the fourth–order correction, that is given by the last two contributions explicitly given in Eq. (2.21) and a further sixth–order term of similar structure which will be spelled out in section 5.3. A systematic further development of the CBF–perturbation series must therefore include higher-order terms in the perturbation series and *simultaneously* include contributions with larger $d$ in the lower-order terms.

### 5.2. *Correlated coupled cluster theory*

There are various ways to derive a perturbative expansion in a correlated basis; one "pedestrian" approach has been mentioned in section 2.2; others may be found in the original literature. [113–18] These approaches become evidently quite cumbersome when one goes to higher orders; one would therefore like to develop methods to generate CBF perturbative corrections to the energy by integral–equation methods. One possible route is the so–called "correlated coupled cluster method". [36] This method uses the algorithms and ideas of the original coupled cluster theory due to Coester and Kümmel [37,38], which have been discussed in Navarro's lectures. [39]

The advantage of using the coupled cluster technique is that it provides a vehicle for generating large classes of CBF diagrams by integral equations. These will be used in the later sections of this chapter in the analysis of certain, well–known sub–classes of diagrams and the possible refinement of our description of many–body systems in situations where simple correlation operators already provide a reasonable overall picture. The original intention [36] in developing these methods has been somewhat different: Originally, one was aiming for a method for including state–dependence into the problem, with the goal of a better understanding of the nuclear matter problem. This goal has not fully materialized: Instead, from the insights derived in the "twisted chain" analysis of Ref. 40, one would come to the somewhat disappointing insight that much *is not well* with the nuclear matter problem, and new techniques are needed which are not necessarily supplied by coupled cluster theory — neither in its original, nor in its "correlated" version.

Our intention here is to formulate a theory that is as close as possible to the original coupled–cluster (or $\exp(S)$) many–body theory, but formulated in terms of the effective interactions introduced in chapter 3. Among others, we want a theory that is formulated entirely in terms of two–body operators. As a matter of course, one could also deal with three–body operators. However, we feel that there is still much to explore at the two–body level, and three–body effects are quite appropriately described at the level of the variational theory as described in Refs. 41 and 42.

The coupled cluster theory writes the ground–state wave function in the $\exp(S)$ form

$$|\Psi_o\rangle = \exp S|\Phi_o\rangle\,, \tag{5.2}$$

where

$$S = \sum_{n\geq 2} S_n \tag{5.3}$$

is a sum of $n$–particle, $n$–hole $(p - h)$ operators.

$$S_n = \frac{1}{n!} \sum_{p_1 \ldots h_n} S_{p_1 \ldots p_n; h_1 \ldots h_n} a^\dagger_{p_1} \ldots a^\dagger_{p_n} a_{h_n} \ldots a_{h_1} \,. \tag{5.4}$$

The $S_n$ may be determined either by a variational procedure

$$\frac{\delta}{\delta S^\dagger_n} \frac{\langle \Phi_o | e^{S^\dagger} \hat{H} e^S | \Phi_o \rangle}{\langle \Phi_o | e^{S^\dagger} e^S | \Phi_o \rangle} = 0 \tag{5.5}$$

or, more conveniently, by writing the Schrödinger equation in the form

$$e^{-S} \hat{H} e^S | \Phi_o \rangle = E | \Phi_o \rangle, \tag{5.6}$$

and observing that for all $n$–particle $n$–hole states $|\Phi_{\mathbf{m}}\rangle$

$$\langle \Phi_{\mathbf{m}} | e^{-S} \hat{H} e^S | \Phi_o \rangle = 0 \qquad\qquad (m > 0). \tag{5.7}$$

The latter method is known as the coupled cluster or exp($S$)–method. The exp($S$) method is formally equivalent to the variational equations (5.5) which might seem more natural within the present formulation of the many–body problem. But the formulation (5.7) is more advantageous in practical applications since in each step only a finite number of diagrams is generated. For a thorough discussion of this point see Navarro's lectures [39] and Ref. 38

Both formulations, the variational view (5.5) or the coupled–cluster equations (5.7), are readily extended to correlated wave functions. Instead of (5.2), write the exact ground–state wave function in the form

$$|\Psi_o\rangle = |e^S o\rangle\,, \tag{5.8}$$

where $S$ is again of the form (5.3), (5.4), but now written in terms of the creation and destruction operators (2.4), (2.5) for *correlated* states.

The *ansatz* (5.8) evidently builds redundancy into the formulation of the many–body problem; we should therefore discuss the philosophy behind our procedure. If we take, for example, $F$ in the Jastrow–Feenberg form, then we are able to sum large classes of diagrams which could not be summed for more complicated operators. However, the Jastrow–Feenberg operator misses much important physics especially in the vicinity of the Fermi surface. If, on the other hand, we omit the $F$ entirely and build the whole particle–hole structure into the $S$–operator, then we can sum much smaller classes of diagrams, but sum them more accurately. Using *both* types of correlations gives us the "best of both worlds": We maintain the diagrammatic richness of the FHNC–EL theory on the one side, and still have the option of improving the accuracy by which certain classes of diagrams are summed when we deem this to be necessary.

The sharing of the tasks between the $F$ and the $S$ operator is seen quite explicitly in the discussion at the end of the past section: The optimized $F$ has built the average correlations in the system, as a consequence the Fermi–sea average (4.58) vanishes, and the $S$ operator builds in gradient corrections. We shall see these connections more completely below and in the next section.

One can determine the operator $S$ either by the variational prescription

$$\frac{\delta}{\delta S_n^\dagger}\frac{\langle o\, e^{S^\dagger} |\, \hat{H}\, | e^S\, o\rangle}{\langle o\, e^{S^\dagger} | e^S\, o\rangle} = 0 \tag{5.9}$$

or by the Schrödinger equation

$$\hat{H}\big|e^S\, o\big\rangle = E\big|e^S\, o\big\rangle\,, \tag{5.10}$$

where $F$ is always determined beforehand. Projection on complete sets of correlated states $\langle m\, e^{-S}|$ leads, after the elimination of the ground–state energy through

$$E = \frac{\langle o\, e^{-S} |\, \hat{H}\, | e^S\, o\rangle}{\langle o\, e^{-S} | e^S\, o\rangle} \tag{5.11}$$

and proper normalization, to a set of correlated coupled–cluster equations

$$\frac{\langle m\, e^{-S} | \hat{H} | e^S\, o\rangle}{\langle o\, e^{-S} | e^S\, o\rangle} = \frac{\langle o\, e^{-S} |\, \hat{H}\, | e^S\, o\rangle}{\langle o\, e^{-S} | e^S\, o\rangle}\frac{\langle m\, e^{-S} | e^S\, o\rangle}{\langle o\, e^{-S} | e^S\, o\rangle}\,. \tag{5.12}$$

Some considerations are in order before we proceed to establish cluster expansions for the ground–state energy expression (5.11) and the coupled–cluster equation (5.12). We aim at the systematic improvement of correlations that have already been treated, albeit in an approximate form, in the Jastrow–Feenberg correlation function. Therefore, one should, in the operator $S$, not go beyond correlations that have been included in $F$ since an overall estimate of multi–particle correlations is considerably easier by introducing and optimizing a multi–particle correlation function. [41,42] Therefore, we restrict ourselves, with a qualification to be discussed below, to two–body operators, *i.e.* we assume that

$$S = S_2 = \frac{1}{2!}\sum_{p_1,p_2;h_1,h_2} S_{p_1,p_2;h_1,h_2}\alpha^\dagger_{p_1}\alpha^\dagger_{p_2}\alpha_{h_2}\alpha_{h_1}\,. \tag{5.13}$$

In analogy to the nomenclature introduced in the coupled cluster theory, we shall refer to this as to the "C-SUB(2)" approximation. For further reference, we also define the operator

$$\hat{S} = \hat{S}_2 = \frac{1}{2!}\sum_{p_1,p_2;h_1,h_2} S_{p_1,p_2;h_1,h_2} a^\dagger_{p_1} a^\dagger_{p_2} a_{h_2} a_{h_1}\,, \tag{5.14}$$

which is an operator acting on *uncorrelated* states.

A second approximation is made consistent with the C-SUB(2) approximation (5.13): We shall include only those many–body matrix elements of the Hamiltonian and the unit operator which may be written in terms of the effective two–body operators $\mathcal{W}^{(2)}(12) \equiv \mathcal{W}(12)$ and $\mathcal{N}^{(2)}(12) = \mathcal{N}(12)$ and *disconnected* products thereof. Two types of higher–order contributions occur beyond these:

(1) Off–diagonal matrix elements of the connected parts of four–body operators $\mathcal{W}^{(d)}(1,\ldots,d)$ and $\mathcal{N}^{(d)}(1,\ldots,d)$ for $d \geq 3$. Inclusion of these sets of many –body contributions will not cause any qualitative change in the cluster expansions to be developed below.

(2) Differences of $\mathcal{W}^{(d)}[\mathbf{m}]$ or $\mathcal{N}^{(d)}[\mathbf{m}]$ for different reference sets of plane–wave orbitals $\mathbf{m}$, e.g.,

$$\mathcal{N}^{(d)}[\mathbf{m}] - \mathcal{N}^{(d)}[\mathbf{o}] = \mathcal{O}(N^{-1}), \tag{5.15}$$

which enter our expansion through the cancellation of unlinked diagrams. These may be written as matrix elements of $r$–body operators, which are off–diagonal in $d$ ($> 2$) states and diagonal in $(r-d)$ states. The omission of this type of contribution (to the energy or the coupled–cluster equations) corresponds also to our concept of including only effective two–body operators. It allows us, moreover, to omit the explicit specification of the reference state by the argument $m$ as mentioned above.

Some manipulations on the ground–state energy (5.11) are needed before we go on to derive an expansion in terms of powers of $S$. We note that $\langle o\, e^{-S}| = \langle o|$ and write

$$E = \frac{\langle o|\hat{H}|e^{S} o\rangle}{\langle o|e^{S} o\rangle} = H_{\mathbf{oo}} + \frac{\langle o|\hat{H}'|e^{S} o\rangle}{1 + \langle o|(e^{S}-1)\, o\rangle}, \tag{5.16}$$

where $\hat{H}' = \hat{H} - H_{oo}$. We recognize in (5.16) again the effective perturbation $H'_{\mathbf{mn}}$ — see Eq. (2.15) — entering the CBF perturbation correction formula (2.21). Expression (5.16) will turn out to be well behaved in the large $N$ limit, since the disconnected portions (5.1), etc., will be canceled by corresponding denominator diagrams. An expansion in powers of $S$ is now readily derived using the power-series expansion method

$$\begin{aligned} E &= H_{oo} + \langle o|\hat{H}'|S\, o\rangle + \left[\frac{1}{2!}\langle o|\hat{H}'|S^2\, o\rangle - \langle o|\hat{H}'|S\, o\rangle\langle o|S\, o\rangle\right] + \dots \\ &= H_{oo} + (\delta E)_1 + (\delta E)_2 + \dots . \end{aligned} \tag{5.17}$$

With the *ansatz* (5.13) for $S$, the first order term of the expansion (5.17) is

$$(\delta E)_1 = \langle o|\hat{H}'|S\, o\rangle = \frac{1}{(2!)^2}\sum_{pp'hh'} \langle hh'|\mathcal{H}(1,2)|pp'\rangle_a\, S_{pp',(hh')_a}\, . \tag{5.18}$$

The second–order term contains 2$p$–2h and 4$p$–4$h$ matrix elements. In the 4$p$–4$h$ case, we label the particle states with $p_1, \dots, p_4$ and the hole states with $h_1, \dots, h_4$. Keeping only the disconnected terms (5.1), we obtain

$$\begin{aligned} (\delta E)_2 &= \frac{1}{2!}\langle o|\hat{H}'|S^2\, o\rangle - \langle o|\hat{H}'|S\, o\rangle\langle o|S\, o\rangle \\ &= \frac{1}{16}\sum_{p_i,h_i}\Big[\frac{1}{2}\langle h_1h_2h_3h_4|\mathcal{H}^{(4)}|p_1p_2p_3p_4\rangle_a - \langle h_1h_2|\mathcal{H}(1,2)|p_1p_2\rangle_a \\ &\quad \times \langle h_3h_4|\mathcal{N}(1,2)|p_3p_4\rangle_a\Big] S_{p_1p_2,(h_1h_2)_a} S_{p_1p_2,(h_1h_2)_a}\, . \end{aligned} \tag{5.19}$$

We now keep, according our strategy outlined above, in $\mathcal{H}^{(4)}$ only those terms that can be written as disconnected products of two–body operators, for example in the form $\mathcal{H}(1,2)\mathcal{N}(3,4)$. Keeping only these terms we first of all show that the disconnected terms cancel. Rearranging the state sums allows us finally to write the energy as in the form

$$E = H_{oo} + \frac{1}{4} \sum_{pp'hh'} \langle hh'|\mathcal{H}|pp'\rangle_a \mathcal{S}_{pp',(hh')_a} \tag{5.20}$$

with

$$\mathcal{S}_{pp',(hh')_a} = S_{pp',(hh')_a} + \frac{1}{4} \sum_{p_1p_2h_1h_2} \langle h_1h_2|\mathcal{N}|p_1p_2\rangle_a \times \\ \times \left[(S^2)_{pp'p_1p_2,(hh'h_2h_2)_a} - S_{pp',(hh')_a} S_{p_1p_2,(h_1h_2)_a}\right] + \ldots \tag{5.21}$$

Here, we mean with $(S^2)_{pp'p_1p_2,(hh'h_2h_2)_a}$ the product of two sets of particle–hole amplitudes, where the hole states are fully antisymmetrized.

The factorization of the final result (5.20) is not unexpected; note that the expression (5.16) contains off–diagonal matrix elements of the interaction only between the ground state and an $n$–particle–$n$–hole state. Since we have *chosen* to keep only those operators that factor into pairs of disconnected two–body operators, *one* of these operators is an interaction, and all the others are non–orthogonality correction that have, at this level, been defined into the renormalized $\mathcal{S}$ operator. Carrying our expansion (5.17) to higher orders in $S$ would lead to higher–order contributions to the $\mathcal{S}$ operator and includes vast classes of nonorthogonality corrections. (Note that $\mathcal{S}$ is identical to $S$ in the limit $F \to 1$.)

For the rather simple form (5.21) to be useful, we must also show that the correlated coupled cluster equations (5.12) can be formulated in terms of the renormalized operator $\mathcal{S}$. For that purpose, we must study the expansion (5.21) in some detail and formulate the general rules according to which the operator $\mathcal{S}$ is calculated from $S$. It is again convenient to use a diagrammatic language which is here the Goldstone–type graphical notation of conventional exp $(S)$ theory. [39]. We need the additional specification that we represent $\mathcal{N}(1,2)$ by a dashed, horizontal line. Some typical diagrams are shown in Fig. 10, from which the general construction principle of $\mathcal{S}$ becomes quite obvious:

$\mathcal{S}$ is represented by the sum of all diagrams that can be constructed from $S$ and $\mathcal{N}(1,2)$ according to the following rules:

(i) the external lines enter only $S$;
(ii) only internal lines may enter $\mathcal{N}(1,2)$,
(iii) no two $\mathcal{N}(1,2)$ operators may be connected directly by a particle or a hole line.

The definition of $\mathcal{S}$ is now readily extended to the inclusion of $S_d$ and $\mathcal{N}^{(d)}$ for $d > 2$. We define $\mathcal{S}$ to be the sum of all diagrams formed according to the rules given above which contribute to the energy through the effective two–body interaction $\mathcal{H}(1,2)$. An example of an $S_4$ term contributing to $\mathcal{S}$ is shown in Fig. 11. The introduction of the operator $\mathcal{S}$ is, at this point, of course, of only esthetic appeal as long as we are bound to derive an algorithm for calculating $\mathcal{S}$ from a given

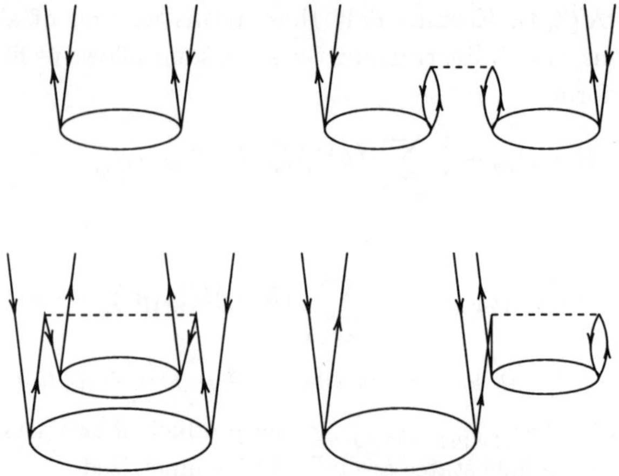

Fig. 10. Some typical diagrams contributing to the expansion of the renormalized operator $\mathcal{S}$ in terms of the original operator $S$ and non–orthogonality corrections.

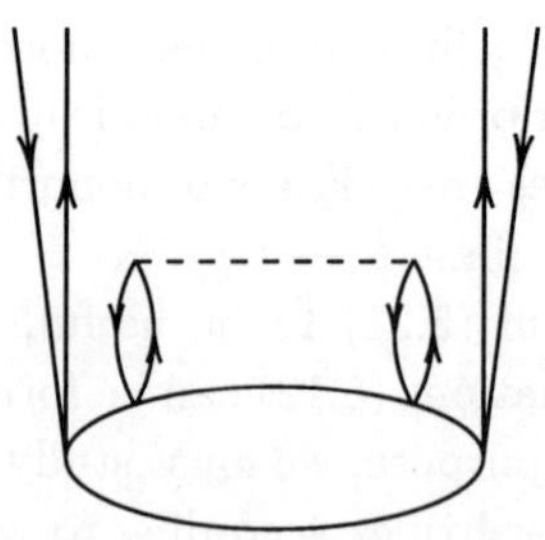

Fig. 11. An example for an $S_4$ diagram contributing to $\mathcal{S}$.

$S$. We will find, however, that the same quantity can be introduced to re–sum our correlated coupled–cluster Eqs. (5.12) and eliminate the "bare" $S$ entirely from our theory.

We proceed now to an expansion of the coupled–cluster equations (5.12) in powers of $S$. Since we aim only at the determination of $\mathcal{S}$, it is sufficient to choose $|\Phi_{\rm m}\rangle$ to be a 2p–2h state. This simplifies, in turn, our considerations due to

$$\langle me^{-S}| = \delta_{mo}\langle o| - \langle mS| \,. \tag{5.22}$$

Eq. (5.12) is rewritten as

$$\frac{\langle m|\hat{H}'|e^{S}o\rangle}{1+\langle o|(e^{S}-1)o\rangle} = \frac{\langle o|\,\hat{H}'\,|e^{S}o\rangle}{1+\langle o|(e^{S}-1)o\rangle}\frac{\langle m|e^{S}o\rangle}{1+\langle o|(e^{S}-1)o\rangle}\,. \tag{5.23}$$

We now expand in powers of $S$

$$\begin{aligned}
0 &= \langle m|\hat{H}'|o\rangle \\
&+ \langle m|\hat{H}'|So\rangle - \langle m|\hat{H}'|o\rangle\langle o|S\,o\rangle - \langle m|o\rangle\langle o|\hat{H}'|So\rangle \\
&+ \frac{1}{2!}\langle m|\hat{H}'|S^2 o\rangle - \langle m|\hat{H}'|o\rangle\langle o|S^2\,o\rangle - \langle m|o\rangle\langle o|\hat{H}'|S^2 o\rangle \\
&- 2\langle m|\hat{H}'|So\rangle\langle o|S\,o\rangle - 2\langle o|\hat{H}'|So\rangle\langle m|S\,o\rangle + \langle m|\hat{H}'|o\rangle\langle m|S\,o\rangle^2 \\
&+ \dots
\end{aligned} \tag{5.24}$$

and separate the diagonal terms:

$$\begin{aligned}
0 &= \langle m|\hat{H}'|o\rangle + (H_{mm} - H_{oo})\langle \Phi_m|\hat{S}|\Phi_o\rangle \\
&+ \sum_{n\neq m} \left[\langle m|H'|n\rangle - \langle m|H|o\rangle\langle o|n\rangle - \langle m|o\rangle\langle o|H'|n\rangle\right] \langle \Phi_n|\hat{S}|\Phi_o\rangle \\
&+ \frac{1}{2!}\sum_{n\neq m} \left[\langle m|H'|n\rangle - \langle m|H'|mo\rangle\langle o|n\rangle - \langle m|o\rangle\langle o|H'|n\rangle\right] \langle \Phi_n|(\hat{S}^2)|\Phi_o\rangle \\
&- \sum_{nn'\neq m} \left[2\langle m|\hat{H}'|n\rangle\langle o|n'\rangle + 2\langle o|\hat{H}'|n\rangle\langle m|n'\rangle - \langle m|H'|o\rangle\langle o|n\rangle\langle o|n'\rangle\right] \\
&\qquad \times\langle \Phi_n|\hat{S}|\Phi_o\rangle\langle n'|\hat{S}|\Phi_o\rangle + \dots .
\end{aligned} \tag{5.25}$$

Eq. (5.25) is an implicit, non–linear equation for the particle–hole amplitudes $S$. For example, if we ignore all but the first two terms, we obtain the second–order CBF correction for the energy.

$$S_{p,p';h,h'} = -\frac{\langle p,p'|\,\mathcal{H}(1,2)\,|h,h'\rangle}{e_p + e_{p'} - e_h - e_{h'}} \tag{5.26}$$

which we have used in section 4.3 for the analysis of momentum–dependent correlations, *cf.* Eq. (4.56). The further analysis of the expansion (5.25) goes along the same lines as the one for the expansion of the ground–state energy (5.17). It is, however, quite tedious and purely technical so that we pass over the details and give only the general calculational recipe below.

By construction, the states $|m\rangle$ and $|n\rangle$ are $2p$–$2h$ states. To be definite, we write

$$\begin{aligned}
|m\rangle &= \alpha^\dagger_{p_1}\alpha^\dagger_{p_2}\alpha_{h_2}\alpha_{h_1}|o\rangle, \\
|n\rangle &= \alpha^\dagger_{p_3}\alpha^\dagger_{p_4}\alpha_{h_4}\alpha_{h_3}|o\rangle .
\end{aligned} \tag{5.27}$$

Matrix elements of effective two–body operators of the type specified in section 3.3 arise when $|m\rangle$ and $|n\rangle$ differ by two or four orbitals. States differing by two orbitals may be generated by coincidence of

- the particle orbitals in $|m\rangle$ with the particle orbitals in $|n\rangle$,
- the hole orbitals in $|m\rangle$ with the hole orbitals in $|n\rangle$, or
- a particle–hole pair in $|m\rangle$ with a particle–hole pair in $|n\rangle$.

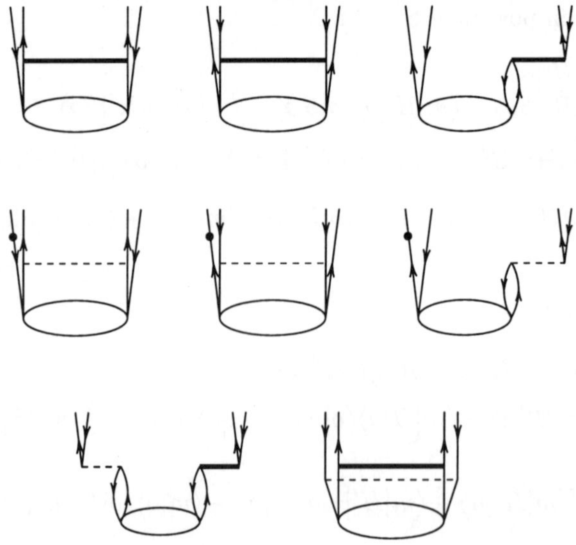

Fig. 12. Some typical diagrams contributing to the correlated coupled cluster equation for $S$

In the $d = 4$ contribution we take again all terms that can be written as matrix elements of unlinked products of two–body operators. An inspection and classification of the distinct contributions is again most efficiently performed using a graphical language. We have to supplement the graphical elements introduced above by the effective two–body interaction $\mathcal{H}(1,2)$ (depicted as a heavy solid line) and the CBF–single–particle (or hole) energies depicted as a heavy dot on a particle or hole line. Some typical diagrams are shown in Fig. 12.

The first three diagrams are known from the conventional $\exp(S)$ theory with the bare interaction replaced by the tamed effective interaction $\mathcal{H}(1,2)$. They generate (upon solving the $\exp(S)$ equations) particle ladders, hole ladders, and ring diagrams, respectively. The next three diagrams show non–orthogonality corrections which are generated from the first diagram by replacing the effective interaction by an $\mathcal{N}$ line, and substituting a single–particle (or hole) energy on one of the outgoing (or incoming) particle (hole) lines directly attached to $S$. Finally we show some diagram arising from the $d = 4$ portions of Eq. (5.25) which represent ladder and ring diagrams containing an interaction line and a normalization correction. We discover already the diagrammatic construction scheme according to which further diagrams contributing to our expansion (5.25) are generated. The expansion (5.25) of the coupled–cluster equations is represented graphically by the sum of all diagrams which have the following properties:

(i) Two hole lines entering and two particle lines exiting at the top of each diagram,
(ii) an arbitrary number of $S$ elements,
(iii) an arbitrary number of $\mathcal{N}(1,2)$ elements,

(iv) one effective interaction operator or one CBF single–particle (or hole) energy, and obey the rules (v) and (vi):

(v) the $S$ elements have only incoming hole lines and outgoing particle lines,

(vi) no $\mathcal{N}(1,2)$ line and no $\mathcal{H}(1,2)$ or $e$ element may be connected directly to another $\mathcal{N}(1,2)$ element.

The further explicit construction of higher–order contributions to Eq. (5.25) serve essentially to confirm the rules (i) to (vi). It suffices to sketch the general way the calculation goes. First, we classify the distinct off–diagonal quantities $H_{mn}$, $J_{mn}$ according to the number of plane–wave orbitals in which the states $|m\rangle$ and $|n\rangle$ (one of which will occasionally be identified with the ground state $|o\rangle$) differ. Products of more than one off–diagonal quantity are sorted in such a way that all terms with the same sum of $d$ values are kept together. Due to our choice (5.13) for $S$, the maximum $d$ value for an $n^{\text{th}}$–order contribution in $S$ will be $d_{\text{max}} = 2n+2$. Next, we retain only those portions of the $d$–body operators, which factorize into products of two body ($\mathcal{H}(1,2)$ or $\mathcal{N}(1,2)$) operators. Upon cancellation of all unlinked diagrams we arrive at our final linked expression for a certain $n^{\text{th}}$–order contribution to our equations (5.25). It is not necessary to present all diagrams that are obtained by this procedure explicitly; progress is as usual accelerated by studying the structure of the expansion and certain subclasses of diagrams of the same topology.

First, we study all diagrams of second order (in $S$) which contain one $\mathcal{N}(1,2)$ element and have a common factor $e_p + e_{p'} - e_h - e_{h'}$. It turns out that (removing the stated common factor) these are identical with the corresponding diagrams appearing in the second–order term of our "dressed" $2p$–$2h$ operator $\mathcal{S}$ (Eq. (5.21)). Identical sub–series arise also from diagrams forming extensions of the "particle–ladder", "hole–ladder", etc., diagrams shown in Fig. 12 (see Fig. 13).

In fact, in all other cases we have studied, we found the building up of the series (5.21). This statement may, of course, also be obtained directly from our construction rules (i) to (vi); as already mentioned, further explicit elaboration of the series (5.21) serves to confirm these rules. Consequently, we may re–sum all $2p$–$2h$ sub–diagrams formed from $S$ and $\mathcal{N}(1,2)$ objects according to the rules spelled out earlier in this section, into the renormalized $2p$–$2h$ operator $\mathcal{S}$. Since, according to the graphical rules defined above, this re–summation can be performed in any place where a bare $S$ element appears, we can re–sum vast classes of nonorthogonality corrections by simply replacing everywhere $S$ by $\mathcal{S}$ and omitting the diagrams summed in the latter object. By this resummation procedure we can eliminate $S$ entirely from our equations in favor of $\mathcal{S}$, and achieve a considerable simplification of the equations. The rules according to which graphical contributions to our (new) coupled–cluster equations in terms of $\mathcal{H}(1,2)$, $\mathcal{N}(1,2)$ and $\mathcal{S}$ are constructed are identical with the rules (i)–(vi) given above, with the additional provision

(vii) no $2p$–$2h$ sub–diagrams occur in which all external lines enter an $\mathcal{S}$ operator, with the trivial exception of the single $\mathcal{S}$ operator.

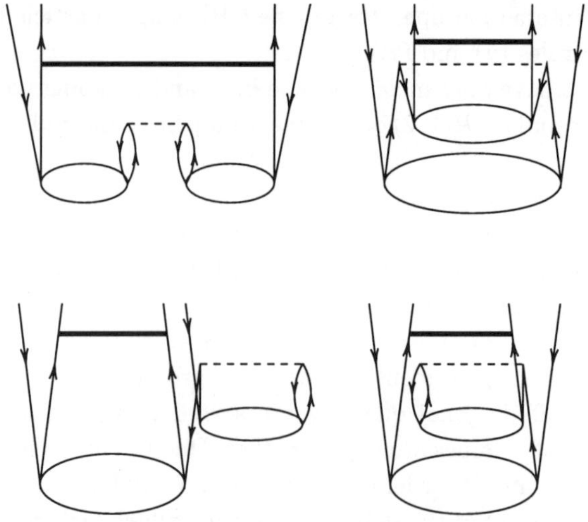

Fig. 13. Diagrams contributing to the correlated coupled cluster equations which are re–summed by the introduction of $\mathcal{S}$

The basic purpose of the exercise of deriving coupled cluster equations in a correlated basis is to establish a method by means of which whole classes of CBF diagrams can be generated by integral equations. The issue appears to be a purely technical one at the first glance, and we can rely on the interpretation of different classes of diagrams (rings, ladders, self–energy corrections) from the ordinary coupled cluster theory discussed in Navarro's lectures. [39] One can, for example, select specific approximations and thus derive a "Bethe–Goldstone", a "Galitszkii" or an "RPA" equation in a correlated basis. However, this view is somewhat superficial, and we shall encounter in the next section a completely different interpretation of the procedure when we focus on a specific class of diagrams, namely the ring–diagrams of CBF theory.

## 5.3. *CBF ring diagrams*

We now take a view that is slightly different from the integral–equation approach of the preceding section by focusing on a specific class of CBF diagrams, namely the ring–diagrams. We will carry out the analysis in a constructive manner, writing down explicitly the relevant diagrams, and rearranging them in intuitive patterns. An alternative procedure would be to follow the elegant analytic proof of Bishop and Lührmann [43] of the fact that the RPA–ring diagrams are a proper subset of the SUB2 approximation of coupled cluster theory, and extend that proof to the CBF version of the coupled cluster theory derived in the preceding section.

This section will display an important new facet of the Jastrow–Feenberg/CBF approach to the many–body problem, namely the inter–relationships between CBF perturbation theory and the diagrammatics of the Jastrow–Feenberg variational method. Whereas we have, in the previous section, considered each CBF matrix element mostly as a "black box", we will now examine the *interplay* between CBF theory and the FHNC summations in the sense that we will see that diagrams that are *topologically* similar in the sense that they have the same momentum flux have also similar physical meaning and can be interpreted as approximations of each other.

We go back to the CBF perturbation expansion (2.21) of the ground–state energy and the systematic procedure, derived in the previous chapter, to generate CBF diagrams to a satisfactorily high order. The result of these derivations is a perturbation expansion that is structurally very similar to an ordinary Goldstone expansion. There are differences, or course, due to the fact that we always deal with many–body wave functions. This is reflected in the appearance of effective three–, four–,... $n$–body interactions and in the normalization corrections which we have encountered above.

We will, in this section, take a very pragmatic point of view and restrict our considerations to an analytically simple case *by ignoring all exchange diagrams.* This is not an absolute necessity and many of the results will also hold in much more general situations. [44] However, if we want to derive the relationship between CBF theory and, for example, the ordinary random phase approximations, we should include only such diagrams where a clean connection can be expected.

Let us look at the first term in the perturbation expansion (2.21).

$$(\Delta E)_2 = -\sum_m{}' \frac{H'_{om}H'_{mo}}{H_{mm} - H_{oo}} \,. \tag{5.28}$$

For ease of writing we abbreviate

$$e_{ph} = e_p - e_h \,. \tag{5.29}$$

Taking $d = 2$, we can write $(\delta E)_2$ as

$$(\Delta E)_2 = -\frac{1}{4}\sum \frac{\left|\langle hh'|\mathcal{H}|pp'\rangle_a\right|^2}{e_{ph} + e_{p'h'}} \approx -\frac{1}{2}\sum \frac{\left|\langle hh'|\mathcal{H}|pp'\rangle\right|^2}{e_{ph} + e_{p'h'}} \,. \tag{5.30}$$

The structure (3.40) of the effective two–body interaction suggests the expansion

$$(\Delta E)_2 = (\Delta E)_2^{(1)} + (\Delta E)_2^{(0)} \,, \tag{5.31}$$

$$(\Delta E)_2^{(1)} = -\frac{1}{4}\sum \frac{\left|\langle hh'|\mathcal{W}|pp'\rangle\right|^2}{e_{ph} + e_{p'h'}} \,, \tag{5.32}$$

$$\begin{aligned}(\Delta E)_2^{(0)} = -\frac{1}{4}\sum\Big[&\langle pp'|\mathcal{N}|hh'\rangle\langle hh'|\mathcal{W}|pp'\rangle + \langle pp'|\mathcal{W}|hh'\rangle\langle hh'|\mathcal{N}|pp'\rangle \\ &+ e_{ph}\langle pp'|\mathcal{N}|hh'\rangle\langle hh'|\mathcal{N}|pp'\rangle\Big] \,.\end{aligned} \tag{5.33}$$

Due to the absence of energy denominators in the term $(\Delta E)_2^{(0)}$, all hole line summations reduce to exchange functions $\ell(rk_F)$, and all particle–line summations to a $\delta$–function *minus* an exchange function. Consequently, the diagrammatic elements to describe these terms are identical to those used in the variational upper bound $H_{oo}$ of the ground–state energy. These are the terms we are looking for: They should *cancel* terms that are generated by the FHNC theory, and replace them by Goldstone–type diagrams.

To see the generic rules, we must develop the expansions to higher order. Formally, we start from the series

$$E = \sum_n (\Delta E)_n \tag{5.34}$$

of all CBF ring–diagrams in terms of the interaction $\langle ij|\mathcal{H}|kl\rangle$ and normalizations $\langle ij|\mathcal{N}|kl\rangle$. The index $n$ counts the number of energy denominators. We seek for a rearranged series (5.34)

$$E = \sum_n (\delta E)_n \tag{5.35}$$

obtained by cancellation of all energy numerator terms and subsequent rearrangement according to the number of remaining energy denominators. In second order, the identification of the two distinct contributions of $(\Delta E)_2$ to $(\delta E)_0$ and $(\delta E)_1$ was given in Eq. (5.31).

The explicit construction of the CBF ring diagrams and the cancellation between the energy–denominator and the energy–numerator terms of the CBF–ring diagrams is still reasonably straightforward in third order. Going back to the general expansion (2.21), we restrict the states $|m\rangle$ and $|n\rangle$ to two–particle, two–hole states. Moreover, to make sure that the matrix elements $H'_{mn}$ are "particle–hole" matrix elements, the states $|m\rangle$ and $|n\rangle$ may differ only by one particle–hole pair, *i.e.*

$$|m\rangle = |\alpha_p^\dagger \alpha_{p''}^\dagger \alpha_{h''} \alpha_h \, o\rangle \qquad |n\rangle = |\alpha_p^\dagger \alpha_{p'}^\dagger \alpha_{h'} \alpha_h \, o\rangle \,. \tag{5.36}$$

For this pair of states, we have therefore

$$\begin{aligned} H'_{mn} &= W_{mn} + \frac{1}{2}\left(H_{mm} + H_{nn} - H_{oo}\right) J_{mn} \\ &= \langle h'p''|\mathcal{W}|p'h''\rangle + \frac{1}{2}\left(2e_{ph} + e_{p'h'} + e_{p''h''}\right)\langle h'p''|\mathcal{N}|p'h''\rangle \\ &= \langle h'p''|\mathcal{H}|p'h''\rangle + e_{ph}\langle h'p''|\mathcal{N}|p'h''\rangle \,. \end{aligned} \tag{5.37}$$

It is again convenient to use a diagrammatic language: The diagrammatic conventions used are analogous to those of the correlated coupled cluster equations: Particle and hole lines are drawn as up– and down going arrows; matrix elements of the two–body operators $\mathcal{H}(12)$ and $\mathcal{N}(12)$ are drawn as a horizontal heavy solid line and a dashed line, respectively. Energy denominators are drawn as horizontal bars, and a heavy dot on a particle– or a hole line represents a single–particle energy $e_k$. These single particle energies appear normally as pairs of a particle– and a hole energy in the form $e_p - e_h$ associated with a particle–hole loop. Note especially the third diagram in Fig. 14: This diagram comes from the last term of Eq. (5.37) and

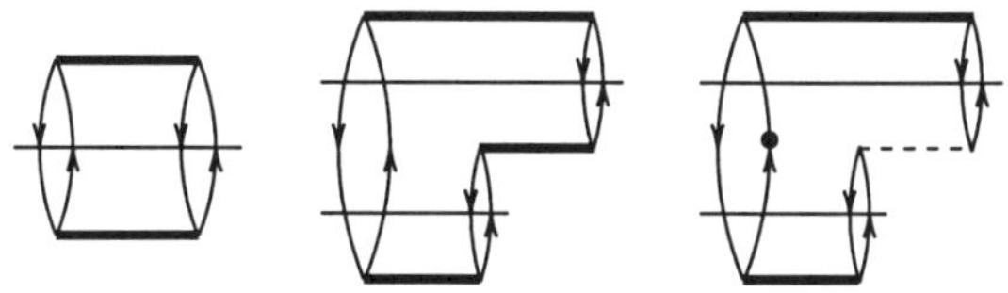

Fig. 14. Diagrammatic representation of second– and third– order CBF ring–diagrams. The heavy solid line represents an effective interaction $\mathcal{H}(1,2)$, and the dashed line a non–orthogonality correction $\mathcal{N}(1,2)$. particle– and hole– lines are drawn as up– and down–going arrows. The horizontal bar indicates an energy denominator, and a filled circle on a particle– or hole line an energy $e_k$; these come normally in the combinations $e_{ph} = e(p) - e(h)$ on a particle–hole loop.

reflects the fact that we are, rigorously speaking, always dealing with many–body matrix elements although they can, occasionally, be written as onr– or two–body operators.

The second and third order diagrams are shown in Fig. 14, and the fourth order diagrams in 15

Writing the third order CBF–correction in terms of the two–body matrix elements $\langle ij|\mathcal{W}|kl\rangle$ and $\langle ij|\mathcal{N}|kl\rangle$, we find for the expansion

$$(\Delta E)_3 = (\Delta E)_3^{(2)} + (\Delta E)_3^{(1)} + (\Delta E)_3^{(0)}\,, \tag{5.38}$$

$$(\Delta E)_3^{(2)} = \sum \frac{\langle pp'|\mathcal{W}|hh'\rangle\langle h'p''|\mathcal{W}|p'h''\rangle\langle hh''|\mathcal{W}|pp''\rangle}{(e_{ph}+e_{p'h'})(e_{ph}+e_{p''h''})} \tag{5.39}$$

$$\begin{aligned}(\Delta E)_3^{(1)} = &\frac{1}{2}\sum\frac{\langle pp'|\mathcal{W}|hh'\rangle}{e_{ph}+e_{p'h'}}\Bigg[\frac{1}{2}e_{p''h''}\langle h'p''|\mathcal{N}|p'h''\rangle\langle hh''|\mathcal{N}|pp''\rangle\\ &+\langle h'p''|\mathcal{W}|p'h''\rangle\langle hh''|\mathcal{N}|pp''\rangle+\langle h'p''|\mathcal{N}|p'h''\rangle\langle hh''|\mathcal{W}|pp''\rangle\Bigg]\\ &+\frac{1}{2}\sum\frac{\langle hh''|\mathcal{W}|pp''\rangle}{e_{ph}+e_{p''h''}}\Bigg[\frac{1}{2}e_{p'h'}\langle pp'|\mathcal{N}|hh'\rangle\langle h'p''|\mathcal{N}|p'h''\rangle\\ &+\langle pp'|\mathcal{W}|hh'\rangle\langle h'p''|\mathcal{N}|p'h''\rangle+\langle pp'|\mathcal{N}|hh'\rangle\langle h'p''|\mathcal{W}|p'h''\rangle\Bigg]\,,\end{aligned} \tag{5.40}$$

$$\begin{aligned}(\Delta E)_3^{(0)} = &\frac{1}{8}\sum\Big[3\langle pp'|\mathcal{W}|hh'\rangle\langle h'p''|\mathcal{N}|p'h''\rangle\langle hh''|\mathcal{N}|pp''\rangle\\ &+3\langle pp'|\mathcal{N}|hh'\rangle\langle h'p''|\mathcal{N}|p'h''\rangle\langle hh''|\mathcal{W}|pp''\rangle\\ &+2\langle pp'|\mathcal{N}|hh'\rangle\langle h'p''|\mathcal{W}|p'h''\rangle\langle hh''|\mathcal{N}|pp''\rangle\\ &+(2e_{ph}+e_{p'h'}+e_{p''h''})\langle pp'|\mathcal{N}|hh'\rangle\langle h'p''|\mathcal{N}|p'h''\rangle\langle hh''|\mathcal{N}|pp''\rangle\Big]\,.\end{aligned} \tag{5.41}$$

The analysis becomes lengthy in fourth order due to the feature of the CBF–perturbation expansion mentioned above: The four–body operators $\mathcal{H}(1234)$ and

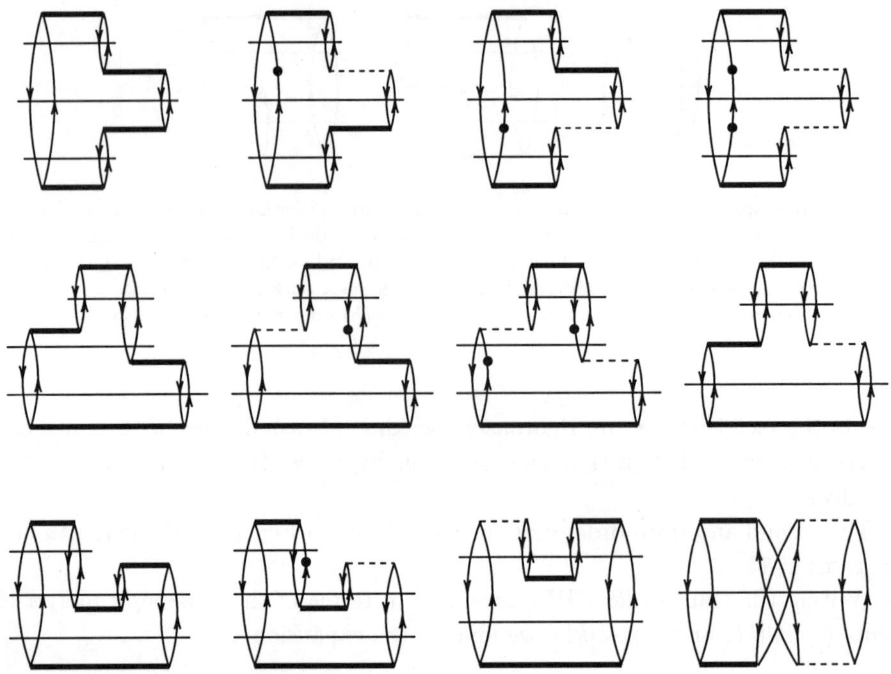

Fig. 15. Fourth–order CBF ring–diagrams. See Fig. 14 for further explanations, note that the diagrams with one or two energy denominators originate from the second and the third order term in Eq. (2.21).

$\mathcal{N}(1234)$ contain unlinked contributions of the form (5.2). In order to construct an irreducible expansion, we have to isolate and cancel all of these disconnected terms. The remaining diagrams are shown in Fig. 15.

We are now ready to read off the third–order contributions to $(\delta E)_0$, $(\delta E)_1$, and $(\delta E)_2$. The contribution to $(\delta E)_2$ is given by the first the term $(\Delta E)_3^{(2)}$ defined in Eq. (5.39). The first contribution to $(\delta E)_1$, *i.e.* the term $(\Delta E)_2^{(1)}$, is supplemented by the term $(\Delta E)_3^{(1)}$ given in Eq. (5.40). Finally, $(\Delta E)_3^{(0)}$ supplements $(\Delta E)_2^{(0)}$ in the expansion of $(\delta E)_0$. Combining $(\Delta E)_2^{(1)}$ with $(\Delta E)_3^{(1)}$ suggests that we renormalize the effective interaction $\mathcal{W}$

$$\mathcal{V}(1,2) = \mathcal{W}_0(1,2) + \mathcal{W}_1(1,2) + \dots \tag{5.42}$$

with $\mathcal{W}_0(1,2) = \mathcal{W}(1,2)$ and

$$\begin{aligned}\langle pp'|\mathcal{W}_1|hh'\rangle = -\sum\Big[&\langle ph''|\mathcal{W}|hp''\rangle\langle p''p'|\mathcal{N}|h''h'\rangle + \langle ph''|\mathcal{N}|hp''\rangle\langle p''p'|\mathcal{W}|h''h'\rangle \\ &+ \frac{1}{2}e_{p''h''}\langle ph''|\mathcal{N}|hp''\rangle\langle p''p'|\mathcal{N}|h''h'\rangle\Big]\,.\end{aligned} \tag{5.43}$$

Anticipating the final result, we evaluate the individual terms in momentum space and express them in terms of $\tilde{V}_{\rm p-h}(q)$ and correction terms that are powers of $\tilde{X}_{dd}(q)$. We find

$$\begin{aligned}
\tilde{\mathcal{W}}_0(q) &= \tilde{\Gamma}'_{dd}(q) - \frac{\hbar^2 q^2}{4m}\tilde{\Gamma}_{dd}(q) \\
&= \tilde{V}_{\rm p-h}(q)\left(1 + 2S_F(q)\tilde{X}_{dd}(q) + 3S_F^2(q)\tilde{X}_{dd}^2(q)\right) + \frac{\hbar^2 q^2}{4m}\tilde{X}_{dd}^2(q) + \mathcal{O}(\tilde{X}_{dd}^3(q)) \\
\tilde{\mathcal{W}}_1(q) &= -2S_F(q)\tilde{\Gamma}_{dd}(q)\tilde{\mathcal{W}}(q) - \frac{\hbar^2 q^2}{4m}\tilde{\Gamma}_{dd}^2(q) \\
&= -2\tilde{V}_{\rm p-h}(q)\left(S_F(q)\tilde{X}_{dd}(q) + 3S_F^2(q)\tilde{X}_{dd}^2(q)\right) - \frac{\hbar^2 q^2}{4m}\tilde{X}_{dd}^2(q) + \mathcal{O}(\tilde{X}_{dd}^3(q))
\end{aligned}$$

and, hence

$$\tilde{\mathcal{V}}(q) = \tilde{\mathcal{W}}_0(q) + \tilde{\mathcal{W}}_1(q) + \ldots = \tilde{V}_{\rm p-h}(q) + \mathcal{O}(\tilde{X}_{dd}^3(q)) . \tag{5.44}$$

These findings are confirmed by extending the rearrangement to the fourth-order ring-diagrams of Fig. 15. For those terms that contain matrix elements of the type $\langle pp'|\mathcal{W}|hh'\rangle$, we find the next correction term to the series (5.42) which cancels the $\mathcal{O}(\tilde{X}_{dd}^3(q))$ term above and shows that $\tilde{\mathcal{V}}(q)$ agrees with $\tilde{V}_{\rm p-h}(q)$ to at least fourth order in $\tilde{X}_{dd}(q)$. We therefore conclude that we can write

$$(\delta E)_2 = -\frac{1}{2}\sum \frac{\left|\langle hh'|V_{\rm p-h}|pp'\rangle\right|^2}{e_{ph} + e_{p'h'}} . \tag{5.45}$$

We also find first correction terms to matrix elements of the type $\langle ph'|\mathcal{W}|hp'\rangle$, which confirm the anticipated result that the sum of all diagrams that contain energy denominators can be written in terms of ordinary RPA ring diagrams in the "particle–hole" interaction $\tilde{V}_{\rm p-h}(q)$.

The expansion also gives sufficient information to identify the leading terms in $(\delta E)_0$. Things simplify significantly if we assume optimized correlations which translates, in our approximation, to using Eq. (4.53) for the effective interactions, in other words we assume that the correlations have been optimized. Then, we find

$$\begin{aligned}
(\Delta E)_2^{(0)} &= \frac{1}{4}\sum_{\mathbf{q}} \frac{\hbar^2 q^2}{2m}\tilde{\Gamma}_{dd}^2(q)S_F(q) , \\
(\Delta E)_3^{(0)} &= -\frac{1}{2}\sum_{\mathbf{q}} \frac{\hbar^2 q^2}{2m}\tilde{\Gamma}_{dd}^3(q)S_F^2(q) .
\end{aligned} \tag{5.46}$$

From the first two terms, (and by verification at fourth order) we conclude that

$$\begin{aligned}
(\delta E)_0 &= \frac{1}{4}\sum_{\mathbf{q}} \frac{\hbar^2 q^2}{2m} S_F(q)\tilde{X}_{dd}^2(q) \\
&= \frac{1}{4}\sum_{\mathbf{q}} \frac{\hbar^2 q^2}{2m} S_F(q)\tilde{\Gamma}_{dd}^2(q)\left[1 - 2S_F(q)\tilde{\Gamma}_{dd}(q) + \ldots\right] .
\end{aligned} \tag{5.47}$$

To interpret our result, we must go back to elementary many–body physics. The Feynman–Hellman theorem says that the ground–state energy per particle can be obtained by coupling constant integration

$$\frac{E}{N} = \frac{E_{HF}}{N} + \frac{1}{2}\int \frac{d^3q}{(2\pi)^3\rho}\tilde{v}(q)\int_0^1 \left[S_\lambda(q) - S_F(q)\right] , \tag{5.48}$$

where $E_{HF}$ is the Hartree–Fock energy, and $S_\lambda(q)$ is the $S(q)$ for a potential $\lambda\tilde{v}(q)$. One can use this theorem also for finding approximate expressions for the energy from an approximate expression for the static structure function $S_\lambda(q)$. For example, the sum of all ring diagrams for a given "particle–hole" potential $\tilde{V}_{\rm p-h}(q)$ is

$$\frac{E^{\rm RPA}}{N} = \frac{E_{HF}}{N} + \frac{1}{2}\int \frac{d^3q}{(2\pi)^3\rho}\tilde{v}(q)\int_0^1 \left[S_\lambda^{\rm RPA}(q) - S_F(q)\right] , \tag{5.49}$$

where

$$S_\lambda^{\rm RPA}(q) = -\int_0^\infty \frac{d(\hbar\omega)}{\pi}\Im m\, \chi^{\rm RPA}(q,\omega) ,$$
$$\chi^{\rm RPA}(q,\omega) = \frac{\chi_0(q,\omega)}{1-\lambda\tilde{v}(q)\chi_0(q,\omega)} . \tag{5.50}$$

Eqs. (5.49) and (5.50) can be used to calculate the sum of all CBF ring diagrams, $\sum_{n=1}^\infty(\delta E)_n$. The first term, $(\delta E)_0$, can be obtained by replacing, in Eq. (5.50), the Lindhard function $\chi_0(q,\omega)$ by a "collective" or "mean spherical" approximation

$$\chi_0^{\rm MSA}(q,\omega) \equiv \frac{2t(q)}{(\omega+i\eta)^2 - \left(\dfrac{t(q)}{S_{\rm F}(q)}\right)^2} . \tag{5.51}$$

This approximation for $\chi_0(q,\omega)$ replaces the particle–hole band by an effective collective mode which is chosen such that the first two energy weighted sumrules are fulfilled:

$$\Im m \int d\omega\, \chi_0^{\rm MSA}(q,\omega) = \Im m \int d\omega\, \chi_0(q,\omega) ,$$
$$\Im m \int d\omega\, \omega\chi_0^{\rm MSA}(q,\omega) = \Im m \int d\omega\, \omega\chi_0(q,\omega) . \tag{5.52}$$

Within the mean spherical approximation, the energy integration can be carried out analytically, leading to the expression (4.51) for $S(k)$. The coupling constant integration can also be carried out, and the resulting energy is exactly *minus* $(\delta E)_0$. The result of all of these considerations is, therefore, quite simple: The Jastrow–Feenberg approximation for the wave function amounts to replacing the the particle–hole propagator by an effective collective mode. Summing all CBF ring diagrams simply removes this approximation, in other words we can write

$$E_{\rm CRPA-Rings} = E_{\rm rings} - E_{\rm rings}^{\rm MSA}$$
$$= \frac{1}{2}\int \frac{d^3q}{(2\pi)^3\rho}\tilde{V}_{\rm p-h}(q)\int_0^1 d\lambda\left(S_\lambda(q) - S_\lambda^{\rm MSA}(q)\right) . \tag{5.53}$$

Thereby, we have established a direct correspondence between the ring diagrams summed by the RPA equations and the chain diagrams of the variational description of the ground state wave function. The degree to which the correspondence can be exploited depends, of course, on the technology available to produce the ingredients of the theory. In the present discussion, we have omitted all exchange diagrams and tacitly also identified the CBF single–particle energies with free kinetic energies. These simplifications have been convenient and have allowed the derivation of closed–form and plausible expressions. We stress that these simplifications are not necessary. However, a very general analysis is not the point of the present discussion: The reason for that is that we have not calculated *all* CBF diagrams, but rather a specific class. Therefore, we should at most expect cancellations between the retained CBF diagrams and a specific class of FHNC diagrams. Since we are dealing with CBF ring diagrams, we must also restrict the analysis to FHNC ring diagrams. If we furthermore omit CBF–exchange diagrams, one should also omit FHNC–exchange diagrams of corresponding topology to remain consistent. One can keep CBF–exchange terms and arrives at similar large–scale cancellations against FHNC–exchange terms, but the calculation is considerably more tedious.[44]

## 6. Dynamics in correlated basis functions

Let us, in the concluding section, turn to the treatment of *dynamics* in correlated basis functions. The ideology is the same as before, but the potential physical outcome is immensely richer than the ground state properties. Historically, one faces the same dilemma as in ground state theories: Intuitively appealing discussions of excitations exist for *weakly interacting* systems, but there is practically no many–body system in the world that can legitimately be called "weakly interacting", and the execution of any theory of excitations without a microscopic foundation relies on semi– or pseudo–phenomenological effective interactions.

As before, CBF theory will be shown to provide a means of mapping the "strongly interacting" problem onto a "weakly interacting" one and, thus, establishes to some extent the legitimacy of phenomenological theories. Depending on the physical system under consideration — electrons, helium liquids, or nucleons — different questions arise. Electron systems are, as usual, the most lenient ones towards simplistic approximations. The real problem of electronic many–particle systems is, with few exceptions, not the short–ranged structure but rather the contamination of the electron system by a non–uniform ionic background. Next in line are nuclear systems. Due to their relatively low density, single–particle pictures are physically reasonable for many purposes; the major problem being the complicated nature of the nucleon–nucleon interaction. At the top of the list of difficulties is again $^3$He. The methods discussed in this section provide an entry point mostly for the study of excitations in $^3$He. However, we know from Saarela's lectures [19] that the simple assumption of one–body excitations misses some essential physics that is recovered only if one also allows for time–dependence of the pair correlations.

We begin by assuming that the essential physics of excited states may be described in the subspace of (correlated) $1p - 1h$ excitations of the Hilbert space. In the absence of correlations, one is led to the time–dependent Hartree–Fock (TDHF) theory [6] which maintains the full antisymmetry of the excited states. A familiar looking expression for the density–density response function like the random–phase approximation (RPA) [22]

$$\chi(q,\omega) = \frac{\chi_0(q,\omega)}{1 - \tilde{V}_{\rm p-h}(q)\chi_0(q,\omega)}, \tag{6.1}$$

where $\chi_0(q,\omega)$ is the familiar Lindhard function, [45] results only if all exchange matrix elements are ignored.

We will derive here the analog of the TDHF equation within the framework of CBF theory. We proceed in three steps:

- First, we embed the small–amplitude limit of the time–dependent Hartree–Fock theory in the CBF context. [46,47] We will draw the connection with the Feynman theory of collective excitations by showing that this theory follows if all matrix elements are replaced by their Fermi sea averages.
- Next, we will introduce a transformation which will bring our equations into the usual form of the time–dependent Hartree–Fock equations with an effective, energy dependent interaction. We will show that the "direct" part of this effective interaction is exactly the "direct" particle–hole interaction of the optimization problem discussed in section 4.3, the quasiparticle interaction discussed in section 4.1, and the effective interaction that is used in calculating the CBF ring diagrams derived in section 5.3. With this, we shall recover the usual formulation of the RPA theory, with definite expressions for effective interactions suitable for strongly interacting systems.
- Finally, we will formulate a theory that can be built upon the best calculations provided by an optimized ground state calculation and replaces those exchange diagrams, that were omitted above, by an "optimal" local approximation.

For the derivation of the CBF version of the TDHF theory, we assume that our system is subjected to a weak, local external field $U_{\rm ext}(\mathbf{r};t)$, and that the excited states can be described as a time–dependent superposition of correlated particle–hole states

$$\begin{aligned} |\Psi(t)\rangle &= \exp\left[-\frac{i}{\hbar}H_{oo}t\right]|\Psi_0(t)\rangle\,, \\ |\Psi_0(t)\rangle &= \frac{1}{I^{1/2}(t)}\exp\left[\sum_{ph} u_{ph}(t)\alpha_p^\dagger\alpha_h\right]|o\rangle, \\ I(t) &= \left\langle o\,\exp\left[\sum_{ph} u_{ph}^*(t)\alpha_h^\dagger\alpha_p\right]\right|\left.\exp\left[\sum_{ph} u_{ph}(t)\alpha_p^\dagger\alpha_h\right]\,o\right\rangle. \end{aligned} \tag{6.2}$$

The weakly–interacting limit $F \to 1$ leads to the general superposition of plane–wave states, which, by virtue of the Thouless theorem [6], may be written in the form

(6.2) by replacing the creation and annihilation operators $\alpha_i^\dagger$, $\alpha_i$ of the correlated states by the usual creation and destruction operators $a_i^\dagger$, $a_i$ of uncorrelated single–particle states. We also can recover the Feynman theory of excitations by assuming that the $u_{ph}(t)$ describe *coherent* states; we will come back to this point further below.

### 6.1. *Equations of motion for correlated states*

A variety of methods to derive equations of motion for small–amplitude excitations is known; the one that is closest to the spirit of the variational theory and that fits naturally into the formulation of our theory is the one already discussed in Saarela's lectures, [19] namely to invoke a least–action principle [48,49]

$$\delta\mathcal{S}[u_{ph}^*(t), u_{ph}(t)] = \delta\int dt\,\mathcal{L}(t) = 0, \tag{6.3}$$

$$\mathcal{L}(t) = \left\langle \Psi(t) \left| \hat{H} + U_{\rm ext}(t) - i\hbar\frac{\partial}{\partial t} \right| \Psi(t) \right\rangle \tag{6.4}$$

for deriving the CBF-analog of the TDHF equations. The least action principle (6.3–6.4) is now applied to a class of correlated, time–dependent states of the form (6.2).

Our immediate task is to cast the Lagrangian defined in Eq. (6.4) in a form that facilitates the evaluation of the stationarity principle (6.3). We first observe that the time–dependent portion of the Lagrangian can be written as

$$\left\langle \Psi(t) \left| -i\hbar\frac{\partial}{\partial t} \right| \Psi(t) \right\rangle = -H_{oo} + \hbar\,\Im m \sum_{ph} \dot{u}_{ph} \left\langle \Psi_o(t) \left| \alpha_p^\dagger \alpha_h \right| \Psi_o(t) \right\rangle . \tag{6.5}$$

We are concerned with small amplitude oscillations; it is therefore sufficient for deriving a non–trivial equation of motion to expand the Lagrangian to second order in the particle–hole amplitudes $u_{ph}$ and $u_{ph}^*$. As a matter of course, one must postulate stationarity of the ground state, in other words that the first variation of the action integral with respect to the particle–hole amplitudes vanishes,

$$\left.\frac{\partial\mathcal{L}}{\partial u_{ph}}\right|_{u_{ph}=0} = \left.\frac{\partial\mathcal{L}}{\partial u_{ph}^*}\right|_{u_{ph}=0} = 0. \tag{6.6}$$

Noting that the external field $U_{\rm ext}(t)$ should be considered of the same order than the particle–hole amplitudes $u_{ph}$, the stationarity conditions (6.6) take the form of a *generalized Brillouin condition*

$$H'_{ph,0} = 0 \tag{6.7}$$

for any $1p - 1h$–state $\left|ph\right\rangle$. If we restrict our discussion to the infinitely extended, homogeneous and isotropic quantum liquid, the condition (6.7) is trivially satisfied due to momentum conservation.

We are now ready to expand the Lagrangian (6.4) to second order in the particle–hole amplitudes:

$$\left\langle \Psi_o(t) \mid \hat{H} - H_{oo} \mid \Psi_o(t) \right\rangle$$

$$= \frac{1}{2} \sum_{pp'hh'} \left[ H'_{o,pp'hh'} u_{ph} u_{p'h'} + \sum_{pp'hh'} H'_{p'h',ph} u^*_{p'h'} u_{ph} + c.c. \right] \tag{6.8}$$

and

$$\left\langle \Psi_o(t) \mid \hat{U}_{\rm ext} \mid \Psi_o(t) \right\rangle = \sum_{ph} \left[ U'_{o,ph} u_{ph} + {\rm c.c.} \right] \tag{6.9}$$

where, in keeping with our usual definitions, $U'_{o,ph} = \left\langle o \middle| U_{\rm ext}(t) - U_{oo} \middle| ph \right\rangle$. The time–derivative term can be further simplified noting that the time–averages of $u_{ph}\dot{u}_{p'h'}$ and $u^*_{ph}\dot{u}^*_{p'h'}$ should be zero, thus

$$i\hbar \int dt\, \Im m \sum_{ph} \dot{u}_{ph} \left\langle \Psi_o(t) \left| \alpha^\dagger_p \alpha_h \right| \Psi_o(t) \right\rangle$$

$$= i\hbar \int dt\, \Im m \sum_{ph,p'h'} \dot{u}_{ph} \left[ \langle 0 \mid pp'hh' \rangle\, u_{p'h'} + \langle p'h' \mid ph \rangle\, u^*_{p'h'} \right]$$

$$= i\hbar \int dt\, \Im m \sum_{ph,p'h'} \dot{u}_{ph} u^*_{p'h'} \langle p'h' \mid ph \rangle\,. \tag{6.10}$$

A set of linear equations of motion is now obtained by taking the first variation of the action integral (6.4)

$$(H_{ph,ph} - H_{oo})\, u_{ph} + \sum_{p'h'} \left[ H'_{ph,p'h'} u_{p'h'} + H'_{pp'hh',o} u^*_{p'h'} \right]$$
$$= U'_{ph,o} + i\hbar \left[ \dot{u}_{ph} + \sum_{p'h'} J_{ph,p'h'} \dot{u}_{p'h'} \right] \tag{6.11}$$
$$(H_{ph,ph} - H_{oo})\, u^*_{ph} + \sum_{p'h'} \left[ H'_{ph,p'h'} u^*_{p'h'} + H'_{o,pp'hh'} u_{p'h'} \right]$$
$$= U'_{ph,o} - i\hbar \left[ \dot{u}^*_{ph} + \sum_{p'h'} J_{ph,p'h'} \dot{u}^*_{p'h'} \right].$$

In the coefficients of the linear equations (6.11) we recover our (off–) diagonal CBF matrix elements:

$$\begin{aligned} H_{ph,ph} - H_{oo} &= e_p - e_h\,, \\ H'_{ph,p'h'} &= \left\langle ph' \middle| \mathcal{H}(1,2) \middle| hp' \right\rangle_a, \\ H'_{pp'hh',0} &= \left\langle pp' \middle| \mathcal{H}(1,2) \middle| hh' \right\rangle_a, \\ J_{ph,p'h'} &= \left\langle ph' \middle| \mathcal{N}(1,2) \middle| hp' \right\rangle_a. \end{aligned} \tag{6.12}$$

Assuming (without loss of generality) harmonic time dependence, we introduce the Fourier decomposition of $u_{ph}(t)$ in the form

$$u_{ph}(t) = x_{ph}e^{-i\omega t} + y^*_{ph}e^{i\omega t}\,. \tag{6.13}$$

One then finds a set of RPA–type equations (we shall refer to these equations as the "correlated RPA (CRPA) equations") of the form

$$\begin{bmatrix} \mathbf{A} & \mathbf{B} \\ \mathbf{B}^* & \mathbf{A}^* \end{bmatrix} \begin{bmatrix} x \\ y \end{bmatrix} = \hbar\omega \begin{bmatrix} \mathbf{M} & 0 \\ 0 & -\mathbf{M}^* \end{bmatrix} \begin{bmatrix} x \\ y \end{bmatrix} + \begin{bmatrix} \mathbf{U} \\ \mathbf{U}^* \end{bmatrix}, \tag{6.14}$$

where

$$\mathbf{A} = (A_{ph,p'h'}) = ([e(p) - e(h)]\delta_{pp'}\delta_{hh'} + \langle ph'|\mathcal{H}(12)|hp'\rangle_a)\,, \tag{6.15}$$

$$\mathbf{B} = (B_{ph,h'p'}) = (\langle pp'|\mathcal{H}(12)|hh'\rangle_a)\,, \tag{6.16}$$

$$\mathbf{M} = (M_{ph,p'h'}) = (\delta_{pp'}\delta_{hh'} + \langle ph'|\mathcal{N}(12)|hp'\rangle_a)\,, \tag{6.17}$$

$$\mathbf{U} = (U_{ph}). \tag{6.18}$$

The normal modes of the system are those solutions of the CRPA equation (6.14) that prevail in the limit of zero external potential, *i.e.* the solutions of the generalized eigenvalue problem

$$\begin{bmatrix} \mathbf{A} & \mathbf{B} \\ \mathbf{B}^* & \mathbf{A}^* \end{bmatrix} \begin{bmatrix} x \\ y \end{bmatrix} = \hbar\omega \begin{bmatrix} \mathbf{M} & 0 \\ 0 & -\mathbf{M}^* \end{bmatrix}. \tag{6.19}$$

## 6.2. *Coherence and the Feynman theory of excitations*

This eigenvalue problem looks quite similar to an ordinary RPA eigenvalue problem [6], see also Eq. (7.36) of Bertsch's lectures. [51] The only new aspect is the *metric matrix* appearing on the right hand side of Eq. (6.19). This metric matrix is obviously due to the non–orthogonality of the correlated single particle states. However, a different interpretation of the equations can be given in terms of the "Feynman" view of excitations. For that purpose, we go back to our analysis of the relationship between the optimization problem and the average–zero property (4.45) of off–diagonal matrix elements. We note that the Fermi–sea averages of both the metric matrix $\mathbf{M}$ and the $\mathbf{A}$ and $\mathbf{B}$ matrices are

$$\begin{aligned} \sum_{hh'} \frac{I_{mm}}{I_{oo}} M_{ph,p'h'} &= \langle \rho_{-q}\rho_q \rangle = S(q)\,, \\ \sum_{hh'} \frac{I_{mm}}{I_{oo}} B_{ph,p'h'} &= \langle o \mid H - H_{oo} \mid \rho_{-q}\rho_q \rangle = 0\,, \\ \sum_{hh'} \frac{I_{mm}}{I_{oo}} A_{ph,p'h'} &= \langle \rho_{-q} \mid H - H_{00} \mid \rho_q \rangle = \frac{1}{2}\langle [\rho_{-q}[T,\rho_q]] \rangle = \frac{\hbar^2 q^2}{2m}\,, \end{aligned} \tag{6.20}$$

the last two lines assuming optimized correlations. If the matrix elements in Eq. (6.19) are replaced by their average values (6.20), the matrix equation reduces to

$$\frac{\hbar^2 q^2}{2m} x(q) = \hbar\omega S(q) x(q)\,, \tag{6.21}$$

where $x(q) = \sum_h x_{ph}$, *i.e.* one obtains the Feynman dispersion relation. The same result would have been obtained if we had not made the approximations on the interaction matrix elements, but on the nature of excitations instead: If we restrict the particle–hole amplitudes to coherent states (*i.e.* assume that the particle–hole amplitudes $u_{ph}$ depend only on the momentum transfer $q = |\mathbf{p} - \mathbf{h}|$, then

$$\sum_{ph} u_{ph} a_p^\dagger a_h = \sum_{ph} u(\mathbf{p} - \mathbf{h}) a_p^\dagger a_h = \sum_q u(q)\hat{\rho}_q\,, \tag{6.22}$$

then the we arrive immediately at the derivation of the Feynman dispersion relation of the preceding section.

Introducing correlations has shifted the emphasis of the equations from the interaction part to the correlation part. The non–trivial part of the interaction matrix elements is not exactly zero as in the boson case, but still averages to zero. The "correlation" part contains now the static structure function, which describes the dynamics of the system. This shows, of course, that much caution must be exercised when applying Eqs. (6.19); any approximations for the matrix elements should be chosen such that the average property is satisfied. Making small mistakes on either side can have disastrous effects on the accuracy of the resultant dispersion relation.

### 6.3. *Diagrammatic reduction*

One thing that is completely missing in the above simplification is the particle–hole continuum. This is evidently a consequence of the averaging procedure, we therefore need to go back to the original equations (6.19) and see how an ordinary RPA theory comes out. The next question is therefore what it takes to get the correct particle–hole continuum, arrive at something like a "local" RPA equation in the usual form (6.1), *and* develop a theory that gives, for a specific excitation operator, the Feynman dispersion relation which we know to be an exact feature. We seek a reformulation of the equations in terms of effective interactions similar to the one carried out in the preceding section. We aim again directly for a density–density response function.

The matrix elements (6.18) of the external potential $U_{\rm ext}$ can be written in terms of matrix elements of the density fluctuation operator $\hat{\rho}(\mathbf{r})$,

$$U_{ph} = \int d^3r \langle ph|\hat{\rho}(\mathbf{r})|o\rangle U_{\rm ext}(\mathbf{r})\,. \tag{6.23}$$

Likewise, the induced density fluctuation is

$$\delta\rho(\mathbf{r}) = \sum_{ph} \left[\langle o|\hat{\rho}(\mathbf{r})|ph\rangle u_{ph} + c.c.\right]. \tag{6.24}$$

The matrix elements of the density operator in momentum space can also be expressed in terms of our correlated–state matrix elements:

$$\begin{aligned}\rho_{ph}(q) &= \langle ph|\sum_k a^\dagger_{k+q}a_k|o\rangle = \sum_k \left[\frac{I_{k+q,k}}{I_{00}}\right]^{1/2}\langle ph \mid k+q,k\rangle \\ &= \sum_k \left[\frac{I_{k+q,k}}{I_{00}}\right]^{1/2}\left[\delta_{h,k}\delta_{p,h+q} + \langle p,k\mid\mathcal{N}\mid h,k+q\rangle_a\right]. \end{aligned} \tag{6.25}$$

Note that the $\rho_q$, being a local operator, commutes with a local $F$, hence the sums go over hole states only and $k+q$ must be a particle state. In the translationally invariant system, momentum conservation in the second term above would also implies that $p = h + q$.

For further reference, we also formulate the static structure function as a sum over specific off–diagonal matrix elements:

$$\begin{aligned}S(q) &= \sum_{k,k'}\langle o|a^\dagger_k a_{k+q}a^\dagger_{k'+q}a_{k'}|o\rangle \\ &= \sum_{k,k'}\left[\frac{I_{k+q,k}I_{k'+q,k}}{I^2_{00}}\right]^{1/2}\left[\delta_{k,k'}\delta_{k+q,k'+q} + \langle k+q,k'\mid\mathcal{N}\mid k,k'+q\rangle_a\right]. \end{aligned} \tag{6.26}$$

We can now formally write down the density–density response function by looking at its definition

$$\delta\rho(\mathbf{r},\omega) = -\int d^3r'\chi(\mathbf{r},\mathbf{r}';\omega)U_{\rm ext}(\mathbf{r}',\omega). \tag{6.27}$$

and, going back to Eq. (6.14), as

$$\chi(\mathbf{r},\mathbf{r}';\omega) = \rho^\dagger(\mathbf{r})\begin{bmatrix}\mathbf{A}-\hbar\omega\mathbf{M}-i\eta & \mathbf{B}\\ \mathbf{B}^* & \mathbf{A}^*+\hbar\omega\mathbf{M}^*+i\eta\end{bmatrix}^{-1}\rho(\mathbf{r}), \tag{6.28}$$

where we think of $\rho^\dagger(\mathbf{r})$ in terms of the particle–hole matrix elements (6.24).

It is feasible to deal with the generalized eigenvalue problem (6.19) or the response function (6.28) numerically [50], provided sufficient care is exercised in the calculation of the matrix elements. Yet, we would like to formulate the equations of motion in a form that emphasizes the interaction, and eliminates the metric matrix and, upon sufficient simplifications, ultimately see how an expression of the form (6.1) comes out. For that purpose, we have to look more closely at the structure of the CBF matrix elements (6.15)–(6.17). To write the equations in a reasonably compact form, define

$$\Omega = \begin{pmatrix}\epsilon_p-\epsilon_h-\hbar\omega-i\eta & 0\\ 0 & \epsilon_p-\epsilon_h+\hbar\omega+i\eta\end{pmatrix} \tag{6.29}$$

and

$$\mathbf{N} = \begin{pmatrix}\langle p'h\mid\mathcal{N}\mid h'p\rangle_a & \langle pp'\mid\mathcal{N}\mid hh'\rangle_a\\ \langle hh'\mid\mathcal{N}\mid pp'\rangle_a & \langle ph'\mid\mathcal{N}\mid p'h\rangle_a\end{pmatrix} \tag{6.30}$$

as well as a corresponding matrix $\mathbf{W}$ consisting of the matrix elements of the operator $\mathcal{W}(1,2)$ introduced in section (3.3). Then the CRPA matrix equation assumes the form

$$\begin{bmatrix} \mathbf{A} - \hbar\omega\mathbf{M} - i\eta & \mathbf{B} \\ \mathbf{B}^* & \mathbf{A}^* + \hbar\omega\mathbf{M}^* + i\eta \end{bmatrix} = \left[\Omega + \frac{1}{2}\Omega\mathbf{N} + \frac{1}{2}\mathbf{N}\Omega + \mathbf{W}\right]$$
$$= \left[\mathbf{1} + \frac{1}{2}\mathbf{N}\right]\left[\Omega + \mathbf{V}_{\mathrm{p-h}}(\omega)\right]\left[\mathbf{1} + \frac{1}{2}\mathbf{N}\right], \tag{6.31}$$

which defines a new, energy–dependent interaction matrix

$$\mathbf{V}_{\mathrm{p-h}}(\omega) = \left[1 + \frac{1}{2}\mathbf{N}\right]^{-1}\left[\mathbf{W} - \frac{1}{4}\mathbf{N}\Omega\mathbf{N}\right]\left[1 + \frac{1}{2}\mathbf{N}\right]^{-1}. \tag{6.32}$$

With this transformation, the calculation of the excited states of the system by solving the CRPA (6.19) is equivalent to solving an ordinary RPA with an energy dependent interaction, where one finds the eigenvalues of

$$\Omega + \mathbf{V}_{\mathrm{p-h}}(\omega) = 0\,. \tag{6.33}$$

With these manipulations and the definition (6.32) we have brought the correlated RPA into a form identical to the conventional RPA, with an energy–dependent effective interaction. Superficially, it looks as we have not accomplished much else but a formal rearrangement. However, the notation $\mathbf{V}_{\mathrm{p-h}}(\omega)$ is of course intended, and it remains to be seen in what sense (or in what approximation) this matrix can be identified with the effective interactions of the preceding sections.

There are two possible ways to proceed. The most general route is a diagrammatic analysis of the expression (6.32). In fact, this is the only way to proceed if one wishes to maintain full generality. The analysis is quite tedious, [44] but interestingly the inverse of the operator $\left[1 + \frac{1}{2}\mathbf{N}\right]$ can be formulated in terms of a *subset* of the diagrams needed to express the off–diagonal matrix elements of the unit operator in section 3.3. Likewise, the elements of the matrix $\mathbf{V}_{\mathrm{p-h}}$ can be expressed in terms of *subsets* of diagrams needed to express $\mathbf{W}$.

### 6.4. *Local approximations*

The second method to proceed is algebraic, but it needs approximations. These are

(1) Take only the direct terms,
(2) Use the local approximation for the operators $\mathcal{N}(1,2) \approx \Gamma_{dd}(r)$ and $\mathcal{W}(1,2) = \mathcal{W}_{\mathrm{loc}}(1,2)$,
(3) Leave out the $z$–factors.

Let us first see the consequences of these approximations for the representations (6.25) and (6.26) of $\rho_{ph}(q)$ and $S(q)$. From their definitions, the $k$ and $k'$ must correspond to hole–states, and $k+q$ as well as $k'+q$ to particle–states.

Then, we get from Eq. (6.25)

$$\begin{aligned}\rho_{ph}(q) &= \sum_k \left[\delta_{h,k}\delta_{p,h+q} + n(k)\bar{n}(k+q)\tilde{\Gamma}_{dd}(q)\right] \\ &= \delta_{p,h+q}n(k)\bar{n}(p)\left(1 + S_F(q)\Gamma_{dd}(q)\right) \\ &\equiv \rho_{ph}^{\rm loc}(q) \end{aligned} \tag{6.34}$$

and from Eq. (6.26)

$$S(q) = S_F(q)\left(1 + S_F(q)\tilde{\Gamma}_{dd}(q)\right) . \tag{6.35}$$

*i.e.* we get the simplest FHNC approximation for the $S(q)$. There may still be more complicated substructures in $\Gamma_{dd}(q)$ containing multiple exchange terms, but exchanges attached to the *external* points of $S(q)$ are due to non–local and exchange matrix elements of the $\mathcal{N}(1,2)$–operator. Details on these structures may be found in Ref. 44.

The simplest FHNC equations consistent with optimization has been described in section 4.3, Eqs. (4.46)-(4.52), we continue to work at this level. The local potential is, in this approximation

$$\langle ph'|\mathcal{W}_{\rm loc}|hp'\rangle = \langle pp'|\mathcal{W}_{\rm loc}|hh'\rangle = \frac{1}{N}\left[\Gamma'_{\rm dd}(q) - \frac{\hbar^2q^2}{4m}\tilde{\Gamma}_{\rm dd}(q)\right] , \tag{6.36}$$

$$\begin{aligned}\langle ph'|\mathcal{H}_{\rm loc}|hp'\rangle &= \langle pp'|\mathcal{H}_{\rm loc}|hh'\rangle \\ &= \frac{1}{N}\left[\tilde{\mathcal{W}}_{\rm loc}(q) + \frac{1}{2}(\epsilon_p + \epsilon_{p'} - \epsilon_h - \epsilon_{h'})\tilde{\Gamma}_{\rm dd}(q)\right] ,\end{aligned} \tag{6.37}$$

where $q = |\mathbf{p} - \mathbf{h}|$ is the momentum transfer. Let us, for the time being, not assume that we have optimized the correlation functions.

The $\mathbf{N}$ matrix is then just a function of momentum transfer $q$,

$$\mathbf{N} = \frac{1}{N}\begin{pmatrix}\tilde{\Gamma}_{\rm dd}(q) & \tilde{\Gamma}_{\rm dd}(q) \\ \tilde{\Gamma}_{\rm dd}(q) & \tilde{\Gamma}_{\rm dd}(q)\end{pmatrix} , \tag{6.38}$$

which enters the equations as a parameter. All summations over particle–hole labels required by the matrix products simply reduce to factors $S_F(q)$. Hence, the inverse

$$\left[1 + \frac{1}{2}\mathbf{N}\right]^{-1} = \mathbf{1} + \frac{1}{2N}\begin{pmatrix}\tilde{X}_{\rm dd}(q) & \tilde{X}_{\rm dd}(q) \\ \tilde{X}_{\rm dd}(q) & \tilde{X}_{\rm dd}(q)\end{pmatrix} . \tag{6.39}$$

can be represented in closed form. One can also carry out the operations (6.32) leading to the matrix $\mathbf{V}_{\rm p-h}(\omega)$ which turns out to be energy independent and of the form

$$\mathbf{V}_{\rm p-h}(\omega) = \frac{1}{N}\begin{pmatrix}\tilde{V}_{\rm p-h}(q) & \tilde{V}_{\rm p-h}(q) \\ \tilde{V}_{\rm p-h}(q) & \tilde{V}_{\rm p-h}(q)\end{pmatrix} \tag{6.40}$$

with (*cf.* Eq. (4.52))

$$\tilde{V}_{\rm p-h}(q) \equiv \tilde{X}'_{\rm dd}(q) - \frac{\hbar^2q^2}{4m}\tilde{X}_{\rm dd}(q) . \tag{6.41}$$

Note that this does not yet assume optimization. With the above transformation, we have indeed brought the "correlated" time–dependent Hartree–Fock equations into a form that is identical to the ordinary RPA equations of motion discussed, for example, in Bertsch's lectures. [51]

### 6.5. *Averaged CRPA equations*

The collective approximation outlined in section 6.2 has led to the Feynman dispersion relation, but did not yield the particle–hole continuum. Section 6.4 has used, on the other hand, an approximate treatment of the integral equations which ignored exchange terms. We were led to the definition of a local "particle–hole" interaction which is the same as the one found in the Euler equations. However, when we calculate the static structure function from a density–density response function obtained from the excitation theory of that section, we will *not* recover the variational result, but rather one that is diagrammatically less complete.

The next question is therefore what it takes to get the correct particle–hole continuum, arrive at something like a "local" RPA equation in the usual sense, *and* develop a theory that gives, for a specific excitation operator, a static structure function that is as good, or better, than the one obtained in FHNC–EL. Evidently, it is far too complicated to calculate all non–local terms if the effective interactions with an accuracy such that the particle–hole averages (6.26) recover the $S(k)$ that comes from the optimization of the ground state. Nevertheless, one would like to combine the benefits of local approximations without compromising the accuracy of the results that had already been obtained. That will, in turn, also provide some feedback on the quality of the Jastrow–Feenberg wave function.

From the way the calculation went for the simplified theory, we conclude that one must not make any approximations for the energy numerator terms appearing in Eq. (3.40) or (6.37), whereas approximations for $\mathcal{N}$ and $\mathcal{W}$ are allowed. The easiest way to generate consistent expressions is to define the Fermi–sea averaging procedure for the overlap matrix elements and then derive the corresponding averages for the interaction matrix elements by the "priming" operation:

$$\mathcal{N}_{\rm loc}(q) \equiv \frac{\sum_{hh'} \langle p'h \mid \mathcal{N} \mid h'p\rangle_a}{\sum_{hh'} 1} = \frac{S(q) - S_F(q)}{S_F^2(q)},$$

$$\mathcal{W}_{\rm loc}(q) \equiv \frac{\frac{\partial}{\partial\beta} \sum_{hh'} \langle p'h \mid \mathcal{N} \mid h'p\rangle_a \Big|_{\beta=0}}{\sum_{hh'} 1} = \frac{S'(q) - S'_F(q)}{S_F^2(q)}. \tag{6.42}$$

We can now simplify the last expression by using the Euler equation, and obtain

$$\mathcal{W}_{\rm loc}(q) = -\frac{\hbar^2 q^2}{4m} \frac{S(q) - S_F(q)}{S_F^2(q)}, \tag{6.43}$$

which defines $\mathcal{W}_{\rm loc}(q)$. The matrix $\mathbf{N}$ is then simply given by Eq. (6.38), with $\tilde{\Gamma}_{\rm dd}(q)$ replaced by $(S(q) - S_F(q))/S_F^2(q)$, and the matrix

$$\left[1+\frac{1}{2}C\right]^{-1} = \mathbf{1} - \frac{1}{2N}\begin{pmatrix} \dfrac{S(q)-S_F(q)}{S(q)S_F(q)} & \dfrac{S(q)-S_F(q)}{S(q)S_F(q)} \\ \dfrac{S(q)-S_F(q)}{S(q)S_F(q)} & \dfrac{S(q)-S_F(q)}{S(q)S_F(q)} \end{pmatrix}. \tag{6.44}$$

Both expressions (6.43) and (6.44) are identical to the expressions of (6.39) and (6.36) of section 6.4 if the simplified FHNC equations of section 4.3 are used.

We can now go back to Eq. (6.32) and calculate the corresponding interaction matrix. As an additional approximation, we need to replace the CBF single–particle energies by a free spectrum. Then we recover, again, a local and energy–independent form (6.40), but with

$$\tilde{V}_{\rm p-h}(q) = \frac{\hbar^2 q^2}{4m}\left[\frac{1}{S^2(q)} - \frac{1}{S_F^2(q)}\right] \tag{6.45}$$

instead of (6.41).

### 6.6. *Response function and dynamic structure function*

To complete the derivations of this section and to justify a number of issues that had already been mentioned in section 5.3, we need to derive the density–density response function. It is not anticipated that we obtain anything else but what is known already with the exception that our theory provides unambiguous definitions and clear approximation schemes for the effective interactions entering the the theory.

We can directly go back to the general definition of the response function (6.28). Using the transformation (6.31), we can also write the density–density response function as

$$\chi(\mathbf{r},\mathbf{r}';\omega) = \rho^\dagger(\mathbf{r})\left[\mathbf{1}+\frac{1}{2}\mathbf{N}\right]^{-1}\left[\mathbf{V}_{\rm p-h}(\omega) - \Omega\right]^{-1}\left[\mathbf{1}+\frac{1}{2}\mathbf{N}\right]^{-1}\rho(\mathbf{r})\,. \tag{6.46}$$

Unless one is prepared to make further approximations, one must now derive a diagrammatic expansion of the correlated matrix elements of the density operator. Again, calculating the combination $\left[\mathbf{1}+\frac{1}{2}\mathbf{N}\right]^{-1}\rho(\mathbf{r})$ simplifies the task by eliminating large classes of diagrams. To show this, we resort again to our local approximations, or, in other words, to keeping the direct terms only. Then we get simply

$$\left[\mathbf{1}+\frac{1}{2}\mathbf{N}\right]^{-1}\rho(\mathbf{r})\rho_{ph}(q) = \delta_{p,h+q}n(k)\bar{n}(p) = \rho^0_{ph}(q)\,, \tag{6.47}$$

in other words the *uncorrelated* matrix elements of the density operator.

With that, we have also derived a density–density response function for the fully correlated theory that has the familiar form (6.1). Of course, the derivation relied on approximations, but also the derivation of (6.1) in a weakly interacting theory required approximations. The most immediate result of our derivations is that a workable form of the driving interaction of an RPA theory has been derived from a microscopic interaction, and the level of implementation is a matter of choice for the user.

Our two different ways to deal with exchange contributions — either keeping only "direct" matrix elements or dealing with the exchange terms in an "averaged" manner, provide *estimates* for the importance of these terms. Quantitative discussions [52] have indicated that these terms are not small. *Whether* the localization spelled out here is appropriate for all purposes is an entirely different question that can only be answered by dealing with the full set of CRPA equations.

A second result arising from the present treatment is, of course, the possibility to calculate an improved static structure function by frequency integration *without* employing the collective approximation for the Lindhard function. It turns out that the differences between the two calculational procedures is generally small, of the percent level. This is one of the reasons for the success of variational wave functions, but the reader is reminded that in both $^3$He and in $^3$He–$^4$He–mixtures, these corrections are not negligible. Beyond the scope of the present treatment we should mention that he correct treatment of the frequency integrations leads in Fermi systems at finite temperatures to a totally different physics.

## Acknowledgments

The work was supported, in part, by the Austrian Science fund under grants No. P12832–TPH and P11098–PHY, the U. S. National Science Foundation under grants PHY–9108066, INT–9014040, and DMR–9509743, as well as by the North Atlantic Treaty Organization and the Austrian Academic Exchange Service. Discussions with numerous senior and junior colleagues are gratefully acknowledged. Parts of this manuscript originate from notes that were written in collaboration with J. W. Clark. I would like to thank Karl Schörkhuber for a critical reading of the manuscript and numerous suggestions.

## References

1. E. Feenberg, *Theory of Quantum Fluids* (Academic, New York, 1969).
2. L. D. Landau, Sov. Phys. JETP **3**, 920 (1957).
3. L. D. Landau, Sov. Phys. JETP **5**, 101 (1957).
4. J. Bardeen, L. N. Cooper, and J. R. Schrieffer, Phys. Rev. **108**, 1175 (1957).
5. S. T. Beliaev, in *Lecture Notes of the 1957 Les Houches Summer School*, edited by C. DeWitt and P. Nozières (Dunod, Paris, 1959), pp. 343–374.
6. D. J. Thouless, *The quantum mechanics of many-body systems*, 2nd ed. (Academic Press, New York, 1972).
7. A. Polls and F. Mazzanti, *Microscopic description of quantum liquids*, this volume.
8. A. D. Jackson, A. Lande, and R. A. Smith, Physics Reports **86**, 55 (1982).
9. A. D. Jackson, A. Lande, and R. A. Smith, Phys. Rev. Lett. **54**, 1469 (1985).
10. E. Krotscheck, A. D. Jackson, and R. A. Smith, Phys. Rev. A **33**, 3535 (1986).
11. V. Apaja *et al.*, Phys. Rev. B **55**, 12925 (1997).

12. E. Feenberg, Phys. Rev. **74**, 206 (1948).
13. P. M. Morse and H. Feshbach, *Methods of Theoretical Physics* (McGraw-Hill, New York - Toronto - London, 1953), Vol. II.
14. J. W. Clark and E. Feenberg, Phys. Rev. **113**, 388 (1959).
15. C.-C. Lin, Ph.D. thesis, Washington University, St. Louis, 1959.
16. P.-O. Löwdin, J. Chem. Phys. **18**, 365 (1950).
17. E. Feenberg, Ann. Phys. (NY) **81**, 154 (1974).
18. R. H. Kulas and W. J. Mullin, J. Low Temp. Phys. **11**, 301 (1973).
19. M. Saarela, *Elementary excitations and dynamic structure of quantum fluids*, this volume.
20. J. W. Clark *et al.*, Nucl. Phys. A **328**, 45 (1979).
21. E. Krotscheck and J. W. Clark, Nucl. Phys. A **328**, 73 (1979).
22. D. Pines and P. Nozieres, *The Theory of Quantum Liquids* (Benjamin, New York, 1966), Vol. I.
23. G. E. Brown, *Many Body Problems* (North Holland, Amsterdam, 1972).
24. G. Baym and C. Pethick, *Landau Fermi Liquid Theory* (Wiley, New York, 1991).
25. S. Babu and G. E. Brown, Ann. Phys. (NY) **78**, 1 (1973).
26. E. Krotscheck *et al.*, Phys. Rev. B **58**, 12282 (1998).
27. A. D. Jackson, E. Krotscheck, D. Meltzer and R. A. Smith, Nucl. Phys. A **386**, 125 (1982).
28. J. C. Wheatley, Rev. Mod. Phys. **47**, 415 (1975).
29. M. Hoffberg, A. E. Glassgold, R. W. Richardson, and M. Ruderman, Phys. Rev. Lett. **24**, 775 (1970).
30. C.-H. Yang and J. W. Clark, Nucl. Phys. A **174**, 49 (1971).
31. E. Krotscheck and J. W. Clark, Nucl. Phys. A **333**, 77 (1980).
32. E. Krotscheck, R. A. Smith, and A. D. Jackson, Phys. Rev. B **24**, 6404 (1981).
33. S. Fantoni, Nucl. Phys. A 363, 381 (1976).
34. C. H. Aldrich and D. Pines, J. Low Temp. Phys. **25**, 677 (1976).
35. R. P. Feynman and M. Cohen, Phys. Rev. **102**, 1189 (1956).
36. E. Krotscheck, H. Kümmel, and J. G. Zabolitzky, Phys. Rev. A **22**, 1243 (1980).
37. F. Coester, in *Lectures in Theoretical Physics: Quantum Fluids and Nuclear Matter* (Gordon and Breach, New York, 1969), Vol. XI B.
38. H. Kümmel, K. H. Lührmann, and J. G. Zabolitzky, Physics Reports **36**, 1 (1978).
39. J. Navarro, R. Guardiola and I. Moliner, *The Coupled Cluster Method and its applications*, this volume.
40. E. Krotscheck, Nucl. Phys. A **482**, 617 (1988).
41. E. Krotscheck, Phys. Rev. B **33**, 3158 (1986).
42. E. Krotscheck, J. Low Temp. Phys. **119**, 103 (2000).
43. R. F. Bishop and K. H. Lührmann, Phys. Rev. B **17**, 3757 (1978).
44. E. Krotscheck, Phys. Rev. A **26**, 3536 (1982).
45. A. L. Fetter and J. D. Walecka, *Quantum Theory of Many-Particle Systems* (McGraw-Hill, New York, 1971).
46. J. W. Clark, in *The many-body problem, Jastrow correlations versus Brueckner Theory*, Vol. 138 of *Lecture Notes in Physics* (Springer, Berlin, Heidelberg, and New York, 1981), p. 184.
47. J. M. C. Chen, J. W. Clark, and D. G. Sandler, Z. Physik A **305**, 223 (1982).
48. P. Kramer and M. Saraceno, *Geometry of the time-dependent variational principle in quantum mechanics*, Vol. 140 of *Lecture Notes in Physics* (Springer, Berlin, Heidelberg, and New York, 1981).
49. A. K. Kerman and S. E. Koonin, Ann. Phys. (NY) **100**, 332 (1976).

50. N.-H. Kwong, Ph.D. thesis, California Institute of Technology, 1982.
51. G. F. Bertsch and K. Yabana, *Density functional theory*, this volume.
52. E. Krotscheck, Ann. Phys. (NY) **155**, 1 (1984).

CHAPTER 8

# THE MAGNETIC SUSCEPTIBILITY OF LIQUID $^3$He

H. Godfrin

*Centre de Recherches sur les Très Basses Températures,*
*Laboratoire Associé à l'Université Joseph Fourier,*
*B.P. 166, 38042 Grenoble Cedex 09, France*
*E-mail: godfrin@labs.polycnrs-gre.fr*

Liquid $^3$He is a model system which allows the investigation of Fermi liquids in a large variety of experimental conditions. In order to understand the effect of the interactions, its properties can be studied as a function of the particle density and temperature. Pure bulk liquid has been extensively investigated, and it constitutes nowadays the canonical example of a three-dimensional Fermi liquid. In addition, liquid $^3$He films adsorbed on a solid substrate offer the possibility to study two-dimensional Fermi fluids where the interactions can be varied in a much wider range than in the bulk. The nuclear magnetism of these systems reveals the subtle properties associated to the large zero point motion of the light $^3$He atoms and to their interactions. We discuss in this lecture NMR measurements performed on these Fermi Liquids and the consequences of the results on the value of the Landau parameters which will be required as an input by future microscopic theories.

## 1. Introduction

Many experimental works have shown that the properties of Liquid $^3$He, a pure, homogeneous and isotropic system of interacting fermions (spin 1/2), can be described by the Landau theory of Fermi Liquids. [1,2] This phenomenological theory associates the excitations of the interacting system (i-particles) to the particles of a non-interacting Fermi gas of the same density. As a consequence, the Fermi fluid shares with the Fermi gas some simple thermodynamic properties: its specific heat is proportional to the temperature, and the low temperature susceptibility (degenerate regime) is independent on the temperature. The interactions lead to a renormalization of the mass and to an enhancement of the T=0 susceptibility. These effects constitute remarkable predictions of Landau's theory, verified by the experiments. [3,4,5,6] However, the magnitude of these effects cannot be inferred from the theory: it is contained in the Landau parameters, which must be deduced from the experiments. This explains the large number of careful determinations of the thermodynamic properties of this system conducted during almost 50 years.

More recently, a large amount of work has been devoted to liquid $^3$He films of atomic thickness. The effect of the dimension is of interest but in addition it is possible to investigate in the two-dimensional (2D) systems a much larger range of densities than in bulk liquid. In the latter, the density range accessible as the pressure is increased (between the saturated vapor pressure and the solidification pressure) is substantial, but restricted to rather dense matter. The absence of a critical point in 2D $^3$He offers in particular the possibility to investigate low and very low density systems, approaching the ideal Fermi gas, as well as very strongly interacting fermionic systems.

In this lecture, the emphasis is placed on the magnetic properties of liquid $^3$He in two and three dimensions. We also discuss the case of $^3$He confined in aerogel.

## 2. The susceptibility of bulk liquid $^3$He

On figure 1 we show the typical behavior of the magnetic susceptibility as a function of temperature. At temperatures well above a few hundred millikelvins, the susceptibility follows quite accurately the Curie Law expected for independent spins. At low temperatures, on the other hand, the susceptibility is constant. The system is in the degenerate regime, and the magnetic behavior is entirely determined by the Fermi statistics. The constant Pauli susceptibility is a well known characteristic of the degenerate Fermi gas. Here, however, we observe the same qualitative behavior for a strongly interacting system, as predicted by Landau's theory. The actual

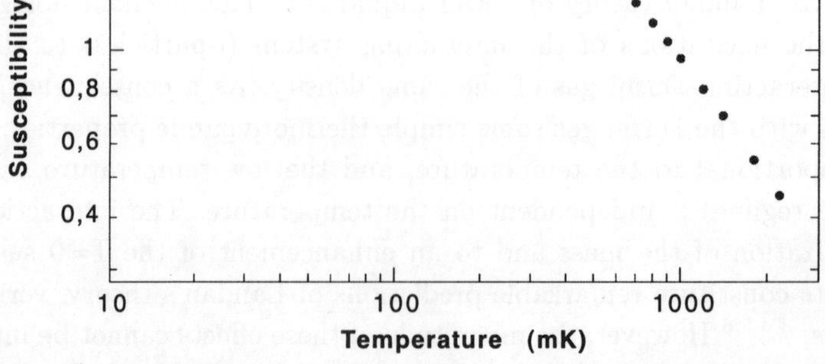


Fig. 1. Susceptibility of liquid $^3$He as a function of temperature, at a pressure of 0.28 $MPa$. Data from Ramm *et al.*. [7] The susceptibility is normalized in such a way that at high temperatures $\chi T = 1$, when $T$ is expressed in kelvins.

value of the Pauli susceptibility is indeed strongly enhanced with respect to that of a Fermi gas of the same particle density.

The data shown here correspond to a low pressure (0.28 MPa). Increasing the pressure leads to a further enhancement of the low temperature susceptibility, as will be seen in the next pages. That is, as the interactions increase, the low temperature susceptibility increases, and at the same time the cross-over temperature between the classical and the degenerate regime is reduced.

This experimental observation has led to two interpretations. In the first approach, the susceptibility increase is considered as a tendency to undergo a ferromagnetic transition, and this has been formally described in a microscopic theory, the "paramagnon model". [8,9,10,11] A different interpretation is provided by the "almost-localized model", [12,13] where the emphasis is placed on the vicinity of a solidification transition.

There is presently no strong experimental argument in favor of either one of these models. In spite of their very different physical grounds, their predictions are qualitatively similar for moderately strong interactions. This controversy has motivated many accurate experimental investigations of the susceptibility of bulk liquid $^3$He.

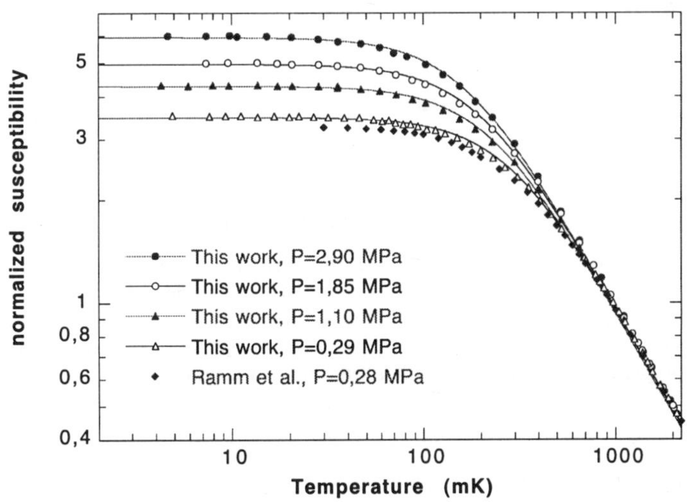


Fig. 2. Normalized susceptibility of liquid $^3$He as a function of temperature, for several pressures (data from Triqueneaux *et al.*. [14] The data from Ramm *et al.* [7] at $P = 0,28\,MPa$ are also shown. The susceptibility is normalized in such a way that at high temperatures $\chi T = 1$, when $T$ is expressed in kelvins.

On figure 2 we show data obtained very recently in my laboratory. [14] The data set includes several pressures reaching that of the melting curve minimum (about $2.9\,MPa$); furthermore, the measurements extend to very low temperatures, in the completely degenerate regime. The results look very similar to the very complete set of data obtained by Ramm *et al.*, [7] which are presently considered as the most accurate reference data of the susceptibility of liquid $^3$He. It was therefore extremely surprizing for us to find a substantial quantitative discrepancy between these experimental results, as seen on figure 2, where the data of Ramm *et al.* are shown for the pressure $0.28\,MPa$. A similar discrepancy is observed at all pressures, as will be shown later on in detail. The main effect is that the susceptibility we measure at low temperatures is larger that that determined by Ramm *et al.* by about 7 %, and even more at high pressures; *i. e.,* the system would be "more magnetic" than is currently believed.

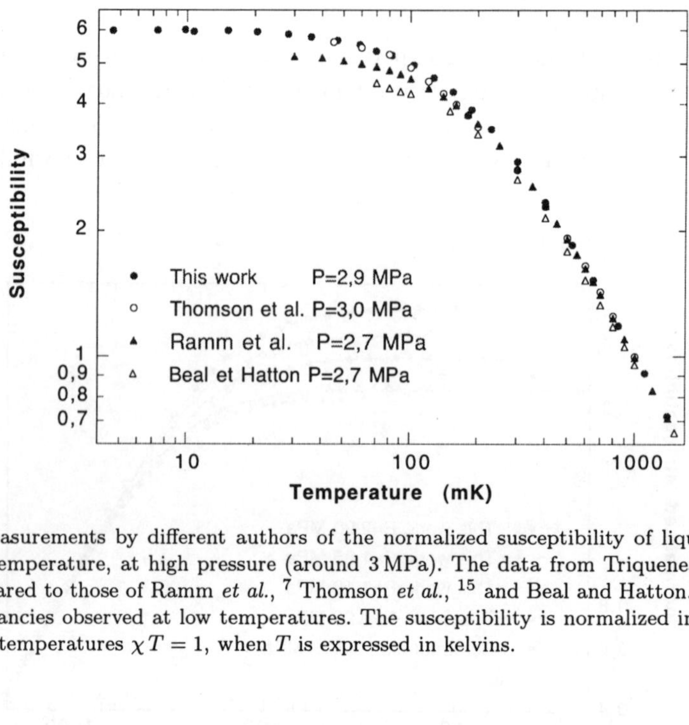


Fig. 3. Measurements by different authors of the normalized susceptibility of liquid $^3$He as a function of temperature, at high pressure (around 3 MPa). The data from Triqueneaux *et al.* [14] can be compared to those of Ramm *et al.*, [7] Thomson *et al.*, [15] and Beal and Hatton. [16] Note the large discrepancies observed at low temperatures. The susceptibility is normalized in such a way that at high temperatures $\chi T = 1$, when $T$ is expressed in kelvins.

This discrepancy is much larger that the quoted accuracy of the experiments (typically 1 to 2 %). It should be pointed out, however, that the data of Ramm *et al.* did already show substantial differences with respect to earlier works. This is evidenced on figure 3, where we plot data from several authors. The low values observed by Beal and Hatton [16] can be explained rather easily, given the fact that substantial corrections due to radiofrequency heating had to be applied. On the other hand, our data agree well with the early results of Thomson *et al.*. [15] It is not the purpose of this lecture to discuss the possible sources of experimental errors or uncertainties, let us briefly say, however, that both Ramm's data and ours have been performed, in principle, in well controlled experimental conditions. The data

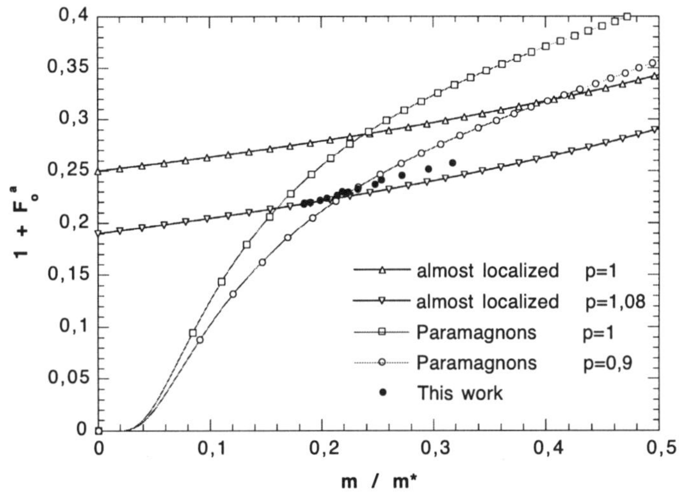


Fig. 4. The magnetic Landau parameters obtained [14] from the susceptibility of liquid $^3$He are plotted, using pressure as an implicit parameter, as a function of the inverse of the normalized effective mass. [17] The predictions of the paramagnon model, [8,9,10] and the almost-localized model, [12,13] for typical adjustable parameters values, are shown.

of Ramm *et al.* have been taken by pulsed NMR methods in a magnetic field of $142\,mT$, while our data have been measured by continuous wave NMR at $29\,mT$; there is no theoretical reason to believe, however, that this can be the origin of the observed discrepancy.

The question is, of course, if such a quantitative difference is significant from the theoretical point of view. The answer is clear when comparing both sets of data to theoretical models, as shown on figure 4.

The paramagnon model [8] assumes a contact interaction $I(\vec{r},\vec{r}') = I\delta(\vec{r}-\vec{r}')$, which leads to a susceptibility enhancement due to spin fluctuations :

$$\frac{\chi_{0,m}}{\chi_{0,m}^{id}} = \frac{1}{1-\bar{I}} \tag{2.1}$$

where $\bar{I} < 1$ is the normalized interaction potential :$\bar{I} = In(E_f)$.

The theory leads to a mass enhancement: [11]

$$\frac{m^\star}{m} = 1 + \frac{9}{2}\,\bar{I}\,\ln\left(1 + \frac{\bar{p}^2\bar{I}}{12\,(1-\bar{I})}\right) \tag{2.2}$$

where $\bar{p} = p/p_f$, of order 1, corresponds to a cut-off in momentum space. [9]

Eliminating the interaction parameter using both equations,2.1 and 2.2, we obtain a relation which can be directly compared to the experimental values :

$$\frac{m^\star}{m} = 1 + \frac{9}{2}\left(1 - \frac{\chi_{0,m}^{id}}{\chi_{0,m}}\right)\ln\left\{1 + \frac{\bar{p}^2\left(1 - \frac{\chi_{0,m}^{id}}{\chi_{0,m}}\right)}{12\,\frac{\chi_{0,m}^{id}}{\chi_{0,m}}}\right\} \tag{2.3}$$

The almost-localized model assumes a fictitious lattice for the fermionic system. This Hubbard-like theory introduces a hard core repulsion as an interaction $U$ for doubly occupied sites. A transition to a localized state is expected for a critical value $U_c$ of the interaction parameter. Defining $I = U/U_c$, the theory yields :

$$\frac{m^\star}{m} = \frac{1}{1-I^2} \tag{2.4}$$

and

$$F_0^a = p\left(\frac{1}{(1+I)^2} - 1\right) \tag{2.5}$$

where $p$, of order 1, depends on the shape of the density of states. [13]

The model gives an effective mass which diverges as the transition to the localized state is approached ($U \to U_c$). Combining the expressions 2.4 et 2.5 we obtain $F_0^a$ as a function of the effective mass by eliminating the parameter $I$ :

$$F_0^a = p\left(\frac{1}{\left(1+\sqrt{1-m/m^\star}\right)^2} - 1\right) \tag{2.6}$$

This formula also allows a direct comparison with experimental values. This is done on figure 4, where the magnetic Landau parameters $1 + F_0^a$ obtained from our measurement [14] of the susceptibility of liquid $^3$He are plotted, using pressure as an implicit parameter, as a function of the inverse of the normalized effective mass obtained by Greywall. [17] This plot allows a direct comparison with the predictions of the paramagnon model [8,9,10] and the almost-localized model, [12,13] for typical adjustable parameters values given on the figure. Note that the experimental values do not show a plateau for small inverse effective masses. The rather constant value $F_0^a = -0.75$ seen in former data, and interpreted as an experimental support to the almost-localized model, must be carefully reconsidered.

It is clear from figure 4 that both models fail to describe accurately the experimental results. The latter display a rather simple linear behavior, which can be naively interpreted as a combination of localization and spin fluctuation effects.

## 3. Liquid $^3$He confined in aerogel

Similar measurements were performed on liquid $^3$He confined in aerogel of 98% porosity. The large specific area of the aerogel gives a large paramagnetic contribution from the solid layers adjacent to the substrate, which can be adequately subtracted from the total susceptibility. The remaining signal should correspond to the susceptibility of bulk liquid.

As seen on figure 5, using the data obtained by Ramm *et al.* [7] leads to a very poor fit, while those of Triqueneaux *et al.* [14] give a good agreement. Note that these measurements are done independently of those shown in the previous section.

It should be noted that the aerogel is not expected to modify substantially the magnetic properties of liquid $^3$He , due to the small value of the Fermi wavelength

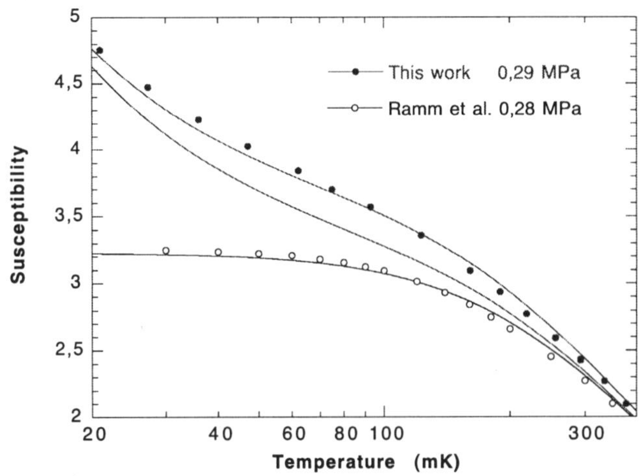


Fig. 5. Analysis of data (dots) taken at a pressure of 0.29 MPa on liquid $^3$He confined in aerogel. [14] A paramagnetic contribution due to the solid $^3$He layers is clearly visible. The circles correspond to the bulk liquid contribution. The middle curve is the expected signal when using the Fermi Liquid parameter $T_f^{\star\star} = 307\,mK$ determined by Ramm *et al.*. [7]

compared to the mean free path of the $^3$He quasiparticles in the very open matrix of aerogel. This is indeed what is found in this experiment.

## 4. Two-dimensional liquid $^3$He

$^3$He atoms adsorbed on a highly homogeneous substrate, like graphite, form two-dimensional Fermi fluids. Their areal density is controlled by the experimentalist simply by choosing the "coverage", *i. e.*, the number of atoms adsorbed on a substrate of a given surface area. The absence of a critical point in two-dimensional $^3$He offers the possibility to investigate the properties of these systems in a very broad density range, going continuously from the low density Fermi gas to the strongly correlated regime.

We show on figure 6 typical susceptibility data [18,19] taken at submonolayer coverage, for the second layer fluid (the first monolayer consists of solid $^4$He which preplates the graphite).

The result is very similar to that shown on figure 5 for bulk $^3$He in aerogel. Here, the paramagnetic contribution visible at very low temperatures is due to $^3$He atoms adsorbed at substrate defects. At temperatures above 10 mK, however, the contribution of the 2D Fermi liquid can be determined unambiguously. Although this curve is qualitatively very similar to those shown before for bulk liquid $^3$He, it should be kept in mind that the two-dimensional nature of the fluid leads to quantitative differences.

These can be seen using the analysis described above. We show on figure 7 the Landau parameter obtained from susceptibility data taken in London [20,21,22] and in Grenoble, [18,19,23,24] represented as a function of the effective mass deduced from

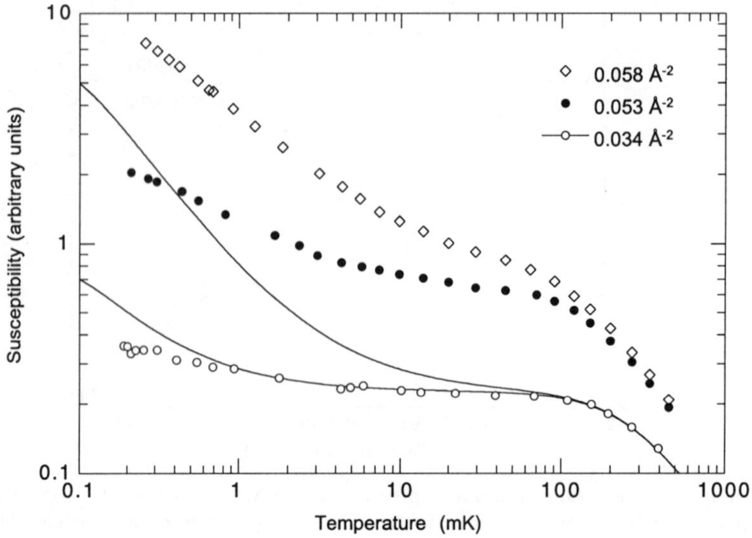


Fig. 6. Magnetic susceptibility [18,19] of liquid $^3$He films of different areal densities, adsorbed on a graphite substrate preplated by a monolayer of $^4$He. A paramagnetic contribution due to $^3$He atoms located at substrate defects is clearly visible at very low temperatures. The bulk liquid contribution is dominant above 10 mK. The lines indicate the signal expected if 0.1 and 1% of the atoms are paramagnetic.

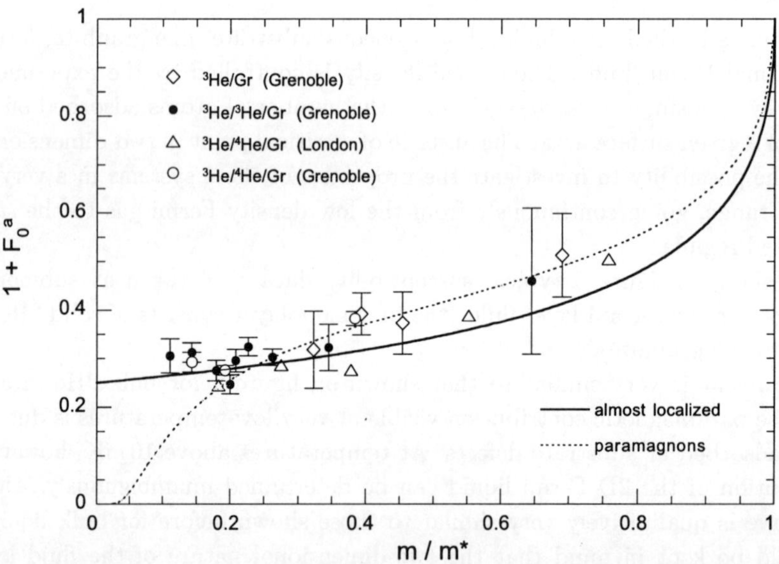


Fig. 7. The magnetic Landau parameters obtained [18,19,20,21,22,23,24] from the susceptibility of 2D liquid $^3$He are plotted, using the areal density as an implicit parameter, as a function of the inverse of the normalized effective mass. [25,26] The predictions of the paramagnon model, [8,9,10,11] and the almost-localized model [12,13] are also shown.

heat capacity data from Seattle [25] and Bell Laboratories. [26] These results correspond to submonolayer coverage: for $^3$He films in the first layer (directly adsorbed on graphite), or for second layer coverage, where a monolayer of solid $^3$He or $^4$He preplates the graphite.

The enhancement of the susceptibility is extremely high in 2D liquid $^3$He near solidification. The figure suggests that the parameter $F_0^a$ tends towards a constant for large interactions (small inverse effective masses). This would be in line with the predictions of the almost-localized model. However, a progressive, rather linear variation similar to that seen for bulk $^3$He in our recent data (see figure 4), cannot be excluded given the relatively large error bars in the 2D-liquid measurements.

## 5. Conclusions

We have presented in this lecture several experimental results concerning the magnetic susceptibility of liquid $^3$He, the archetype of a Fermi Liquid. The Landau theory provides an adequate framework for the understanding of the qualitative properties of three-dimensional and two-dimensional $^3$He, in a very large range of interaction strength. However, the understanding at the microscopic level is still very crude. The simple models described here capture some of the main features of the interacting fermion problem, but an accurate description is still lacking. This is the task of theorists for the next years. I hope that the experimental results presented here will contribute to this goal, or at least motivate the development of such a theory.

## 6. Acknowledgements

The results presented in this lecture have been obtained in collaboration with C. Bäuerle, J. Bossy, Yu.M. Bunkov, A.S. Chen, E. Collin, S.N. Fisher, R. Harakaly, K.-D. Morhard and S. Triqueneaux in the ultra-low temperature group of the CNRS-CRTBT.

## References

1. L. D. Landau, Sov. Phys. JETP **3**, 920 (1956).
2. L. D. Landau, Sov. Phys. JETP **5**, 101 (1957).
3. D. Pines and D. Nozières, *The Theory of Quantum Liquids*, Addison Wesley Publishing Co. (1966).
4. J. Wilks, *Liquid and Solid Helium*, Clarendon Press, Oxford (1967).
5. D. Vollhardt and P. Wölfle, *The Superfluid Phases of Helium 3*, ed. Taylor & Francis (1990).
6. W. P. Halperin and E. Varoquaux, *Helium Three*, edited by W.P. Halperin and L.P. Pitaevskii, Elsevier (1990).
7. H. Ramm, P. Pedroni, J. R. Thompson and H. Meyer, J. Low Temp. Phys. **2**, 539 (1970); see also J. R. Thompson, H. Ramm, J. F. Jarvis and H. Meyer, J. Low Temp. Phys. **2**, 521 (1970).
8. N. F. Berk and J. R. Schrieffer, Phys. Rev. Lett. **17**, 433 (1966).

9. S. Doniach and S. Engelsberg, Phys. Rev. Lett. **17**, 750 (1966).
10. M. T. Béal-Monod, J. Low Temp. Phys. **37** 123 (1979) and erratum J. Low Temp. Phys. **39**, 231 (1980).
11. M. T. Béal-Monod and A. Theumann, *Proceedings of the International Conference on Ordering in Two Dimensions*, edited by S. K. Sinha, North Holland, Amsterdam (1980).
12. M. C. Gutzwiller, Phys. Rev. A **137**, 1726 (1965).
13. D. Vollhardt, Rev. Mod. Phys. **56**, 99 (1984).
14. S. Triqueneaux, *Thesis*, Université J. Fourier, CRTBT-CNRS, Grenoble (1999), unpublished.
15. A. L. Thomson, H. Meyer and E. D Adams, Phys. Rev. **128**, 509 (1962).
16. B. T. Beal and J. Hatton, Phys. Rev. A **139**, 1751 (1965).
17. D. S. Greywall, Phys. Rev. B **33**, 7520 (1986).
18. C. Bäuerle, Yu. M. Bunkov, A. S. Chen, S. N. Fisher and H. Godfrin, J. Low Temp. Phys. **110**, 333 (1998).
19. C. Bäuerle, *Thesis*, Université J. Fourier, CRTBT-CNRS, Grenoble (1996), unpublished.
20. C. P. Lusher, B. Cowan and J. Saunders, Phys. Rev. Lett. **67**, 2497 (1991).
21. J. Saunders, C. P. Lusher and B. Cowan, in *Excitations in Two-Dimensional and Three-Dimensional Quantum Fluids*, edited by A.G.F. Wyatt and H.J. Lauter, Plenum, New York (1991).
22. M. Siqueira, *PhD Thesis*, Royal Holloway, University of London (1995), unpublished.
23. K.-D. Morhard, *Thesis*, Université J. Fourier, CRTBT-CNRS, Grenoble (1995), unpublished.
24. K.-D. Morhard, C. Bäuerle, J. Bossy, Y. Bunkov, S.N. Fisher and H. Godfrin, Phys. Rev. B **53**, 2658 (1996).
25. M. Bretz, J. G. Dash, D. C. Hickernell, E. O. McLean and O. E. Vilches, Phys. Rev. A **8**, 1589 (1973).
26. D. S. Greywall, Phys. Rev. B **41**, 1842 (1990).

## CHAPTER 9

# THE HYPERSPHERICAL HARMONIC METHOD: A REVIEW AND SOME RECENT DEVELOPMENTS

S. Rosati

*Dipartimento di Fisica "E.Fermi", Università di Pisa,*
*and Istituto Nazionale di Fisica Nucleare, Sezione di Pisa,*
*I-56127 Pisa, Italy*
*E-mail: sergio.rosati@df.unipi.it*

Studies of the Schrödinger equation solutions where the wave function is expanded in terms of a basis of hyperspherical harmonic functions started a long time ago and have continued until present times. This approach has been applied to different types of microscopic many-particle systems and its degree of success is strongly determined by the number of particles and by their mutual interactions. For three- and four-particle systems and soft interactions the required number of basis functions is not particularly large. For bigger systems and/or strong interactions this number becomes so great that modifications of the original method need to be devised in order to solve the corresponding numerical problem with present-day computing systems. A few of these modifications, such as using the potential basis, the adiabatic and the correlated expansions are discussed here. Results of the calculations are presented for the bound states of the helium atom and of nuclei with $A = 3$ and 4, interacting through realistic two-nucleon potentials. More recently a lot of effort has been devoted to the extension, and modification, of the hyperspherical harmonic approach to scattering problems involving charged particles. As a consequence, the properties of few-nucleon reactions of astrophysical interest can be studied and the results will be briefly discussed.

## 1. Introduction

Different methods are nowadays available for studying many-particle problems. One of these is based on the so-called hyperspherical formalism where the wave function (w.f.) of the system is expanded in terms of a (complete) set of basis functions, namely the hyperspherical harmonic (HH) functions . In the first step of the technique a set of Jacobi coordinates is chosen and then the hyperspherical coordinates are introduced. The latter coordinates, a hyperradius and some hyperangles whose number is related to that of the particles of the system, are a rather natural extension to the case of multi-particle systems of the polar coordinates $r$, $\theta$, $\phi$ commonly used for three dimensional problems.

The hyperspherical coordinates were first introduced for the three-particle system in a posthumous paper of Gronwall. [1] These coordinates were later used rather extensively for different types of strongly interacting systems. The first important developments of the method were given in the pioneering papers of Delves, [2] Erens *et al.*, [3] Simonov [4] and Fabre de la Ripelle. [5]

In the hyperspherical approach, as will be discussed in Sec. 5, the w.f. satisfying the Schrödinger equation is expanded in terms of products of hyperradial and specific hyperangular functions. Having fixed the number of expansion terms, the quantities to be calculated are the hyperradial functions and, for a bound state problem, the eigenvalues of the hamiltonian. This is usually accomplished by solving a (large) number of coupled second order differential homogeneous equations, or by further expanding the hyperradial functions in terms of an appropriate basis, and then determining all the coefficients of the expansion as the solutions of a set of homogeneous linear equations. The corresponding eigenvalue problem can be solved with the aid of standard numerical techniques.

In the HH approach a large part of the calculation can be done analytically, even though some difficulties can arise when the appropriate symmetry characteristics of the w.f. are imposed. However, a serious drawback of the approach lies in the necessarily large number of terms required in the expansion in order to get sufficiently accurate results. With the presently available computing systems only the w.f. of the bound states of nuclei with $A = 3$ and $A = 4$ can be calculated with high precision. [6,7]

For nuclear systems with larger $A$ values it is still problematic to get comparable accuracy for bound and scattering states. An analogous situation holds also for other many-particle systems. For this reason variations of the technique have been suggested and exploited in order to speed up the convergence process. So, important improvements have been obtained by using symmetry properties, [4] the introduction of an "optimal" subset (potential harmonics ), [8] the adiabatic approximation , or the correlated hyperspherical harmonic (CHH) expansion.

In the latter approach the HH functions are multiplied by appropriate correlation factors, accelerating the convergence process quite considerably by taking into account the effects of singularities or of strongly repulsive terms in the interparticle interactions. [9] In particular, starting from the work of Ref. 10, Haftel and Mandelzweig have shown that a substantial gain in the rate of convergence can be obtained for the helium atom. The CHH basis has been extensively applied by the Pisa group to study both bound and scattering states of nuclear systems interacting through realistic two- and three-body nuclear potentials. Very accurate results have been obtained for the $A = 3$ [11] and $A = 4$ bound states [12] and for the scattering states of three nucleons. [13,14] Recently, a different technique, allowing for a reduction of the number of the HH functions, has been used by the Trento group [15] in the computation of the energy and of some other properties of few nucleon systems. Another useful way for improving the convergence of the three-nucleon system has been exploited by Rosati and Viviani [16] utilizing other types of hyperspherical functions. The application of the HH basis to nuclei with $A > 4$ is now being investigated and some interesting results have already been obtained by Barnea and Viviani. [17]

## 2. Microscopic systems

A number of systems of remarkable physical interest requires a quantum-mechanical description. Here, we will focus our attention on the non-relativistic problem of calculating the bound or scattering state solutions of the Schrödinger equation. The study of atomic and molecular systems started with the development of quantum mechanics and is still actively pursued nowadays as an important objective of chemistry. Since we are interested in the application of the HH techniques, only the instructive case of the helium atom will be discussed in some details. Among other systems of interest we may recall the muonic molecules, the $^4He_3$ molecular cluster, the nuclear and hypernuclear systems and the three-quark states in a non-relativistic approach.

In the atomic case the interaction is electromagnetic. Apart from the difficulty due to its singularity at the $r=0$ interparticle separation, reasonably precise calculations can be made rather easily. However, the spectroscopic measurements do attain very high accuracy, hence the theoretical calculations must have an extremely refined precision in order to allow for significant comparison with experiments.

The electromagnetic interaction is again active in the case of muonic molecules though some important corrections due to relativistic effects must also be taken into account. Most of the work devoted to the study of muonic molecules concerns those of the $ab\mu$ type, where $a, b = p, d, t$ (see, for example, the paper of Abramov *et al.* [18] and references therein). Here again a high degree of precision is needed and the existence of loosely bound states forces to a rather refined treatment.

The difficulty of the calculations on helium clusters has a different origin. A number of helium-helium potentials are available (see, for example, the review article of Jansen and Aziz [19]). All of them are characterized by a very strong repulsion at small distances. A rough idea of this difficulty can be gathered just considering the simple, but reasonably accurate, Lennard-Jones potential, having a repulsive part proportional to $r^{-12}$. Therefore, in the region where the repulsion predominates, halving the distance causes the repulsion to increase by a factor of about $4 \times 10^3$. For this reason the standard HH expansion cannot be successful. This difficulty can be overcome by introducing suitable correlation factors that help to considerably reduce the number of basis functions required. For example, the bound and excited states of the helium trimer can be evaluated [20] using this approach with a precision similar to that of the best available techniques. Finally, the use of a non-relativistic quark-quark potential in a three-quark system does not present difficulties and also the HH approach can be rather easily applied. [21]

Much of our attention will be focused on the application of the HH method, and its variants, to nuclear systems. The difficulties are now due to the strong state dependence and to the large repulsion of the nucleon-nucleon (*NN*) interaction at small distances. Moreover, the nuclear interaction should be derived in a consistent way from quantum chromodynamics (QCD), which still appears to be an impossible task. As a consequence, there are only a few realistic semi-phenomenological

*NN* potentials available which contain some theoretical hints and a number of free parameters, determined by fitting the deuteron properties and a large set of *NN* scattering data for various energies up to the pion production threshold. Moreover, three-nucleon ($3N$) interaction terms also appear to be of some relevance. The situation is similar for hypernuclear systems where the strong interaction is responsible for their structure and again a semi-phenomenological hyperon-nucleon potential must be used. In conclusion, even though the study of nuclear systems represents a very difficult task, significant improvements have been achieved in recent years using different theoretical approaches, thanks above all to the continuously increasing power of the computing systems. A few of these approaches will be briefly mentioned here since their results will be compared with those given by the HH-type approaches.

Methods based on Monte Carlo sampling have shown to be powerful tools in few– and many–body physics and have been applied to many nuclear systems. In the variational approach (VMC) flexible wave functions are constructed and the required multidimensional integrals are calculated by a Monte Carlo technique. An even higher accuracy is obtained by the Green's function Monte Carlo (GFMC) approach where the Monte Carlo technique (see Ref. 22 and the references cited there) is used to calculate the Green's function.

In another approach, the Faddeev equations have been solved for the $A = 3$ bound and continuum states both in momentum and coordinate space representations of realistic nuclear interactions. A review of the main results has been given by Glöckle *et al.*. [23] Much work has also been done to solve the Faddeev-Yakubovsky equations for the $A = 4$ systems and, at present, the situation is very satisfactory for the bound state of the $\alpha$-particle. [24] A variety of scattering states, with $A \geq 4$, is of particular physical interest but their study is very complicated. The extension of the Faddeev-Yakubovsky method to these systems appears at the moment to be extremely difficult and further theoretical effort is necessary.

Variational methods, based on the expansion of the w.f. in terms of gaussian-type functions (allowing for most of the integrations to be made analytically), have proved to be powerful in light nuclei. In particular, interesting results have been obtained by the so-called stochastic variational method. [25,26] Also a very large basis of harmonic oscillator functions can be found to yield significant results. [27]

The various HH-type techniques recently applied to nuclear systems will be discussed in some details in Sec. 8.

## 3. Jacobi coordinates

The choice of the basis is the key point when the w.f. of the system is expanded in a set of basis functions. This is strongly related to the choice of the coordinate set. First of all, for isolated systems the center of mass coordinates should be chosen to be part of that set in order to single out its motion from the problem of the internal structure. Clearly, many choices of the internal coordinates are possible which, to a large extent, determine the strength of the model. The best possible situation occurs when the Schrödinger equation can be solved by separating the variables. Unfortunately, the strong interaction is highly non-separable. So, it would prove useful to be able to satisfy three conditions: *i*) if important collective variables exist, then they should be introduced; *ii*) the basis functions should allow for a convergence as fast and as accurate as possible; *iii*) the important clustering structures should be grossly described by a limited number of functions. A useful choice of the coordinates is one in which the kinetic energy operator is expressed in a separable form of the center of mass and internal coordinates.

Let us consider a system of $A$ particles and indicate by $m_i$, $\boldsymbol{r}_i$, $\boldsymbol{p}_i = -i\hbar\boldsymbol{\nabla}_{\boldsymbol{r}_i}$ $(i = 1, \ldots, A)$ the mass, position and momentum of the $i$-th particle. The kinetic energy operator $T$ can be written as

$$T = \sum_{i=1}^{A} \frac{\boldsymbol{p}_i^2}{2\,m_i} = -\frac{\hbar^2}{2} \sum_{i=1}^{A} \nabla^2_{\boldsymbol{X}_i}\,, \tag{3.1}$$

where $\boldsymbol{X}_i = \sqrt{m_i}\,\boldsymbol{r}_i$. We wish $T$ to have the form

$$T = \frac{\boldsymbol{p}^2_{\mathrm{cm}}}{2\,m_{\mathrm{tot}}} + \frac{1}{2\,M} \sum_{i=1}^{N} \boldsymbol{p}^2_{\boldsymbol{y}_i} = -\frac{\hbar^2}{2\,m_{\mathrm{tot}}} \nabla^2_{\boldsymbol{R}} - \frac{\hbar^2}{2\,M} \sum_{i=1}^{N} \nabla^2_{\boldsymbol{y}_i}\,, \tag{3.2}$$

where $N = A - 1$, $\boldsymbol{p}_{\mathrm{cm}} = \boldsymbol{p}_1 + \ldots + \boldsymbol{p}_A$ is the center of mass momentum, $\boldsymbol{R}$ is the center of mass coordinate, $m_{\mathrm{tot}} = m_1 + \ldots + m_A$, $M$ is a reference mass possibly equal to unity and $\boldsymbol{y}_i$ $(i = 1, \ldots, N)$, the Jacobi coordinates , are linear combinations of $\boldsymbol{X}_1, \ldots, \boldsymbol{X}_A$. Let $\boldsymbol{y}_A = \boldsymbol{R}$ and

$$\boldsymbol{y}_i = \sum_{j=1}^{A} c_{ij} \boldsymbol{X}_j\,, \qquad (i = 1, \ldots, N)\,, \tag{3.3}$$

with

$$c_{Aj} = \frac{\sqrt{m_j}}{m_{\mathrm{tot}}}, \qquad (j = 1, \ldots, A)\,. \tag{3.4}$$

Since $\boldsymbol{\nabla}_{\boldsymbol{X}_i} = \sum_{j=1}^{A} c_{ji} \boldsymbol{\nabla}_{\boldsymbol{y}_j}$, we have

$$\begin{aligned} T &= -\frac{\hbar^2}{2m_{\mathrm{tot}}} \nabla^2_{\boldsymbol{y}_A} - \frac{\hbar^2}{2M} \sum_{i,j,k=1}^{N} c_{ji} c_{ki} \boldsymbol{\nabla}_{\boldsymbol{y}_j} \cdot \boldsymbol{\nabla}_{\boldsymbol{y}_k} \\ &= -\frac{\hbar^2}{2m_{\mathrm{tot}}} \nabla^2_{\boldsymbol{y}_A} - \frac{\hbar^2}{2M} \sum_{i=1}^{N} \nabla^2_{\boldsymbol{y}_i}\,, \end{aligned} \tag{3.5}$$

when the following conditions are satisfied

$$\sum_{i=1}^{A} c_{ji} c_{ji} = \frac{1}{M} \quad (j = 1, \ldots, N)\,, \qquad \sum_{i=1}^{A} c_{ji} c_{ki} = 0 \quad (j \neq k = 1, \ldots, A)\,. \tag{3.6}$$

In general, there are many possibilities for satisfying these conditions. The Jacobi coordinates are those appropriate for the various partitions or cluster configurations. They are given here explicitly for $A = 2$, 3 and 4.

$\underline{A=2}$.
In this case the solution is unique and well-known. The equations are

$$c_{11}^2 + c_{12}^2 = \frac{1}{M}\,, \quad c_{11}c_{21} + c_{12}c_{22} = 0\,. \tag{3.7}$$

The first relation is verified by putting $c_{11} = 1/\sqrt{M}\,\sin\alpha$ and $c_{12} = 1/\sqrt{M}\,\cos\alpha$. From the second equation it follows $\tan\alpha = -\sqrt{m_2/m_1}$ and, in conclusion,

$$\boldsymbol{y}_1 = \sqrt{\frac{m_1}{Mm_{\rm tot}}}\boldsymbol{X}_2 - \sqrt{\frac{m_2}{Mm_{\rm tot}}}\boldsymbol{X}_1 = \sqrt{\frac{m_1m_2}{Mm_{\rm tot}}}(\boldsymbol{r}_2 - \boldsymbol{r}_1)\,, \tag{3.8}$$

$$\boldsymbol{y}_2 = \boldsymbol{R} = \frac{m_1\boldsymbol{r}_1 + m_2\boldsymbol{r}_2}{m_{\rm tot}}\,. \tag{3.9}$$

(2) $\underline{A=3}$.
A possibility is to take $\boldsymbol{y}_2$ proportional to $\boldsymbol{r}_2 - \boldsymbol{r}_1$, and $c_{23}= 0$. One easily gets

$$c_{21} = -\sqrt{\frac{m_2}{M(m_1+m_2)}}\,, \quad c_{22} = \sqrt{\frac{m_1}{M(m_1+m_2)}}\,, \tag{3.10}$$

$$c_{11} = -\sqrt{\frac{m_1m_3}{Mm_{\rm tot}(m_1+m_2)}}\,, \quad c_{12} = -\sqrt{\frac{m_2m_3}{Mm_{\rm tot}(m_1+m_2)}}\,,$$

$$c_{13} = \sqrt{\frac{m_1+m_2}{Mm_{\rm tot}}}\,. \tag{3.11}$$

The Jacobi coordinates for $A = 3$ are represented in Fig. 1. Other choices of

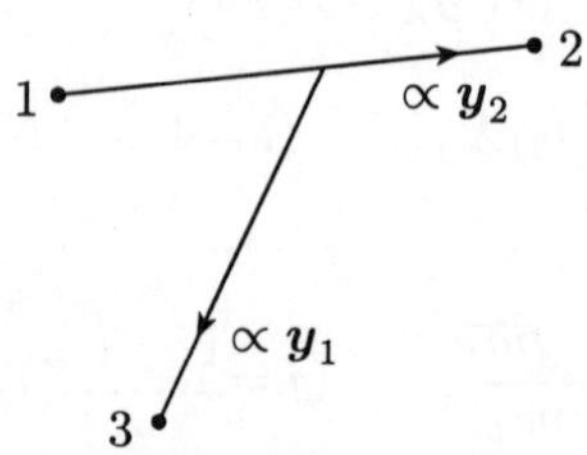


Fig. 1. Jacobi coordinates for three particles.

interest correspond to take $\boldsymbol{y}_2$ proportional to $\boldsymbol{r}_i - \boldsymbol{r}_j$, $i \neq j = 1,2,3$.

(3) $\underline{A=4}$.
In this case there are two possible cluster configurations (partitions).
Choice $a$):

$$c_{31} = -\sqrt{\frac{m_2}{Mm_{12}}}\,, \quad c_{32} = \sqrt{\frac{m_1}{Mm_{12}}}\,, \quad c_{33} = c_{34} = 0\,, \tag{3.12}$$

$$c_{21} = -\sqrt{\frac{m_1m_3}{Mm_{12}m_{123}}}\,, \quad c_{22} = -\sqrt{\frac{m_2m_3}{Mm_{12}m_{123}}}\,,$$

$$c_{23} = \sqrt{\frac{m_{12}}{Mm_{123}}}\,, \quad c_{24} = 0\,, \tag{3.13}$$

$$c_{11} = -\sqrt{\frac{m_1 m_4}{M m_{\rm tot} m_{123}}}\,, \quad c_{12} = -\sqrt{\frac{m_2 m_4}{M m_{\rm tot} m_{123}}}\,,$$
$$c_{13} = -\sqrt{\frac{m_3 m_4}{M m_{\rm tot} m_{123}}}\,, \quad c_{14} = \sqrt{\frac{m_{123}}{M m_{\rm tot}}}\,, \tag{3.14}$$

where $m_{ij} = m_i + m_j$ and $m_{ijk} = m_i + m_j + m_k$. The corresponding coordinates

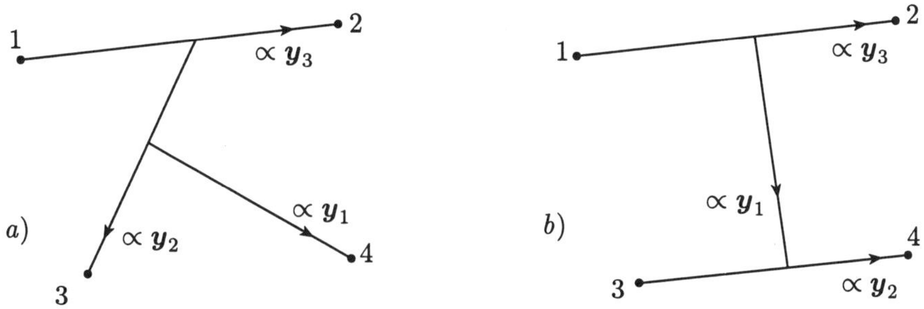


Fig. 2. Jacobi coordinates for four particles.

are indicated in Fig. 2*a*).

Choice *b*):

$$c_{31} = -\sqrt{\frac{m_2}{M m_{12}}}\,, \quad c_{32} = \sqrt{\frac{m_1}{M m_{12}}}\,, \quad c_{33} = c_{34} = 0\,, \tag{3.15}$$

$$c_{21} = c_{22} = 0\,, \quad c_{23} = -\sqrt{\frac{m_4}{M m_{34}}}\,, \quad c_{24} = \sqrt{\frac{m_3}{M m_{34}}}\,, \tag{3.16}$$

$$c_{11} = -\sqrt{\frac{m_1 m_{34}}{M m_{\rm tot} m_{12}}}\,, \quad c_{12} = -\sqrt{\frac{m_2 m_{34}}{M m_{\rm tot} m_{12}}}\,,$$
$$c_{13} = \sqrt{\frac{m_3 m_{12}}{M m_{\rm tot} m_{34}}}\,, \quad c_{14} = \sqrt{\frac{m_4 m_{12}}{M m_{\rm tot} m_{34}}}\,, \tag{3.17}$$

and the corresponding coordinates are indicated in Fig. 2*b*).

## 4. Hyperspherical coordinates

Let $\boldsymbol{y}_1, \ldots, \boldsymbol{y}_N$ be a set of Jacobi coordinates so that the Laplace operator can be written as

$$\nabla^2 = \sum_{i=1}^{N} \nabla^2_{\boldsymbol{y}_i} \equiv \sum_{i=1}^{N} \left[ \frac{\partial^2}{\partial {y_i}^2} + \frac{2}{y_i}\frac{\partial}{\partial y_i} - \frac{\hat{l}^2(\hat{\omega}_i)}{{y_i}^2} \right], \tag{4.1}$$

where $\hat{l}^2(\hat{\omega}_i)$ is the square of the angular momentum associated with the polar angles $\hat{\omega}_i \equiv (\theta_i, \phi_i)$ of $\boldsymbol{y}_i$. For the case $N = 1$ the hyperspherical coordinates are $(r, \hat{\omega}_1) \equiv (y_1, \theta_1, \phi_1)$. If $N = 2$, together with $\hat{\omega}_1$ and $\hat{\omega}_2$, one has the coordinates $y_1$ and $y_2$. It is convenient to introduce just a single "length" $r$ called the hyperradius

$$r = \sqrt{y_1^2 + y_2^2}\,, \tag{4.2}$$

and a second angular variable, the hyperangle, $\Phi$ defined by

$$y_2 = r\cos\Phi\,, \quad y_1 = r\sin\Phi\,, \qquad \left(0 \le \Phi \le \frac{\pi}{2}\right), \tag{4.3}$$

so that Eq. (4.2) is satisfied. The generalization to larger $N$ values is straightforward. In fact, together with the $N$ angular variables $\hat{\omega}_i$, one introduces the hyperradius

$$r = \sqrt{\sum_{i=1}^{N} y_i^2}\,, \tag{4.4}$$

and $N-1$ hyperangles $\Phi_N, \ldots, \Phi_2$ given by the relations

$$\begin{aligned} y_N &= r\cos\Phi_N\,, \\ y_i &= r\sin\Phi_N \cdots \sin\Phi_{i+1}\cos\Phi_i\,, \qquad (i = 2, \ldots, N-1)\,, \\ y_1 &= r\sin\Phi_N \cdots \sin\Phi_2\,, \end{aligned} \tag{4.5}$$

to satisfy Eq.(4.4). Therefore, we have

$$\frac{\partial}{\partial y_i} = \frac{y_i}{r}\frac{\partial}{\partial r} + D_{1i}\,, \tag{4.6}$$

$$\frac{\partial^2}{\partial y_i^2} = \left(\frac{1}{r} - \frac{y_i^2}{r^3}\right)\frac{\partial}{\partial r} + \frac{y_i^2}{r^2}\frac{\partial^2}{\partial r^2} + D_{2i}\,, \tag{4.7}$$

where the operators $D_{1i}$ and $D_{2i}$ imply derivatives with respect to the hyperangular variables. From Eqs. (4.1) and (4.5) it can be shown that

$$\nabla^2 = \sum_{i=1}^{N} \nabla^2_{\boldsymbol{y}_i} = \frac{\partial^2}{\partial r^2} + \frac{3N-1}{r}\frac{\partial}{\partial r} + \frac{\Lambda_N^2(\Omega_N)}{r^2}\,, \tag{4.8}$$

where $\Omega_N \equiv (\hat{\omega}_1, \ldots, \hat{\omega}_N, \Phi_2, \ldots, \Phi_N)$, and the operator $\Lambda_N(\Omega_N)$ is the so-called grandangular momentum. The explicit expression of $\Lambda_N^2(\Omega_N)$ can be obtained by means of the recurrence relation [5]

$$\begin{aligned} \Lambda_i^2(\Omega_i) &= \frac{\partial^2}{\partial \Phi_i^2} + \Big[(3i-2)\cot\Phi_i + 2(\cot\Phi_i - \tan\Phi_i)\Big]\frac{\partial}{\partial \Phi_i} - \frac{\hat{l}^2(\omega_i)}{\cos^2\Phi_i} + \frac{\Lambda_{i-1}^2(\Omega_{i-1})}{\sin^2\Phi_i} \\ &= 4(1-z_i)\frac{\partial^2}{\partial z_i^2} + 6\Big[2 - i(1+z_i)\Big]\frac{\partial}{\partial z_i} - 2\frac{\hat{l}^2(\omega_i)}{1+z_i} + 2\frac{\Lambda_{i-1}^2(\Omega_{i-1})}{1-z_i}\,, \end{aligned} \tag{4.9}$$

where $z_i = \cos 2\Phi_i$. The latter equation is obtained starting from Eq. (4.1) and calculating the expression of $D_{1i}$ and $D_{2i}$ in Eqs. (4.6)-(4.7) with the aid of Eq. (4.5). The volume element $d\tau = d\boldsymbol{y}_1 \ldots d\boldsymbol{y}_N$ can be expressed in terms of the hyperspherical coordinates

$$d\tau = d\hat{\omega}_1 \ldots d\hat{\omega}_N\,(y_1^2 dy_1) \ldots (y_N^2 dy_N) = d\hat{\omega}_1 \ldots d\hat{\omega}_N\, y_1^2 \ldots y_N^2\, J\, dr\, d\Phi_2 \ldots d\Phi_N\,, \tag{4.10}$$

where

$$J = \begin{vmatrix} y_1/r & y_2/r & y_3/r & \dots & y_N/r \\ y_1 \cot \Phi_N & y_2 \cot \Phi_N & y_3 \cot \Phi_N & \dots & -y_N \tan \Phi_N \\ y_1 \cot \Phi_{N-1} & y_2 \cot \Phi_{N-1} & y_3 \cot \Phi_{N-1} & \dots & 0 \\ \vdots & \vdots & \vdots & \dots & \vdots \\ y_1 \cot \Phi_2 & -y_2 \tan \Phi_2 & 0 & \dots & 0 \end{vmatrix}$$

Therefore,

$$J = \frac{1}{r} y_1 \dots y_N \frac{1}{\sin \Phi_2 \cos \Phi_2} \cdots \frac{1}{\sin \Phi_N \cos \Phi_N}, \tag{4.11}$$

so that, after writing $d\tau = r^{D-1} dr\, d\Omega_N$ with $D = 3N$, one obtains

$$d\Omega_N = d\hat{\omega}_1 \dots d\hat{\omega}_N \prod_{j=2}^{N} (\sin \Phi_j)^{3j-4} \cos^2 \Phi_j \, d\Phi_j \,. \tag{4.12}$$

## 5. Hyperspherical functions

Let us write the Laplace operator of Eq. (4.1) in the form

$$\nabla^2 = \sum_{i=1}^{N} \nabla^2_{\boldsymbol{y}_i} = \sum_{i=1}^{D} \frac{\partial^2}{\partial \boldsymbol{x}_i{}^2}, \tag{5.1}$$

where $x_1, \dots, x_D$ are the cartesian components of the vectors $\boldsymbol{y}_1, \dots \boldsymbol{y}_N$. A generic homogeneous polynomial of order $n$ in the $x$-coordinates has the form

$$f_n = \sum_{(n)} a_{(n)} x_1^{n_1} \dots x_D^{n_D}, \tag{5.2}$$

where $a_{(n)}$ is a numerical coefficient, $(n)$ stands for the set $n_1 \dots n_D$ and the summation is extended to various choices of these indices satisfying the condition

$$n_1 + \dots + n_D = n \,. \tag{5.3}$$

It can be verified directly that

$$\sum_{j=1}^{D} x_j \frac{\partial f_n}{\partial x_j} = n f_n \,. \tag{5.4}$$

A given $f_n$ is called a harmonic polynomial and denoted as $h_n$ if, when Eq. (5.3) is satisfied, the coefficients $a_{(n)}$ are such that

$$\nabla^2 h_n = 0 \,. \tag{5.5}$$

Any homogeneous polynomial of order $n$ can be expressed (see, for example, Ref. 28) in the form

$$f_n = h_n + r^2 h_{n-2} + r^4 h_{n-4} + \dots , \tag{5.6}$$

which is called the canonical decomposition of $f_n$ in powers of the hyperradius $r$. If $h_G$ is a harmonic polynomial, the function

$$Y_G(\Omega) = r^{-G} h_G , \tag{5.7}$$

is independent on the hyperradius. Since $\nabla^2 h_G = 0$ and applying Eq. (4.8) for the Laplace operator, one obtains, in terms of the HH coordinates,

$$\begin{aligned} \nabla^2 h_G &= \left\{ \frac{\partial^2}{\partial r^2} + \frac{(D-1)}{r}\frac{\partial}{\partial r} + \frac{\Lambda^2(\Omega)}{r^2} \right\} r^G Y_G(\Omega) \\ &= \left[ G(G+D-2) + \Lambda^2(\Omega) \right] r^{G-2} Y_G(\Omega) = 0 . \end{aligned} \tag{5.8}$$

Therefore,

$$\left[ \Lambda^2(\Omega) + G(G+D-2) \right] Y_G(\Omega) = 0 . \tag{5.9}$$

The function $Y_G(\Omega)$ is an eigenfunction of $\Lambda^2(\Omega)$ and is called a hyperspherical harmonic (HH) function .

Many harmonic polynomials and HH functions can be constructed. For example, in the well-known case $D = 3$ the grandangular momentum reduces to minus the angular momentum operator $\hat{l}^2$. Introducing the index $m$ to label a set of linearly independent eigenfunctions of $\hat{l}^2$ for a given eigenvalue $l(l+1)$, we have

$$\hat{l}^2 Y_{lm}(\hat{\omega}) = l(l+1) Y_{lm}(\hat{\omega}) . \tag{5.10}$$

By requiring furthermore that $Y_{lm}(\hat{\omega})$ be an eigenfunction of $\hat{l}_z$, the index $m$ can be taken equal to the corresponding eigenvalue and the functions $Y_{lm}(\hat{\omega})$ are the well-known spherical harmonic functions.

Let $\{G\}$ indicate the grandangular momentum $G$ together with the other quantum numbers useful to characterize the orthonormal set. One has

$$\int d\Omega \, Y^*_{\{G\}}(\Omega) Y_{\{G'\}}(\Omega) = \delta_{\{G\},\{G'\}} . \tag{5.11}$$

It follows from the expression (4.9) for $\Lambda_i^2(\Omega_i)$ that the functions $Y_{\{G\}}(\Omega)$ can be constructed by using a recursive relation. [29] For $N = 2$ the equation to be solved is

$$\Lambda_2^2(\Omega_2) Y_{\{G\}}(\Omega_2) = -G(G+4) Y_{\{G\}}(\Omega_2) . \tag{5.12}$$

Let us consider

$$Y_{\{G\}}(\Omega_2) = F(\cos 2\Phi_2)(\cos\Phi_2)^{l_2}(\sin\Phi_2)^{l_1} Y_{l_1 m_1}(\hat{\omega}_1) Y_{l_2 m_2}(\hat{\omega}_2) , \tag{5.13}$$

where $F(\cos 2\Phi_2)$ is a function of $z = \cos 2\Phi_2$ to be determined. Using the expression for $\Lambda_i^2(\Omega_i)$ given in Eq. (4.9), one obtains the equation

$$(1-z^2)F'' + (\alpha - \beta z)F' + \gamma F = 0 , \tag{5.14}$$

where

$$\alpha = l_2 - l_1, \quad \beta = l_1 + l_2 + 3, \quad \gamma = G(G+4) - (l_1+l_2)(l_1+l_2+4). \tag{5.15}$$

The Eq. (5.14) is satisfied if $F$ is chosen to be proportional to the Jacobi polynomial $P_n^{l_1+1/2,l_2+1/2}(z)$ for $G = 2n + l_1 + l_2$, with $n$ integer (see, for example, Ref. 30). Therefore, the solution for $N = 2$ is

$$\begin{aligned} Y_{\{G\}}(\Omega_2) = {} & N_n^{l_2,\nu_2}(\cos\Phi_2)^{l_2}(\sin\Phi_2)^{l_1} Y_{l_1 m_1}(\hat{\omega}_1) Y_{l_2 m_2}(\hat{\omega}_2) \\ & \times P_n^{l_1+1/2,l_2+1/2}(\cos 2\Phi_2), \end{aligned} \tag{5.16}$$

with $N_n^{l_2,\nu_2}$ a normalization constant and $\nu_2 = 2n + l_1 + l_2 + 2$. It can be readily verified that $r^G Y_{\{G\}}(\Omega_2)$ is a harmonic polynomial of order $G = 2n + l_1 + l_2$.

For an arbitrary value of $N$ let us write

$$Y_{\{G\}}(\Omega_N) = Y_{\{G\}}(\Omega_{N-1})(\cos\Phi_N)^{l_N}(\sin\Phi_N)^{G_{N-1}} Y_{l_N m_N}(\hat{\omega}_N) F(\cos 2\Phi_N), \tag{5.17}$$

then Eq. (5.9) is satisfied by

$$F(\cos 2\Phi_N) = N_{n_N}^{l_n,\nu_n} P_{n_N}^{\nu_{N-1},l_n+1/2}(\cos 2\Phi_N), \tag{5.18}$$

where

$$G_j = \sum_{i=1}^{j}(l_j + 2n_j), \quad n_1 = 0, \quad G \equiv G_N, \quad \nu_j = G_j + \frac{3j}{2} - 1, \tag{5.19}$$

and $\{G\} \equiv \{l_1, \ldots l_N, m_1, \ldots m_N, n_2, \ldots n_N\}$ for a total of $3N-1$ quantum numbers. In conclusion, the HH function can be cast in the form [5]

$$Y_{\{G\}}(\Omega_N) = \left[\prod_{i=1}^{N} Y_{l_i,m_i}(\hat{\omega}_i)\right]\left[\prod_{j=2}^{N} {}^{(j)}P_{n_j}^{l_j,G_{j-1}}(\Phi_j)\right], \tag{5.20}$$

with

$${}^{(j)}P_{n_j}^{l_j,G_{j-1}}(\Phi_j) = N_{n_j}^{l_j,\nu_j}(\cos\Phi_j)^{l_j}(\sin\Phi_j)^{G_{j-1}} P_{n_j}^{\nu_{j-1},l_j+1/2}(\cos 2\Phi_j). \tag{5.21}$$

The explicit expression for the normalization constant is [5]

$$N_{n_j}^{l_j,\nu_j} = \left[\frac{2\nu_j\,\Gamma(\nu_j - n_j)\,\nu_j!}{\Gamma(\nu_j - n_j - l_j - 1/2)\Gamma(n_j + l_j + 3/2)}\right]^{1/2}. \tag{5.22}$$

Again, it can be verified that $r^G Y_{\{G\}}(\Omega_N)$ is a harmonic polynomial of order $G$.

The functions $Y_{\{G\}}(\Omega_N)$ can be linearly combined to yield eigenfunctions of the total angular momentum $L$, $L_z$. For instance, these eigenfunctions can be constructed by means of the following coupling scheme:

$$\begin{aligned} \mathcal{H}_{\{G\}}(\Omega_N) = \sum_{m_1,\ldots,m_N} & (\ell_1 m_1 \ell_2 m_2|L_2 M_2)(L_2 M_2 \ell_3 m_3|L_3 M_3) \\ & \ldots (L_{N-1} M_{N-1} \ell_N m_N|L L_z)\, Y_{\{G\}}(\Omega_N), \end{aligned} \tag{5.23}$$

where $(\ell_i m_i \ell_j m_j|L_j M_j)$ are Clebsch–Gordan coefficients, $M_i = \sum_{j=1,i} m_j$ and the symbol $\{G\}$ now stands for the set of $3N-1$ quantum numbers

$$\{G\} \equiv \{\ell_1, \ldots, \ell_N,\ L_2, \ldots, L_{N-1}, L, L_z,\ n_2, \ldots, n_N\}. \tag{5.24}$$

To give an example, the explicit expression of the $\mathcal{H}$ function for $A = 4$ is

$$\mathcal{H}_{\ell_1\ell_2\ell_3L_2LL_zn_2n_3}(\Omega_3) = \Big[\big[Y_{\ell_1}(\hat{\omega}_1)Y_{\ell_2}(\hat{\omega}_2)\big]_{L_2}Y_{\ell_3}(\hat{\omega}_3)\Big]_{LL_z} \times {}^{(2)}\mathcal{P}^{\ell_1,\ell_2}_{n_2}(\Phi_2)\,{}^{(3)}\mathcal{P}^{\ell_3,2n_2+\ell_1+\ell_2}_{n_3}(\Phi_3)\,. \tag{5.25}$$

Of course, other coupling schemes can be used.

For systems including Fermi particles, the HH functions must be multiplied by spin-isospin functions and the various momenta must be combined to give definite values of the total angular momentum $J, J_z$. The corresponding functions will be still denoted as $\mathcal{H}_{\{\tilde{G}\}}$, where $\{\tilde{G}\}$ now also includes the spin-isospin quantum numbers.

In general, the w.f. must satisfy certain symmetry relations. Let $(i, j, k, \ldots)$ be a permutation $p$ of the indices $1, 2, \ldots A$, where the first two indices $i$ and $j$ specify the so called "reference pair". The w.f. can be cast in the form

$$\Psi(1,2,\ldots A) = r^{-(D-1)/2}\sum_{\{\tilde{G}\}}\Big[\sum_p \alpha_p \mathcal{H}_{\{\tilde{G}\}}(i,j,k,\ldots)\Big]u_{\{\tilde{G}\}}(r)\,, \tag{5.26}$$

where $\alpha_p = 1$ for identical bosons and $\alpha_p = (-1)^P$ for identical fermions, $P = 0, 1$ according to the parity of the permutation. The summation over $p$ is extended to all the necessary permutations. Putting

$$\mathcal{H}^{(S)}_{\{\tilde{G}\}} = \sum_p \alpha_p \mathcal{H}_{\{\tilde{G}\}}(i,j,k,\ldots)\,, \tag{5.27}$$

we can rewrite Eq. (5.23) in the form

$$\Psi(1,2,\ldots,A) = r^{-(D-1)/2}\sum_{\{\tilde{G}\}}\mathcal{H}^{(S)}_{\{\tilde{G}\}}u_{\{\tilde{G}\}}(r)\,. \tag{5.28}$$

For realistic $NN$ interactions it can be more convenient to classify the $\mathcal{H}$ functions in a different way. The quantum numbers $\{\tilde{G}\}$ can be separated into two sets. The first one contains all the orbital quantum numbers $l_1, \ldots, l_N$, and those specifying their intermediate couplings, the couplings of the spin and isospin of the particles. All these numbers characterize a "channel". A generic $\alpha$-channel is therefore specified by the quantum numbers

$$\alpha \equiv (l_{1\alpha}, l_{2\alpha}, \ldots, L_{12\alpha}, L_{123\alpha}, \ldots, S_{12\alpha}, S_{123\alpha}, \ldots, T_{12\alpha}, T_{123\alpha}, \ldots, T, T_z, J, J_z), \tag{5.29}$$

with an obvious notation. The second set of quantum numbers is $(n_\alpha) \equiv (n_{1\alpha}, \ldots, n_{N\alpha})$.

The important point to notice is that it is possible, in a variational approach, to decide which $\alpha$-channels can be expected to be more important by looking at the operatorial dependence of the interaction. Then the sum over $(n_\alpha)$ can be extended in such a way to accurately describe the dependence on the hyperangular variables for a given $\alpha$-channel.

In conclusion, in place of $\mathcal{H}_{\{\tilde{G}\}}(i,j,k,\ldots)$ and $\mathcal{H}^S_{\{\tilde{G}\}}$ we now have $\mathcal{H}_{\alpha,(n_\alpha)}$ and $\mathcal{H}^{(S)}_{\alpha,(n_\alpha)}$. The w.f., instead of the expression given in Eq. (5.28), is then written as

$$\Psi(1,2,\ldots,A) = r^{-(D-1)/2} \sum_{\alpha,(n_\alpha)} \mathcal{H}^{(S)}_{\alpha,(n_\alpha)} u_{\alpha,(n_\alpha)}(r) \,. \tag{5.30}$$

## 6. The coupled equations

The Schrödinger equation in the center of mass frame can be expressed in terms of the Jacobi coordinates $\boldsymbol{y}_1,\ldots,\boldsymbol{y}_N$ as

$$[T+V-E]\Psi(1,\ldots,A)=\left[-\frac{\hbar^2}{2\,M}\sum_{i=1}^{N}\nabla^2_{\boldsymbol{y}_i}+V(\boldsymbol{y}_1,\ldots,\boldsymbol{y}_N)-E\right]\Psi(1,\ldots,A)=0\,, \tag{6.1}$$

where the dependence of the potential on the spin and isospin of the particles is understood. In the hyperspherical formalism, the w.f. $\Psi$ is expanded in terms of the functions $\mathcal{H}^S_{\{\tilde{G}\}}$ satisfying the appropriate symmetry conditions. Then the problem to be solved is to calculate all the hyperradial functions $u_{\{\tilde{G}\}}(r)$. To this aim it is convenient to apply a variational procedure. The Rayleigh-Ritz principle is appropriate for bound states. First one expands the w.f. limiting the expansion to a finite number of terms. In correspondence with a given choice of $\Psi$ in Eq. (5.28), the quantum numbers $\{\tilde{G}\}$ are limited to a finite set:

$$\Psi_v(1,\ldots,A) = r^{-(D-1)/2} \sum_{\{\tilde{G}\}}^{max} \mathcal{H}^{(S)}_{\{\tilde{G}\}} u_{\{\tilde{G}\}}(r) \,. \tag{6.2}$$

The variational condition can be cast in the form

$$\delta_u < \Psi_v | H - E | \Psi_v >= 0, \tag{6.3}$$

where the symbol $\delta_u$ denotes the functional derivative with respect to a generic hyperradial function. Let us now introduce the quantities

$$N_{\{\tilde{G}\},\{\tilde{G}'\}} = \sum_{spin,isospin} \int d\Omega\, \mathcal{H}^{(S)\,*}_{\{\tilde{G}\}} \mathcal{H}^{(S)}_{\{\tilde{G}'\}} \,, \tag{6.4}$$

$$T_{\{\tilde{G}\},\{\tilde{G}'\}} = \sum_{spin,isospin} \int d\Omega\, \mathcal{H}^{(S)\,*}_{\{\tilde{G}\}} T\, \mathcal{H}^{(S)}_{\{\tilde{G}'\}} \,, \tag{6.5}$$

$$V_{\{\tilde{G}\},\{\tilde{G}'\}} = \sum_{spin,isospin} \int d\Omega\, \mathcal{H}^{(S)\,*}_{\{\tilde{G}\}} V\, \mathcal{H}^{(S)}_{\{\tilde{G}'\}} \,, \tag{6.6}$$

where $T$ ($V$) is the total kinetic (potential) energy operator. It should be noted that $N_{\{\tilde{G}\},\{\tilde{G}'\}}$ is a numerical quantity while $T_{\{\tilde{G}\},\{\tilde{G}'\}}$, and also in general $V_{\{\tilde{G}\},\{\tilde{G}'\}}$, are operators. From Eq. (6.3) one obtains the following set of second order differential equations

$$\sum_{\{\tilde{G}'\}} \left[T_{\{\tilde{G}\},\{\tilde{G}'\}} + V_{\{\tilde{G}\},\{\tilde{G}'\}} - E N_{\{\tilde{G}\},\{\tilde{G}'\}}\right] u_{\{\tilde{G}'\}}(r) = 0 \,. \tag{6.7}$$

Without loss of generality, the functions $\mathcal{H}^{(S)}_{\{\tilde{G}\}}$ can be chosen as the eigenfunctions of a given grandangular momentum $G$ and Eq. (6.7) can be written in the form

$$\sum_{\{G'\}}\left[-\frac{\hbar^2}{2M}N_{\{\tilde{G}\},\{\tilde{G}'\}}\left\{\frac{d^2}{dr^2}+\frac{\mathcal{L}(\mathcal{L}+1)}{r^2}-E\right\}+V_{\{\tilde{G}\},\{\tilde{G}'\}}\right]u_{\{\tilde{G}'\}}(r)=0\,, \quad (6.8)$$

where $\mathcal{L}=G+(D-3)/2$. In general $N_{\{\tilde{G}\},\{\tilde{G}'\}}\neq\delta_{\{G\},\{G'\}}$ since the $\mathcal{H}^{(S)}_{\{\tilde{G}\}}$ functions are combinations of the same $Y_{\{G\}}$ functions expressed in the various coordinate sets $\Omega^p$. For the bound states the conditions $u(r)\to 0$, both for $r\to 0$ and for $r\to\infty$, must be satisfied.

In order to solve Eq. (6.7) one can use standard numerical techniques, as Gauss integration and Monte Carlo or Quasi Random Number (QRN) methods. Alternatively, a function $\mathcal{H}_{\{\tilde{G}\}}(\Omega^p)$ for a generic permutation $p$ can be expressed in terms of the $\mathcal{H}_{\{\tilde{G}\}}$ functions pertaining to a given permutation, say $p_1\equiv 1,2,\ldots A$. For $A=3$ the coefficients of the transformation have been obtained by Reynal and Revai. [31] For $A\geq 4$ the problem is more involved and has been studied by just a few researchers. [32–34]

The main difficulty in calculations involving the HH expansion is that, due to the enormous degeneracy of the basis as $G$ increases, as many basis functions as possible (typically up to a few thousands) need to be treated. Indeed the convergence problem is in general quite serious in the case of realistic interactions. The expansion is successful for $A$=3 and 4, while appropriate modifications of the HH method can be useful for larger $A$ values. On the contrary, the solution of the corresponding eigenvalue problem does not present severe difficulties due to the available numerical codes.

## 7. The hyperspherical harmonic expansion in momentum space

Let $\boldsymbol{y}_1,\ldots,\boldsymbol{y}_A$ be a set of Jacobi coordinates and $\hbar\boldsymbol{k}_1,\ldots,\hbar\boldsymbol{k}_A$ the conjugate Jacobi coordinates in momentum space. The expressions for $\hbar\boldsymbol{k}_1,\ldots,\hbar\boldsymbol{k}_N$ in terms of $\boldsymbol{p}_1,\ldots,\boldsymbol{p}_A$ are the same as those for $\boldsymbol{y}_1,\ldots,\boldsymbol{y}_N$ in terms of $\boldsymbol{r}_1,\ldots,\boldsymbol{r}_A$ and $\hbar\boldsymbol{k}_A=\sum_{i=1}^A\boldsymbol{p}_i$. From the condition $(H-E)\,|\Psi\rangle$=0 follows that

$$\langle\boldsymbol{k}_1,\ldots,\boldsymbol{k}_A\,|\,H-E\,|\,\boldsymbol{k}'_1,\ldots,\boldsymbol{k}'_A\rangle\,\langle\boldsymbol{k}'_1,\ldots,\boldsymbol{k}'_A\,|\,\Psi\rangle=0\,, \quad (7.1)$$

where integration over all the $\boldsymbol{k}'$ variables is understood. The following relation may be shown to hold (see, for example, Ref. 28)

$$\exp\left(i\sum_{j=1}^A\boldsymbol{k}_j\cdot\boldsymbol{y}_j\right)=e^{i\boldsymbol{k}_A\cdot\boldsymbol{y}_A}\frac{(2\pi)^{D/2}}{(kr)^{D/2-1}}\sum_{\{G\}}i^G Y_{\{G\}}(\Omega)Y^*_{\{G\}}(\Omega_k)J_{\mathcal{L}+1/2}(kr)\,, \quad (7.2)$$

where $\Omega$ and $\Omega_k$ denote the sets of the coordinate and momentum space hyperangles, respectively, and

$$k=\sqrt{k_1^2+\ldots+k_N^2}\,, \quad (7.3)$$

is the hypermomentum, playing in momentum space the role corresponding to $r$ in coordinate space. Let

$$\Psi(1,\ldots,A)=\frac{e^{i\boldsymbol{k}_{\rm tot}\cdot\boldsymbol{R}}}{(2\pi)^{3/2}}\sum_{\{G\}}Y_{\{G\}}(\Omega)u_{\{G\}}(r)\,, \tag{7.4}$$

where $\hbar\boldsymbol{k}_{\rm tot}$ is the total momentum and $\boldsymbol{R}=\boldsymbol{y}_A$. Then

$$\begin{aligned}\langle\boldsymbol{k}'_1,\ldots,\boldsymbol{k}'_A\mid\Psi\rangle&=\int d\boldsymbol{y}_1\ldots d\boldsymbol{y}_A\frac{e^{-i\sum_{j=1}^{A}\boldsymbol{k}'_j\cdot\boldsymbol{y}_j}}{(2\pi)^{3A/2}}\Bigg[\sum_{\{G\}}Y_{\{G\}}(\Omega)u_{\{G\}}(r)\Bigg]\frac{e^{i\boldsymbol{k}_{\rm tot}\cdot\boldsymbol{R}}}{(2\pi)^{3/2}}\\&=\delta(\boldsymbol{k}'_A-\boldsymbol{k}_{\rm tot})\sum_{\{G\}}Y_{\{G\}}(\Omega_{k'})v_{\{G\}}(k')\,,\end{aligned} \tag{7.5}$$

where

$$v_{\{G\}}(k')=\frac{(-i)^G}{(2\pi)^3}\int dr\,\frac{r^{D-1}}{(k'r)^{D/2-1}}J_{\mathcal{L}+1/2}(k'r)u_{\{G\}}(r)\,. \tag{7.6}$$

Therefore, we get

$$\begin{aligned}&\langle\boldsymbol{k}_1,\ldots,\boldsymbol{k}_A\mid H-E\mid\boldsymbol{k}'_1,\ldots,\boldsymbol{k}'_A\rangle\langle\boldsymbol{k}'_1,\ldots,\boldsymbol{k}'_A\mid\Psi\rangle\\&=\int d\boldsymbol{k}'_1\ldots d\boldsymbol{k}'_A\left[\{T(\boldsymbol{k}'_1,\ldots,\boldsymbol{k}'_A)-E\}\,\delta(\boldsymbol{k}_1-\boldsymbol{k}'_1)\ldots\delta(\boldsymbol{k}_A-\boldsymbol{k}'_A)\right.\\&\quad\left.+\delta(\boldsymbol{k}_A-\boldsymbol{k}'_A)V(\boldsymbol{k}_1,\ldots,\boldsymbol{k}_N;\boldsymbol{k}'_1,\ldots,\boldsymbol{k}'_N)\right]\delta(\boldsymbol{k}'_A-\boldsymbol{k}_{\rm tot})\sum_{\{G\}}Y_{\{G\}}(\Omega_k)v_{\{G\}}(k)\\&=\delta(\boldsymbol{k}_A-\boldsymbol{k}_{\rm tot})\Bigg[\{T(\boldsymbol{k}_1,\ldots,\boldsymbol{k}_A)-E\}\sum_{\{G\}}Y_{\{G\}}(\Omega_k)v_{\{G\}}(k)\\&\quad+\int d\boldsymbol{k}'_1\ldots d\boldsymbol{k}'_NV(\boldsymbol{k}_1,\ldots,\boldsymbol{k}_N;\boldsymbol{k}'_1,\ldots,\boldsymbol{k}'_N)\Bigg]\sum_{\{G\}}Y_{\{G\}}(\Omega_{k'})v_{\{G\}}(k')\,.\end{aligned} \tag{7.7}$$

Taking $\boldsymbol{k}_{\rm tot}$=0, the hyperradial functions $v_{\{G\}}(k)$ must satisfy the following set of integral equations

$$\begin{aligned}&[T(\boldsymbol{k}_1,\ldots,\boldsymbol{k}_N)-E]\sum_{\{G\}}Y_{\{G\}}(\Omega_k)v_{\{G\}}(k)\\&+\int d\boldsymbol{k}'_1\ldots d\boldsymbol{k}'_NV(\boldsymbol{k}_1,\ldots,\boldsymbol{k}_N;\boldsymbol{k}'_1,\ldots,\boldsymbol{k}'_N)\sum_{\{G\}}Y_{\{G\}}(\Omega_{k'})v_{\{G\}}(k')=0\,,\end{aligned} \tag{7.8}$$

which can be solved by standard numerical methods.

The kinetic energy operator can be also taken in its relativistic form,

$$T=\sum_{i=1}^{A}\sqrt{\boldsymbol{p}_i^2c^2+m_i^2c^4}=\sum_{i=1}^{A}m_ic^2+\sum_{i=1}^{A}\frac{\boldsymbol{p}_i^2}{2m_i}+\ldots\,, \tag{7.9}$$

and can be expressed in terms of $\boldsymbol{k}_1,\ldots,\boldsymbol{k}_N$ only, since $\boldsymbol{k}_{\rm tot}$=0. As an example, for identical particles with $A$=3, using Eqs. (3.10) and (3.11) with $m_1=m_2=m_3=m$

and $M = 1$, we have

$$\boldsymbol{y}_3 = \frac{1}{3}(\boldsymbol{r}_1 + \boldsymbol{r}_2 + \boldsymbol{r}_3)\,, \qquad \hbar\boldsymbol{k}_3 = (\boldsymbol{p}_1 + \boldsymbol{p}_2 + \boldsymbol{p}_3) = 0\,, \tag{7.10}$$

$$\boldsymbol{y}_2 = \frac{1}{\sqrt{2}}(\boldsymbol{r}_2 - \boldsymbol{r}_1)\,, \qquad \hbar\boldsymbol{k}_2 = \frac{1}{\sqrt{2}}(\boldsymbol{p}_2 - \boldsymbol{p}_1)\,, \tag{7.11}$$

$$\boldsymbol{y}_1 = \sqrt{\frac{2}{3}}\left[\boldsymbol{r}_3 - \frac{1}{2}(\boldsymbol{r}_1 + \boldsymbol{r}_2)\right]\,, \qquad \hbar\boldsymbol{k}_1 = \sqrt{\frac{2}{3}}\left[\boldsymbol{p}_3 - \frac{1}{2}(\boldsymbol{p}_1 + \boldsymbol{p}_2)\right]\,, \tag{7.12}$$

so that

$$T = \hbar c\left[\sqrt{\frac{1}{6}k_1^2 + \frac{1}{2}k_2^2 + \frac{1}{\sqrt{3}}\boldsymbol{k}_1\cdot\boldsymbol{k}_2 + \frac{m^2c^2}{\hbar^2}} + \sqrt{\frac{1}{6}k_1^2 + \frac{1}{2}k_2^2 - \frac{1}{\sqrt{3}}\boldsymbol{k}_1\cdot\boldsymbol{k}_2 + \frac{m^2c^2}{\hbar^2}}\right.$$
$$\left. + \sqrt{\frac{2}{3}k_1^2 + \frac{m^2c^2}{\hbar^2}}\right] = 3mc^2 + \frac{\hbar^2}{2m}(k_1^2 + k_2^2) + \ldots\,. \tag{7.13}$$

## 8. Results for the $A = 3$, 4 nuclei with the hyperspherical harmonic expansion

As stated in the introduction there is a serious problem to be solved when applying the HH method to the study of strongly interacting systems with large repulsion at small distances. The convergence of the expansion is extremely slow and a very large number of basis functions is required.

Even in the case of central potentials the rate of convergence is satisfactory only for potentials with very soft repulsion, as it has been shown for the three-nucleon system by Ehrens *et al.*. [3] Some of the results obtained by these authors are reported in Table 1 for three simple effective interactions, namely those proposed by Baker, [35] Volkov [36] and Malfliet and Tjon. [37] The first interaction is always attractive; the second one contains a fairly soft repulsion, while the third one has a repulsive part behaving as $r^{-1}$ when $r \to 0$. Inspection of the table reveals that full convergence is attained for the first interaction for a number $k_{max} = 12$ of HH components and for $k_{max} = 18$ in the second case. However, even $k_{max} = 27$ is insufficient for the third interaction, which shows a rather large repulsive part. In nuclear systems the calculation of the $A = 3$, 4 nuclei ground state properties is even more difficult with realistic interactions. All the important channels must to be included and, in general, all the spin-isospin quantum numbers must be considered. However, it can be expected that, due to centrifugal effects, the channels with increasing values of the angular momenta will provide smaller and smaller contributions. As an example, in Table 2 the first twelve channels for the three-nucleon system are listed in order of importance in establishing the structure of the system. An analogous ordering of channels can also be made for the four-nucleon system. [12] In fact, it has been shown that for $A = 3\,(4)$ at least $N_c = 8\,(80)$ channels must be considered. Moreover, for each channel the dependence on the hyperangular variables and on $r$ must be

Table 1. Triton binding energy (in MeV) as a function of the number $k_{max}$ of HH components in the w.f. for the Baker [B], Volkov [V] and Malfliet and Tjon [MT(V)] potentials. The results are taken from Ref. 3 where the parameters of the potentials are also specified.

| $k_{\max}$ | B | V | MT(V) |
|---|---|---|---|
| 0 | 9.2062 | 7.7080 | 0.5660 |
| 2 | 9.6134 | 8.0786 | 1.8386 |
| 3 | 9.7462 | 8.3296 | 4.2234 |
| 4 | 9.7646 | 8.3759 | 4.9824 |
| 5 | 9.7738 | 8.4162 | 5.8854 |
| 6 | 9.7779 | 8.4428 | 6.6316 |
| 9 | 9.7794 | 8.4609 | 7.3762 |
| 12 | 9.7795 | 8.4641 | 7.6217 |
| 15 | 9.7795 | 8.4646 | 7.7136 |
| 18 | 9.7795 | 8.4647 | 7.7521 |
| 21 | | | 7.7697 |
| 24 | | | 7.7785 |
| 27 | | | 7.7831 |

Table 2. Quantum number values for the first 12 channels in the expansion of the three-nucleon ground state. The orbital angular momenta $\ell_{1\alpha}$, $\ell_{2\alpha}$ and $L_\alpha$ are those associated with the Jacobi coordinates $\boldsymbol{y}_1, \boldsymbol{y}_2$ for the channel $\alpha$ and their resultant value, $S_{2\alpha}$ ($T_{2\alpha}$) is the resultant spin (isospin) of the two reference particles and $S_\alpha$ and $T_\alpha$ are the total spin and isospin of the state.

| $\alpha$ | $\ell_{1\alpha}$ | $\ell_{2\alpha}$ | $L_\alpha$ | $S_{2\alpha}$ | $T_{2\alpha}$ | $S_\alpha$ | $T_\alpha$ |
|---|---|---|---|---|---|---|---|
| 1 | 0 | 0 | 0 | 1 | 0 | 1/2 | 1/2 |
| 2 | 0 | 0 | 0 | 0 | 1 | 1/2 | 1/2 |
| 3 | 0 | 2 | 2 | 1 | 0 | 3/2 | 1/2 |
| 4 | 2 | 0 | 2 | 1 | 0 | 3/2 | 1/2 |
| 5 | 2 | 2 | 0 | 1 | 0 | 1/2 | 1/2 |
| 6 | 2 | 2 | 2 | 1 | 0 | 3/2 | 1/2 |
| 7 | 2 | 2 | 1 | 1 | 0 | 1/2 | 1/2 |
| 8 | 2 | 2 | 1 | 1 | 0 | 3/2 | 1/2 |
| 9 | 1 | 1 | 0 | 1 | 1 | 1/2 | 1/2 |
| 10 | 1 | 1 | 1 | 1 | 1 | 1/2 | 1/2 |
| 11 | 1 | 1 | 1 | 1 | 1 | 3/2 | 1/2 |
| 12 | 1 | 1 | 2 | 1 | 1 | 3/2 | 1/2 |

accurately described. When these two conditions are satisfied the HH approach allows us to treat those nuclear systems with high precision. In order to illustrate this, the results obtained in Ref. 6 for the AV14 [38] and AV18 [39] potentials are reported in Table 3. For the case $A = 3$ these are compared with the corresponding estimates obtained with other accurate approaches, namely the Faddeev equations (FE) method, [40,41] the coupled-rearrangement-channel (CRC) [25] and the GFMC techniques. [42]

Recently a benchmark test of the accuracy attainable in the calculation of the four-nucleon bound state in the case of the AV14 interaction has been performed [7].

Table 3. Binding energy ($B$), mean kinetic energy ($T$), radius ($R$) and $S'$- $P$- and $D$-wave percentages for the triton obtained by different methods for the AV14 and AV18 potentials.

| Potential | Method | $B$(MeV) | $T$(MeV) | $R$(fm) | $P_{S'}$(%) | $P_P$(%) | $P_D$(%) |
|---|---|---|---|---|---|---|---|
| | HH [6] | 7.6844 | 45.678 | 1.776 | 1.126 | 0.0761 | 8.968 |
| | FE [40] | 7.680 | | | | | |
| AV14 | FE [41] | 7.670 | | | 1.12 | 0.08 | 8.96 |
| | CRC [25] | 7.6844 | | | 1.126 | 0.076 | 8.968 |
| | GFMC [42] | 7.670(8) | | | | | |
| AV18 | HH [6] | 7.6181 | 46.707 | 1.769 | 1.294 | 0.0658 | 8.511 |
| | FE [24] | 7.623 | | | | | |

Table 4. Binding energy ($B$), mean kinetic energy ($T$), radius ($R$) and percentages of total angular momentum components for the four-nucleon ground state. All the data reported in the Table are taken from Ref. 7.

| Method | $B$(MeV) | $T$(MeV) | $R$(fm) | $P_S$(%) | $P_P$(%) | $P_D$(%) |
|---|---|---|---|---|---|---|
| FE | 25.94(5) | 102.39(5) | 1.485(3) | 85.71 | 0.38 | 13.91 |
| CRCGV | 25.90 | 102.25 | 1.482 | 85.73 | 0.37 | 13.90 |
| SVM | 25.92 | 102.35 | 1.486 | 85.72 | 0.368 | 13.91 |
| HH | 25.90(1) | 102.44 | 1.483 | 85.72 | 0.369 | 13.91 |
| GFMC | 25.93(2) | 102.3(1.0) | 1.490(5) | | | |
| NCSM | 25.80(20) | 103.35 | 1.485 | 86.73 | 0.29 | 12.98 |
| EIHH | 25.944(10) | 100.8(9) | 1.486(1) | 85.73 | 0.370(1) | 13.89(1) |

To this end, several efficient methods were developed: the Faddeev-Yakubovsky (denoted in the Table as FE), the CRC Gaussian basis variational (CRCGV), the stochastic variational (SVM), the HH variational, the GFMC, the no-core shell model (NCSM) and the effective hypersperical harmonic interaction (EIHH) methods. The details of the various approaches can be found in the original papers. [7] The results are given in Table 4.

## 9. Modified hyperspherical harmonic expansions

As discussed in the preceding section, we came to the conclusion that it is possible to obtain very satisfactory results for the $A = 3$, 4 systems and realistic interactions using the HH approach. However, a large number of basis functions must be included in the variational w.f. and the corresponding calculations are laborious. Therefore, it would be very useful to devise criteria *i*) for identifying the HH components which give the major contributions and *ii*) for finding appropriate modifications of the expansion in order to speed up the convergence rate. This problem is discussed in the following three subsections.

### 9.1. *The potential basis*

Because of the large degeneracy of the HH expansion it is important to determine some rules for selecting those components which give the most significant contributions. For a system of particles with $A \geq 3$ a possible approach is to establish, first of all, a complete description of the two-body correlations, then of the three-body ones and so on. It is reasonable to expect that the two-body correlations will give the largest contributions to the structure of the system. This point has been discussed in detail by Navarro. [43] A way for constructing these correlations in the HH method was proposed a few years ago by Fabre de la Ripelle [8] and the corresponding set of HH functions constitutes the so-called potential basis (PB) . The underlying idea is to consider those terms in the expansion of the w.f. which nearly exhaust the effect of the particle interaction leaving only a "weak" residual interaction. Obviously, an accurate description of the effects of the residual interactions requires larger and larger sets of HH functions. Let us refer to the form of the w.f. function as given in Eq. (5.28). A drastic limitation on the channels to be included is achieved by imposing the following conditions:

*i*) for each spin-isospin channel, consider only the minimum angular momentum value of the reference pair $i, j$ and take all the others equal to zero;
*ii*) allow only for a dependence on the hyperangle $\Phi_N^{(p)}$ for each amplitude corresponding to the permutation $p$ [see Eq. (5.26)].

Since $r_{ij}^{(p)}$ is proportional to $r\cos\Phi_N^{(p)}$, the best w.f. of the form

$$\Psi_{\rm PB}(1,\ldots,A) = \sum_p \psi(r_{ij}^{(p)}, r), \tag{9.1}$$

can be constructed using the potential basis with an increasing value of the grand-angular momentum $G$.

The first proof that the PB can represent a good approximation to the full HH expansion was given in connection with the solution of the trinucleon bound state for the completely space-symmetric $S$-state. [3,8,44] While the PB was initially devised for a purely central potential, its extension to realistic potentials is rather straightforward. In that case too, the PB functions give the largest contributions and a first order approximation to the problem is obtained correspondingly.

### 9.2. *The correlated expansion*

If the particle interaction is strongly repulsive at small distances the convergence of the HH expansion can be terribly slow. Thus, for the calculation of the trinucleon ground state with $NN$ realistic potentials, HH functions with $G$ up to 180 must be considered [6] in order to get a very close agreement with other accurate techniques. [24,25,45,46] For larger systems the situation is even worse and the number of basis elements that are needed becomes extremely large. Consequently, it is quite difficult to extend the calculation to scattering states or to include $3N$ interaction terms.

In order to overcome this difficulty, in Ref. 47 the radial dependence of each amplitude was modified by the inclusion of a suitably chosen correlation factor. The role of these correlation factors is to speed up the convergence of the expansion by improving the description of the system when two particles are close to each other. In such configurations, there are cancellations between large contributions from the kinetic and the potential energy terms which require describing all the important details of the w.f. The CHH angular momentum-spin-isospin basis is taken to have the form

$$\Phi^{\rm CHH}_{\alpha,(n_\alpha)}(i,j,k,\ldots) = F_\alpha(r_{ij},r_{ik},r_{jk},\ldots)\mathcal{H}_{\alpha,(n_\alpha)}(i,j,k,\ldots)\,, \tag{9.2}$$

where $F_\alpha$ is the correlation factor. Of course, putting $F_\alpha = 1$ takes us back to the standard HH expansion. The Jastrow form is the usual choice for $F_\alpha$. However, for each pair of particles the correlation function should depend on the state of the pair and, as a consequence, the problem of choosing $F_\alpha$ requires some extra attention. Let us start by discussing the simplest case, namely $A = 3$. Two interesting choices possible for $F_\alpha$ are the Jastrow form (CHH expansion) or a function of the distance between the reference pair particles. In the latter case, the expansion is called the pair-correlated hyperspherical harmonic (PHH) expansion .

*i)* Jastrow correlation factor:

$$F_\alpha(r_{ij},r_{ik},r_{jk}) = f_\alpha(r_{ij})g_\alpha(r_{ik})g_\alpha(r_{jk})\,. \tag{9.3}$$

*ii)* PHH correlation factor:

$$F_\alpha(r_{ij},r_{ik},r_{jk}) = f_\alpha(r_{ij})\,. \tag{9.4}$$

A simple but successful way for selecting the correlation functions is as follows. Let us again consider the case of a nuclear system. The total w.f. of the coordinates of a given pair of nucleons when they are rather close to each other and all the remaining nucleons are far from them, depends mainly on their mutual interaction. Therefore, the radial w.f. of the relative motion of that pair in the angular–spin state $[\beta \equiv {}^{2S_\beta+1}(\ell_\beta)_{j_\beta}]$ can be approximately described by the solution of an equation of the form

$$\sum_{\beta'}\left\{-\frac{\hbar^2}{m}\left[\frac{d^2}{dx^2}+\frac{2}{x}\frac{d}{dx}-\frac{\ell_\beta(\ell_\beta+1)}{x^2}\right]\delta_{\beta\beta'} + V_{\beta\beta'}(x) + \lambda_{\beta\beta'}(x)\right\}\phi_{\beta'}(x) = 0\,, \tag{9.5}$$

where $m$ is the nucleon mass, $x$ is the interparticle separation, $V_{\beta\beta'}(x) = \langle\beta' \mid V(i,j) \mid \beta\rangle$ and $V(i,j)$ is the $NN$ potential. Depending on the quantum numbers, the state $\beta$ can be coupled or not to the other ones. An additional term $\lambda_{\beta\beta'}(x)$ has been included in Eq. (9.5) to simulate the average effect of the other nucleons

Table 5. Binding energy ($B$), kinetic energy mean value ($T$) and $S'$- $P$- and $D$-wave percentages for triton in terms of the number of channels $N_c$.

| $N_c$ | $B$(MeV) | $T$(MeV) | $P_{S'}(\%)$ | $P_D(\%)$ | $P_P(\%)$ |
|---|---|---|---|---|---|
| 8 | 7.660 | 45.551 | 1.128 | 8.926 | 0.066 |
| 12 | 7.678 | 45.645 | 1.127 | 8.962 | 0.076 |
| 18 | 7.683 | 45.671 | 1.126 | 8.967 | 0.076 |
| Ref. 25 | 7.684 | 45.677 | 1.126 | 8.968 | 0.076 |

on the pair. The only important condition to be satisfied is $|\lambda_{\beta\beta'}(x)| \ll |V_{\beta\beta'}(x)|$, when $x$ is small. In Ref. 47 it was found that a satisfactory choice is

$$\lambda_{\beta\beta'}(x) = \Lambda_\beta e^{-\gamma x}\, \delta_{\beta\beta'}\,. \tag{9.6}$$

The precise value of $\gamma$ has been found to be unimportant and the range of values $(0.2 \div 0.4)$ fm$^{-1}$ is adequate. Correspondingly, the depth $\Lambda_\beta$ can be chosen so that $\phi_\beta(x)$ satisfies the healing condition $\phi_\beta(x) \to x^{\ell_\beta}$ when $x \to \infty$. The functions $\phi_\beta(x)/x^{\ell_\beta}$ have then been used to construct the correlation factor for a given channel.

Let us again consider the $A = 3$ case. The functions $f_\alpha(x)$ are related to the reference pair characterized by definite values of the angular momentum, spin and isospin for each channel. Therefore, these functions can be taken as the solutions of Eq. (9.5). However, since the total w.f. has been constructed in the $LS$ coupling scheme, the total angular momentum $j$ of the reference pair does not have a definite value, in general. For the first three channels of Table 2 such a problem does not exist because $\ell_{1\alpha} = 0$ and, since $J = 1/2$, one easily obtains $j = S_{2\alpha}$. Consequently, the correlations $f_\alpha$, $\alpha$ =1–3, correspond to the states $^3S_1$, $^1S_0$ and $^3D_1$, respectively.

The channels with $\alpha > 3$ are less important in producing the structure of the system than the first three channels. For the channels with $\ell_{2\alpha} = 0$, $S_{2\alpha} = 1$, $T_{2\alpha} = 0$ one has $f_\alpha(x) = \phi_{^3S_1}(x)$; for the channels with $\ell_{2\alpha} = 0$, $S_{2\alpha} = 0$, $T_{2\alpha} = 1$, $f_\alpha(x) = \phi_{^1S_0}(x)$; and for the channels with $\ell_{2\alpha} = 2$, $S_{2\alpha} = 1$, $T_{2\alpha} = 0$, $f_\alpha(x) = \phi_{^3D_1}(x)/x^2$. Finally, for the channels with $\ell_{2\alpha} = 1$, the solution of Eq. (9.5) has taken into account only the central part of the pair potential in the state $^3P_j$.

For the PHH expansion $g_\alpha(x) = 1$. In the CHH case the functions $g_\alpha(x)$ have been taken equal to $\bar{\phi}(r)$, solution of Eq. (9.5) with $\ell_\beta = 0$ and

$$V_{\beta,\beta'}(x) = \frac{1}{2}\left[V_S^{(c)}(x) + V_T^{(c)}(x)\right]\,, \tag{9.7}$$

where $V_S^{(c)}(x)$ $[V_T^{(c)}(x)]$ is the projection of the central part of the nuclear potential on the state $^1S_0$ $[^3S_1]$. For large interparticle separation distances the correlation factors goes to unity in order to recover the HH expansion which is well suited for describing such configurations. To give an idea of the convergence of the PHH expansion we report in Table 5 only the results obtained in Ref. 11 for the triton using the AV14 potential. From the table one can see that the PHH expansion converges rapidly yielding results which are in complete agreement with those obtained with the HH expansion in Ref. 6 and reported in Table 3. The quantum numbers which specify the channels with $\alpha > 12$ can be found in Ref. 25.

For systems with $A > 3$ analogous criteria can be helpful for constructing appropriate correlation factors. The case $A$=4 has been studied in detail using the correlated expansion in Ref. 17.

### 9.3. *The adiabatic approximation*

The so-called adiabatic approximation (AA), was first used for studying nuclear systems by Macek [48] and Fabre de la Ripelle. [49] It was originally proposed many years ago by Born and Oppenheimer for calculating the structure of a diatomic molecule. First, for a fixed internucleon distance $R$ the Schrödinger equation for the electronic motion was solved and the corresponding eigenvalue $U(R)$ determined. The set of eigenvalues corresponding to different values of $R$ was then used as an effective potential for determining the vibrational and rotational levels of the molecule (for a discussion of the method see the paper by Ballhausen and Hansel. [50]) Putting

$$\Psi(r,\Omega_N) = r^{-(D-1)/2}\Phi(r,\Omega_N)\,, \tag{9.8}$$

the Schrödinger equation becomes

$$\left[-\frac{\hbar^2}{2M}\left\{\frac{\partial^2}{\partial r^2}+\frac{\Lambda^2-(D-1)(D-3)/4}{r^2}\right\}+V(r,\Omega_N)-E\right]\Phi(r,\Omega_N)=0. \tag{9.9}$$

The adiabatic hyperspherical harmonic (AHH) functions are the solutions of the following equation

$$\left[-\frac{\hbar^2}{2M}\left\{\frac{\Lambda_N^2}{r^2}-\frac{(D-1)(D-3)}{4r^2}\right\}+V(r,\Omega_N)\right]\phi_m(r,\Omega_N)=U_m(r)\phi_m(r,\Omega_N), \tag{9.10}$$

where the index $m = 0,1,2,\dots$ labels the various eigenfunctions and the corresponding "eigenpotentials" $U_m(r)$. Clearly, the eigenfunctions reduce to the usual HH functions $\mathcal{H}(\Omega_N)$ and $U_m(r)\sim r^{-2}$ if the potential term is dropped out. When the AHH functions have been calculated, the complete w.f. can be expanded as:

$$\Psi(r,\Omega_N) = r^{-(D-1)/2}\sum_{m=0}^{M}\phi_m(r,\Omega_N)\,w_m(r)\,. \tag{9.11}$$

The following set of $M+1$ coupled equations is then obtained from the Schrödinger equation

$$-\frac{\hbar^2}{2M}\left\{w_m''(r)+\sum_{n=0}^{M}\left[B_{mn}w_n'(r)+C_{mn}w_n(r)\right]\right\}+\left[U_m(r)-E\right]w_m(r) = 0, \tag{9.12}$$

with

$$B_{mn}(r) = \left\langle \phi_m(r,\Omega_N) \middle| 2\frac{\partial}{\partial r} \middle| \phi_n(r,\Omega_N) \right\rangle_\Omega , \tag{9.13}$$

$$C_{mn}(r) = \left\langle \phi_m(r,\Omega_N) \middle| \frac{\partial^2}{\partial r^2} \middle| \phi_n(r,\Omega_N) \right\rangle_\Omega , \tag{9.14}$$

where the symbol $\langle\ \rangle_\Omega$ implies the integration over all the angular and hyperangular variables. The crucial point is the calculation of the AHH functions. A possibility is to expand $\phi_m(r,\Omega_N)$ in terms of the HH functions . Obviously a large number of HH functions is necessary. To take care of this fact, the AHH functions can be expanded in terms of a more suitable basis, such as that of the PHH one. The spline technique has also been used to solve Eq. (9.12) for the $\mu$-atom and atomic systems [51] with quite accurate results. In general, only a few terms in the expansion given in Eq. (9.11) are necessary to obtain a good estimate of the ground state energy of the system.

Table 6. Convergence of the binding energy (in MeV) for the $A=3$, 4 systems in the case of the adiabatic approximation when using the AV14 potential. $M$ is the number of functions $\phi_m(r,\Omega_N)$ used in the expansion of the w.f. The number $K$ of the PHH functions used for expanding each function $\phi_m(r,\Omega)$ is $K=48$ for $A=3$ and $K=81$ for $A=4$.

| $M$ | $B(A{=}3)$ | $B(A{=}4)$ |
|---|---|---|
| 1 | 7.44 | 20.01 |
| 5 | 7.61 | 21.05 |
| 9 | 7.65 | 21.08 |
| 13 | 7.66 | 21.09 |

The application of the AHH expansion to the ground states of $A=3$, 4 nuclei with realistic interactions has been made by Kievsky and Viviani. [52] The results are reported in Table 6 and show a rapid convergence with the number $M$ of AHH functions included.

### 9.4. *The extended hyperspherical harmonic expansion*

An alternative way for taking into account the correlation effects in the w.f. , known as the extended hyperspherical harmonic (EHH) expansion, has been recently proposed [16,53] . Let us consider the case of a three-body system and indicate with $\boldsymbol{y}_1^{(p)}$ and $\boldsymbol{y}_2^{(p)}$ the Jacobi coordinates corresponding to the permutation $p$. It can be observed that

$$r^{2n+\ell_{1\alpha}+\ell_{2\alpha}}\ {}^{(2)}\mathcal{P}_n^{\ell_{1\alpha},\ell_{2\alpha}}(\Phi_2^{(p)}) = (y_1^{(p)})^{\ell_{1\alpha}}(y_2^{(p)})^{\ell_{2\alpha}} \sum_{m=0}^{n} a_{nm}\ (y_2^{(p)})^{2m} r^{2(n-m)} , \tag{9.15}$$

where $\Phi_2^{(p)}$ is the hyperspherical angle for the permutation $p$ and $a_{nm}$ are numerical coefficients. The radial structure $R_\alpha(y_1^{(p)}, y_2^{(p)})$ of a $\alpha$–channel in the expansion of the w.f. is the following:

$$R_\alpha(y_1^{(p)}, y_2^{(p)}) = \sum_{n=N_\alpha^0}^{N_\alpha} \frac{u_{\alpha,n}(r)}{r^{\frac{5}{2}}} \, {}^{(2)}\mathcal{P}_n^{\ell_{1\alpha},\ell_{2\alpha}}(\Phi_2^{(p)}) \,. \tag{9.16}$$

The role of $N_\alpha^0$ is to avoid the inclusion of the linearly dependent states (see Refs. 16, 53 for more details). Therefore, apart from the kinematical factor $(y_1^{(p)})^{\ell_{1\alpha}}(y_2^{(p)})^{\ell_{2\alpha}}$, only even powers of the distance $y_2^{(p)} \equiv r_{ij}$ enter the expansion. Let us now consider a configuration of the system where the particles $i$ and $j$ are close to each other. The dependence on $r_{ij}$ of the system w.f. for small values of this distance is expected to be linear, in general. Odd powers of $r_{ij} = r\cos\Phi_2^{(p)}$ can be linearly expressed in terms of even powers of $\cos\Phi_2^{(p)}$, with $\Phi_2^{(p)}$ ranging from zero to $\pi/2$, but an infinite expansion is required. Therefore, if the linear dependence on the corresponding hyperangle $\Phi_2^{(p)}$ is important, the resulting expansion is expected to be very slowly convergent. As previously discussed, in fact, when the pair correlations are important (as in the case of strongly repulsive *NN* potentials), the HH expansion converges very poorly. The inclusion of suitably chosen correlation functions is a rather effective way to take care of "odd" powers in the expansion of each amplitude. Alternatively, there is also the possibility of including explicitly some odd powers in the expansion of the radial dependence by considering, in place of an expression such as Eq. (9.16), the following:

$$\begin{aligned} R_\alpha(y_1^{(p)}, y_2^{(p)}) = &\sum_{n=N_\alpha^0}^{N_\alpha} \frac{u_{\alpha,n}(r)}{r^{\frac{5}{2}}} \, {}^{(2)}\mathcal{P}_n^{\ell_{1\alpha},\ell_{2\alpha}}(\Phi_2^{(p)}) \\ &+ \cos\Phi_2^{(p)} \sum_{n'=0}^{N'_\alpha} \frac{w_{\alpha,n'}(r)}{r^{\frac{5}{2}}} \, {}^{(2)}\mathcal{P}_n^{\ell_{1\alpha},\ell_{2\alpha}}(\Phi_2^{(p)}) \,. \end{aligned} \tag{9.17}$$

This expansion basis is overcomplete, *i.e.* a few states can be linearly dependent on the others. As an example, the application of the EHH approach to the helium atom is discussed in the next subsection.

### 9.5. *Application of the EHH expansion to the helium atom*

The problem of calculating the energy levels of the helium atom was tackled by Hylleraas [54] in 1930. The technique used is discussed in many textbooks of Quantum Mechanics as a significative example of the application of the variational method. Later on Pekeris [55] improved the accuracy of the calculation. The w.f. was constructed using the "perimetric coordinates"

$$u = \varepsilon(r_{32} - r_{31} + r_{12}) \,, \quad v = \varepsilon(r_{31} - r_{32} + r_{12}) \,, \quad z = 2\varepsilon(r_{31} + r_{32} + r_{12}) \,, \tag{9.18}$$

where the indices 3,2,1 stand for the nucleus and the two electrons, respectively, and $\varepsilon = \sqrt{B}$, with $B$ the binding energy of the atom. The radial part of the w.f. has the form

$$\phi = e^{-\frac{1}{2}(u+v+z)} G(u,v,z)\,,$$
$$G(u,v,z) = \sum_{i=0}^{I}\sum_{j=0}^{J}\sum_{k=0}^{K} A_{ijk} L_i(u) L_j(v) L_k(z)\,. \tag{9.19}$$

Here $L_m(\xi)$ is a Laguerre polynomial of order $m$ and the coefficients $A_{ijk}$ are determined by the variational principle. Such a form has been generalized by a few authors. In particular, the choice in the paper of Drake *et al.* [56] is

$$\phi = \sum_{i=0}^{I}\sum_{j=0}^{J}\sum_{k=0}^{K} \left[a_{ijk}^{(1)}\varphi_{ijk}(\alpha_1,\beta_1) + a_{ijk}^{(2)}\varphi_{ijk}(\alpha_2,\beta_2)\right] \pm \text{exchange term}\,, \tag{9.20}$$

where the $a^{(1)}, a^{(2)}$ are trial parameters and

$$\varphi_{ijk}(\alpha,\beta) = r_{31}^i r_{32}^j r_{12}^k e^{-\alpha r_{31} - \beta r_{32}}\,. \tag{9.21}$$

The pairs $(\alpha_1,\beta_1)$, $(\alpha_2,\beta_2)$ are non-linear trial parameters. The exchange term in Eq. (9.20) has the same functional form as the first term but with $r_{31} \leftrightarrow r_{32}$. The results of Refs. 55, 56 are reported in Table 7. They refer to the case where the mass of the helium nucleus is considered to be infinite with respect to the electron mass (for the case of finite mass value see the cited references).

Table 7. Ground state of the helium atom in atomic mass units (amu). $N$ is the number of linear parameters. In the final line the extrapolated value is indicated.

| | Ref. 55 | | Ref. 56 |
|---|---|---|---|
| $N$ | $B$(amu) | $N$ | $B$(amu) |
| 95 | 2.903723389 | 44 | 2.903724131 |
| 125 | 2.903723878 | 67 | 2.903724351 |
| 203 | 2.903724228 | 135 | 2.90372437655 |
| 210 | 2.903724311 | 182 | 2.90372437696 |
| ⋮ | ⋮ | ⋮ | ⋮ |
| extr. | 2.903724354 | 236 | 2.90372437702 |

It should be noted that there are other important corrections to be taken into account when calculating the energy levels, such as the relativistic and QED corrections.

The HH method has been applied to the helium atom problem by several researchers (see, for example, the paper of Krivec [57] and the references therein). The best HH estimate of the ground state energy has been obtained by Fabre de la Ripelle *et al.* [58] with the result $B$= 2.9030674 amu where only the first four digits are correct. The convergence rate of the HH expansion is slow due to the singularity of the Coulomb potential when $r \to 0$. Haftel and Mandelzweig [59] modified the HH expansion basis multiplying the basis functions by an appropriate correlation factor to improve the convergence. This is known as the correlation function hyperspherical harmonic method (CFHHM) . The w.f. is written as

$$\Psi = \Theta\,\Phi\,, \tag{9.22}$$

where

$$\Theta = e^{-\beta r_{12} - \delta(r_{31}+r_{32})}\,, \qquad \beta = -\frac{1}{2}\,, \qquad \delta = \frac{2m_\alpha}{m_\alpha + 1}\,, \tag{9.23}$$

where $m_\alpha$ is the $\alpha$-particle mass. With such a choice of $\beta$ and $\delta$ the behaviour of the Coulomb interaction at $r \to 0$ is correctly taken into account. The function $\Phi$ is then expanded in terms of the HH basis. The equation to be solved is written in the form

$$\Theta^{-1}[H - E]\Theta\,\Phi = (\Theta^{-1}T\,\Theta + V - E)\,\Phi = 0\,, \tag{9.24}$$

where $T$ and $V$ are the total kinetic and potential energies. The details of the calculation can be found in the original papers. The conclusion reached is that the convergence is strongly improved. The convergence pattern is shown in Table 8.

Table 8. Ground state binding energy of the helium atom in atomic mass unit with the CFHHM of Ref. 59 and the EHH approach. $N_{\rm HH}$ is the number of HH basis functions.

| $N_{\rm HH}$ | CFHHM | EHH |
|---|---|---|
| 1 | 2.855504862 | 2.90347103 |
| 4 | 2.902870977 | 2.90369776 |
| 9 | 2.903701425 | 2.90371936 |
| 16 | 2.903718577 | 2.90372322 |
| 25 | 2.903723654 | 2.90372388 |
| 36 | 2.903723987 | 2.90372415 |

More recently the HH expansion has been revisited. [60] The w.f. has the form

$$\Psi = \chi_{S,S_z}(1,2)\,\psi\,, \qquad \psi = \varphi(\boldsymbol{y}_1^{(1)}, \boldsymbol{y}_2^{(1)}) + \varphi(\boldsymbol{y}_1^{(2)}, \boldsymbol{y}_2^{(2)}) + \varphi_3(\boldsymbol{y}_1^{(3)}, \boldsymbol{y}_2^{(3)})\,, \tag{9.25}$$

where $\boldsymbol{y}_1^{(p)}, \boldsymbol{y}_2^{(p)}$ are the Jacobi coordinates corresponding to the even (odd) permutation $p$ of $1,2,3$ for total spin $S = 0\,(1)$. The amplitudes $\varphi, \varphi_3$ are then expanded in terms of the HH basis. It should be noted that:

*i*) in principle, the amplitude $\varphi_3(\boldsymbol{y}_1^{(3)}, \boldsymbol{y}_2^{(3)})$ should be, by itself, sufficient to produce the exact w.f. . However, the convergence of the expansion is dramatically slow in such a case ;

*ii*) the three amplitudes $\varphi$ in Eq. (9.25) are not orthogonal to each other. Therefore, some care is needed to avoid duplication of basis elements.

With $\Psi$ as in Eq. (9.25) and not a very large number of basis functions the estimate $B = 2.9028$ amu is obtained and therefore the result is not satisfactory. As a possible remedy we can consider the EHH expansion of the form

$$\Psi = \chi_{S,S_z}(1,2) \sum_{p=1}^{3} \left[ \varphi_p^{\rm HH}(\boldsymbol{y}_1^{(p)}, \boldsymbol{y}_2^{(p)}) + \cos\Phi_2^{(p)}\, \varphi_p^{\rm EHH}(\boldsymbol{y}_1^{(p)}, \boldsymbol{y}_2^{(p)}) \right] , \tag{9.26}$$

where $\cos\Phi_2^{(p)} = y_2^{(p)}/r$ and both $\varphi^{\rm HH}$ and $\varphi^{\rm EHH}$ are expanded in terms of the HH basis with the hyperradial functions expressed as a combination of the Laguerre functions $e^{-\gamma r} L_n(\gamma r)$. The results obtained are shown in Table (8) and are compared with the results of Ref. 59. As it can be seen from the table and by comparing the results with the value $B$=2.9037243770 amu given by Drake *et al.* [56], both methods provide seven correct digits for the ground state energy. For the first excited level an accuracy of six digits has been obtained by means of the EHH technique. Further improvements in the EHH approach are presently under investigation.

## 10. Variational calculations for three- and four-nucleon scattering processes

The study of few-nucleon systems has reached a promising state–of–the–art in the last few years. Several methods are now available for computing static and dynamic observables. In particular, a remarkable accuracy can be obtained using the Faddeev equations technique, [23,45,46,61] the GFMC method [62] and variational techniques. [11,25,63] In the preceding sections the results for the bound states of strongly interacting systems have been reported. In this section the application of the approach to studying three- and four-nucleon scattering processes will be discussed. As in the case of bound states, the approach is variational and is based on the Kohn variational principle. Only a brief description of a three-body process formalism is given here. More details can be found in Refs. 11, 64.

### 10.1. *N − d scattering*

The scattering wave function $\Psi$ for the *N*-*d* process is written as a sum of two terms $\Psi = \Psi_C + \Psi_A$. The first term $\Psi_C$ describes the system when the three nucleons are close to each other. For large interparticle separations and energies below the deuteron breakup threshold (DBT) it goes to zero, whereas for higher energies it must reproduce a three outgoing particle state. It is written as a sum of three Faddeev–like amplitudes $\psi_C(\boldsymbol{y}_1^{(p)}, \boldsymbol{y}_2^{(p)})$, where $\boldsymbol{y}_1^{(p)}$ and $\boldsymbol{y}_2^{(p)}$ are the Jacobi coordinates corresponding to an even permutation $p \equiv (i,j,k)$ of the particle indices 1, 2 and 3. Each amplitude $\psi_C(\boldsymbol{y}_1^{(p)}, \boldsymbol{y}_2^{(p)})$, has total angular momentum $JJ_z$ and total isospin $TT_z$ and it is decomposed into channels using $LS$ coupling, namely

$$\psi_C(\boldsymbol{y}_1^{(p)}, \boldsymbol{y}_2^{(p)}) = \sum_{\alpha=1}^{N_c} R_\alpha(y_1^{(p)}, y_2^{(p)}) \mathcal{H}_\alpha^{(p)} , \tag{10.1}$$

$$\mathcal{H}_\alpha^{(p)} = \left\{ \left[ Y_{\ell_{1\alpha}}(\hat{y}_1) Y_{\ell_{2\alpha}}(\hat{y}_2) \right]_{\Lambda_\alpha} \left[ s_\alpha^{jk} s_\alpha^i \right]_{S_\alpha} \right\}_{JJ_z} \left[ t_\alpha^{jk} t_\alpha^i \right]_{TT_z} , \tag{10.2}$$

where $y_1^{(p)}, y_2^{(p)}$ are the absolute values of the Jacobi coordinates and $\mathcal{H}_\alpha$ is the angular-spin-isospin function for each channel. The two-dimensional amplitude $R_\alpha$ is expanded in terms of the PHH basis

$$R_\alpha(y_1^{(p)}, y_2^{(p)}) = r^{\ell_{1\alpha}+\ell_{2\alpha}} f_\alpha(y_2^{(p)}) \left[ \sum_{n_\alpha} u_{\alpha,n_\alpha}(r) \, {}^{(2)}P_{n_\alpha}^{\ell_{1\alpha},\ell_{2\alpha}}(\Phi_2^{(p)}) \right] . \tag{10.3}$$

Here again the hyperspherical variables are defined by the relations $y_2^{(p)} = r \cos \Phi_2^{(p)}$ and $y_1^{(p)} = r \sin \Phi_2^{(p)}$ and $f_\alpha(y_2^{(p)})$ is a pair correlation function.

The second term $\Psi_A$ describes the asymptotic motion of a deuteron relative to the third nucleon. It can be written as a sum of three amplitudes with the generic one of the form

$$\Omega_{LSJ}^\lambda(\boldsymbol{y}_1^{(p)}, \boldsymbol{y}_2^{(p)}) = \sum_{l_\alpha=0,2} w_{l_\alpha}(y_2^{(p)}) \mathcal{R}_L^\lambda(y_1^{(p)})$$
$$\times \left\{ \left[ [Y_{l_\alpha}(\hat{y}_2^{(p)}) s_\alpha^{jk}]_1 s^i \right]_S Y_L(\hat{y}_1^{(p)}) \right\}_{JJ_z} [t_\alpha^{jk} t^i]_{TT_z} , \tag{10.4}$$

where $w_{l_\alpha}(y_2^{(p)})$ is the deuteron component in the state $l_\alpha = 0, 2$ and $L$ is the relative angular momentum of the deuteron and the incident nucleon. The superscript $\lambda$ indicates the regular ($\lambda \equiv R$) or the irregular ($\lambda \equiv I$) solution. In the *p-d* (*n-d*) case the $\mathcal{R}^\lambda$ are related to the regular and irregular Coulomb (spherical Bessel) functions, respectively. The function $\Omega^\lambda$ can be combined to form the general asymptotic state

$$\Omega_{LSJ}^+(\boldsymbol{y}_1^{(p)}, \boldsymbol{y}_2^{(p)}) = \Omega_{LSJ}^0(\boldsymbol{y}_1^{(p)}, \boldsymbol{y}_2^{(p)}) + \sum_{L'S'} {}^J\mathcal{L}_{LL'}^{SS'} \, \Omega_{L'S'J}^1(\boldsymbol{y}_1^{(p)}, \boldsymbol{y}_2^{(p)}) , \tag{10.5}$$

with the following asymptotic functions

$$\Omega_{LSJ}^0(\boldsymbol{y}_1^{(p)}, \boldsymbol{y}_2^{(p)}) = u_{00} \Omega_{LSJ}^R(\boldsymbol{y}_1^{(p)}, \boldsymbol{y}_2^{(p)}) + u_{01} \Omega_{LSJ}^I(\boldsymbol{y}_1^{(p)}, \boldsymbol{y}_2^{(p)}) , \tag{10.6}$$
$$\Omega_{LSJ}^1(\boldsymbol{y}_1^{(p)}, \boldsymbol{y}_2^{(p)}) = u_{10} \Omega_{LSJ}^R(\boldsymbol{y}_1^{(p)}, \boldsymbol{y}_2^{(p)}) + u_{11} \Omega_{LSJ}^I(\boldsymbol{y}_1^{(p)}, \boldsymbol{y}_2^{(p)}) . \tag{10.7}$$

The matrix elements $u_{ij}$ are selected in accordance with the four different possible choices of the matrix $\mathcal{L}$: the $K$-matrix, $K^{-1}$-matrix, $S$-matrix or the $T$-matrix. [65] The three-nucleon scattering w.f. for an incident state with relative angular momentum $L$, spin $S$ and total angular momentum $J$ is

$$\Psi_{LSJ}^+ = \sum_p \left[ \Psi_C(\boldsymbol{y}_1^{(p)}, \boldsymbol{y}_2^{(p)}) + \Omega_{LSJ}^+(\boldsymbol{y}_1^{(p)}, \boldsymbol{y}_2^{(p)}) \right] , \tag{10.8}$$

and its complex conjugate is $\Psi^-_{LSJ}$. A variational estimate of the trial parameters in the w.f. $\Psi^+_{LSJ}$ can be obtained by requiring, in accordance with the generalized Kohn variational, that the functional

$$[{}^J\mathcal{L}^{SS'}_{LL'}] = {}^J\mathcal{L}^{SS'}_{LL'} - \frac{2}{\det(u)} \langle \Psi^+_{LSJ} \,|\, H - E \,|\, \Psi^+_{L'S'J} \rangle \,, \qquad (10.9)$$

be stationary. It is convenient here to use the complex form [65] of the Kohn variational principle since the numerical instabilities, normally present in practical applications of this principle [66], are then avoided.

The variation of the functional with respect to the hyperradial functions leads to a set of coupled differential equations of the form

$$\sum_{\alpha', n'_\alpha} \left[ A^{\alpha\alpha'}_{n_\alpha n'_\alpha}(r) \frac{d^2}{dr^2} + B^{\alpha\alpha'}_{n_\alpha n'_\alpha}(r) \frac{d}{dr} + C^{\alpha\alpha'}_{n_\alpha n'_\alpha}(r) \right.$$
$$\left. + \frac{2M}{\hbar^2} E N^{\alpha\alpha'}_{n_\alpha n'_\alpha}(r) \right] u_{\alpha', n'_\alpha}(r) = D^\lambda_{\alpha n_\alpha}(r) \,. \qquad (10.10)$$

For each asymptotic state ${}^{(2S+1)}L_J$ two different inhomogeneous terms $D^\lambda_{\alpha n_\alpha}$ can be constructed in correspondence to the asymptotic $\Omega^\lambda_{LSJ}$ functions with $\lambda \equiv 0, 1$. Boundary conditions must be specified for the hyperradial functions. For energies below the DBT they must go to zero when $r \to \infty$. Above the DBT energy they must describe the breakup configuration asymptotically.

The convergence of the expansion of the internal part $\Psi_C$ is studied by first considering the channels having values as low as possible for the orbital angular momenta $\ell_{1\alpha}$ and $\ell_{2\alpha}$, and then including a few channels of higher $\ell_{1\alpha}$, $\ell_{2\alpha}$ values in the expansion. The number of the PHH functions belonging to those channels is increased and higher order channels are included until the desired degree of convergence is obtained.

## 10.2. *Results for the N − d scattering*

To give an idea of the accuracy attainable in the construction of $N$-$d$ scattering states, a few of the results obtained in a benchmark calculation on $n$-$d$ scattering below DBT will be reported here. In Refs. 67, 68 a detailed comparison has been made of the mixing parameter values obtained by the Bochum group by solving the Faddeev equations in momentum space with those of the Pisa group using the PHH method. In Table 9 the eigenphase shifts and mixing parameters for the state with $J = 1/2$ and positive parity given in Ref. 67 are reported. Results for the phase shift parameters up to $J = 9/2$ and some other scattering observables can be found in the original paper. The conclusion of the benchmark, as can also be seen from the table, is that there is a strict agreement between the results obtained by the two different methods.

High quality measurements for $p$-$d$ scattering have been presented in Ref. 69. Therefore a significant comparison can be made between the experimental data and

Table 9. Phase shifts $\delta$ and mixing parameters $\eta$ (in degrees) for $J^\Pi = \frac{1}{2}^+$ at three different energy values for the AV14 potential. $\lambda$ is the angular momentum of the projectile nucleon, $\Sigma$ is the sum of deuteron and nucleon spins. The calculations of the Bochum group are in momentum space, those of the Pisa group in configuration space.

| | $E_{lab} = 1\,\text{MeV}$ | | $E_{lab} = 2\,\text{MeV}$ | | $E_{lab} = 3\,\text{MeV}$ | |
|---|---|---|---|---|---|---|
| $\delta_{\Sigma\lambda}$ | Bochum | Pisa | Bochum | Pisa | Bochum | Pisa |
| $\delta_{\frac{3}{2}2}$ | -0.999 | -1.00 | -2.57 | -2.58 | -3.91 | -3.91 |
| $\delta_{\frac{1}{2}0}$ | -17.8 | -17.7 | -28.0 | -27.9 | -34.9 | -34.9 |
| $\eta$ | 1.03 | 1.04 | 1.20 | 1.21 | 1.25 | 1.26 |

the theoretical results. The validity of the Kohn principle for the charged particle $p$-$d$ scattering below DBT has been proved in Ref. 70. Applications of the PHH expansion to this process can be found in Refs. 41, 71, 72.

### 10.3. *Results for the low energy $n$—$^3H$ and $p$—$^3He$ scattering*

Let us first consider the $n$-$^3$H scattering at zero energy. The singlet $a_s$ and triplet $a_t$ scattering lengths can be deduced from the experimental values of the total cross section $\sigma_T$ and the coherent scattering length $a_c$:

$$\sigma_T = \pi(|a_s|^2 + 3|a_t|^2)\,, \qquad a_c = \frac{1}{4}a_s + \frac{3}{4}a_t\,. \tag{10.11}$$

The $n$-$^3$H cross section has been accurately measured over a wide range of energies and the extrapolation to zero energy does not present any problem. The value obtained is $\sigma_T = 1.70 \pm 0.03\,\text{b}$. [73] The coherent scattering length has been measured by neutron–interferometry techniques. The most recent values reported in the literature have been obtained by the same group: they are $a_c = 3.82 \pm 0.07\,\text{fm}$ [74] and $a_c = 3.59 \pm 0.02\,\text{fm}$, [75] the latter value being obtained with a more advanced experimental setup. Recently, the $a_c = 3.607 \pm 0.017\,\text{fm}$ estimate has been obtained from $p$-$^3$He data by using an approximated Coulomb–corrected $R$-matrix theory. [76]

The calculated singlet and triplet $n$-$^3$H scattering lengths for the different potential models are plotted versus the corresponding $^3$H binding energy in Fig. 3. The experimental values [75,76] of $a_s$ and $a_t$ have also been reported. The models including only *NN* forces are the AV14, [38] AV8 [77] and AV18 [39] potentials. Adding $3N$ forces we have: the AV14 + Urbana model VIII (AV14UR), [78] AV18 + Urbana model IX (AV18UR), [79] AV14 + Brazil with $\Lambda = 5.6m_\pi$ (AV14BR1) and AV14 + Brazil with $\Lambda = 5.8m_\pi$ (AV14BR2). [80] In the AV14UR and AV18UR models, one of the parameters of the $3N$ potential was chosen to reproduce the experimental $^3$H binding energy value $B_3 = 8.48$ MeV. The AV14BR1 and AV14BR2 models have been chosen to give slightly larger $B_3$ values. It should be noted that all the results for the singlet (triplet) scattering length fall essentially on a straight line. However, the experimental values extracted from the data do not lie on the theoretical curves.

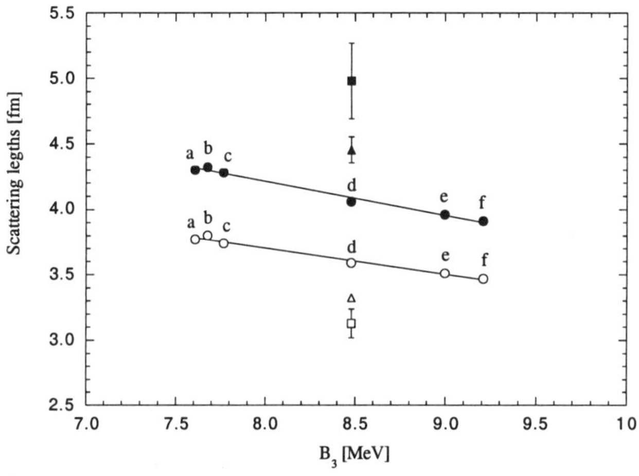


Fig. 3. Singlet (full symbols) and triplet (open symbols) $n$-$^3$H scattering lengths plotted against the corresponding trinucleon binding energy for different interaction models. The circles labelled by a, b, c, d, e, f correspond to the AV18, AV14, AV8, AV18UR, AV14BR1 and AV14BR2 models, respectively. The AV14UR and AV18UR model predictions are almost coincident. The squares (triangles) are the experimental values of Ref. 75 (Ref. 76). The straight lines are linear fits of the theoretical results.

This disagreement is caused by a rather small discrepancy between the calculated and measured coherent scattering lengths.

A preliminary study of the $p$-$^3$He differential cross section and some polarization observables has been presented in Ref. 81 . In fact, rather accurate experimental data are available at low energies and it is therefore possible to perform detailed comparisons. As a result it appears that a proper treatment of the Coulomb repulsion is necessary in order to have a correct description of the observables at small angles.

## 11. Electro-weak reaction on few-nucleon systems

In the preceding section it has been seen that the wave functions that describe the scattering of projectiles of different energies by the $A = 2$ and 3 bound systems below and above their breakup threshold can be espressed with high precision using the correlated hyperspherical harmonic (CHH) method. The wave functions can then be used to study a number of reactions, some of them being of relevant astrophysical interest.

The status of *ab initio* calculations in the low-energy region of proton radiative capture [82] $^2$H$(p,\gamma)\,^3$He and weak capture [83] $^3$He$(p,e^+\nu_e)\,^4$He reactions will be briefly reviewed in the present section. The latter process is also known as the *hep* reaction. These studies provide examples of some useful applications of the CHH method.

### 11.1. *The p − d radiative capture*

Proton radiative capture calculations offer the possibility of testing the models of nuclear interactions and currents, thanks to the numerous experimental data of the TUNL and Wisconsin groups. [84]. The theoretical description of this process requires the knowledge of the nuclear bound- and scattering-state wave functions, for the given hamiltonian model.

The calculation reviewed here [82] is based on PHH wave functions obtained with the AV18UR potential. The electromagnetic current and charge operators include one- and two-body components. The one-body terms can be obtained in the standard way from a non-relativistic reduction of the covariant single-nucleon current, including terms proportional to $1/m^2$ (where $m$ is the nucleon mass). The two-body electromagnetic current consist of "model-independent"(MI) and "model-dependent"(MD) terms (see Refs. 42, 85 for a review). The MI terms, which give the leading two-body contributions, are obtained from the *NN* interaction and, by construction, satisfy current conservation with it. The MD two-body currents are purely transverse and therefore cannot be linked directly to the underlying *NN* interaction. Among the MD currents, those associated with excitation of $\Delta$ isobars are the most important ones in the momentum-transfer regime discussed here. These currents have been treated using the transition-correlation operator (TCO) scheme (originally developed in Ref. 86 and further extended in Ref. 85), which explicitly includes the $\Delta$ degrees of freedom in the nuclear wave functions. The $\Delta$-currents, although the most important MD currents, still give much smaller contributions than the leading MI terms.

While the main two-body contributions to the electromagnetic current are linked to the form of the *NN* interaction through the continuity equation, the most important two-body electromagnetic charge operators are model dependent and should be viewed as relativistic corrections. The model commonly used [87] for the two-body charge operators includes the $\pi$-, $\rho$-, and $\omega$-meson exchange terms with both isoscalar and isovector components, as well as the (isoscalar) $\rho\pi\gamma$ and (isovector) $\omega\pi\gamma$ charge transition couplings. At moderate values of the momentum transfer ($q < 5\,\mathrm{fm}^{-1}$), the contribution due to the $\pi$-meson exchange charge operator has been found to be typically one order of magnitude larger than that of any of the remaining two-body mechanisms and of the one-body relativistic corrections. [85]

The available experimental data on the *p-d* reaction includes differential cross sections, vector and tensor analyzing powers, photon polarization coefficients, as well as the astrophysical *S*-factor up to energies of 10 MeV in the center of mass (c.m.) reference frame. [84] The comparison between theory and experiment [82] has led to the following conclusions: *i*) the theoretical predictions obtained by including only one-body currents are in strong disagreement with the data; *ii*) the differences between the theory and experiment disappear, to a large extent, when two-body currents are taken into account; *iii*) some discrepancies remain in the c.m. energy range 0–100 keV, particularly for the differential cross sections and the tensor analyzing powers at small angles. The origin of these discrepancies has been investigated in detail in Ref. 82.

The results for the astrophysical $S$-factor in the c.m. energy range 0–2 MeV, are summarized in Fig. 4.

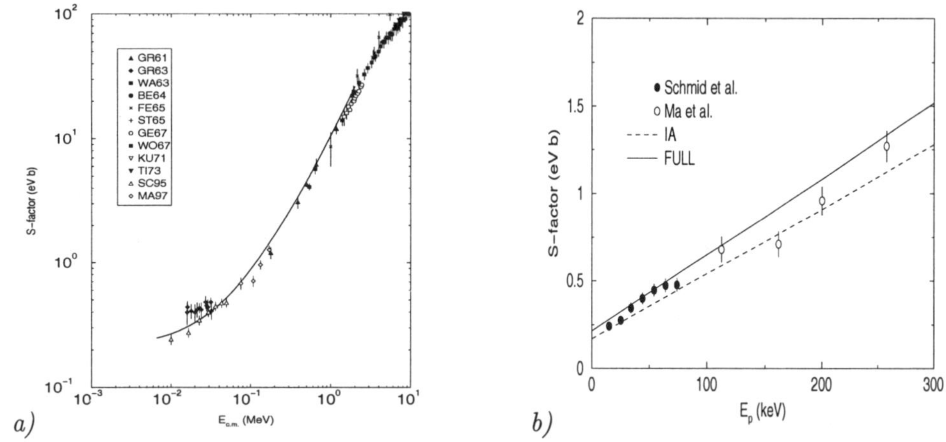


Fig. 4. The $S$-factor for the $^2$H$(p,\gamma)^3$He reaction in the c.m. range 0–2 MeV, obtained with the AV18UR hamiltonian model and one-body currents only (dashed line) or both one- and two-body currents (solid line). In *a)* only the results of the full calculation are shown. The experimental data are taken from the web site `http://pntpm.ulb.ac.be/nacre.htm`.

### 11.2. *The hep reaction*

The weak proton capture on $^3$He (the *hep* reaction) is one of the nuclear reactions of the solar *p*-*p* chain which produces neutrinos. The particularity of the *hep* neutrinos is that they are the most energetic ones: their end-point energy is even larger than that of the $^8$B neutrinos. However, the *hep* neutrino flux is typically much smaller than that of the $^8$B neutrinos. Therefore a significant distortion can be provoked by the *hep* neutrinos in the high-energy part of the $^8$B neutrino spectrum. In fact, the first results which the Super-Kamiokande (SK) collaboration presented in the late 1990's for the $^8$B neutrino spectrum showed an enhancement of events in the highest-energy bin. [88,89] The situation in 1999, after 825 days of data acquisition, was the following: [90] *i*) the ratio between the number of measured and predicted events on the basis of the Standard Solar Model (SSM98) [91] was approximately 0.47, as obtained from the low-energy part of the spectrum; *ii*) the excess of events in the high-energy part of the spectrum could be explained by an enhancement of the *hep* SSM98 flux by a factor of about 17.

A direct measurement of the *hep* cross section in the low-energy regime cannot be performed with the available experimental techniques, since the rate is too low. Therefore, the Standard Solar Model can estimate the *hep* neutrino flux only from theoretical studies. In particular, the SSM98 estimate is based on the calculation of Ref. 86 for the astrophysical $S$-factor at zero energy in the c.m. frame. This study, the last of a lengthy series[a], uses variational Monte Carlo (VMC) wave functions obtained with the AV14UR potential and, in a partial wave expansion of the initial $p-{}^3$He state, includes only the ${}^3S_1$ channel and neglects any dependence on the momentum transfer $q$ of the reaction. Based on this calculation, the SSM98 value for the *hep* astrophysical $S$-factor is $2.3 \times 10^{-20}$ keV b.

The *hep* $S$-factor has been recalculated using CHH wave functions obtained with the AV18UR hamiltonian model, and including all $S$- and $P$-wave capture states. The nuclear weak current and charge operators have vector and axial-vector parts, with corresponding one- and two-body components. All the one-body operators have been obtained from the non-relativistic reduction of the single-nucleon covariant current, as it has been done for the electromagnetic current and charge operators.

The two-body vector current and charge operators have been constructed from the isovector components of the corresponding electromagnetic operators in accordance with the conserved-vector-current hypothesis; the model used is therefore the same as the one discussed for the $p$-$d$ reaction. However, in the weak vector charge only the "$\pi$-like" and "$\rho$-like" terms have been retained, since they give the most important contributions at low momentum-transfer.

The two-body axial charge operator includes a pion-range term, which follows from the soft-pion theorem and current algebra arguments, [92,93] and short-range terms associated with scalar- and vector-meson exchanges. The latter are obtained consistently with the $NN$ interaction model following a procedure [94] similar to that used to derive the MI two-body electromagnetic current operators. [83] The $\Delta$-excitation terms have also been included, but they have been found to be unimportant. [83]

In contrast to the electromagnetic case, the axial current operator is not conserved. Thus, its two-body components cannot be linked to the $NN$ interaction and so should be viewed as model dependent. Among the two-body axial current operators, those due to $\pi$- and $\rho$-meson exchanges and to the $\rho\pi$-transition mechanism have been included. However, the leading two-body terms in the axial current are due to $\Delta$-isobar excitation. These contributions have again been treated non-perturbatively using the TCO scheme.

Due to the poor knowledge of the axial coupling constants for the $N \to \Delta$ and $\Delta \to \Delta$ transitions, the largest model dependence is on the axial current. In particular, the $N \to \Delta$ transition plays a crucial role in those weak processes where the leading one-body operators are suppressed. To minimize this model dependence, the $N$-$\Delta$ axial coupling constant has been adjusted so as to reproduce the experimental value of the Gamow-Teller matrix element in tritium $\beta$-decay. [83,95] While this procedure is model dependent, its actual model dependence is in fact very weak, as it has been shown in Refs. 83, 95.

[a]For a historical review of the different calculations for the *hep* reaction see Ref. 83.

Table 10. The *hep* $S$-factor, in units of $10^{-20}$ keV b, calculated with CHH wave functions corresponding to the AV18UR hamiltonian model, at $p$-$^3$He c.m. energies $E$=0, 5, and 10 keV. The rows labelled "one-body" and "full" list the contributions obtained by retaining the one-body only and both one- and two-body terms in the nuclear weak current. The contributions due to the $^3S_1$ channel only and all $S$- and $P$-wave channels are listed separately.

| | $E$=0 keV | | $E$=5 keV | | $E$=10 keV | |
|---|---|---|---|---|---|---|
| | $^3S_1$ | $S+P$ | $^3S_1$ | $S+P$ | $^3S_1$ | $S+P$ |
| one-body | 26.4 | 29.0 | 25.9 | 28.7 | 26.2 | 29.3 |
| full | 6.38 | 9.64 | 6.20 | 9.70 | 6.36 | 10.1 |

The results for the *hep* $S$-factor are summarized in Table 10. By inspection of the table, it can be noted that: *i*) the energy dependence is rather weak: the value at 10 keV is only about 4% larger than that at 0 keV; *ii*) the $P$-wave capture states are found to be important, contributing about 40% of the calculated $S$-factor. However, the contributions from the $D$-wave channels are expected to be very small; [83] *iii*) the many-body axial currents play a crucial role in the (dominant) $^3S_1$ capture, where they reduce the $S$-factor by more than a factor of four. These results have been shown to be weakly model dependent. [83] In fact, the zero-energy $S$-factor calculated with the older AV14UR interaction is $10.1 \times 10^{-20}$ keV b, only about 4% larger than the AV18UR value of $9.64 \times 10^{-20}$ keV b listed in Table 10.

In conclusion, the new estimate of $10.1 \times 10^{-20}$ keV b for the $S$-factor at 10 keV (the Gamow-peak energy is 10.7 keV) is about 4.5 times larger than the SSM98 value.

To study the implications of these results for the SK energy spectrum of recoil electrons scattered from solar neutrinos, the parameter $\alpha \equiv (S_{\rm new}/S_{\rm SSM98}) \times P_{\rm osc}$ has been introduced, where $P_{\rm osc}$ is the *hep* neutrino suppression constant. At the present time, $\alpha = (10.1 \times 10^{-20}\,{\rm keV\,b})/(2.3 \times 10^{-20}\,{\rm keV\,b}) = 4.4$ if *hep* neutrino oscillations are ignored ($P_{\rm osc}$=1). The SK data available up to 1999, after 825 days of data acquisition, were presented as a ratio of the measured electron spectrum to what is expected in the SSM98 with no neutrino oscillations. The high energy results are represented in Fig. 5 by the filled points. The error bars denote the combined statistical and systematic errors. Also shown in Fig. 5, with opaque squares, are the more recent 1117-day data. [96] The result of choosing three different values of $\alpha$ is shown by the three lines. The values of $\alpha$ considered are 4.4 and 2.2, which correspond to $P_{\rm osc}$ = 1 and 0.5, respectively, and 20, close to the factor of 17 mentioned in Ref. 90. From Fig. 5 it appears that, given the prediction of Ref. 83, the theory is in agreement with the latest SK data.

Finally, it is interesting to note that the SSM98 has been recently revisited [97] (SSM00) and the new value for the *hep* flux has been renormalized following the calculation reviewed here. Consequently, the most recent SK 1258-day data [98] are

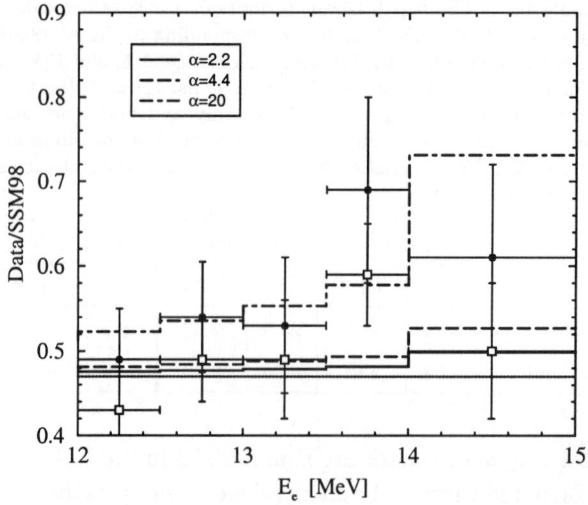


Fig. 5. The highest energy part of the SK electron energy spectrum. The 825-day data were extracted graphically from Fig. 8 of Ref. 90, and are shown as points, while the 1117-day data were extracted from Fig. 8 of Ref. 96 and are shown by the squares. The curves correspond respectively to no *hep* contribution (dotted line) and to $\alpha = 2.2$ (solid line), 4.4 (long-dashed line) and 20 (dot-dashed line).

presented as the ratio between the measured and the SSM00 predicted events. Considering also the effects of a new measurement of the $^8$B spectrum,[99] no high energy enhancement is evident any more.

## 12. Conclusions

In this paper a few of the important characteristics of the HH expansion have been discussed. The analysis is necessarily not exhaustive and some interesting related topics are not investigated in detail or even considered at all.

In the Introduction and in Sec. 2 the general status of the research on a number of microscopic systems of wide interest in physics has been examined. Those systems can also be studied in the framework of the HH method. In Sec. 3 the so-called Jacobi variables have been defined and explicit expressions for systems of three and four particles with different masses have been given. In Sec. 4 the hyperspherical coordinates have been introduced and the related expression for the total kinetic energy operator has been obtained. Sec. 5 is devoted to the definition of the hyperspherical harmonic functions and to an analysis of their most important properties. The problem of constructing trial HH wave functions satisfying the required symmetry conditions has been then analyzed. In particular, the expansion of the w.f. in terms of channels has been introduced. The Rayleigh-Ritz variational principle has then been used in Sec. 6 in order to determine the form of the trial wave function; this requires the solution of a system of coupled second order differential equations.

The conclusion is that high precision results can be obtained for the bound states of three- and four-particle systems. In the case of strong interactions this can be accomplished in the HH approach only by using very large sets of basis functions and solving the corresponding variational problem. For larger systems the situation is worse and modifications of the HH technique are required. For this purpose, in Sec. 9 the so-called potential basis, the adiabatic, the extended and the correlated HH expansions have been briefly revisited.

At present different ways for studying atomic systems are available. The principal elements of the HH expansion in momentum space have been discussed in Sec. 7 without specific numerical applications. Particular attention has been devoted to the calculation of the energy levels of the helium atom. With this in mind, the HH and CHH approaches can also be successful as has been discussed in Sec. 7. In Sec. 8 some of the results obtained by applying HH-type techniques to three- and four-nucleon bound state systems with realistic interactions have been reported and compared with those of other accurate methods. The conclusion is that these methods are quite powerful for obtaining an accurate description of the structure of these systems.

In recent times a lot of attention has been paid to studying scattering processes involving a few particles. The Kohn variational principle, in conjuction with the CHH expansion, has been applied with remarkable success. Very accurate results have been obtained in the case of three- and four-nucleon scattering. It is important to notice that the validity of the variational principle has been proved for the case where two or more charged particles are present. This point has been rather briefly examined in Sec. 10. One important result is that quite accurate wave functions can be obtained for scattering processes involving three or four nucleons using realistic interactions. This fact is of particular interest since these wave functions are a relevant ingredient in calculations of important nuclear reaction parameters. In this context, Sec. 11 is devoted to a short review of some recent results obtained from the study of a number of electro-weak reactions on few-nucleon systems.

As fas as the use of the HH method in studies of systems with $A > 4$ is concerned, much of the discussion given in this paper is still valid. However, the numerical applications require a notable computing effort and much work must be done in order to attain descriptions that are as accurate as those of the three- and four-particle systems.

**Acknowledgements**

The author wishes to thank L. Lovitch for a careful and critical checking of the manuscript, L.E. Marcucci for discussions on its contents and A. Kievsky and M. Viviani for their contributions to the elaboration of many of the results presented in this article.

## References


1. T. H. Gronwall, Phys. Rev. **51**, 655 (1937).
2. L. M. Delves, Nucl. Phys. **9**, 391 (1959); *Advances in Nuclear Physics*, Vol. 5, eds. M. Baranger and E. Vogt (New York: Plenum Press 1972). p. 126.
3. G. Erens, J.L. Visschers and R. van Wageningen, Ann. Phys. (N.Y.) **67**, 461 (1971).
4. Yu. Simonov, Sov. J. Nucl. Phys. (Engl. Transl.) **3**, 461 (1966).
5. M. Fabre de la Ripelle, Ann. Phys. (N.Y.) **147**, 281 (1983).
6. A. Kievsky, L. E. Marcucci, S. Rosati and M. Viviani, Few-Body Sys. **22**, 1 (1997).
7. K. Kamada *et al.*, Phys. Rev. C **64**, 044001 (2001).
8. M. Fabre de la Ripelle, "Proceedings of the International School on Nuclear Theoretical Physics" (Predeal, 1969) (A. Corciovei, Ed.), Inst. Atom. Phys., Bucarest, 1969; Rev. Roum. Phys. **14**, 1215 (1969); M. Beiner and M. Fabre de la Ripelle, Lett. Nuovo Cim. **1**, 584 (1971).
9. A. M. Gorbatov, A. V. Bursak, Yu. N. Krilov and B. V. Rudak, Sov. J. Nucl. Phys. (Engl. Transl.) **40**, 233 (1984).
10. V. B. Mandelzweig, Phys. Lett. A **80**, 36 (1980); **82**, 47 (1981).
11. A. Kievsky, M. Viviani and S. Rosati, Nucl. Phys. A **577**, 511 (1994).
12. M. Viviani, A. Kievsky and S. Rosati, Few-Body Sys. **18**, 25 (1995).
13. A. Kievsky, S. Rosati, M. Viviani and W. Tornow, Nucl. Phys. A **607**, 402 (1996).
14. A. Kievsky, M. Viviani and S. Rosati, Phys. Rev. C **64**, 024002 (2001).
15. N. Barnea, W. Leidemann and G. Orlandini, Nucl. Phys. A **650**, 427 (1999).
16. S. Rosati and M. Viviani, Few-Body Sys. **27**, 73 (1999).
17. N. Barnea and M. Viviani, Phys. Rev. C **61**, 034003 (2000).
18. D. J. Abramov, L. N. Bogdanova, V. V. Gusev and L. I. Ponomarev, Physics of Atomic Nuclei **61**, 457 (1998).
19. A. R. Jansen and R. A. Aziz, J. Chem. Phys **103**, 9626 (1995).
20. P. Barletta and A. Kievsky, Phys. Rev. A **64**, 042514 (2001).
21. M. Ferraris, M. Giannini, M. Pizzo, E. Santopinto and L. Tiator, Phys. Lett. B **364**, 231 (1995).
22. R. B. Wiringa, S. C. Pieper, J. Carlson and V. R. Pandharipande, Phys. Rev. C **62**, 014001 (2000).
23. W. Glöckle, H. Witala, H. Kamada and J. Golak, Phys. Rep. **274**, 107 (1996).
24. A. Nogga, H. Kamada and W. Glöckle, Phys. Rev. Lett. **85**, 944 (2000).
25. H. Kameyama, M. Kamimura and Y. Fukushima, Phys. Rev. C **40**, 974 (1989); Nucl. Phys. A **508**, 17c (1990).
26. Y. Suzuki and K. Varga, *Stochastic variational approach to quantum-mechanical few-body problems*, (Springer-Verlag, 1998).
27. P. Navrátil and P. R. Barret, Phys. Rev. C **59**, 1906 (1999); P. Navrátil, G. P. Kamuntavicius and P. R. Barret, Phys. Rev. C **61**, 044001 (2000); P. Navrátil, J. P. Vary and P. R. Barret, Phys. Rev. Lett. **84**, 5728 (2000).
28. J. Avery, *Hyperspherical Harmonics*, (Kluwer Academic Publishers, Dordrecht, 1989).
29. F. Zernike and H. C. Brinkman, *Proc. Kon. Ned. Acad. Wetensch.* **33**, 3 (1935).
30. M. Abramowitz and I. Stegun, *Handbook of Mathematical Functions*, (Dover Publications Inc., New York, 1970).
31. J. Raynal and J. Revai, Nuovo Cimento **68**, 682 (1970).
32. V. D. Efros, Few-Body Sys. **19**, 167 (1995).
33. M. Viviani, Few-Body Sys. **25**, 177 (1998).
34. N. Barnea and A. Novoselsky, Ann. Phys. (N.Y.) **256**, 192 (1997).
35. G. A. Baker, Jr., J. L. Gamel, B. J. Hill and J. G. Wills, Phys. Rev. **125**, 1754 (1962).
36. A. B. Volkov, Nucl. Phys. **74**, 33 (1965).

37. R. A. Malfliet and J. A. Tjon, Nucl. Phys. A **127**, 161 (1969).
38. R. B. Wiringa, R. A. Smith and T. L. Ainsworth, Phys. Rev. C **29**, 1207 (1984).
39. R. B. Wiringa, V. G. J. Stoks and R. Schiavilla, Phys. Rev. C **51**, 38 (1995).
40. W. Glöckle *et al.*, Few-Body Sys. Suppl. **8**, 9 (1995).
41. J. L. Friar, B. F. Gibson and G. L. Payne, Phys. Rev. C **36**, 1138 (1987).
42. J. Carlson and R. Schiavilla, Rev. Mod. Phys. **70**, 743 (1998).
43. J. Navarro, in this volume.
44. J. L. Ballot, M. Beiner and M. Fabre de la Ripelle, in "The Nuclear Many Body Problem", (F. Calogero and C. Ciofi degli Atti, Eds.), Vol. 1, Bologna, 1973.
45. C. R. Chen, G. L. Payne, J. L. Friar and B. F. Gibson, Phys. Rev. C **31**, 266 (1985); **33**, 1740 (1986).
46. S. Ishikawa, T. Sasakawa, T. Sawada and T. Ueda, Phys. Rev. Lett. **53**, 1877 (1984); T. Sasakawa and S. Ishikawa, Few-Body Sys. **1**, 3 (1986).
47. S. Rosati, M. Viviani and A. Kievsky, Few-Body Sys. **9**, 1 (1990).
48. J. M. Macek, J. Phys. B **1**, 831 (1968).
49. M. Fabre de la Ripelle, *C.R. Acad. Sc. Paris* **274**, 104 (1972).
50. C. J. Ballhausen and Aage E. Hansel, *Annual Rev. Phys. Chem.* **23**, 15 (1972).
51. Y. Zhou, C. D. Lin and J. Shertzen, J. Phys. B **26**, 3937 (1993).
52. A. Kievsky and M. Viviani, Few-Body Sys. Suppl. **8**, 44 (1995).
53. S. Rosati and M. Viviani, Proceedings of the "Workshop on Microscopic Approaches to the Structure of Light Nuclei", June 1998, in *Advances in Quantum Many Body Theory*, **2**, Ed. R.F. Bishop.
54. E. A. Hylleraas, Zeit. Phys. **65**, 209 (1930).
55. C. L. Pekeris, Phys. Rev. **126**, 1470 (1962).
56. C. W. F. Drake *et al.*, Phys. Rev. Lett. **77**, 171 (1996).
57. R. Krivec, Few-Body Sys. **25**, 7084 (1998).
58. M. Fabre de la Ripelle *et al.*, Phys. Rev. **44**, 7084 (1991).
59. M. I. Haftel and V. Mandelzweig, Ann. Phys. (N.Y.) **189**, 29 (1989).
60. F. Paterni, Thesis, Dipartimento di Fisica, University of Pisa, 2000.
61. F. Ciesielsky and J. Carbonell, Phys. Rev. C **58**, 58 (1998).
62. B. S. Pudliner *et al.*, Phys. Rev. C **56**, 1720 (1997).
63. K. Varga, Y. Ohbayashi and Y. Suzuki, Phys. Lett. B **396**, 1 (1997).
64. A. Kievsky, M. Viviani, S. Rosati, Phys. Rev. C **56**, 2987 (1997).
65. R. R. Lucchese, Phys. Rev. A **40**, 6879 (1989).
66. C. Schwartz, Phys. Rev. **124**, 1468 (1961).
67. D. Hüber *et al.*, Phys. Rev. C **51**, 1100 (1995).
68. A. Kievsky, M. Viviani, S. Rosati, D. Hüber, W. Glöckle, H. Kamada, H. Witala and J. Golak Phys. Rev. C **58**, 3085 (1998).
69. S. Shimizu *et al.*, Phys. Rev. C **52**, 1193 (1995).
70. M. Viviani, A. Kievsky and S. Rosati, Few-Body Sys. **30**, 39 (2001).
71. C. R. Brune *et al.*, Phys. Rev. C **63**, 044013 (2001).
72. A. Kievsky, S. Rosati and M. Viviani, Phys. Rev. C **64**, 041001 (2001).
73. T. W. Phillips, B. L. Berman and J. D. Seagrave, Phys. Rev. C **22**, 384 (1980).
74. S. Hammerschmied, H. Rauch, H. Clerc and U. Kischko, Zeit. Phys. A **302**, 323 (1981).
75. H. Rauch, D. Tuppinger, H. Woelwitsch and T. Wroblewski, Phys. Lett. B **165**, 39 (1985).
76. G. M. Hale *et al.*, Phys. Rev. C **42**, 438 (1990).
77. R. B. Wiringa (unpublished).
78. R. B. Wiringa, Phys. Rev. C **43**, 1585 (1991).
79. B. S. Pudliner *et al.*, Phys. Rev. Lett. **74**, 4396 (1995).

80. H. T. Coelho, T. K. Das and M. R. Robilotta, Phys. Rev. C **28**, 1812 (1983).
81. M. Viviani, A. Kievsky, S. Rosati, E. A. George and L. D. Knutson, Phys. Rev. Lett. **86**, 3739 (2001).
82. M. Viviani *et al.*, Phys. Rev. C **61**, 064001 (2000).
83. L. E. Marcucci *et al.*, Phys. Rev. Lett. **84**, 5959 (2000); L. E. Marcucci *et al.*, Phys. Rev. C **63**, 015801 (2001).
84. G. J. Schmid *et al.*, Phys. Rev. Lett. **76**, 3088 (1996); L. Ma *et al.*, Phys. Rev. C **55**, 588 (1997); M. K. Smith and L. D. Knutson, Phys. Rev. Lett. **82**, 4591 (1999).
85. L. E. Marcucci, D. O. Riska and R. Schiavilla, Phys. Rev. C **58**, 3069 (1998).
86. R. Schiavilla *et al.*, Phys. Rev. C **45**, 2628 (1992).
87. R. Schiavilla, D. O. Riska and V. R. Pandharipande, Phys. Rev. C **41**, 309 (1990).
88. Y. Fukuda *et al.*, Phys. Rev. Lett. **82**, 2430 (1999).
89. M. Smy, hep-ex/9903034.
90. Y. Suzuki, contribution to Lepton-Photon Symposium 99 (1999), `http://www-sk.icrr.u-tokyo.ac.jp/doc/sk/pub/index.html`.
91. J. N. Bahcall, S. Basu and M. H. Pinsonneault, Phys. Lett. B **433**, 1 (1998).
92. K. Kubodera, J. Delorme and M. Rho, Phys. Rev. Lett. **40**, 755 (1978).
93. I. S. Towner, Nucl. Phys. A **542**, 631 (1992).
94. M. Kirchbach, D. O. Riska and K. Tsushima, Nucl. Phys. A **542**, 616 (1992).
95. R. Schiavilla *et al.*, Phys. Rev. C **58**, 1263 (1998).
96. R. Svoboda, contribution to Tau 2000 (2000), `http://www-sk.icrr.u-tokyo.ac.jp/doc/sk/pub/index.html`.
97. J. N. Bahcall *et al.*, Astrophys. J. **555**, 990 (2001).
98. S. Fukuda *et al.*, Phys. Rev. C **86**, 5651 (2001).
99. C. E. Ortiz *et al.*, Phys. Rev. Lett. **85**, 2909 (2000).

# CHAPTER 10

# THE NUCLEAR MANY–BODY PROBLEM

S. Fantoni
*International School for Advanced Studies, SISSA,*
*Via Beirut 2/4, I-34014 Trieste, Italy*
*International Centre for Theoretical Physics, ICTP,*
*Strada Costiera 32, I-34014 Trieste, Italy*
*E-mail: fantoni@sissa.it*

The most recent developments in the nuclear many–body problem are briefly reviewed, trying to focus on the major advances reached in the last few years and on the still open problems. Particular attention is devoted to quantum simulations for nuclear and neutron matter, which have recently received much attention, and, given the rapid increase in computer power, pose themeselves amongst the best candidates for a new generation of *ab initio* calculations in nuclear physics. The newly developed Auxiliary Field Diffusion Monte Carlo method, or other stochastic methods using auxiliary field ideas, appear to embody all the main features to perform simulations of large nuclear systems, solving, under this respect, the long standing *spin problem*. The use of the Auxiliary Field Diffusion Monte Carlo method to compute properties of many particle systems interacting via spin-isospin dependent nuclear potentials is described. By combining diffusion Monte Carlo for the spatial degrees of freedom and auxiliary field Monte Carlo to separate the spin-isospin operators, ground state energies and other properties can be calculated for medium size nucleon systems. Recent applications of the method to obtain the equation of state and the compressibilty of neutron matter are presented and discussed. These calculations use realistic interactions such as the Argonne $v_8'$ and $v_6'$ two–nucleon potentials plus the Urbana IX three-nucleon potential. Other properties of astrophysical interest such as the spin susceptibility of neutron matter and the symmetry energy are also discussed. A new homework problem is proposed to test the modern many–body techniques with a nuclear potential which includes tensor, spin–orbit and three–body interactions.

## 1. Introduction

In this contribution I will focus my attention on the recent developments of quantum simulations for nuclear physics. This is a line of research, which, in spite of the fact of being currently carried out by a relatively small number of nuclear theorists, looks very promising for a better quantitative understanding of the nuclear interaction, the nuclear structure and the nuclear phenomenology.

The quantum simulations I am referring to in this contribution address the non relativistic many–body problem underlying what has been recognized in the last few years as the Standard Model of Nuclear Physics (SMNP), namely a model which embodies in a realistic way the main features of an hadronic system at low energy. Deviations from this model should indicate the presence of new physics or the need of introducing more fundamental degrees of freedom.

Account of what is meant by this model and of its growing up in the last twenty years can be found, for instance, in the proceedings of Elba's workshops on electron–nucleus scattering[a], starting from the first one in 1988 [1] or in the review book of Ref. 2

Let me here outline the line of thoughts which has brought to the SMNP. A first important characterization is that nucleons are the only explicit degrees of freedom to be considered in describing the structure of nuclei and nuclear matter. A second assumption is that these nuclear systems can be safely treated by using non relativistic quantum theory, so that the interaction amongst the nucleons is provided by a many-body potential depending upon nucleon spatial coordinates and their spin–isospin variables.

Obviously, the above assumptions do not mean that the role of other degrees of freedom or of relativity are considered to be inessential for the understanding of nuclear reactions, like, for instance, those produced at HERA or at Jefferson Laboratory, in which the external probe has intermediate energies (few GeV). It means that, while the scattering mechanism on the *nucleon* has to be treated within a fully relativistic scheme, the structure of the target nucleus and the behavior of the spectator nucleons can be safely treated by means of a non relativistic theory.

Following the above statements, the SMNP considers the many–body Schrödinger equation as the basis of our understanding of the structure of ordinary and esotic nuclei, as well as of nuclear matter in the extreme condition of low temperature and high density. Relativistic corrections are then added perturbatively, but they are not expected to change the non relativistic picture in any significant way.

The main goals of the research on the SMNP model have been and still are the following.

**(i)** Search for a many–body nuclear interaction, capable to describe all the nuclear systems of interest, from the deuteron to nuclear matter up to densities of $\sim 0.5 fm^{-3}$ and temperatures of few tenths of $MeV$.

**(ii)** Search for the current operators which are consistent with the above effective interaction.

**(iii)** developments of powerful and efficient many–body methods to solve the corresponding Schrödinger equation through *ab-initio* calculations, namely those with no approximations on the nuclear interaction.

[a]Reference to the series can be found in http://www.eipc.it

The need of sophisticated many–body techniques in nuclear physics is dictated by the fact that realistic effective nuclear forces give rise to short range correlations amongst the nucleons, even at moderately low density, such as equilibrium density of nuclear matter $\rho_0 = 0.16\ fm^{-3}$. Typical effects of these correlations, which can hardly be explained by any standard mean field picture, are the following:

- the occurence of a dip in the short range behavior of the two–body distribution function, which has measurable effects on the inclusive electron scattering cross section off nuclei, particularly, at low missing energy;
- the depletion of the single particle occupancy $n(\epsilon)$ for states below the Fermi sea, and the consequent appearence of a tail, extending up to very high excited states ($\epsilon \gg \epsilon_F$);
- the quenching and broadening of spectral function and the longitudinal response function at high momentum transfer.

The above features, first predicted by Correlated Basis Function (CBF) calculations, based upon Fermi Hypernetted Chain (FHNC) theory, [3,4] have received much attention in recent years from both theorists and experimentalists. [5] Given these features, nuclear matter and nuclei have to be considered typical strongly correlated fermion systems, like those familiar in condensed matter physics. In addition to the presence of strong correlations, here, one is also facing with the so called *spin problem*, coming from the strong spin–isospin dependence of the nuclear interaction.

## 1.1. *The nuclear interaction*

The search for a realistic nuclear interaction, together with a consistent definition of the current operators, has been a long debated issue and has involved the collective efforts of several nuclear theorists. As a proof of this, practically all the modern nuclear interactions have city names, like Urbana, Nijmegen, Bonn.

There is a generalized consensus on the fact that the problem of determining a realistic two–body NN potential, $V_2$, is substantially solved. The so called *modern two–body potentials* [6,7,8] are phase–shift equivalent and all fit the Nijmegen data [9] below 350 $MeV$ with a $\chi^2/N_{data} \sim 1$. They seem to give not too different results for the properties of both light nuclei and nuclear matter. Evidence of this can be drawn from Fig. 1, which displays the results of a recent calculation [10] of the equation of state of neutron matter, performed by using two–hole line Brueckner Hartree Fock (BHF) theory for practically all the modern two–body NN potentials (Argonne $v_{18}$ ($A18$) [6], Nijmegen I (Nij-I), [7] Nijmegen-II (Nij-II), [7] Reid93 [7] and CD-Bonn. [8] Contrary to the case of light nuclei, where techniques like Faddeev, Green Function Monte Carlo or Correlated Hyperspherical method agree to very high accuracy, [11] the theoretical uncertainty on the equation of state of nuclear matter is unsatisfactory, being larger than the differences amongst the modern NN potentials. This is clearly indicated in Fig. 2, where Brueckner–Hartree–Fock and

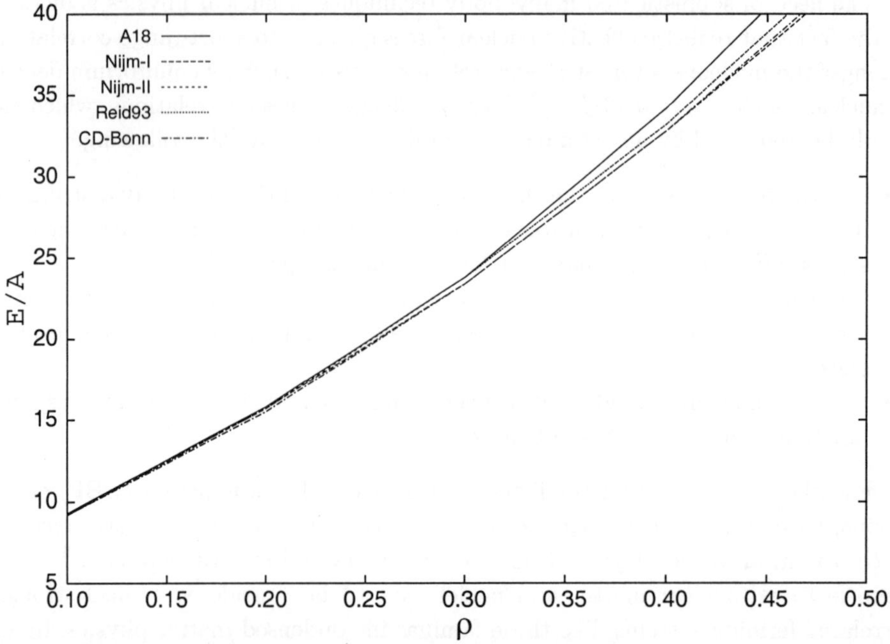


**Fig. 1.** Equivalence of modern two–body NN potentials: BHF results [10] for the equation of state of neutron matter with various *modern* two–body potentials. The energy per particle is in $MeV$, and the density $\rho$ in fm$^{-3}$.

Fermi Hypernetted Chain (FHNC/SOC) calculations are compared for the $A18$ model of neutron matter.

On the basis of our present knowledge of the SMNP, the two–body NN interaction $V_2$ alone cannot account for the properties of neither nuclear matter nor light nuclei. Nuclear matter saturates at too high density and $^3$He nucleus is underbound. The nuclear interaction must include at least a three-body term $V_3$.

However, finding a realistic three–body potential for ordinary matter and estimating its contribution in high density nuclear matter is still an open and debated issue. The structural forms of $V_3$ considered so far are limited to the three–nucleon processes with a minimal number of intermediate states, and its actual determination heavily relies on the requirement that $V_2 + V_3$ reproduce the ground state and the low lying states energies of light nuclei, [14,15,16] which constitutes a relatively small set of experimental data.

In order to show the relevance of this problem, Fig. 3 gives the results of a preliminary calculations [17] performed by using the Auxiliary Field Diffusion Monte Carlo (AFDMC) method [18] for the equation of state of neutron matter. Results obtained with three different $V_3$, the Urbana IX [14,15] (UIX) and the recent Illinois II and IV [16] (IL2 and IL4) (see section 2.2) are compared. Already at twice the nuclear matter density, the energy contributions from the three–body potentials are large and very different from each other, in spite of the fact that all of them provide

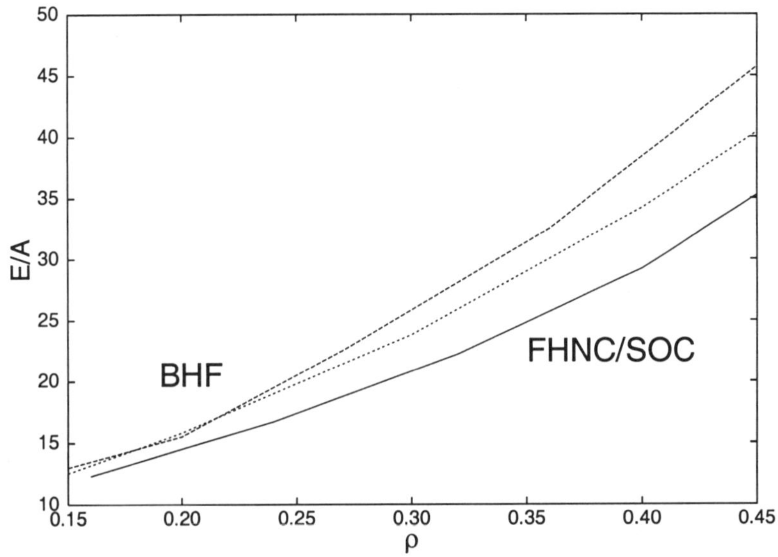


Fig. 2. FHNC/SOC [12] and BHF results for the neutron matter equation of state with $A18$ potential. The lower of the two BHF dashed lines corresponds to the calculation of Ref. 10. The upper curve has been obtained by Baldo *et al.* [13] in a BHF calculation, with the continuous choice for the single particle spectrum, which includes also the three hole line diagrams. Differences between the two BHF calculations cannot be ascribed to the contributions from the three hole line diagrams. The energy per particle is in $MeV$, and the density $\rho$ in fm$^{-3}$.

a satisfactory fit to the ground state and the low energy spectrum of nuclei with $A \leq 8$, IL2 and IL4 being only marginally better than UIX.

### 1.2. *Quantum simulations*

Significative advances have been made in traditional many–body techniques during the last couple of decades. Report on these advances can be found in this volume as well as in Ref. 19. The most recent *ab initio* calculations, carried out on medium–heavy nuclei and nuclear matter, along CBF, BHF or Self Consistent Green's Function theories, show a semiquantitative agreement, and they all recognize the fundamental role of NN correlations, as previously discussed.

However, the remaining discrepancies amongst them (see Fig. 2) put serious limitations on both determining the many–body character of the nuclear interaction and making quantitative studies on the nuclear properties at the accuracy needed by the present nuclear phenomenology.

More recently, there has been a significative effort to export quantum simulations methods from Condensed Matter to Nuclear Physics. While the stochastic methods are not yet flexible enough to fully substitute the more traditional many–body methods, they may provide important benchmarks for any given model hamiltonian, so that the validity of approximations, still present in these methods, can be fully ascertained. For instance, one would like to estimate the effect of neglecting the

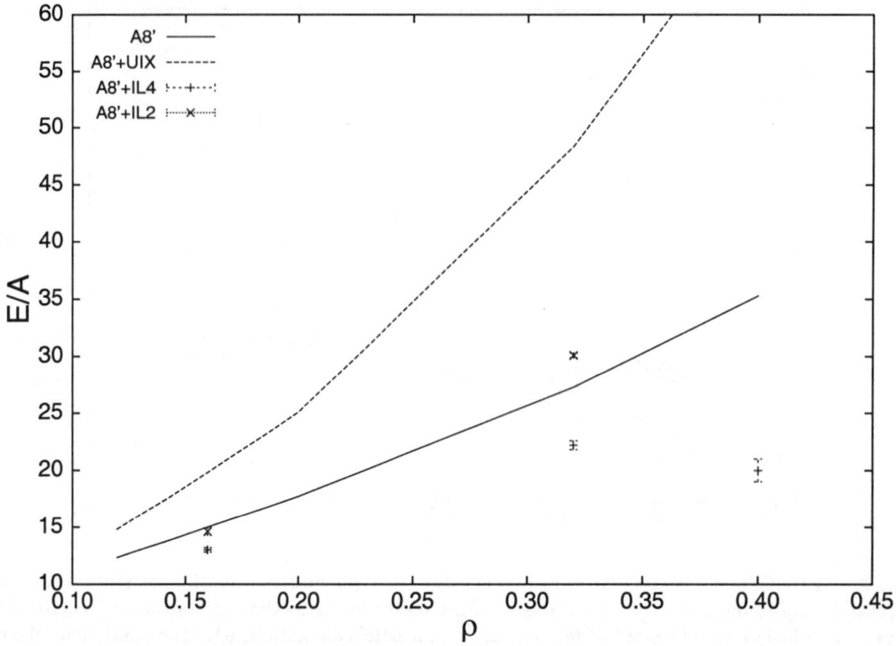


Fig. 3. The three–body potential problem: the equation of state of neutron matter is calculated by using the two–body potential $A8'$ and different three–body potentials. The calculation has been performed [17] by using the AFDMC method with 14 neutrons in a periodic box and a time step $\Delta\tau = 5 \times 10^{-5} MeV^{-1}$. The error bars indicates the statistical errors of the simulations, and do not include finite size errors. The energy per particle is in $MeV$, and the density $\rho$ in $\mathrm{fm}^{-3}$.

*elementary diagrams* and/or the *commutator terms* in FHNC/SOC theory, [20,21] as well as to check the *convergence* of the perturbative corrections of the various many–body theories. [22]

Quantum simulations have been very successful in calculating the properties of strongly interacting systems of interest in condensed matter physics. They are substantially exact, apart from statistical errors, finite size effects and the well known *sign problem* [23] for Fermi systems. These methods have been extended to perform quantum simulations for nuclear systems, where the particles interact via a potential having a strong spin-isospin dependence. [24,25] However, the exponential growth of the number of spin-isospin states with the number of nucleons $A$, the so called *spin problem*, have limited the application of these methods to systems of up to $\approx 10$ nucleons. [16]

Recently, it has been proposed an alternative Quantum Monte Carlo method, named Auxiliary Field Diffusion Monte Carlo [18] to circumvent this problem and allow for quantum simulations in heavy nuclear systems. It is based on sampling the spin-isospin states rather than performing a complete spin sum. In this approach the scalar parts of the Hamiltonian are propagated as in standard Diffusion Monte Carlo (DMC). Auxiliary fields are introduced to replace the spin dependent interaction by using Hubbard–Stratonovich transformations. The method consists of a Monte Carlo sampling of the auxiliary fields and then propagating the spin variables at the sampled values of the auxiliary fields. This propagation results in a rotation of each particle's spinors. The fermion sign problem is taken care of by applying a path-constraint approximation analogous to the fixed-node approximation. The AFDMC method for the spin–isospin calculations is essentially an application of the method of Zhang *et al.* [26] used in condensed matter lattice systems to the spin-isospin states of nucleon systems.

The AFDMC method has already demonstrated that energies per particle can be calculated for spin–isospin dependent interactions, in the cases of a neutron drop [27,28] with $A = 7, 8$, of neutron matter, simulated with up to 66 neutrons in a periodic box, [27,28,29,30] of $^4$He nucleus and of nuclear matter with up to 76 nucleons. [31] It has beeen found that (i) the neutron matter calculations scale in particle number roughly like fermion Monte Carlo calculations with central forces; and (ii) the results obtained compare very well with the existing Green Function Monte Carlo estimates.

### 1.3. *Plan of the paper*

The Hamiltonian of the SMNP is described in the next Section. Section 3 is devoted to a brief description of the AFDMC method, and of a new method recently developed to estimate the finite size corrections and based on FHNC/SOC. Section 4 will present some recent results for neutron and nuclear matter, like the equation of state, the spin susceptibility and the symmetry energy. Conclusions and perspectives are given in the last Section.

## 2. The Hamiltonian

The Hamiltonian of the SMNP is given by

$$\begin{aligned} H &= T + V_2 + V_3 \\ &= -\frac{\hbar^2}{2m} \sum_{i=1,N} \nabla_i^2 + \sum_{i<j} v_{ij} + \sum_{i<j<k} V_{ijk} \ , \end{aligned} \tag{2.1}$$

In the following of this section I will limit my attention to effective interactions of the Urbana–Argonne type. Most of the *ab initio* calculations performed in nuclear matter and heavy nuclei have been considering this type of interaction, for both hystorical and practical reasons.

### 2.1. *Two–body potential*

The two–body part of the Argonne–Urbana potential is usually denoted with $v_l$ where the subscript $l$ indicates the number of operatorial components included in the model interaction: $l = 1$ stands for a purely scalar interaction; $l = 4$ corresponds to a spin–isospin dependent potential, but only central; $l = 6$ includes the tensor components, and $l = 8$ includes also the spin–orbit components. More specifically,

$$v_l = \sum_{i<j} v_{ij} = \sum_{i<j} \sum_{p=1}^{l} v_p(r_{ij}) O^{(p)}(i,j) \ , \tag{2.2}$$

where $i$ and $j$ label the two nucleons, $r_{ij}$ is the internucleon distance, and the spin–isospin dependent operators $O^p(i,j)$ for $p = 1,8$ are given by

$$O^{p=1,8}(i,j) = \left(1, \vec{\sigma}_i \cdot \vec{\sigma}_j, S_{ij}, \vec{L}_{ij} \cdot \vec{S}_{ij}\right) \otimes (1, \vec{\tau}_i \cdot \vec{\tau}_j) \ . \tag{2.3}$$

$S_{ij} = 3(\hat{r}_{ij} \cdot \vec{\sigma}_i)(\hat{r}_{ij} \cdot \vec{\sigma}_j) - \vec{\sigma}_i \cdot \vec{\sigma}_j$ is the two–nucleon tensor operator, and $\vec{L}_{ij}$ and $\vec{S}_{ij}$ are the relative angular momentum and the total spin, given by

$$\vec{L}_{ij} = \frac{1}{2\imath}(\vec{r}_i - \vec{r}_j) \times (\vec{\nabla}_i - \vec{\nabla}_j) \ , \tag{2.4}$$

$$\vec{S}_{ij} = \frac{1}{2}(\vec{\sigma}_i + \vec{\sigma}_j) \ . \tag{2.5}$$

A two–body potential with the 8 components given in Eq. (2.3) can provide a very good fit to NN scattering data up the meson treshold. However, the modern potentials include more terms. The Argonne $v_{18}$ ($A18$) potential consists of $l = 18$ components. [6] Besides the above 8 components it includes six other charge independent terms ($L^2$, $L^2\ \vec{\sigma}_i \cdot \vec{\sigma}_j$, $(\vec{L} \cdot \vec{S})^2$) $\otimes(1, \vec{\tau}_i \cdot \vec{\tau}_j)$, as well as other 4 charge–symmetry–breaking and charge–dependent components. It fits both *pp* and *np* scattering data up to 350 MeV with a $\chi^2 \sim 1$ per degree of freedom.

Most of the quantum simulations carried out for nuclear and neutron matter have used a simplified and isoscalar version of the $A18$ potential, the so called $v_8'$ two–body potential ($A8'$). [15] This potential has been obtained with a new fit to the N–N data, by including the eight spin–dependent operators of Eq. (2.3) only. It equals the isoscalar part of $v_{18}$ in all $S$ and $P$ waves as well as in the $^3D_1$ wave and its coupling to the $^3S_1$. It has been used in a number of GFMC calculations in light nuclei, [15] as well as FHNC/SOC calculations in $^{16}$O, $^{40}$Ca and nuclear matter. [32] Differences with the $A18$ potential give very small contributions that can safely be estimated either by perturbation theory or from FHNC/SOC calculations.

The potential obtained from $A8'$ by dropping the two spin–orbit components is denoted as $A6'$. It has been used to estimate the effect of the spin–orbit component of the interaction.

A third even more simplified model interaction, which has been often used in *ab initio* calculations in light nuclei and nuclear matter, is the MS3 potential. It is a semirealistic interaction of $v_4$ type with a repulsive core that was introduced to reproduce both the $S$-wave scattering data up to about 60 MeV and the binding energy of the deuteron and the alpha particle. [33,34] This interaction has been shown to provide a reasonable qualitative description of light and medium nuclei. [11,35]

### 2.2. *Three-body interaction*

The three-body potentials of the Urbana–Argonne interactions have the following form [14]

$$V_{ijk} = V^{SI}_{ijk} + V^{SD}_{ijk} \ , \tag{2.6}$$

where the structure of the spin dependent part, $V^{SD}_{ijk}$, is derived from the three–nucleon processess characterized by two– and three–pion exchanges and only one $\Delta$ in the intermediate states. It is attractive and fixes the nuclear matter saturation density at the experimental value. However, it gives too much binding. The spin independent part $V^{SI}_{ijk}$ is purely phenomenological, and it simulates higher order three–nucleon processes which correspond to an effective repulsive potential. It has been introduced to balance the overbinding of nuclear matter given by $V^{SD}_{ijk}$.

The Urbana IX three–body potential has a spin dependent part given only by the three–nucleon two–pion exchange process, the so called Fujita–Miyazawa term:

$$\begin{aligned} V^{SD}_{ijk} &= B_{2\pi} \sum_{cyclic} \{\vec{\tau}_i \cdot \vec{\tau}_j, \vec{\tau}_i \cdot \vec{\tau}_k\}\{X^{\pi}_{ij}, X^{\pi}_{ik}\} \\ &+ \frac{B_{2\pi}}{4} \sum_{cyclic} [\vec{\tau}_i \cdot \vec{\tau}_j, \vec{\tau}_i \cdot \vec{\tau}_k] \left[X^{\pi}_{ij}, X^{\pi}_{ik}\right] \ , \end{aligned} \tag{2.7}$$

with the operator $X^{\pi}_{ik}$ given by

$$X^{\pi}_{ik} = Y(m_\pi, c_3; r_{ik})\vec{\sigma}_i \cdot \vec{\sigma}_k + T(m_\pi, c_3; r_{ik})S_{ik} \ . \tag{2.8}$$

The spin independent part is given by

$$V^{SI}_{ijk} = U_0 \sum_{cyclic} T^2(m_\pi, c_3; r_{ij})T^2(m_\pi, c_3; r_{ik}) \ . \tag{2.9}$$

The analytic expressions of $T$ and $Y$ can be found in the Appendix, together with those of the $A8'$ two–body potential. The values of the strengths $B_{2\pi}$, $U_0$ and of the cutoff $c_3$ have been fitted to reproduce the ground state and the low lying states of light nuclei ($A \leq 8$), [14,16] and are listed in the Appendix.

The interaction $A18$ plus the Urbana IX will be denoted as $AU18$, $A8'$ plus Urbana IX as $AU8'$ and $A6'$ plus Urbana IX as $AU6'$.

New three–body potentials have been recently proposed [16] to include the S-wave term and the three–pion diagrams with only one $\Delta$ in each intermediate states. The potentials denoted with IL2 and IL4 in Fig. (3) correspond to two different parametrizations of these Feynman diagrams and provide a better description of light nuclei with respect to UIX. [16]

## 3. The AFDMC method

The novelty of this method relies on the way of sampling the spin states of the nucleons instead of summing over all of them as in previous applications of GFMC method on nuclear problems, avoiding a direct sampling in the usual spin up/down basis, which is known to give high variance.

Therefore, AFDMC is a method for sampling the states produced when operated on by the imaginary time propagator $\exp(-(V_2^{SD}+V_3^{SD})\Delta t)$ to compute

$$\Psi(R,S) = \int dR'dS'G_0(R,R')e^{-V\Delta t+E_0\Delta t}P_{LS}\Psi(R',S') \,, \tag{3.1}$$

where $R \equiv \vec{r}_1,\ldots,\vec{r}_N$ and $S \equiv \eta_1,\ldots,\eta_N$ generically denote the spatial coordinates and the spin–isospin states for all the particle in the system and $\Psi(R,S) =< \Psi|R,S>$ is the ground state wave function. $G_0(R,R')$ is the free propagator, which is diagonal in spin space and given by

$$\begin{aligned} G_0(R,R') &= \left(\frac{m}{2\pi\hbar^2\Delta t}\right)^{3/2} \exp\left(-\frac{m|R-R'|^2}{2\hbar^2\Delta t}\right) \\ &= \left(\frac{m}{2\pi\hbar^2\Delta t}\right)^{3/2} \exp\left(-\frac{m|\Delta R|^2}{2\hbar^2\Delta t}\right) \,, \end{aligned} \tag{3.2}$$

where

$$\Delta\vec{r}_j = \vec{r}_j - \vec{r}_j' \,. \tag{3.3}$$

The unprimed coordinates denote the new positions, and the primed coordinates the old ones. $P_{LS}$ is the spin–orbit propagator, which will be discussed in section (3.3).

The way AFDMC deals with the spin–isospin operators is to break up these operators into terms that give a single new coherent state when they operate on a coherent state. By sampling these terms the spin-isospin sums are effectively sampled. In addition, the propagation is local in both space and spin. More explicitely, as $\Delta t \to 0$ the propagator goes smoothly to the identity and the walker remains the same.

Therefore, the spin part of the propagator should be viewed as a sum of terms like

$$\prod_{i=1}^{A}[a_i + \vec{b}_i \cdot \vec{\sigma}_i] \,. \tag{3.4}$$

While this coherent state basis introduces unwanted components that have to be averaged out, it avoids the *all or nothing* sampling that occurs in the $z$ component basis.

In the following of this section, I will give a schematic description of the method for the case of a neutron systems. The inclusion of the isospin variables is straightforward.

### 3.1. The auxiliary field breakup for a $v_6$ two–body potential

For $N$ neutrons, the $v_6$ two-body interaction of Eq. (2.2) can be written as

$$\sum_{i<j}\sum_{p=1}^{3} v_p^{PNM}(r_{ij})O^{(p)}(i,j) = V_2^{SI} + V_2^{SD}$$
$$= V_2^{SI} + \frac{1}{2}\sum_{i,\alpha,j,\beta} \sigma_{i,\alpha}A_{i,\alpha,j,\beta}\sigma_{j,\beta} \,. \tag{3.5}$$

The $A_{i,\alpha,j,\beta}$ matrix is taken to be zero when $i = j$ and it is real and symmetric. It therefore has real eigenvalues and eigenvectors. The eigenvectors and eigenvalues are defined by

$$\sum_{j,\beta} A_{i,\alpha,j,\beta}\psi_n^\beta(j) = \lambda_n\psi_n^\alpha(i)\,. \tag{3.6}$$

The matrices are more conveniently written in terms of their eigenvectors and eigenvalues to give the spin–dependent part of the potential

$$V_2^{SD} = \frac{1}{2}\sum_{i,j,n} \vec{\sigma}_i \cdot \vec{\psi}_n(i)\lambda_n\vec{\psi}_n(j)\cdot\vec{\sigma}_j$$
$$= \frac{1}{2}\sum_{n=1}^{3A}(O_n)^2\lambda_n \,, \tag{3.7}$$

with

$$O_n = \sum_i \vec{\sigma}_i \cdot \vec{\psi}_n(i)\,. \tag{3.8}$$

The Hubbard-Stratonovich transformation is then used to write the spin part of the propagator in the form of Eq. (3.4). It is given by

$$e^{-\frac{1}{2}\lambda_n O_n^2\Delta t} = \left(\frac{\Delta t|\lambda_n|}{2\pi}\right)^{\frac{1}{2}}\int_{-\infty}^{\infty} dx e^{-\frac{1}{2}\Delta t|\lambda_n|x^2 - \Delta t s\lambda_n O_n x} \tag{3.9}$$

where $s$ is 1 for $\lambda < 0$, and $s$ is $i$ for $\lambda > 0$.

Notice that the AFDMC makes use of the Hubbard–Stratonovich transformation to linearize only the spin part of the propagator and not the full two–body interaction, as attempted in previous quantum simulation of lattice systems.

Each of the $O_n$ is a sum of one–body operators as required above. They do not commute, so the time step $\Delta\tau$ has to be kept small to ignore the commutator terms.

### 3.2. Break–up for the three–body potential

For a neutron system the spin–dependent part of Urbana IX potential, given in Eqs. (2.6) and (2.7) reduces to a sum of terms containing only two-body spin operators but with a form and strength that depends on the positions of three particles. As it will be seen below, for a fixed position of the particles, the inclusion of three-body potentials of the Urbana IX type in the Hamiltonian does not add any additional complications within the AFDMC framework.

The anticommutator of Eq. (2.7) can be written as

$$\{X^{\pi}_{ik}, X^{\pi}_{kj}\} = 2\, x^{\mu\nu}_{ijk}\sigma^{\mu}_{i}\sigma^{\nu}_{j} \;, \tag{3.10}$$

where

$$x^{\mu\nu}_{ijk} = y_{ik}y_{kj}\delta_{\mu\nu} + y_{ik}t^{\mu\nu}_{kj} + t^{\mu\nu}_{ik}y_{kj} + t^{\mu\alpha}_{ik}t^{\alpha\nu}_{kj} \;, \tag{3.11}$$

and

$$\begin{aligned} y_{ik} &= Y(m_\pi, c_3, r_{ik}) - T(m_\pi, c_3, r_{ik}), \\ t^{\mu\nu}_{ik} &= 3\, T(m_\pi, c_3, r_{ik})\hat{r}^{\mu}_{ik}\hat{r}^{\nu}_{ik} \;. \end{aligned} \tag{3.12}$$

The spin–dependent part of the three–body interaction $V_3^{SD}$ can then be easily incorporated in the matrix $A_{i,\alpha,j,\beta}$ of Eq. (3.5), by the following substitution

$$A_{i,\alpha,j,\beta} \to A_{i,\alpha,j,\beta} + 2 \sum_k B_{2\pi} x^{\alpha\beta}_{ijk} \;. \tag{3.13}$$

For the case of nuclear matter the break–up is much more complicated, because the commutator term in Eq. 2.7, which vanishes for neutron matter, cannot be easily incorporated in the matrix $A_{i,\alpha,j,\beta}$.

### 3.3. *The spin-orbit propagator*

A first approximation to the spin–orbit propagator $P_{LS}$ can be obtained by operating the derivative of the $\vec{L}_{jk} \cdot \vec{S}_{jk}$ operator on the free propagator $G_0$ given in Eq. (3.2),

$$(\vec{\nabla}_j - \vec{\nabla}_k)G_0(R,R') = -\frac{m}{\hbar^2 \Delta t}(\Delta\vec{r}_j - \Delta\vec{r}_k)G_0(R,R') \;, \tag{3.14}$$

and substituting this expression into the propagator,

$$P_{LS} = \exp\left(\sum_{j\neq k} \frac{m v_{LS}(r_{jk})}{4i\hbar^2}[\vec{r}_{jk} \times (\Delta\vec{r}_j - \Delta\vec{r}_k)] \cdot \vec{\sigma}_j\right) \;. \tag{3.15}$$

This approximate expression for the spin-orbit operator already has the form of Eq. (3.4). Propagation by the spin orbit terms is given by rotating the spinors around the relative angular momentum vector which can be calculated from the new and old positions. However this term also includes some extra incorrect contributions.

These contributions can be identified by expanding Eq. (3.1) up to first order in $\Delta t$ by using the $P_{LS}$ given above and comparing with the correct propagation provided by the Schrödinger equation. The spurious terms can be eliminated by introducing in the propagation the proper counter terms which have the form of the following two- plus three- body potential. [30]

$$V_2^{add} = -\sum_{j<k} \frac{m r_{jk}^2 v_{LS}^2(r_{jk})}{8\hbar^2} \left[2 + \vec{\sigma}_j \cdot \vec{\sigma}_k - \vec{\sigma}_j \cdot \hat{r}_{jk} \vec{\sigma}_k \cdot \hat{r}_{jk}\right] . \tag{3.16}$$

and

$$\begin{aligned} V_3^{add} = - \sum_{j<k<p} \sum_{\text{cyclic}} & \frac{m r_{jk} r_{jp} v_{LS}(r_{jk}) v_{LS}(r_{jp})}{16\hbar^2} \\ & \{\hat{r}_{jk} \cdot \hat{r}_{jp} \left[2 + \vec{\sigma}_k \cdot \vec{\sigma}_j + \vec{\sigma}_p \cdot \vec{\sigma}_j + \vec{\sigma}_k \cdot \vec{\sigma}_p\right] \\ & - \vec{\sigma}_j \cdot \hat{r}_{jk} \vec{\sigma}_k \cdot \hat{r}_{jp} - \vec{\sigma}_p \cdot \hat{r}_{jk} \vec{\sigma}_j \cdot \hat{r}_{jp} - \vec{\sigma}_p \cdot \hat{r}_{jk} \vec{\sigma}_k \cdot \hat{r}_{jp}\} . \end{aligned} \tag{3.17}$$

In Ref. 30 it is shown that the $V_2^{add}$ and $V_3^{add}$ terms are necessary to get full agreement between the extrapolated mixed and growth energies. [23]

### 3.4. *Trial wave function and path constraint*

In all the AFDMC calculations performed so far the following simple trial function given by a Slater determinant of one-body space-spin orbitals multiplied by a central Jastrow correlation has been used:

$$|\Psi_T\rangle = \left[\prod_{i<j} f(r_{ij})\right] A \left[\prod_i |\phi_i, s_i>\right] . \tag{3.18}$$

The overlap of a walker with this wave function is the determinant of the space-spin orbitals, evaluated at the walker position and spinor for each particle, and multiplied by a central Jastrow product.

For unpolarized neutron matter in a box of side $L$, the orbitals are plane waves that fit in the box times up and down spinors. The usual closed shells are 2, 14, 38, 54, 66, ... particles.

The Jastrow correlation function $f(r)$ has been taken as the first component of the FHNC/SOC correlation operator $\hat{F}_{ij}$

$$\hat{F}_{ij} = \sum_{p=1}^{l} f_p(r_{ij}) O^{(p)}(ij), \tag{3.19}$$

which minimizes the FHNC/SOC energy per particle of nucleon matter at the desired density, as in Ref. 36

The nodal structure of the trial function is fully determined by the Slater Determinant. A more complex correlation factor, such as that used in the FHNC/SOC variational calculations would be highly desirable, and could in principle be used. However straightforward evaluation of this wave function would require an exponentially growing number of operations as the particle number increases as discussed in the Introduction. AFDMC can realistically deal with trial functions having a combination of only a relatively small number of Jastrow correlated Slater Determinants, so that each of them can be rotated in the spin-isospin space as explained above.

Alternatively one can try to construct determinants which incorporate the main spin–dependent correlations, similar to *backflow correlations* which can be fully included in the Slater Determinant. This is for instance the case of the spin–orbit correlations, which can be partially taken into account into the trial wave function by adding a back–flow into the orbitals of the Slater Determinant:

$$e^{\imath\vec{k}\cdot\vec{r}_1}\begin{pmatrix} a \\ b \end{pmatrix} \longrightarrow e^{\imath\vec{k}\cdot\vec{r}_1+1/2\sum_{j\neq 1} f_b(r_{1j})(\vec{r}_{1j}\times\vec{k}_1)\cdot\vec{\sigma}_1}\begin{pmatrix} a \\ b \end{pmatrix}$$

$$\longrightarrow \begin{pmatrix} e^{\imath\vec{k}\cdot\vec{r}_1} \left\{a[\cosh(A_1)+\widehat{A_{1z}}\sinh(A_1)]+b[(\widehat{A_{1x}}+\imath\widehat{A_{1y}})\sinh(A_1)]\right\} \\ e^{\imath\vec{k}\cdot\vec{r}_1} \left\{b[\cosh(A_1)-\widehat{A_{1z}}\sinh(A_1)]+a[(\widehat{A_{1x}}-\imath\widehat{A_{1y}})\sinh(A_1)]\right\} \end{pmatrix} \quad (3.20)$$

where $a$ and $b$ are spinor components, and

$$\vec{A}_1 = \frac{1}{2}\sum_{j\neq 1} f_b(r_{1j})\vec{r}_{1j}\times\vec{k}_1 \ . \quad (3.21)$$

The $\vec{L}\cdot\vec{S}$ correlation in FHNC/SOC calculations is given by

$$F_B(12) = \frac{1}{4\imath} f_b(r_{12})\ [\vec{r}_{12}\times(\vec{\nabla}_1-\vec{\nabla}_2)\cdot(\vec{\sigma}_1+\vec{\sigma}_2)] \ , \quad (3.22)$$

where the subscript $b$ stands for the the seventh component ($p = 7$) of $\hat{F}(12)$. From the results of the expectation value of the potential energy at the two-body level of the FHNC cluster expansion, one observes that the leading spin-orbit terms are those in which the spin–orbit potential couples the scalar correlation to the spin–orbit one (usually denoted as bbc and cbb terms), as shown in Table 1 for the case of the $A8'$ potential. One can easily verify that the above terms are exactly reproduced by the following spin–orbit correlation:

$$\widetilde{F_B}(12) = \frac{1}{2\imath} f_b(r_{12})\ [\vec{r}_{12}\times\vec{\nabla}_1\cdot\vec{\sigma}_1 - \vec{r}_{12}\times\vec{\nabla}_2\cdot\vec{\sigma}_2], \quad (3.23)$$

which is fully equivalent to the backflow correlation of Eq. (3.20).

As in standard fermion Diffusion Monte Carlo, the AFDMC method has the usual sign problem. In this case the overlap of the walkers with the trial function is complex. Therefore, the usual fermion sign problem is here a phase problem. To deal with it, the path of the walkers is constrained to regions where the real part of

Table 1. Spin–orbit terms in the two-body cluster expansion of $< V_2 >$ for neutron matter at $\rho_0$ and $2\rho_0$. The potential used is $A8'$, set to zero beyond half the side $L = (14/\rho)^{1/3}$ of the periodic box for 14 neutrons. The energies are in MeV.

| term | $F_B(0.16)$ | $F_B(0.32)$ | $\widetilde{F_B}(0.16)$ | $\widetilde{F_B}(0.32)$ |
|---|---|---|---|---|
| bcc+cbb | -7.556 | -11.107 | -7.556 | -11.107 |
| Total | -6.305 | -9.908 | -5.114 | -8.614 |

the overlap with our trial function is positive. For spin independent potentials this reduces to the fixed-node approximation. It is straightforward to show that if the sign of the real part is that of the correct ground state, one gets the correct answer and small deviations give second order corrections to the energy. It has not been proved that this constraint always gives an upper bound to the ground state energy although it appears to do so for the calculations performed so far.

### 3.5. *Tail corrections*

Monte Carlo simulations are often performed calculating expectation values only within a sphere of radius $L/2$ around a particle, where $L$ is the length of the box side. Tail corrections are then estimated by integrating out the spin–independent part of the two–body potential from $L/2$ up to infinity. The AFDMC calculations have been performed within the full simulation box, and, in order to include also the contribution from the neighbor cells, the Jastrow factor $f(r)$ and the components $v_p(r)$ of the potential have been tabulated in the following form

$$F(x,y,z) = \prod_{mno} f(|(x+mL_x)\hat{x} + (y+nL_y)\hat{y} + (z+oL_z)\hat{z}|)$$
$$V_p(x,y,z) = \sum_{mno} v_p(|(x+mL_x)\hat{x} + (y+nL_y)\hat{y} + (z+oL_z)\hat{z}|)\,. \tag{3.24}$$

It has been found always adequate to include only the 26 additional neighbor cells corresponding to $m$, $n$, and $o$ taking the values $-1$, 0, and 1.

The AFDMC results reported in this paper are therefore already tail corrected.

### 3.6. *The AFDMC algorithm*

The algorithm used in the AFDMC simulations of neutron and nuclear matter is given by:

**(1)** Sample the $|R,S>$ initial walkers from $| < \Psi_T|R,S > |^2$ using Metropolis Monte Carlo, where $< \Psi_T|R,S >$ is a given trial function.

**(2)** Propagate in the usual Diffusion Monte Carlo way with a drifted gaussian for half a time step.

**(3)** Diagonalize the potential matrix $A_{i,\alpha,j,\beta}$ for each walker, to get the eigenvectors $\lambda_n$ and the eigenvectors $\psi_n^\alpha(i)$. 3A auxiliary field variables are required for the $\sigma$ terms. Therefore, each walker requires the diagonalization of a $3A$ by $3A$ matrix.

**(4)** Sample a value of $x$ for each of the $3A$ auxiliary field variables from the Hubbard–Stratonovich transformation of Eq. (3.9). A discrete version of this transformation is given by the following three–point formula. [37]

$$\begin{aligned} e^{-\frac{1}{2}\lambda_n O_n^2 \Delta t} &= \int_{-\infty}^{\infty} dx f(x)\, e^{-\Delta t s \lambda_n O_n x} + O(\Delta t^3)\ , \\ f(x) &= \frac{1}{6}\left[\delta(x-h) + \delta(x) + \delta(x+h)\right]\ , \\ h &= \left(\frac{3}{|\lambda_n|\Delta t}\right)^{\frac{1}{2}}\ . \end{aligned} \tag{3.25}$$

Other expression with 5 or more interpolation points are possible, or one can use a gaussian distribution. For all the simulations already performed, Eq. (3.25) has been found always accurate enough.

**(5)** Make a rotation in the spin space, by applying the following $2 \times 2$ matrices $M_k(i,j)$, to the $k$–th particle spinor.

$$M_k(i,j) = \begin{pmatrix} [\cosh(A_n) + \sinh(A_n)\psi_n^z(k)] & [\sinh(A_n)(\psi_n^x(k) - \imath\ \psi_n^y(k))] \\ [\sinh(A_n)(\psi_n^x(k) + \imath\ \psi_n^y(k))] & [\cosh(A_n) - \sinh(A_n)\psi_n^z(k)] \end{pmatrix}, \tag{3.26}$$

where

$$A_n = \Delta t |\lambda_n| x_n \sqrt{[(\psi_n^x(k))^2 + (\psi_n^y(k))^2 + (\psi_n^z(k))^2]}\ , \tag{3.27}$$

and $x_n$ is the sampled Hubbard–Stratonovich value. For positive values of $\lambda_n$, one has a similar set of equations, in which $\sinh(A_n)$ is substituted with $\imath\ \sin(-A_n)$.

**(6)** Propagate the spin–orbit component of the potential as discussed in section (3.3).

**(7)** Repeat the propagations in the opposite order to produce a reversible propagator and lower the time step error.

**(8)** Combine all weight factors and evaluate a new value of $< \Psi_T|R,S >$. If the real part is negative, enforce constrained path by dropping the walker, but keeping it in the calculation of the mixed energy.

**(9)** Evaluate the averages of $\langle\Psi_T|R,S\rangle$ and of $\langle\Psi_T|H|R,S\rangle$ to calculate the mixed energy. Evaluate also the growth energy from the normalization. [23]

**(10)** Repeat as necessary.

Table 2. Finite size corrections for symmetrical nuclear matter: [31] PBFHNC results for the MS3 potential at $\rho = 0.16 fm^{-3}$. The PBFHNC calculations have been performed with a Jastrow correlated wave function, whereas the FHNC/SOC result has been obtained with a correlation operator of the type $F_4$. PBFHNC and FHNC/SOC calculations include the basic four–point elementary diagram $E_4$.

| $A$ | $PB-FHNC$ | $FHNC/SOC$ | $AFDMC$ |
|---|---|---|---|
| 28 | -13.6 | - | -16.17(6) |
| 76 | -15.6 | - | -18.08(3) |
| 2060 | -14.0 | - | ↓ |
| ∞ | -14.0 | -14.9 | -16.5(1) |

### 3.7. *Finite size effects: The periodic box FHNC method*

The FHNC method has been recently reformulated [38,31] so that it can be used to calculate the expectation values for the same system as that used in quantum simulations: a fixed number of fermions $A$ in a periodic box (PBFHNC theory). One can carry out such type of calculation for any given finite number of fermions. Therefore, it provides a very efficient method to estimate the finite size corrections to quantum simulations.

The integral equations to sum the FHNC diagrams have been derived in Ref. 38 for the case of a Jastrow correlated wave function of the type given in Eq. (3.18). Table 2 give the results for symmetrical nuclear matter with the MS3 two–body potential and Jastrow correlated wave function. The PBFHNC results are compared with the full FHNC/SOC results and the corresponding AFDMC. The finite size corrections to AFDMC calculations are extracted from the corresponding PBFHNC results. A more realistic estimate would require the extension of the PBFHNC theory to treat spin–isospin dependent correlation operators of the type given in Eq. (3.19), as in FHNC/SOC. One can see that for this potential (with no tensor force) spin–isospin dependent correlations are not extremely important. However, a different behaviour is expected for potential of the $v_6$ or $v_8$ type, which have tensor components.

Results obtained with spin–isospin dependent correlation operator, at the second order of the cluster expansion, [30] are reported in Table 3. These and other similar results could be used to estimate finite size corrections in the case of simulations carried out with two–body potentials of the $A18$ type. However, the inclusion of three–body force will drastically change the behaviour in the number of particles given in the Table.

Table 3. Finite size correction for neutron matter. Kinetic, potential and total energy per particle (in $MeV$) at the second order of the FHNC cluster expansion, calculated within the PBFHNC theory and with the $A6'$ and $A8'$ interactions and correlation operators of the $F_6$ and $F_8$ type respectively. The density considered is nuclear matter experimental saturation density $\rho_0$.

| $A$ | $T_F$ | $< T6 >_2$ | $< V6 >_2$ | $< H6 >_2$ | $< T8 >_2$ | $< V8 >_2$ | $< H8 >_2$ |
|---|---|---|---|---|---|---|---|
| 14 | 35.600 | 44.47 | -29.41 | 15.06 | 46.36 | -36.58 | 9.78 |
| 38 | 33.703 | 42.41 | -29.43 | 12.98 | 44.28 | -36.28 | 8.00 |
| 66 | 34.917 | 43.64 | -29.07 | 14.57 | 45.55 | -36.07 | 9.48 |
| 114 | 35.646 | 44.40 | -28.87 | 15.53 | 46.32 | -35.94 | 10.38 |
| 1030 | 35.139 | 43.88 | -28.95 | 14.92 | 45.79 | -35.97 | 9.82 |
| $\infty$ | 35.094 | 43.84 | -28.96 | 14.88 | 45.75 | -35.97 | 9.78 |

## 4. AFDMC applications to nucleon matter

In this section some of the most recent AFDMC applications to nuclear and neutron matter are discussed and critically compared with the corresponding results obtained within CBF anf BHF theories.

### 4.1. *Equation of state of neutron matter*

Neutron matter simulations have been carried out for the $AU8'$ interaction up 66 neutrons in a periodical box at various densities ranging from $0.75\rho_0$ up to $2.5\rho_0$.

Let me first discuss some results obtained with 14 neutrons interacting via the two–body potential $A8'$, mainly to point out a possible problem connected with the treatment of the spin–orbit interaction. Simulations with a larger number of neutrons and using $A18$ instead of $A8'$ will not change the main conclusions. The results are displayed in Fig. 4, together with the variational FHNC/SOC results for the same interaction, obtained by using correlation operators of the $F_6$ and $F_8$ forms, and the BHF results for the $A18$ potential. One can see that SOC(F6) and SOC(F8) in the figure give quite different equations of state, particularly at high density. This is due to the intrinsic approximations present in FHNC/SOC theory to treat the spin–orbit correlations. Given the fact that second order perturbative corrections will necessarily lower the FHNC/SOC energies, the limited number of neutrons considered in the AFDMC simulations and the not totally negligible differences between $A8'$ and $A18$, the SOC(F6), AFDMC and BHF results show a satisfactory agreement. This indicates that the treatment of the spin–orbit correlations in FHNC/SOC needs to be improved upon. Higher order terms beyond the two–body cluster diagrams included in the present treatment can not be neglected. Moreover, the contribution from elementary diagrams is not so small as it was believed, particularly in the case of neutron matter. The results of PBFHNC and

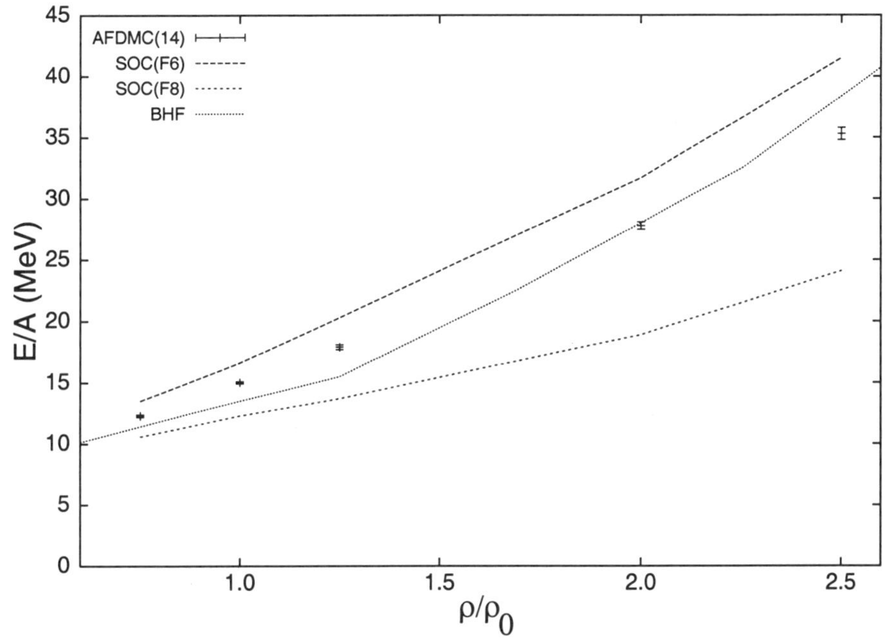


Fig. 4. The spin-orbit problem. The FHNC/SOC variational results obtained with correlation functions of type $F6$ or $F8$ are compared with the BHF of Ref. 13 and AFDMC results. [30] The interaction is $A8'$ for SOC(F6), SOC(F8) and AFDMC, and $A18$ for BHF. The AFDMC results have been obtained with 14 neutrons and the reported error includes also the uncertainty due to finite size errors.

FHNC/SOC calculations with Jastrow, $F_4$ and $F_6$ correlations, reported in this paper, include four–body elementary diagrams, which generally give a not negligible repulsive contribution to the energy per particle. [31]

Table 4 shows the dependence of the AFDMC simulations on the number of neutrons considered in the simulation, and gives an estimate of the spin–orbit contribution to the energy per particle. The following comments are in order: (i) except for the low density cases, the size dependence is dominated by the three–body force and, because of this, it results to be quite different from that shown by PBFHNC (see Table 3); this feature urgently asks for the extension of PBFHNC theory to treat spin dependent correlations of the type $F_6$; (ii) the spin–orbit contribution to the energy per particle is much smaller than in FHNC/SOC with $F_8$ correlations, and in good agreement with FHNC/SOC with $F_6$, confirming the indications coming from Fig. 4.

In Fig. 5 the AFDMC equation of state of pure neutron matter is compared with that obtained with CBF theory and $F_6$ correlations. [39] The equation of state obtaind by Akmal *et al.* [12] with the $AU18$ potential by using FHNC/SOC and $F_8$ correlations is also reported.

Table 4. AFDMC energies per particle in MeV of 14 and 66 neutrons in a periodic box for $AU6'$ and $AU8'$ interaction models at various densities. [30] Error bars for the last digit are shown in parentheses.

| $\rho(fm^{-3})$ | $AU6'(14)$ | $AU8'(14)$ | $AU6'(66)$ | $AU8'(66)$ |
|---|---|---|---|---|
| 0.12 | 14.96(6) | 14.80(9) | 14.93(4) | |
| 0.16 | 19.73(5) | 19.76(6) | | |
| 0.20 | 25.29(6) | 25.13(8) | 26.51(6) | |
| 0.32 | 48.27(9) | 48.4(1) | 53.11(9) | 54.3(6) |
| 0.40 | 69.9(1) | 70.3(2) | 79.4(2) | |

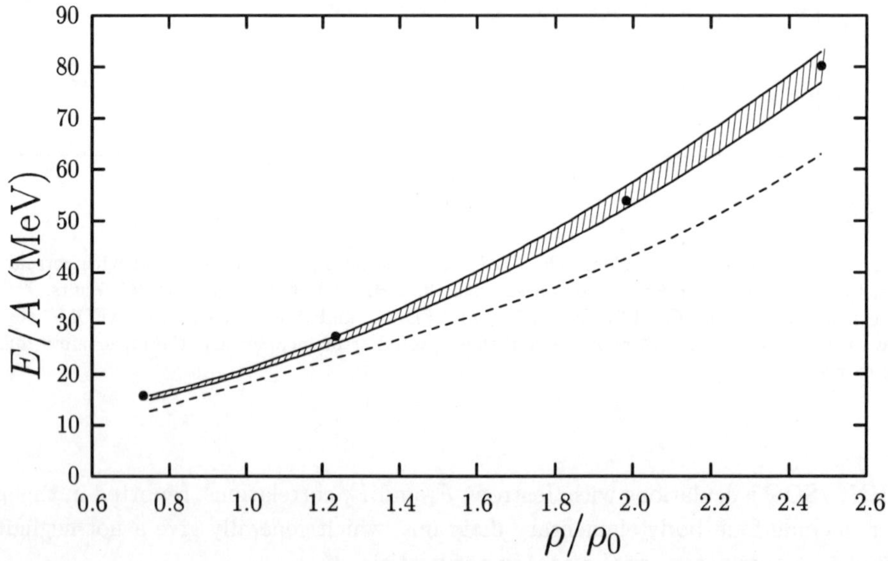


Fig. 5. AFDMC equation of state of the $AU6'$ model of neutron matter (dots); [29] CBF theory [39] results for the same interaction model are in the shaded area where the highest values correspond to the variational estimate. The equation of state obtained in Ref. 12 for the $AU18$ interaction by using FHNC/SOC theory is given by dashes. The statistical errors of the AFDMC estimates are smaller than the symbols.

The compressibility $\mathcal{K}$, given by

$$\frac{1}{\mathcal{K}} = \rho^3 \, \frac{\partial^2 E_0(\rho)}{\partial \rho^2} + 2\rho^2 \frac{\partial E_0(\rho)}{\partial \rho} \, , \tag{4.1}$$

can be estimated from the equation of state by taking $E_0 = E/A$. For a Fermi gas the compressibility is $\mathcal{K}_0 = 9\pi^2 m/(k_f^5 \hbar^2)$. The AFDMC results for $\mathcal{K}/\mathcal{K}_0$ are shown in Fig. 6. They are compared with the corresponding CBF estimates [39] and other existing BHF calculations performed with the old Reid potential [40] and the FHNC/SOC calculations of Ref. 12 with the $AU18$ potential.

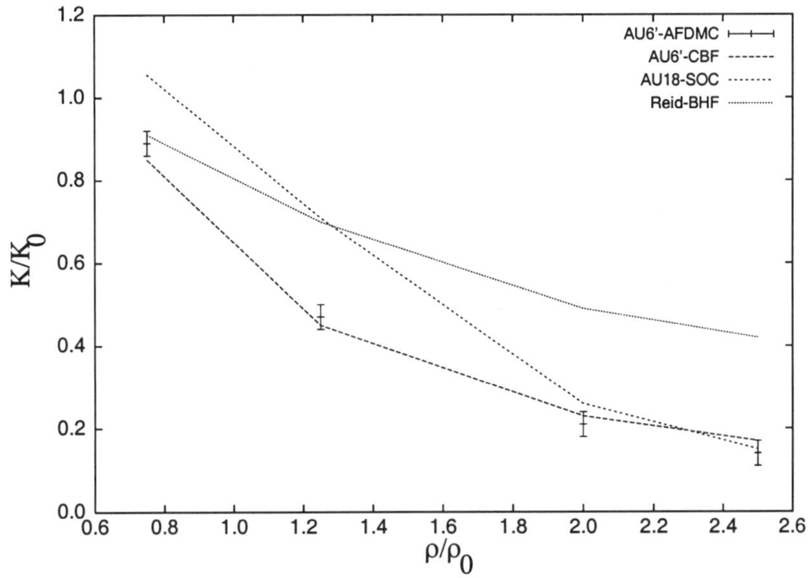


Fig. 6. Neutron matter compressibility. The AFDMC results [29] are compared with the CBF results based on $F_6$ correlations, [39] the BHF results for the Reid potential [40] and the variational FHNC/SOC reults of Akmal *et al.* [12] for the $AU18$ interaction.

### 4.2. *Symmetry energy of nuclear matter*

The AFDMC can deal with $N \neq Z$ systems, and it has been applied to compute the symmetry coefficient of the mass formula for the semirealistic two–body potential MS3 which has no tensor force. The resulting values of $E/A$ at $\rho_0$ for symmetrical nuclear matter are given in Table 2, where they are compared with the FHNC/SOC and PBFHNC results. The finite size correction is estimated from the corresponding PBFHNC results.

Fig. 7 shows the dependence of the energy per particle on the asymmetry parameter $\alpha = (N - Z)/(N + Z)$. [31]

The figure also reports the FHNC/SOC results for symmetrical nuclear matter ($\alpha = 0$) and pure neutron matter ($\alpha = 1$). The dahed line corresponds to a quadratic fit for the FNHC/SOC results giving a symmetry energy of 41.59 $MeV$ (FHNC/SOC can compute only $N = Z$ or $N = A$ matter). The function $E(\alpha)$ provided by the AFDMC results is not fully quadratic in $\alpha$, and corresponds to a symmetry energy of $\sim 36.4$ $MeV$. However, considering the possibility that the nodal surface adopted for the neutron matter is better than that of nuclear matter, as well as the fact that the AFDMC energies for $N \neq Z$ are not finite size corrected, one cannot expect an accuracy better than $\sim 10\%$.

One would like to perform similar quantum simulations with more realistic interaction than MS3.

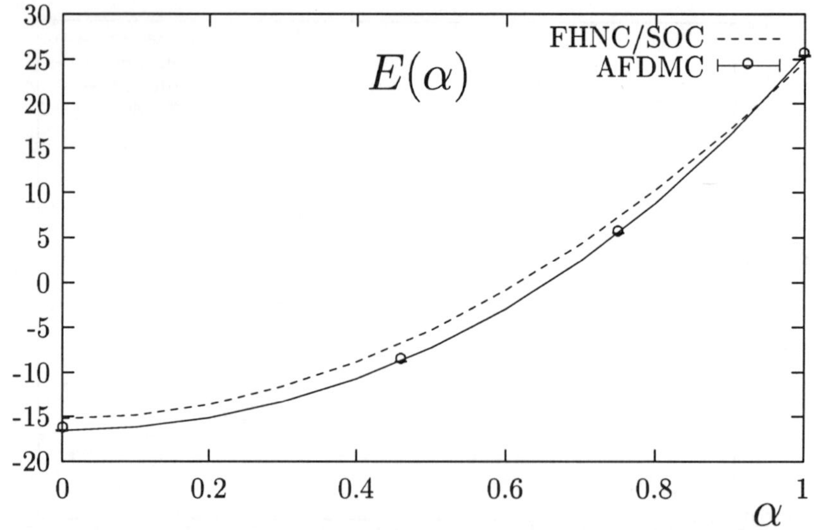


Fig. 7. AFDMC and FHNC/SOC energy per particle of nuclear matter for several values of the asymmetry parameter. [31] The lines correspond to polinomial fits of the calculated energies.

### 4.3. *Spin susceptibility of neutron matter*

The spin susceptibility of neutron matter is an important quantity to estimate the mean free path of a neutrino in dense neutron matter, which is a relevant information for the understanding of the mechanisms underlying the sopernovae explosion and the cooling process of neutron stars. [41,42,43,44]

The Hamiltonian for the spin susceptibilty in a magnetic field, ignoring any orbital effects, is given by

$$H = H_0 - \mu \sum_i \vec{\sigma}_i \cdot \vec{B} \, , \tag{4.2}$$

where $\mu = 6.03 \times 10^{-18}$MeV/Gauss is the neutron magnetic moment. The susceptibility is defined as

$$\chi = -\rho\mu^2 \frac{\partial^2 E_0(b)}{\partial b^2}\bigg|_{b=0} \, , \tag{4.3}$$

where $\vec{b} = \mu\vec{B}$.

The AFDMC method can treat spin polarized systems and compute the energy per particle for a given density $\rho$ and a given polarization $J_z = (N\uparrow - N\downarrow)/(N\uparrow + N\downarrow)$. The spin susceptibility has been recently calculated for neutron matter interacting via $AU6'$ and $AU8'$ potentials, which give the same results within the statistical accuracy. [29] The results, normalized with the spin susceptibility of the Fermi free gas $\chi_0 = \mu^2 m k_f/(\hbar^2\pi^2)$, are plotted in Fig. 8. They are also compared with the results extracted from the Landau parameters calculations by Backmann *et al.* [40] performed with BHF and the Reid potential, and with those by Jackson

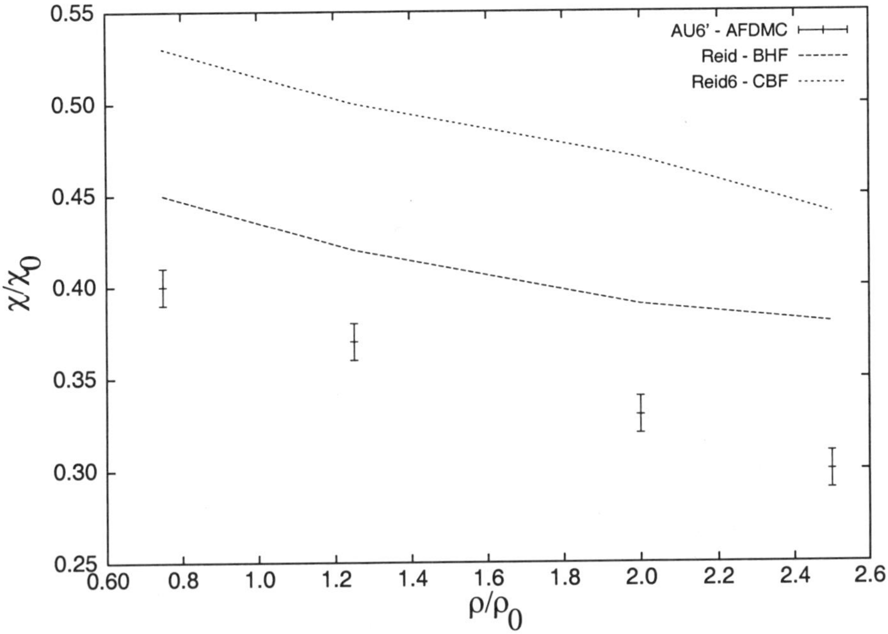


Fig. 8. Spin susceptibility ratio $\chi/\chi_F$ of neutron matter. The AFDMC results [29] for the interactions $AU6'$, $AU8'$ are compared with those extracted from the Landau parameters calculated with BHF [40] and CBF [45] many–body methods. The AFDMC estimate for the Reid6 potential at $\rho/\rho_0 = 1.25$ is $\chi/\chi_0 = 0.36(1)$, a bit smaller than the corresponding CBF result.

*et al.*, [45] performed with CBF perturbation theory and a $v_6$ version of the Reid potential[b] [c].

Clearly, there is a strong effect of NN correlations which reduces $\chi$ of about a factor 3 at twice the equilibrium density of nuclear matter. The spin susceptibility remains rather small, $\chi \sim 0.4\chi_F$, at a low density value, such as $\rho = 0.12\ fm^{-3}$. These results imply that NN correlations strongly reduce (of about a factor ten) the neutrino absorption cross section coming from the vector–axial current, which couples to spin–density fluctuations of neutron matter. [43]

One would like to extend this study to a nucleon matter system having a small fraction of protons and to compute the full spin response at zero and low temperature.

[b]Notice that extracting $\chi/\chi_0$ from the Landau parameters $G_0$ and $F_1$, as done in Fig. 8, may not be very accurate in the presence of a tensor NN force.

[c]A recent BHF calculation[46] of the spin susceptibility of neutron matter with $A18$ potential provides results which are very close to the AFDMC results displayed in the figure.

## 5. Outlook and Conclusions

Some of the most recent developments in quantum simulations for nuclear and neutron matter have been reviewed. Particular attention has been devoted to the Auxiliary Field Diffusion Monte Carlo method, which can be used for *ab initio* calculations of nuclear systems with a large number of nucleons interacting via realistic potentials which include tensor, spin–orbit and three–body force. Calculations with up to 76 nucleons in a periodical box have already been performed. The method has been succesfully applied to symmetrical and asymmetrical nuclear matter, and to unpolarized and polarized neutron matter for a wide range of densities.

While there is much work to be done to validate the results obtained, I believe that this method or one based on the auxiliary field ideas should be able to produce accurate Monte Carlo calculations of the structure of a wide variety of nuclear systems. While previous Monte Carlo calculations have been severely restricted in particle number by the spin-isospin sums, that restriction is lifted by using the auxiliary field break up of the spin-isospin part of the Hamiltonian, while using standard diffusion Monte Carlo for the spatial degrees of freedom.

Work in progress includes calculating with isospin dependent tensor, spin–orbit and three–body forces, so that a full range of nuclear systems can be realistically attacked; calculating the properties of light nuclei to compare with exact GFMC calculations; and investigating pion condensation. In addition, including explicit meson degrees of freedom can also be attempted. In the language of this paper, each meson field mode corresponds to an auxiliary field. [18]

A pressing need is to find trial functions other than the simplest Slater determinant that can be evaluated efficiently. This would allow us to both lower the variance of our calculations as is usual when better guiding functions are used in the importance sampling of the random walk, as well as to obtain a better path constraint.

Another important need is to generalize the PBFHNC method to include spin–dependent correlations as in FHNC/SOC calculations. This would allow for realistic estimates of the finite size effects.

Finally, it would be extremely useful to compare AFDMC simulations of neutron and nuclear matter with other calculations, like for instance BHF or CBF calculations. This is not possible today, because the various many–body methods have been applied with different nuclear interactions. A realistic potential like $AU8'$ [15] seems to be the ideal interaction to test the above many–body techniques. Calculating the ground state of light and medium nuclei, nuclear matter, neutron matter and neutron drops with this interaction would be an useful homework problem. Standard GFMC or other powerful few–body techinques could also be attempted for a system of 14 neutrons in a periodic box. For this reason, a detailed description of the $AU8'$ potential can be found in the Appendix.

## Acknowledgements

The unpublished results reported in this paper have been obtained in collaboration with Kevin Schmidt, Antonio Sarsa and Adelchi Fabrocini. Vijay Pandharipande and Christopher Pethick are also aknowledged for useful discussions. Portions of this work were supported by MURST-National Research Projects and CINECA computing center.

## Appendix

The various equations defining the $AU8'$ potential, originally proposed by Pudliner *et al.* [15] are reported in the following, together with the values of the various parameters entering the two– and three–body parts.

The conventional value of $\hbar^2/(2m)$ (even in pure neutron matter calculations) is

$$\frac{\hbar^2}{2m} = 20.73554 MeV fm^2 , \tag{5.1}$$

which corresponds to the $n - p$ averaged mass.

The various components of the two–body $A8'$ potentials can be written in the form

$$v_p(r) = \sum_{m=1}^{8} A_{p,m} F_m(r) , \tag{5.2}$$

where the odd and the even components refer to the $\tau$–independent and $\tau$–dependent operators respectively. The spin–independent part $v_{ij}^{SI}$ of the two–body potential is given by the first component $v_{p=1}(r_{ij})$ only. The constants $A_{p,m=1,4}$ are reported in Table 5, while the remaining $A_{p,m=5,8}$ vanish except $A_{4,5} = A_{6,7} = 1/3$ and $A_{4,6} = A_{6,8} = 2/3$. The functions $F_m(r)$ are given by

$$\begin{aligned}
F_1(r) &= T^2(\mu, c; r) , \\
F_2(r) &= (1 + a_0 r) W(r) , \\
F_3(r) &= \mu r W(r) , \\
F_4(r) &= (\mu r)^2 W(r) , \\
F_5(r) &= a_1 Y(m_0, c; r) - a_2 r W(r) , \\
F_6(r) &= a_3 Y(m_c, c; r) - a_4 r W(r) , \\
F_7(r) &= a_1 T(m_0, c; r) , \\
F_8(r) &= a_3 T(m_c, c; r) ,
\end{aligned} \tag{5.3}$$

Table 5. Argonne $v8'$ two–body potential. Matrix $A_{p,m=1,4}$ appearing in Eq. (5.2).

| p | $A_{p,1}$ | $A_{p,2}$ | $A_{p,3}$ | $A_{p,4}$ |
|---|---|---|---|---|
| 1 | -7.52251741 | 2616.39024949 | 0 | 147.79390526 |
| 2 | -0.12318501 | 84.20118403 | 0 | -61.22868919 |
| 3 | 0.48726001 | -82.48240972 | 0 | 49.26463509 |
| 4 | 0.65399916 | -107.98800762 | 0 | -20.40956306 |
| 5 | 0.94963459 | -2.91931242 | -424.28015518 | -398.23289299 |
| 6 | -0.17865545 | -0.97310414 | 234.18526077 | -256.12175941 |
| 7 | -0.71193373 | -373.43774331 | 0 | 653.08534247 |
| 8 | -0.28568125 | -201.79028547 | 0 | 354.25604242 |

where the coeficients $a_0 - a_4$ are given by

$$\begin{aligned} a_0 &= 0.37929090 fm^{-1} \,, \\ a_1 &= 3.15588245 \,, \\ a_2 &= 10.48427302 fm^{-1} \,, \\ a_3 &= 3.48918764 \,, \\ a_4 &= 11.21004425 fm^{-1} \,, \end{aligned} \tag{5.4}$$

and the masses $m_0$, $m_c$ and $\mu$ and the cutoff parameter $c$ are given by

$$\begin{aligned} m_0 &= 0.68401113 fm^{-1} \,, \\ m_c &= 0.70729025 fm^{-1} \,, \\ \mu &= 0.69953054 fm^{-1} \,, \\ c &= 2.1 fm^{-2} \,. \end{aligned} \tag{5.5}$$

The Tensor, Yukawa and Wood–Saxon functions are given by

$$\begin{aligned} T(m,c;r) &= (1 + \frac{3}{mr} + \frac{3}{(mr)^2}) \frac{e^{-mr}}{mr} (1 - e^{-cr^2})^2 \,, \\ Y(m,c;r) &= \frac{e^{-mr}}{mr} (1 - e^{-cr^2}) \,, \\ W(r) &= \frac{1}{1 + \exp(\ 5(r - 0.5))} \,. \end{aligned} \tag{5.6}$$

In the case of pure neutron matter (PNM), the isospin exchange operators can be replaced with the identity. Therefore, the Argonne $v_8'$ two-body potential will have only the four components $v_l^{PNM} = v_{2l-1} + v_{2l}$, with $l = 1, 4$.

The Urbana IX three–body potential is given in section 2.2 in Eqs. (2.6), (2.7), (2.8) and (2.9). The values of the various parameters are given by

$$\begin{aligned} B_{2\pi} &= -0.0293MeV \ , \\ U_0 &= 0.0048MeV \ , \\ m_\pi &= 0.69953054fm^{-1} \ , \\ c_3 &= 2.1fm^{-2} \ . \end{aligned} \tag{5.7}$$

In the case of pure neutron matter the anticommutator term in Eq. 2.7 vanishes and $\{\vec{\tau}_i \cdot \vec{\tau}_j, \vec{\tau}_i \cdot \vec{\tau}_k\} = 2$.

## References


1. *Electron–Nucleus Scattering,* edited by A. Fabrocini, S. Fantoni, S. Rosati, M. Viviani, (World Scientific, 1989).
2. *Modern Topics in Electron Scattering,* edited by B. Frois and I. Sick, (World Scientific, 1991).
3. S. Fantoni, V. R. Pandharipande, Nucl. Phys. A **427**, 473 (1984).
4. O. Benhar, A. Fabrocini and S. Fantoni, Nucl. Phys. A **505**, 267 (1989); Phys. Rev. C **41**, R24 (1989).
5. V. R. Pandharipande, I. Sick, Peter K. A. deWitt Huberts, Rev. Mod. Phys. **69**, 981 (1997).
6. R. B. Wiringa, V. G. J. Stocks and R. Schiavilla, Phys. Rev. C **51**, 38 (1995).
7. V. G. J. Stocks, R. A. M. Klomp, C. P. F. Terheggen, and J. de Swart, Phys. Rev. C **49**, 2950 (1994).
8. R. Machleidt, F. Sammarruca, and Y. Song, Phys. Rev. C **53**, R1483 (1996).
9. V. G. J. Stocks, R. A. M. Klomp, M. C. M. Rentmeester, and J. de Swart, Phys. Rev. C **48**, 792 (1993).
10. L. Engvik, M. Hjorth–Jensen, R. Machleidt, H. Müther and A. Polls, Nucl. Phys. A **627**, 85 (1997).
11. S. Rosati, *The hyperspherical harmonic method,* this volume.
12. A. Akmal,V. R. Pandharipande and D. G. Ravenhall, Phys. Rev. C **58**, 1804 (1998).
13. M. Baldo, G. Giansiracusa, U. Lombardo and H. .Q. Song, Phys. Lett. B **473**, 1 (2000).
14. B. S. Pudliner, V. R. Pandharipande, J. Carlson and R. B. Wiringa, Phys. Rev. Lett. **74**, 4396 (1995).
15. B. S. Pudliner, V. R. Pandharipande, J. Carlson, S. C. Pieper, R. B. Wiringa, Phys. Rev. C **56**, 1720 (1997).
16. S. C. Pieper, V. R. Pandharipande, R. B. Wiringa and J. Carlson, Phys. Rev. C **64**, 014001 (2001).
17. K. E. Schmidt, A. Sarsa and S. Fantoni, in preparation.
18. K. E. Schmidt and S. Fantoni, Phys. Lett. B **446**, 99 (1999).
19. *Microscopic Quantum Many–Body Theories and Their Applications,* editors J. Navarro, A. Polls, Lecture Notes in Phys. Vol. 510, (Springer, 1998).
20. A. Polls and F. Mazzanti *Microscopic description of quantum liquids,* this volume.
21. S. Fantoni and A. Fabrocini, Lecture Notes in Phys. Vol. **510**, 119 (1998).
22. E. Krotscheck, *Theory of Correlated Basis Functions,* this volume.
23. K. E. Schmidt and M. H. Kalos, in *Monte Carlo Methods in Statistical Physics,* Vol. II (Springer-Verlag, Berlin, 1984), p. 125.

24. J. Lomnitz-Adler and V. R. Pandharipande, Nucl. Phys. A **361**, 399 (1981).
J. Carlson, Phys. Rev. C **36**, 2026 (1987); **38**, 1879 (1988).
26. S. Zhang, J. Carlson, and J. Gubernatis, Phys. Rev. Lett. **74**, 3652 (1995); Phys. Rev. B **55**, 7464 (1997).
27. S. Fantoni, A. Sarsa and K. E. Schmidt, Prog. Part. Nucl. Phys. **44**, 63 (2000).
28. K. E. Schmidt, A. Sarsa and S. Fantoni, Int. J. Mod. Phys. B **15**, 1510 (2001).
29. S. Fantoni, A. Sarsa and K. E. Schmidt, Phys. Rev. Lett. **87**, 181101 (2001).
30. S. Fantoni, F. Pederiva, A. Sarsa, K. E. Schmidt and J. Carlson, to be submitted to Phys. Rev. C.
31. S. Fantoni, A. Sarsa and K. E. Schmidt, in *Advances in Quantum Many–Body Theories* Vol. 5, edited by R. F. Bishop, K. A. Gernoth and N. R. Walet, (World-Scientific, Singapore 2001).
32. A. Fabrocini, F. Arias de Saavedra and C. Co' Phys. Rev. C **61**, 044302 (2000).
33. I. A. Afnan and Y. C. Tang Phys. Rev. **175** (1968) 1337.
34. R. Guardiola 1981, in *Recent Progress in Many-Body Theories* Lecture Notes in Physics 142, (Springer-Verlag, Berlin, 1981), p. 398.
35. F. Arias de Saavedra, G. Co' and A. Fabrocini Phys. Rev. C **63**, 064308 (2001).
36. R. B. Wiringa, V. Ficks, and A. Fabrocini, Phys. Rev. C **38**, 1010 (1988).
37. S. E. Koonin, D. J. Dean and K. Langanke, Phys. Rept. **278**, 1 (1997).
38. S. Fantoni and K. E. Schmidt, Nucl. Phys. A **690**, 456 (2001).
39. A. Fabrocini, private communication.
40. S. O. Bäckmann and C. G. Källman, Phys. Lett. B **43**, 263 (1973).
41. G. G. Raffelt, *Stars as Laboratories for Fundamental Physics*, The University of Chicago Press, Chicago & London (1996).
42. R. F. Sawyer, Phys. Rev. D **11**, 2740 (1975); Phys. Rev. C **40**, 865 (1989).
43. N. Iwamoto and C. J. Pethick, Phys. Rev. D **25**, 313 (1982).
44. S. Reddy, M. Prakash, James M. Lattimer and Jose A. Pons, Phys. Rev. C **59**, 2888 (1999).
45. A. D. Jackson, E. Krotscheck, D. E. Meltzer and R. A. Smith, Nucl. Phys. A **386**, 125 (1992).
46. M. Baldo, private communication.

# INDEX


$\Delta$ isobar, 370, 372
$\beta$–derivative, 210
$^3$He droplets, 173
$^3$He impurity in liquid $^4$He, 54, 93, 97
  two–body radial distribution function, 95
$^4$He droplets
  collective states, 168
  ground state, 166
$^4$He nucleus, 156
$^4$He phase diagram, 53
$^{16}$O nucleus, 156
*hep* reaction, 372, 373
$m_0$ sum rule, 247
$m_{-1}$ sum rule, 247
$n$–particle current, 220
$n$–particle densities, 208
$n$–particle density
  impurity, 255
  time dependent, 219
$n$–particle distribution, 208
$n$-body correlations, 123

A=3 nuclear systems
  bound state, 340, 355
  scattering state, 340
A=4 nuclear systems
  bound state, 340, 355
Abe diagrams, 75
ACA, 96
action integral, 217
  kinetic energy, 218
  time derivative, 218
action principle
  impurity, 255
anomalous dispersion, 233
astrophysical $S$-factor, 370
average correlation approximation, 96

backflow, 206
backflow correlations, 98, 114, 298
BCS theory, 265
  pairing interaction, 294
  trial state, 289
  variational, 289
benzene, 42
BGY–equations
  linearized , 239
BH molecule, 139
Bloch oscillations, 180
Bogoljubov amplitudes, 289
Born–Green–Yvon (BGY) equations, 241
Bose–Einstein condensation, 179, 191
Bose–Einstein condensation
  in finite systems, 192
  in helium droplets, 195
  in homogeneous fluids, 192
  in liquid helium, 194
  in non interacting systems, 192
  temperature dependence, 193
  in magnetic traps, 179
bra ground state, 133
Brueckner Hartree Fock, 381

$C_9$, 42
$C_{60}$, 22
CBF theory, 207, 267
  diagonal matrix elements, 270, 274, 275
  effective interactions, 283
  infinite order, 298
  off–diagonal matrix elements, 270, 274
  overlap matrix elements, 270, 274, 278
  particle–hole spectrum, 277
CC2, 143, 145
CC3, 143
CC4, 143
CCn, 139
CCn approximation, 134
CCSD, 139
CCSDT, 139

CCSDTQ, 139
center-of-mass motion, 144
chemical potential, 80, 94
  impurity, 94
classification of diagrams
  Fermi systems, 109
cluster expansion, 58, 63, 80, 103
  classification of diagrams, 65
  composite diagrams, 67
  diagonal matrix elements, 275
  diagrammatic rules, 67
  diagrams, 63
  elementary diagrams, 72
  Fermi systems, 107, 109
  impurity problem, 95
  nodal diagrams, 67
  overlap matrix elements, 278
  simple diagrams, 67
collective excitations
  density, 206
column density, 200
complex Kohn principle, 367
condensate fraction, 53, 192
condensate photoionization, 186
configuration interaction method, 122, 126
continuity equation, 220
  asymptotic, 248
  impurity, 255, 257
  one–particle, 228, 229
  two–particle, 227
convolution approximation
  impurity, 256
Cooper pairs, 293
correlated basis functions, 93, 99
correlated coupled cluster theory, 299
correlated RPA, 319
correlation energy, 199
correlation function
  one–body, 216
  one–particle, 229
  time dependent, 216
  two–body, 217
correlation functions
  three–body (triplet), 228
  two–body, 228
Coulomb plasma, 141
coupled cluster equations, 129
critical frequency, 232
CRPA, 319
currents, 221
  Feynman, 227
  Feynman approximation, 222
  full theory, 240
  one–particle, 229
  two–particle, 226
cusp condition, 212

density
  Feynman approximation, 222
  one–body, 228
  two particle, 226
density fluctuation operator, 115, 207, 320
density fluctuations, 240
  two–particle, 240
density–density response function, 321, 325
diagrams, 63
  articulation points, 65
  classification, 65
  elementary, 72
  Fermi systems, 109
  irreducible, 65
  linked, 65
  nodal, 67
  reducible, 65
  rules, 67
  statistical correlations in Fermi systems, 109
  subdiagrams, 68
  symmetry factors, 65, 68
  unlinked, 65
diamond, 42
dielectric function, 33, 42
direct correlation function, 211
distribution function
  three–particle, 225
  two–body, 58
driving terms, 221
  convolution approximation, 226
  two–particle, 226
dynamic structure function, 90, 207, 325
  Fermi systems, 114
  Feynman, 224
  free Fermi gas, 114
  sum rules, 92

ECCM, 134
effective mass, 98, 102
effective potential
  elementary diagrams, 211

electron gas, 141
elementary diagrams, 113
elementary excitation modes, 215
equation of state, 80
Euler–Lagrange equation, 57, 81, 210
  momentum space, 213, 214
  optimum correlation factor, 57
  paired phonon analysis, 84
  particle–hole interaction, 83
  Schrödinger-like, 211
  stability conditions, 83
Euler–Lagrange equations
  particle–hole interaction, 316
evaporative cooling, 181
exchange interaction, 20, 45
exchange interaction
  Slater approximation, 13
excitation operator, 224
excitations
  roton, 224
exp(S) formalism, 122, 123
extended CCM, 134
external potential, 221, 222

f-sum rule, 28, 248
Faddeev equations, 342
Faddeev-Yakubovsky method, 342
Feenberg, 207
Feenberg ansatz, 62
Feenberg effective potential, 209, 210
Fermi hypernetted chain equations, 102, 109, 110
Feshbach resonance, 199
few–body systems
  coupled-rearrangement-channel, 355
  CRC gaussian basis variational method, 356
  effective HH interaction method, 356
  Faddeev equations, 342, 355, 365
  Faddeev-Yakubovsky method, 342, 356
  Green's function Monte Carlo, 342, 356, 365
  no-core shell model, 356
  stochastic variational method, 342, 356
  variational Monte Carlo, 342, 372
  Green's function Monte Carlo, 355
Feynman, 206
Feynman diagrams, 267
FHNC, 111
fluctuation–dissipation theorem, 215
Fourier transform, 221

gap equation, 291
gas parameter , 194
Gauss integration, 352
gaussian expansion, 153, 155, 165
generating functional, 273
GGA (generalized gradient approximation), 16, 20, 22
Goldstone diagrams, 267
grandangular momentum, 346, 348
Green's function, 247
Gross–Pitaevskii energy functional, 199
Gross–Pitaevskii equation, 199
ground state hamiltonian, 217
GW approximation, 46

$H_2O$ molecule, 139
Hückel model, 41
half–density radius, 201
hamiltonian
  impurity, 255
  time dependent, 217
hard core potential, 142
harmonic polynomial, 347, 348
Hartree-Fock theory, 3–6, 7, 12, 20
  time–dependent Hartree–Fock theory, 265, 316
HCSUB(n) approximation, 142
helium atom, 341, 362
helium droplets, 163
helium properties, 53
helium–helium interaction, 341
  Aziz *II* potential, 55
  Aziz potential, 55
  Lennard–Jones potential, 55, 341
HF molecule, 139
HNC, 68
HNC correlated Hartree equation, 199
HNC/0, 69
HNC/4, 72
Hohenberg-Kohn theorem, 9
hydrodynamic effective mass, 261
hypernetted chain equations, 63, 68, 69, 210
  two–body radial distribution function, 69
hyperpolarizability, 45
hyperspherical coordinates, 339, 345
hyperspherical coordinates

hyperangle(s), 339, 346
hypermomentum, 353
hyperradius, 339, 345
hyperspherical harmonics (HH), 339, 341, 348, 361, 374
hyperspherical harmonics (HH)
adiabatic HH expansion (AHH), 340, 360
correlated HH expansion (CHH), 340, 358
correlation function HH method (CFHHM), 364
extended HH expansion (EHH), 361, 365
pair-correlated HH expansion (PHH), 358, 366
potential basis expansion (PB), 340, 357

incompressibility, 223
independent $n$-body correlations, 151
independent pairs, 124
independent triplets, 124
independent two-body correlations, 151
induced potential, 211, 213
intermediate normalization, 151
intermediate normalization condition, 129
IP (ionization potential), 15
iron, 22
isothermal compressibility, 80

J-TICI2, 162, 163, 166, 168
J-TICI3, 164, 166, 168, 170, 172–174
Jackson–Feenberg identity, 59, 61, 274
Jacobi coordinates
in coordinate space, 339, 343, 345, 351, 352, 355
in momentum space, 352
Jastrow ansatz, 209
Jastrow correlation, 151, 162–164, 166, 168, 194, 358
Jastrow wave function
variational wave function, 57
Jastrow–Feenberg wave function, 265
jellium, 7, 40, 141

kinetic energy, 209
Kohn-Sham theory, 13, 16
Kramers–Kronig relations, 231

Laguerre polynomial, 363, 365
lambda transition, 53
Landau, 206
Landau parameters, 265, 285
Landau parameters in neutron matter, 400
Landau's excitation spectrum, 88
Landau's quasiparticle theory, 284
Landau–Pomeranchuk spectrum, 94
Landau-Zener tunneling, 184
LDA (local density approximation), 13–15, 20, 21
least–action principle, 217, 219, 317
light nuclei, 143
linear response, 31, 36, 316
linear response function, 207, 215, 224, 242
Feynman, 223
impurity, 257
random phase, 215
linked cluster expansion, 152
linked diagrams, 65
linked operators, 127
local density approximation, 197
long range order, 192
low density expansion, 197
low excited states
Fermi systems, 114
Feynman ansatz, 88
impurity problem, 98
multiphonon state, 100
low excited states in a Bose fluid, 87
LSDA (local spin density approximation), 14, 22

many–body methods
auxiliary field diffusion Monte Carlo, 382, 383, 388
Brueckner Hartree Fock, 381
correlated basis functions, 195, 381
diffusion Monte Carlo, 195
Fermi hypernetted chain, 381
Green's function Monte Carlo, 195, 383
hypernetted chain, 194
path integral Monte Carlo, 195
periodic box FHNC, 395
single operator chain approximation, 381
variational Monte Carlo, 194
many–body problem , 339
nuclear, 379

maxon, 206
metal clusters, 39
microscopic theory
  variational, 208
momentum distribution, 192
Monte Carlo, 207, 352
MOT, 181
multireference state, 140

natural orbits, 195
NCCM, 133
neutron matter
  compressibility, 398
  equation of state, 396
  neutrino mean free path, 400
  spin susceptibility, 400
neutron scattering, 88, 89, 94
NMR, 24
nodal surfaces, 266
nonlinear polarizability, 45
normal CCM, 133
normalization factor, 217
normalization integral, 209
nuclear current and charge operators, 370, 372
nuclear interaction
  $3N$ potential, 358
  $NN$ potential, 341
  Argonne $v_8'$, 385
  Argonne $v_{14}$, 355, 360, 368
  Argonne $v_{14}$a , 355
  Argonne $v_{18}$, 355, 368, 381, 385
  Argonne $v_8$, 368
  AU8', 387, 403
  Baker, 354
  Brazil 1, 368
  Brazil 2, 368
  CD–Bonn, 381
  Illinois II, 382
  Illinois IV, 382
  Malfliet and Tjon, 354
  modern two–body potentials, 381
  Nijmegen I, 381
  Nijmegen II, 381
  Reid93, 381
  Urbana IX, 368, 382
  Urbana VIII, 368
  Volkov, 354
nuclear matter
  symmetry energy, 399
nuclear physics
  AU8' homework problem, 402
numerical methods
  fast Fourier transform, 17
  Kohn-Sham , 17–19
  matrix, 35
  multigrid, 18
  order-$N$, 20
  real-time, 37
  Sternheimer, 37
  TDDFT , 34–39

one–body density matrix, 192
one–particle density fluctuation, 220
optical lattice, 179
optical rotatory power, 44
optimized ground state, 217
Ornstein–Zernike equation, 211
oscillator length, 192
oscillator strength, 28, 41

paired phonon analysis, 84
parquet theory, 267
particle–hole effective interaction, 212
path constraint, 393
perimetric coordinates, 362
perturbation theory, 99, 265
  Brillouin–Wigner, 100, 271
  Feenberg–Feshbach, 271
  non–orthogonal, 269
  Rayleigh–Schrödinger, 271
phonons, 206
polarizability, 27, 32
polarized $^4$He, 114
pole strength, 215
polyenes, 42
pressure, 80
proton radiative capture, 370
proton radiative capture $^2\text{H}(p,\gamma)\,^3\text{He}$, 369, 370
proton weak capture $^3\text{He}(p,e^+\nu_e)\,^4\text{He}$ (the *hep* reaction), 369
proton weak capture $^3\text{He}(p,e^+\nu_e)\,^4\text{He}$ (the *hep* reaction), 371

quadratic form, 208
quantum many body problem, 50
quasi random number (QRN) integration, 352
quasiparticles, 284

effective mass, 284
interaction, 284
occupation numbers, 284
spectrum, 284
variational interaction, 288

random–phase approximation (RPA), 316
reference spectrum, 243
reference state, 131, 132
roton, 88, 206
RPA, 316
rubidium, 179

$Sc^+$, 140
scaling approximation, 73
scattering length, 194
Schiff–Verlet correlation factor
variational wave function, 60
self–energy
CBF–approximation, 229
full theory, 243
imaginary part, 216, 231
impurity, 257
numerical calculation, 230
real part, 231
sequential relation
current, 221
density, 220
shadow wave function, 207
size-extensivity, 122, 128, 129
Slater function, 104, 273
SOC, 381
speed of sound, 214
spin–orbit backflow correlation, 392
spin–orbit propagator, 390
standard model in nuclear physics, 380
standard solar model (SSM), 371, 373
static response function, 223
static structure function, 71, 86, 321
Fermi systems, 112
free Fermi gas, 106
structure function, 212
Coulomb gas, 214
SUB(2), 145, 147
SUB(n), 139
SUB(n) approximation, 134, 142
sum rule, 246
compressibility, 248
sum rules, 168, 170
Super-Kamiokande, 371
superconductivity, 265
superfluidity, 50, 51
superposition approximation, 75
surface plasmon, 39

TDDFT (time-dependent density functional theory), 25–46
thermodynamic limit, 193
Thomas-Fermi approximation, 9, 11, 12, 200
three–body correlations, 73, 78, 80, 166, 168
variational wave function, 73
three–body potential, 382
three–phonon coupling, 230
TICC2, 147, 148, 150, 151, 161
TICC2
configuration representation, 145
coordinate representation, 150
TICI2, 148, 150, 156, 158, 161
TICI3, 165
time dependent wave function, 216
TOP, 179
total energy, 208
Jastrow wave function, 60
transition currents, 256
transition-correlation operator scheme, 370
translational invariance, 144
trap frequency, 192
triplet distribution
convolution approximation, 225
two–body radial distribution function, 58
Fermi systems, 103, 111
free Fermi gas, 107
two–particle current, 250
two-body correlations, 156

uniform limit approximation, 226
unlinked diagrams, 65
unlinked operators, 127

V4 interactions, 156, 157, 162
V6 interactions, 160, 162
Van der Waals interaction, 24
variational principle, 57, 363
variational principle
Kohn principle, 365
Rayleigh-Ritz principle, 351
variational wave function, 56, 94